Data Engineering Best Practices

Architect robust and cost-effective data solutions
in the cloud era

Richard J. Schiller

David Larochelle

Data Engineering Best Practices

Copyright © 2024 Packt Publishing

Group Product Manager: Apeksha Shetty

Publishing Product Manager: Nilesh Kowadkar

Book Project Manager: Hemangi Lotlikar

Senior Editor: David Sugarman

Technical Editor: Sweety Pagaria

Copy Editor: Safis Editing

Proofreader: David Sugarman

Indexer: Manju Arasan and Tejal Soni

Production Designer: Alishon Mendonca

DevRel Marketing Coordinator: Nivedita Singh

First published: September 2024

Production reference: 1060924

Published by Packt Publishing Ltd.

Grosvenor House

11 St Paul's Square

Birmingham

B3 1RB, UK

ISBN 978-1-80324-498-3

www.packtpub.com

Contributors

About the authors

Richard J. Schiller is a chief architect, distinguished engineer, and startup entrepreneur with 40 years of experience delivering real-time large-scale data processing systems. He holds an MS in computer engineering from Columbia University's School of Engineering and Applied Science and a BA in computer science and applied mathematics. He has been involved with two prior successful startups and has coauthored three patents. He is a hands-on systems developer and innovator.

David Larochelle has been involved in data engineering for startups, Fortune 500 companies, and research institutes. He holds a BS in computer science from the College of William & Mary, a Masters in computer science from the University of Virginia, and a Master's in communication from the University of Pennsylvania. David's career spans over 20 years, and his strong background has enabled him to work in a wide range of organizations, including startups, established companies, and research labs.

About the reviewers

Kamal Baig has over 19 years of experience within the IT space. He has a solid background in data and application development integration and seamlessly transitioned into the Azure solutions architect role. Throughout his career, Kamal has consistently demonstrated a deep understanding of data architecture principles and best practices, leveraging Azure technologies to design and implement cutting-edge solutions that meet the complex needs of modern enterprises. His expertise spans data analytics modernization, data warehouses, data mesh, and data products. Coming from CPG, hospitality, and education domains, he has designed scalable data solutions to ensure security, compliance, and regulatory requirements to align with organizational goals.

John Bremer has 20 years of experience in the market research and data science space. A pioneer creating impactful innovation and value for clients and stakeholders, John has successfully designed and executed research and data strategies and projects for various industries and sectors, leveraging his expertise in data analysis, data mining, and data science. As the President of Phantom 4 Solutions, he provides on-demand support and consulting for organizations in many roles, including Chief Research Officer, Chief Data Science Officer, or Chief Data Analytics Officer. John has a proven track record of managing and transforming high-performance quant teams, and is a respected and valued consultant and decision-maker on data-related matters.

Lindsey Nix is an experienced product manager with a demonstrated history of working in the aerospace, finance, and semiconductors industries. Lindsey is skilled in management, system requirements, software documentation, technical writing, business development, strategic planning, and information assurance. She is a strong consulting professional with a Master's degree in business administration, systems engineering, and data analytics from San Jose State University.

Shanthababu Pandian has over 23 years of IT experience, specializing in data architecting, engineering, analytics, DQ&G, data science, ML, and Gen AI. He holds a BA in electronics and communication engineering, three Master's degrees (M.Tech, MBA, M.S.) from a prestigious Indian university, and has completed postgraduate programs in AIML from the University of Texas and data science from IIT Guwahati. He is a director of data and AI in London, UK, leading data-driven transformation programs focusing on team building and nurturing AIML and Gen AI. He helps global clients achieve business value through scalable data engineering and AI technologies. He is also a national and international speaker, author, technical reviewer, and blogger.

Marianna Petrovich brings over 30 years of experience to the table. Her passion for software engineering, cloud and data intricacies, quality, and governance is evident in her work. Marianna's expertise in data engineering has made her a sought-after consultant and advisor. Trusted for her knowledge of modern data platforms and cloud tools, she guides clients with her exceptional skills in both data and engineering. Currently, she heads the enterprise data engineering team at Circana. Holding a Master's degree in big data from ASU, Marianna resides in Northern California with her husband and eight children. Her aspiration is to inspire the next generation by teaching data engineering to children.

Bill Sun is a senior IT enterprise and solutions architect with expertise in cloud computing, big data, AI/ML, and DevOps. Known for his strong communication skills and leadership, Bill has driven significant projects at Fortune 500 companies. His accomplishments include cloud migrations, data pipeline optimizations, and the development of unified platform services. Bill holds a Master's in computer science from Johns Hopkins, BA degrees from Tsinghua University, and multiple certifications, including Azure and AWS.

Table of Contents

3

A Data Engineer's Journey – IT's Vision and Mission   63

4

Architecture Principles   99

5

Architecture Framework – Conceptual Architecture Best Practices   129

6

Architecture Framework – Logical Architecture Best Practices   181

7

8

9

Key Considerations for Agile SDLC Best Practices

10

Key Considerations for Quality Testing Best Practices

11

Key Considerations for IT Operational Service Best Practices

12

Key Considerations for Data Service Best Practices   329

13

14

15

12

Key Considerations for Data Service Best Practices   329

13

14

15

16

Machine Learning Pipeline Best Practices and Processes   433

17

Takeaway Summary – Putting It All Together   465

18

Preface

Are you an IT professional, IT manager, or business leader looking for an effective large-scale data engineering solution platform? Have you experienced the pain of slogging through piles of literature? Have you had to implement a series of painful **proofs of concept**? If so, this book is for you.

You will emerge on the other side able to implement correctly architected, data-engineered solutions that address real problems you will face in the development process.

Data engineering is rapidly evolving, and the modern data engineer needs to be equipped with software engineering practices to succeed in today's fast-paced data-driven world. This hands-on book takes a practical approach to applying software and data engineering practices to modern use cases, including the following:

- Migrating to cloud-based storage and processing
- Applying Agile methodologies
- Prioritizing governance, privacy, and security

This book is ideal for data engineers and analytics teams looking to enhance their skills and gain a competitive edge in the industry. While reading the book, you will be prompted with ideas, questions, and plans for implementation that would not have been considered prior to reading.

This book assumes that you have a foundational knowledge of at least one cloud vendor service, in particular, **Amazon Web Services** (**AWS**) or Microsoft's Azure. Additionally, you should be well versed in a scripting language (such as Python) and a primary language (such as Java or C/C++), have encountered concurrent/distributed big data processing, and ideally have some experience with analytic services such as **Azure Analysis Services** (**AAS**), Microsoft Power BI, or other third-party analytic solutions. This book is largely aimed at developers and architects who understand Python and cloud computing but want a complete framework for future-proofing successful solutions.

The book is not proscriptive regarding IT solutions, but it does raise key considerations for evaluation as the technology field evolves. After reading this book, IT architects will be equipped to dialogue with cloud vendors and third-party vendors following best practices, so that any solution developed remains robust, of high quality, and cost-effective over time.

This book's structure is as follows:

- Mission/vision
- Principles
- Architecture
- Best practices
- Design patterns
- Use cases

Where pertinent, vendor selection criteria are presented wherein business value statements affect weighting, so that decisions are correctly made to implement an organization's goals. Real-life examples and lessons sum up key points. The book is structured to enable you to envision a reference architecture for your organization and then see the implementation of the business solution in the context of the reference architecture. As the content of the chapters is absorbed, it is a best practice to organize the solution forming in your mind. This is our first key consideration:

"Envision what it means to my company's goals."

Organize your notes and takeaways from the perspective of *"What does it mean for my goals?"* while building up a reference architecture and solution strawman.

By the end of this book, you will be able to architect, design, and implement end-to-end cloud-based data processing pipelines. You will also be able to provide customers with access to data as a product supporting various machine learning, analytic, and big data use cases… all within a well-architected data framework. You will know how to build or buy logical components aligned to the architected data framework's principles and best practices using Agile software development processes tuned to work for an organization. Although this book will *not* supply all the answers, it will shine a light on the path to success while avoiding the pitfalls encountered by many, including the author's own experiences. It will save you countless hours of frustration and enable more rapid creation of better-architected systems.

Who this book is for

If you are an IT professional, IT manager, or business leader looking to build a large-scale data engineering solution, then this book will provide you with a solid set of best practices. As a data engineer, it will give you details behind the best-practice recommendations so you assess the right approaches for your effort. All this should take many hours of pain out of your engineering efforts. If you have to implement a series of proofs of concepts, then this book points to the technologies and vendors that you should avoid so that the proof of concept does not become a **proof of failure** (**POF**). If all this is of interest to you, then this book is for you.

What this book covers

- *Chapter 1, Overview of the Business Problem Statement*, provides a definition of the business problem faced by the data engineer. It also provides an introduction to the entire book.

- *Chapter 2, A Data Engineer's Journey – Background Challenges*, elaborates on the challenges faced when building a modern data system.

- *Chapter 3, A Data Engineer's Journey – IT's Vision and Mission*, illustrates various mission and vision statements and urges you to develop one if one does not already exist. This way, you can keep your focus on the end and not deviate from your strategy.

- *Chapter 4, Architecture Principles*, elaborates on the need to develop principles that keep you solidly grounded in reality. Many examples are provided and explained because they drive the best practices.

- *Chapter 5, Architecture Framework – Conceptual Architecture Best Practices*, depicts architecture as the framework for design engineering. Too often projects go off the rails because the architecture shifts and the structure of the engineering design falls apart. Architecture is a communication tool to keep consensus, especially when things go wrong – and they always do in any engineering effort.

- *Chapter 6, Architecture Framework – Logical Architecture Best Practices*, describes the need to formally define and document the architecture for all, thus tying the conceptual level to the physical level of the architecture.

- *Chapter 7, Architecture Framework – Physical Architecture Best Practices*, defines what will be built and eventually what was built and where it all operates.

- *Chapter 8, Software Engineering Best Practice Considerations*, elaborates on the software best practices needed for the data engineering effort to succeed.

- *Chapter 9, Key Considerations for Agile SDLC Best Practices*, discusses the project management and development processes needed to deliver a data solution.

- *Chapter 10, Key Considerations for Quality Testing Best Practices*, provides testing best practices for a data factory.

- *Chapter 11, Key Considerations for IT Operational Service Best Practices*, defines operational requirements for a data solution.

- *Chapter 12, Key Considerations for Data Service Best Practices*, elaborates on data services, where the focus is on refining raw data into a gem, like a diamond, with facets. It takes the focus away from servicing data as a blob. Examples are provided to illustrate this important message.

- *Chapter 13, Key Considerations for Management Best Practices*, gets into the details of data factory curation and processing with a focus on difficult problems to solve.

- *Chapter 14, Key Considerations for Data Delivery Best Practices*, continues *Chapter 13*'s theme but addresses difficult problem areas for a business and the impediments that can be overcome with the best practices presented.

- *Chapter 15, Other Considerations – Measures, Calculations, Restatements and Data Science Best Practices*, defines the analysis workbench and various tools and processes for the data consumer. This is what is necessary to deliver data at the end of the data factory.

- *Chapter 16, Machine Learning Pipeline Best Practices and Processes*, dives deeper into machine learning/deep learning, **Generative AI** (**GenAI**), and ways to apply knowledge engineering to cooperatively address the future vision where AI takes center stage.

- *Chapter 17, Takeaway Summary – Putting It All Together*, presents the book's conclusion and parting wishes for the development of your future-proof data engineering designs.

- *Chapter 18, Appendix and Use Cases*, delivers on the promise to elaborate on a few high-level use cases with a primer on the technologies used in those use cases.

To get the most out of this book

This book has been written at an intermediate level for data engineers, architects, and managers. There are no tools that you need on your desktop; however, if you want to become hands-on with the tools and technologies referenced, there will be short links (to the {https://packt-debp.link} domain) that are similar to traditional endnotes in each chapter. The journey toward best practices begins with the business context, the mission, vision, and principles that set the foundation for success, and then the development of an architecture. This is followed by engineering designs across a number of important areas driven by people, process, and technology needs.

As the book progresses, the technical topics get deeper, ending with machine learning, and GenAI with a practical look at how to tune LLMs with **RAG** and **prompt engineering**, and a good exploration of **knowledge engineering**.

Get in touch

Feedback from our readers is always welcome.

General feedback: If you have questions about any aspect of this book, email us at customercare@packtpub.com and mention the book title in the subject of your message.

Errata: Although we have taken every care to ensure the accuracy of our content, mistakes do happen. If you have found a mistake in this book, we would be grateful if you would report this to us. Please visit www.packtpub.com/support/errata and fill in the form.

Piracy: If you come across any illegal copies of our works in any form on the internet, we would be grateful if you would provide us with the location address or website name. Please contact us at copyright@packt.com with a link to the material.

If you are interested in becoming an author: If there is a topic that you have expertise in and you are interested in either writing or contributing to a book, please visit authors.packtpub.com.

Share Your Thoughts

Once you've read *Data Engineering Best Practices*, we'd love to hear your thoughts! Scan the QR code below to go straight to the Amazon review page for this book and share your feedback.

https://packt.link/r/1-803-24498-4

Your review is important to us and the tech community and will help us make sure we're delivering excellent quality content.

Download a free PDF copy of this book

Thanks for purchasing this book!

Do you like to read on the go but are unable to carry your print books everywhere?

Is your eBook purchase not compatible with the device of your choice?

Don't worry, now with every Packt book you get a DRM-free PDF version of that book at no cost.

Read anywhere, any place, on any device. Search, copy, and paste code from your favorite technical books directly into your application.

The perks don't stop there, you can get exclusive access to discounts, newsletters, and great free content in your inbox daily

Follow these simple steps to get the benefits:

1. Scan the QR code or visit the link below

https://packt.link/free-ebook/978-1-80324-498-3

2. Submit your proof of purchase
3. That's it! We'll send your free PDF and other benefits to your email directly

1

Overview of the Business Problem Statement

We begin with the task of defining the business problem statement.

"Businesses are faced with an ever-changing technological landscape. Competition requires one to innovate at scale to remain relevant; this causes a constant implementation stream of **total cost of ownership (TCO)** *budget allocations for refactoring and re-envisioning during what would normally be a run/manage phase of a system's lifespan."*

This rapid rate of change means the goalposts are constantly moving. "Are we there yet?" is a question I heard from my kids constantly when traveling. It came from not knowing where we were or having any idea of the effort to get to where we were going, with a driver (me) who had never driven to that destination before. Thank goodness for Garmin (automobile navigation systems) and Google Maps, and not the outdated paper maps that were used in the past. See how technology even impacted that metaphor? Garmin is being displaced by Google for mapping use cases. This is not always because it is better but because it is free (if you wish to be subjected to data collection and advertising interruptions) and it is hosted on everyone's smart device.

Now, I can tell my grandkids that in exactly 1 hour and 29 minutes, they will walk into their home after spending the weekend with their grandparents. The blank stare I get in response tells it all. Mapped data, rendered with real-time technology, has changed us completely.

Technological change can appear revolutionary when it's occurring, but when looking back over time, the progression of change appears to be a no-brainer series of events that we take for granted, and even evolutionary. That is what is happening today with data, information, knowledge, and analytical data stores in the cloud. The term **DataOps** was popularized by Andy Palmer, co-founder and CEO of Tamr {https://packt-debp.link/MGj4EU}. The data management and analytics world has referenced the term often. In 2015, Palmer stated that DataOps is not just a buzzword, but a critical approach to managing data in today's complex, data-driven world.

> *I believe that it's time for data engineers and data scientists to embrace a similar (to DevOps) new discipline – let's call it DataOps – that at its core addresses the needs of data professionals on the modern internet and inside the modern enterprise. (Andy Palmer {https://packt-debp.link/ihlztK})*

In *Figure 1.1*, observe how data quality, integration, engineering, and security are tied together with a solid DataOps practice:

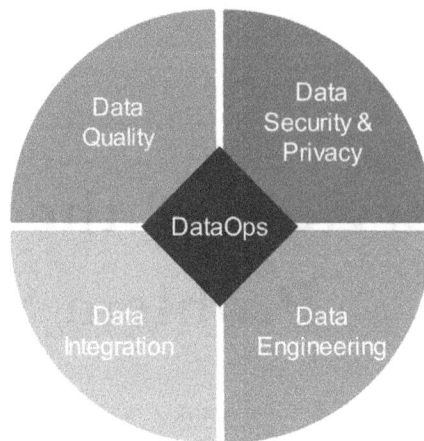

Figure 1.1 – DataOps in the enterprise

The goal of this chapter is to set up the foundation for understanding *why* the best practices presented are structured as they are in this book. This foundation will provide a firm footing to make the framework you adopt in your everyday engineering tasks more secure and well-grounded. There are many ways to look at solutions to data engineering challenges, and each vendor, engineering school, and cloud provider will have its own spin on the formula for success. That success will ultimately depend on what you can get working today and keep working in the future. A unique balance of various forces will need to be obtained. However, this balance may be easily upset if the foundation is not correct. As a reader, you will have naturally formed biases toward certain engineering challenges. These can force you into niche (or single-minded) focus directions – for example, a fixation on robust/highly available multi-region operations with a de-emphasized pipeline software development effort. As a result, you may overbuild robustness and underdevelop key features. Likewise, you can focus on hyper-agile streaming of development changes into production at the cost of consumer data quality. More generally, there is a significant risk from just *doing IT* and losing focus on why we need to carefully structure the processing of data in a modern information processing system. You must not neglect the need to capture data with its semantic context, thus making it true and relevant, instead of the software system becoming the sole interpretation of the data. This freedom makes data and context equal to information that is fit for purpose, now and in the future.

We can begin with the *business problem statement*.

What is the business problem statement?

Data engineering approaches are rapidly morphing today. They will coalesce into a systemic, consistent whole. At the core of this transformation is the realization that data is information that needs to represent facts and truths along with the rationalization that created those facts and truths over time. There must not be any *false facts* in future information systems. That term may strike you as odd. *Can a fact be false?* This question may be a bit provocative. But haven't we often built IT systems to determine just that?

We process data in software systems that preserve business context and meaning but force the data to be served only through those systems. It does not stand alone and if consumed out of context, it would lead to these false facts propagating into the business environment. Data can't stand alone today; it must be transformed by information processing systems, which have technical limitations. Pragmatic programmers' {https://packt-debp.link/zS3jWY} imperfect tools and technology will produce imperfect solutions. Nevertheless, the engineer is still tasked with removing as many as possible, if not all, false facts when producing a solution. That has been elusive in the past.

We often take shortcuts. We also justify these shortcuts with statements like: *"there simply is not enough time!"* or *"there's no way we can get all that data!"* The business *"can't afford to curate it correctly,"* or lastly *"there's no funding for boiling the ocean."* We do not need to boil the ocean.

What we are going to think about is how we are going to turn that ocean directly into steam! This should be our response, not a rollover! This rethinking mindset is exactly what is needed as we engineer solutions that will be future-proof. What is hard is still possible if we rethink the problem fully. To turn that metaphor around – we will use data as the new fuel for the engine of innovation.

> **Fun fact**
>
> In 2006, mathematician Clive Humby coined the phrase "data is the new oil" {https://packt-debp.link/SiG2rL}.

Data systems must become self-healing of false facts to enable them to be knowledge-complete. After all, what is a *true fact*? Is it not just a hypothesis backed up by evidence until such time that future observations disprove a prior truth? Likewise, organizing information into knowledge requires not just capturing semantics, context, and time series relevance but also the asserted reason for a fact being represented as information truth within a dataset. This is what knowledge defines: truth! However, it needs correct representation.

> **Note**
>
> The truth of a knowledge base is composed of facts that are proven by assertions that withstand the test of time and do not hide information context that makes up the truth contained within the knowledge base.

But sometimes, when we do not have enough information, we guess. This guessing is based on intuition and prior experience with similar patterns of interconnected information from related domains. We humans can be very wrong with our guesses. But strongly intuited guesses can lead to great leaps in innovation which can later be backfilled with empirically collected data.

Until then, we often stretch the truth to span gaps of knowledge. Information relationship patterns need to be retained as well as the hypothesis recording accurate *educated guesses*. In this manner, data truths can be guessed. They can also be guessed well! These guesses can even be unwound when proven to be wrong. It is essential that data is organized in a new way to support intelligence. Reasoning is needed to support or refute hypotheses, and the retention of information as knowledge to form truth is essential. If we don't address organizing big data to form knowledge and truth within a framework consumable by the business, we are just wasting cycles and funding on cloud providers.

This book will focus on best practices; there are a couple of poor practices that need to be highlighted. These form **anti-patterns** that have crept into the data engineer's tool bag over time that hinder the mission we seek to be successful in. Let's look into these anti-patterns next.

Anti-patterns to avoid

What are anti-patterns? These are architectural patterns that form blueprints for ease of implementation. Just like when building a physical building, a civil architect will use blueprints to definitively communicate expectations to the engineers. If a common solution is recurring and successful, it is reused often as a pattern, like the framing of a wall or a truss for a type of roofline. Likewise, an anti-pattern is a pattern to be avoided: like not putting plumbing on an outside wall in a cold climate, because the cold temperature could freeze those pipes.

The first anti-pattern we describe deals with retaining *stuff* as data that we think is valuable but can no longer even be understood or processed given how it was stored, and it's contextual meaning gets lost since it was never captured when the data was first retained in storage (such as cloud storage).

The second anti-pattern involves not knowing the business owner's meaning for column-formatted data, nor how those columns relate to each other to form business meaning because this meaning was only preserved in the software solution, not in the data itself. We rely on **entity relationship diagrams** (**ERDs**), that are not worth the paper they were printed on, to gain some degree of clarity that is lost the next time an agile developer does not update them. Knowing what we must avoid in the future as we develop a future-proof, data-engineered solution will help set the foundation for this book.

In order to get a better understanding of the two anti-patterns just introduced, the following specific examples should help illustrate what to avoid.

Anti-pattern #1 – Must we retain garbage?

As an example of *what not to do*, in the past, I examined a system that retained years of data, only to be reminded that the data was useless after three months. This is because the processing code that created that data had changed hundreds of times in prior years and continued to evolve without being noted in the dataset produced by that processing. The assumptions put into those non-mastered datasets were not preserved in the data framework. Keeping that data around was a red herring, just waiting for some future big data analyst to try and reuse it. When I asked, *"Why was it even retained?"* I was told it had to be, according to company policy. We are often faced with someone who thinks piles of stuff are valuable, even if they're not processable. Some data can be the opposite of valuable. It can be a business liability if reused incorrectly. Waterfall gathered business requirements or even loads of agile development stories will not solve this problem without a solid data framework for data semantics as well as data lineage for the data's journey from information to knowledge. Without this smart data framework, the insights gathered would be *wrong*!

Anti-pattern #2 – What does that column mean?

Likewise, as another *not-to-do* example, I once built an elaborate, colorful graphical rendering of web consumer usage across several published articles. It was truly a work of art, though I say so myself. The insight clearly illustrated that some users were just not engaging a few key classes of information that were expensive to curate. However, it was a work of pure fiction and had to be scrapped! This was because I misused one key dataset column that was loaded with data that was, in fact, the inverted rank of users access rather than an actual usage value.

During the development of the data processing system, the prior developers produced no metadata catalog, no data architecture documentation, and no self-serve textual definition of the columns. All that information was retained in the mind of one self-serving data analyst. The analyst was holding the business data hostage and pocketing huge compensation for generating insights that only that individual could produce. Any attempt to dethrone this individual was met with one key and powerful consumer of the insight overruling IT management. As a result, the implementation of desperately needed governance mandated enterprise standards for analytics was stopped. Using the data in such an environment was a walk through a technical minefield.

Organizations must avoid this scenario at all costs. It is a data-siloed, poor-practice anti-pattern. It arises due to individuals seeking to preserve a niche position or a siloed business agenda. In the case just illustrated, that anti-pattern was to kill the use of the governance-mandated enterprise standard for analytics. The problem can be protected from abuse by properly implementing governance in the data framework where data becomes self-explanatory.

Let's consider a real-world scenario that illustrates both of these anti-patterns. A large e-commerce company has many years of customer purchase data that includes a field called `customer_value`. Originally, this field was calculated using the total amount the customer spent, but its meaning has changed repeatedly over the years without updates to the supporting documentation. After a few years, it was calculated as `total_spending - total_returns`. Later, it becomes `predicted_lifetime_value` based on a **machine learning** (**ML**) model. When a new data scientist joins the company and uses the field to segment customers for a marketing campaign, the results are disastrous! High value customers from early years are undervalued while new customers are overvalued! This example illustrates how retaining data without proper context (Anti-pattern #1) and lack of clear documentation for data fields (Anti-pattern #2) can lead to significant mistakes.

Patterns in the future-proof architecture

Our effort in writing this book is to strive to highlight for the data engineer the reality that in our current information technology solutions, we process data as information, when, in fact, we want to use it to inform the business knowledgably.

Today, we glue solutions together with code that manipulates data to mimic information for business consumption. What we really want to do is to retain the business information with the data and make the data smart so that information in context forms knowledge that will form insights for the data consumer. The progression of data begins with just raw data that is transformed into information, and then knowledge, through the preservation of semantics along with context; and finally, the development of analytic derived insights will be elaborated on in future chapters. In *Chapter 18*, we have included a number of use cases that you will find interesting. From my experience over the years, I've learned that making data smarter has always been rewarded.

The resulting *insights* may be presented to the business in new innovative manners when the business requires those insights from data. The gap we see in the technology landscape is that in order for data to be leveraged as an insight generator, its data journey must be an informed one. Innovation can't be pre-canned by the software engineer. It is teased out of the minds of business and IT leaders from the knowledge the IT data system presents from different stages of the data journey. This requires data, its semantics, its lineage, its direct or inferred relationships to concepts, its time series, and its context to be retained.

Technology tools and data processing techniques are not yet available to address this need in a single solution, but the need is clearly envisioned. One monolithic data warehouse, data lake, knowledge graph, or in-memory repository can't solve the total user-originated demand today. Tools need time to evolve. We will need to implement tactically and think strategically regarding what data (also known as *truths*) we present to the analyst.

> **Key thought**
> Implement: *Just enough, just in time.*
> Think strategically: *Data should be smart.*

Applying innovative modeling approaches can bring systemic and intrinsic risk. Leveraging new technologies will produce key advantages for the business. Minimizing the risk of technical or delivery failure is essential. When thinking of the academic discussions debating data mesh versus data fabric, we see various cloud vendors and tool providers embracing the need for innovation… but also creating a new technical gravity that can suck in the misinformed business IT leader.

Remember, this is an evolutionary event and for some it can become an *extinction level event*. Microsoft and Amazon can embrace *well architected* best practices that foster greater cloud spend and greater cloud vendor lock-in. Cloud **platform-as-a-service** (**PaaS**) offerings, cloud architecture patterns, and biased vendor training can be terminal events for a system and its builders. The same goes for tool providers such as the creators of **relational database management systems** (**RDBMS**), data lakes, operational knowledge graphs, or real-time in-memory storage systems. None of the providers or their niche consulting engagements come with warning signs. As a leader trying to minimize risk and maximize gain, you need to keep an eye on the end goal:

"I want to build a data solution that no one can live without – that lasts forever!"

To accomplish this goal, you will need to be very clear on the mission and retain a clear vision going forward. With a well-developed set of principles, best practices, and clear position regarding key considerations, with an unchallenged governance model … the objective is attainable. Be prepared for battle! The field is always evolving and there will be challenges to the architecture over time, maybe before it is even operational. Our suggestion is to always be ready for these challenges and *do not* count on political power alone to enforce compliance or governance of the architecture.

You will want to consider these steps when building a modern system:

- Collect the **objectives and key results** (**OKRs**) from the business and show successes early and often.
- Always have a demo ready for key stakeholders at a moment's notice for key stakeholders.

- Keep those key stakeholders engaged and satisfied as the **return on investment** (**ROI**) is demonstrated. Also, consider that they are funding your effort.

- Keep careful track of the feature to cost ratio and know who is getting value and at what cost as part of the system's **total cost of ownership** (**TCO**).

- Never break a data **service level agreement** (**SLA**) or data contract without giving the stakeholders and users enough time to accommodate impacts. It's best to not break the agreement at all, since it clearly defines the data consumer's expectations!

- Architect data systems that are backwardcompatible and never produce a broken contract once the business has engaged the system to glean insight. Pulling the rug out from under the business will have more impact than not delivering a solution in the first place, since they will have set up their downstream expectations based on your delivery.

You can see that there are many patterns to consider and some to avoid when building a modern data solution. Software engineers, data admins, data scientists, and data analysts will come with their perspectives and technical requirements in addition to **objectives and key results** (**OKRs**) that the business will demand. Not all technical players will honor the nuances that their peers' disciplines require. Yet, the data engineer has to deliver the future-proof solution while balancing on top of a pyramid change.

In the next section, we will show you how to keep the technological edge and retain the balance necessary to create a solution that withstands the test of time.

Future-proofing is …

To future-proof a solution means to create a solution that is relevant to the present, scalable, and cost-effective, and will still be relevant in the future. This goal is attainable with a constant focus on building out a reference architecture with best practices and design patterns.

The goal is as follows:

Develop a scalable, affordable IT strategy, architecture, and design that leads to the creation of a future-proof data processing system.

When faced with the preceding goal, you have to consider that change is *evolutionary* rather than *revolutionary*. That means that a data architecture is solid and future-proof. Making a system 100% future-proof is an illusion; however, the goal of attaining a near future-proof system must always remain a prime driver of your core principles.

The attraction of shiny lights must never become bait to catch an IT system manager in a web of errors, even though cool technology may attract a lot of venture and seed capital or even create a star on one's **curriculum vitae** (**CV**). It may just as well all fade away after a breakthrough in a niche area is achieved by a disrupter. Just look at what happened when OpenAI, ChatGPT, and related **large language model** (**LLM**) technology started to roll out. Conversational **artificial intelligence** (**AI**) has changed many systems already.

After innovation rollout, what was once hard is now easy and often available in open source to become commoditized. Even if a business software method or process-oriented **intellectual property** (**IP**) is locked away with patent protection… after some time – 10, 15, or 20 years – it is also free for reuse. In the filing disclosure of the IP, valuable insights are also made available to the competition. There can only be so many cutting-edge tech winners, and brilliant minds tend to develop against the same problem at the same time until a breakthrough is attained, often creating similar approaches. It is at this stage that data engineering is nearing an inflection point.

There will always be many more losers than winners. Depending on the size of an organization's budget and its culture for risk/reward, there can arise a shiny light idea that becomes a blazing star. 90% of those who pursue the shooting star wind up developing a dud that fades away along with an entire IT budget. Our suggestion is to follow the business's money and develop agilely to minimize the risk of IT-driven failure.

International Data Corporation (**IDC**) and the business intelligence organization Qlik came up with the following comparison:

"Data is the new water."

You can say that data is oil or it is water – a great idea is getting twisted and repurposed, even in these statements. It's essential that data becomes information and that information is rendered in such a way as to create direct, inferred, and derived knowledge. Truth needs to be defined as knowledge in context, including time. We need systems to be not data processing systems but knowledge aware systems that support intelligence, insight, and development of truths that withstand the test of time. In that way, a system may be future-proof. Data is too murky, like dirty water. It's clouded by the following:

- Nonsense structures developed to support current machine insufficiency

- Errors due to misunderstanding of the data meaning and lineage

- Deliberate opacity due to privacy and security

- Missing context or state due to missing metadata

- Missing semantics due to complex relationships not being recorded because of missing data and a lack of funding to properly model the data for the domain in which it was collected

Data life cycle processes and costs are often not considered fully. Business use cases drive what is important (note: we will elaborate a lot more on how use cases are represented by conceptual, logical, and physical architectures in *Chapters 5-7* of this book). Use cases are often not identified early enough. The data services that were implemented as part of the solution are often left undocumented. They are neither communicated well nor maintained well over the data's timeframe of relevancy. The result is that the data's quality melts down like a sugar cube left in the rain. It loses its worth organically as its value degrades in time. Data efficacy loses value over time. This may be accelerated by the business and technical contracts not being maintained, and without that maintenance comes the loss of trust in a dataset's governance. The resulting friction between business silos becomes palpable. A potential solution has been to create business data services with data contracts. These contracts are defined by well-maintained metadata, and describe the dataset at rest (its semantics) as well as its origin (its lineage) and security methods. They also include software service contracts for the timely maintenance of the subscribed quality metrics.

Businesses need to enable datasets to be priced, enhanced as value-added sets, and even sold to the highest bidder. This is driven over time by the cost of maintaining data systems, which can only increase. The data's relevance (correctness), submitted for value-added enrichment and re-integration into commoditized data exchanges, is a key objective:

Don't move data; enrich it in place along with its metadata to preserve semantics and lineage!

The highest bidder builds on the data according to the framework architecture and preserves the semantic domain for which the data system was modeled. Like a ratchet that never loses its grip, datasets need to be correct and hold on to the grip of reality over time. This reality for which the dataset was created can be proposed by the value-added resellers without sacrificing the quality or data service level.

Observe that, over time, the cost of maintaining data correctness, context, and relevance will exceed any single organization's ability to sustain it for a domain. Naturally, it remains instinctual for the IT leader to hold on to the data and produce a silo. This natural reality to hide imperfections for an established system that is literally melting down must be fixed in the future data architecture's approach. Allowing the data to evolve/drift, be value-added, and yet remain correct and maintainable is essential. Imperfect alignment of facts, assertions, and other modeled relationships within a domain would be diminished with this approach.

Too often in today's processing systems, the data is curated to the point where it is considered *good enough for now*. Yet, it is not good enough for future repurposing. It carries all the assumptions, gaps, fragments, and partial data implementations that made it just *good enough*. If the data is right and self-explanatory, its data service code is simpler. The system solution is engineered to be elegant. It is built to withstand the pressure of change since the data organization was designed to evolve and remain 100% correct for the business domain.

"There is never enough time or money to get it right… the first time! There is always time to get it right later… again and again!"

This pragmatic approach can stop the IT leader's search for a better data engineering framework. Best practices could become a bother since the solution just works, and we don't want to fix what works. However, you must get real regarding the current tooling choices available. The cost to implement any solution must be a right fit, yet as part of the architecture due diligence process, you still need to push against the edge of technology to seize on innovation opportunities, when they are ripe for the taking.

Consider semantic graph technology in **OWL-RDF** and its modeling and validation complexities via **SPARQL**, compared to using the labeled property graphs with custom code for the semantic representation of data in a subject domain's knowledge base. Both have advantages and disadvantages; however, neither scales without implementing a change-data-capture mechanism syncing an in-memory analytics storage area for real time analytics use case support. Cloud technology has not kept up with making a one-size-fits all, data store, data lake, or data warehouse. It's better said that one technology solution to fit all use cases and operational service requirements does not exist.

Since one size does not fit all, one data representation does not fit all use cases.

A monolithic data lake, Delta Lake, raw data storage, or data warehouse does not fit the business needs. Logical segmentation and often physical segmentation of data are needed to create the right-sized solution needed to support required use cases. The data engineer has to balance cost, security, performance, scale, and reliability requirements, as well as provider limitations. Just as one shoe size does not fit all... the solution has to be implementable and remain functional over time.

Organization into zone considerations

One facet of the data engineering best practices presented in this book is the need for a primary form of data representation for important data patterns. A **raw ingest zone** is envisioned to hold input **Internet of Things** (**IoT**) data, raw retailer point-of-sale data, chemical property reference data, or web analytics usage data. We are proposing that the concept of the zone be a formalization of the layers set forth in the *Databricks Medallion Architecture* (https://www.databricks.com/glossary/medallion-architecture). It may be worth reading through the structure of that architecture pattern or waiting until you get a chance to read *Chapter 6*, where a more detailed explanation is provided.

Raw data may need data profiling systems processing applied as part of ingest processing, but that is to make sure that any input data is not rejected due to syntactic or semantic incorrectness. This profiled data may even be normalized in a basic manner prior to the next stage of processing in the data pipeline journey. Its transformation involves processing into the bronze zone and later into the silver zone, then the gold zone, and finally made ready for the consumption zone (for real-time, self-serve analytics use cases).

The bronze, silver, and gold zones host information of varying classes. The gold zone data organization looks a lot like a classic data warehouse, and the bronze zone looks like a data lake, with the silver zone being a cache enabled data lake with a lot of derived, imputed, and inferred data drawn from processing data in the bronze zone. This silver zone data supports **online transaction processing** (**OLTP**) use cases but stores processed outputs in the gold zone. The gold zone may also support OLTP use cases directly against information.

The consumption zone is enabled to provide for the measures, calculated metrics, and **online analytic processing** (**OLAP**) needs of the user. Keeping it all in sync can become a nightmare of complexity without a clear framework and best practices to keep the system correct. Just think about the loss of linear dataflow control in an AWS or Azure cloud PaaS solution required to implement this zone blueprint. Without a clear architecture, data framework, best practices, and governance... be prepared for many trials and errors.

Cloud limitations

Data engineering best practices must take into consideration current cloud provider limitations and constraints that drive cost for data movement and third-party tool deployment for analytics when architecting. Consider the ultimate: a zettabyte cube of memory with sub-millisecond access for terabytes of data, where compute code resides with data to support relationships in a massive fabric or mesh. Impossible, today! But wait... maybe tomorrow this will be reality. Meanwhile, how do you build today in order to effortlessly move to that vision in the future? This is the focus of the best practices of this book. All trends point to the eventual creation of big-data, AI enabled data systems.

There are some key trends and concepts forming as part of that vision. Data sharing, confidential computing, and concepts such as *bring your algorithm to the data* must be considered as core approaches to repurposing datasets, their semantics, and their business value as data leaves the enterprise and enters the publicly available domain. With the data, information, and derived knowledge comes the data consumption handle, which is more than loosely defined metadata. It consists of the lineage, semantics, context, and timely value needed to sustain trust so that monetary compensation for the stream of value-added information will be possible. These royalty contracts make data resellable and demystified. Just like a book is published, so can data be published. The best practices of this book will position data to support the development of value-added services over curated information in the course of time. Smart data becomes a value-added ecosystem in and of itself, which is as important as the software data processing systems of past generations, where data was but a snapshot of the processing state for which that system was created.

The Intelligence Age

Future state **data-as-a-service** (**DaaS**) offerings will depend on the new data engineering framework that this book will highlight. Also, we will show the best practice considerations for the development framework required by that structure. The process of curating knowledge from data, and its metadata into information and then insights, will involve the process of transforming data into truths that withstand the test of time. This novel data framework required for the commoditization of information is essential as we enter the **Intelligence Age** and exit the **Information Age**. In the *Intelligence Age*, insights are gathered from knowledge formed from information operated upon by human and AI systems.

Achieving AI goals requires the application of various machine and deep learning algorithms that define the *Intelligence Age*. Along with these algorithms, you can envision the development of extremely advanced quantum computers and zettabyte in-memory storage arrays. These will be needed as part of the new data engineering organization. What is often not discussed is the data engineering framework required to facilitate the algorithms; otherwise, the data lake will become a data swamp in short order. The power of computing will advance in leaps and bounds. What we build today in software systems paves the way for those hardware advances – not the reverse. Software algorithms drive the hardware computing power needed; otherwise, the hardware remains underutilized. Data engineering has been a laggard in the evolutionary process.

For this purpose, this book has been created to future-proof data engineering architectures and designs by providing best practices with considerations for senior IT leadership thinking. Along with that come some practical data architecture approaches with use case justifications for the data architect and data engineer; these will add color to those best practice descriptions.

Use case definitions

Question: Why should the development of any data engineering system be use-case-driven?

Answer: If one cannot develop a solution that integrates with the business's needs, it is irrelevant; it can't be communicated, nor can its efficacy be quantifiable.

Without use cases, a data processing system does not provide the tangible value required to keep up funding continuance. Even if a solution is the best thing since peanut butter, it will quickly devolve into an ugly set of post-mortem discussions when failures to meet expectations start to arise.

The solution needs to be built as a self-documenting, self-demonstrable collection of use cases that support a test-driven approach to delivery.

It's part art and part science, but fully realizable with a properly focused system data architecture. Defining reference use cases and how they will support the architecture is a high bar to achieve. As the use cases are created and layered into the development plans as features of the solution, you must not get lost in the effort. To keep the focus, you need a vision, a strategy, and a clear mission.

The mission, the vision, and the strategy

You should begin with the mission and vision for which this overview section has laid a foundation. These should be aligned with the organization's strategy, and if they are not… then alignment must be achieved. We will elaborate more on this in subsequent sections.

Principles and the development life cycle

Principles govern the choices made in the development of the business strategy defining the architecture, where the technologists apply art and science to fulfill the business needs. Again, alignment is required and necessary; otherwise, there will be difficulties when the first problems arise and they are not easily surmountable. The cost of making mistakes early is far greater than making errors later in the engineering development life cycle. The data engineering **life cycle** begins with architecture.

The architecture definition, best practices, and key considerations

The **architecture** can be developed in many ways, but what we as engineers, architects, and authors have discovered is that the core architecture deliverable needs to have three main components:

- A conceptual architecture
- A logical architecture
- A physical architecture

The conceptual architecture shows the business mission, vision, strategy, and principles, clearly implemented as an upward and outward facing communications tool. In the conceptual architecture's definition, there will be a capabilities matrix that shows all the capabilities needed for your solution and these will be mapped deliverables. This will be the focus of *Chapter 5*, but for now, it is enough to know that the foundation of the solutions' concepts will be your principles that are aligned with the vision, mission, and strategy of your business.

The logical architecture shows the software services and dataflows necessary to implement the conceptual architecture and ties the concepts to the logical implementation of those concepts. The physical architecture defines the deployable software, configurations, data models, ML models, and cloud infrastructure of the solution.

Our **best practices** and **key considerations** are drawn from years of experience with big data processing systems and start-ups in the areas of finance, health, publishing, and science. When working in those areas, projects included analytics of social media, health, and retail analytic data.

Use cases can be created using information contained in the logical as well as the physical architecture:

- Logical use cases:

  - Software service flows

  - Dataflows

- Physical use cases include:

  - Infrastructural configuration information

  - Operational process information

  - Software component inventory

  - Algorithm parameterization

  - Data quality/testing definition and configuration information

  - DevOps/MLOps/TestOps/DataOps trace information

Reusable **design patterns** are groupings of these use cases that have clean interfaces and are generic enough to be repurposed across data domains, therefore reducing the cost to develop and operate these patterns. With the simplification of the software design due to the smart data framework's organization, use cases will coalesce into patterns easily. This will be an accelerator for software development in the future. Dataflows will be represented by these design patterns, which make them more than just static paper definitions. They will be operational design patterns that reflect the data journey through the data framework's engineered solution that is aligned with the architecture.

The DataOps convergence

The data journey is a path from initial raw data ingestion through classification that ultimately positions transformed information for end user consumption. Curated, consumable data progresses through various zones of activity. These are going to be defined better in subsequent chapters, but the zones are bronze, silver, and gold. Datasets are curated in a data factory manner that is logically and physically grouped into these zones of activity. All custom built and configured data factory hosted data pipeline journeys utilize a data engineer's standard process, which you will develop; otherwise, IT operations and the maintenance of service levels through agreeable contracts would be at risk. Data transformation and cataloging activities are centered around what others have coined DataOps.

By 2025, a data engineering team guided by DataOps practices and tools will be 10 times more productive than teams that do not use DataOps. (2022 Gartner Market Guide for DataOps Tools, {https://packt-debp.link/6JRtF4})

DataOps, according to Gartner, is composed of five core capabilities:

- Orchestration capabilities involve the following:

  - Connectivity

  - Scheduling

  - Logging

  - Lineage

  - Troubleshooting

  - Alerting

  - Workflow automation

- Observability capabilities enable the following:

  - Monitoring of live or historic workflows

  - Insights into workflow performance

  - Cost metrics

  - Impact analysis

- Environment management capabilities cover the following:

  - **Infrastructure as code (IaC)**

  - Resource provisioning

  - Credential management

  - IaC templates (for reuse)

- Deployment automation capabilities include the following:

  - Version control

  - Approvals

  - Cloud CI/CD and pipelines

- Test automation capabilities provide the following:

  - Validation

  - Script management

  - Data management

To illustrate how these DataOps principles can be applied, imagine a large retail company deploying an inventory management system. See *Figure 1.2*:

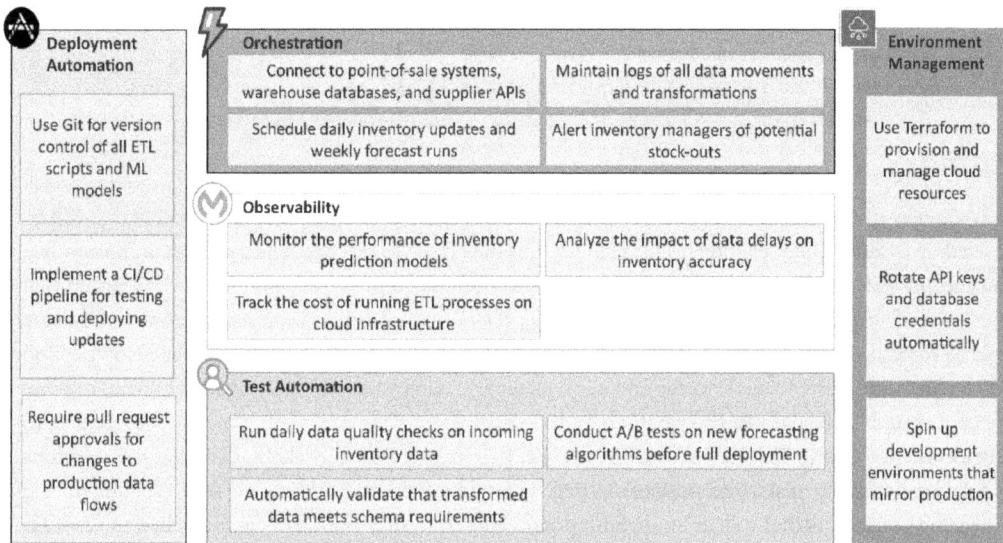

Figure 1.2 – Retail inventory management capabilities

Many third party vendors have jumped on the DataOps hype and produced fantastic tooling to jumpstart the convergence of DevOps, **MLOps**, and **TestOps** practices for modern cloud data systems.

The data engineering best practices of this book will also support the DataOps practices noted by Gartner while remaining neutral to the specific tooling choices. The focus will be on the data engineering framework that the *DataOps* effort will make streamlined, efficient, and future-proof. Refer to *Figure 1.3*:

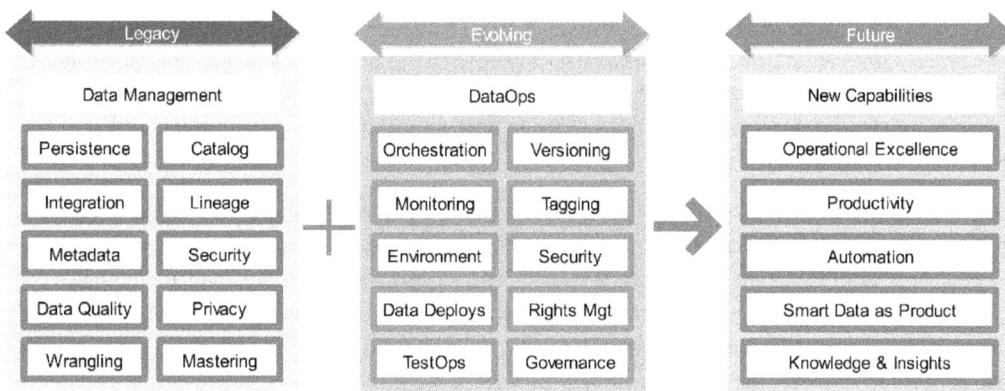

Figure 1.3 – DataOps tools augmenting data management tasks

It is clear that DataOps adds a lot of value to legacy data management processes to enable a future where new capabilities are made possible. In the following quote, you can see how modern DataOps processes will enable faster development:

> *A reference customer quoted that they were able to do 120 releases a month by adopting a DataOps tool that was suitable for their environment, as opposed to just one release every three months a year ago. (Gartner, 2022, {https://packt-debp.link/41DfFu})*

Summary

In this overview of the business problem, you have learned a number of foundational elements that will be elaborated on in subsequent chapters. This chapter introduced the topics needed to gain an understanding of the current state of data engineering and the creation of future-proof designs. You have learned that businesses are faced with an ever-changing technological landscape. Competition requires one to innovate at scale to remain relevant. This causes a constant implementation stream of total-cost-of-ownership (TCO) budget allocations for refactoring and re-envisioning during what would normally be a run/manage phase of a system's lifespan. In this chapter, and in subsequent chapters, we make many references to the engineering solution's TCO. These references will be reminders to all stakeholders that the solutions developed are within the real world business setting. They are not created in some abstract vacuum, devoid of budgeting constraints that will, at times, limit possibilities. It is important to note that when the TCO is clear, yet constrained by budgets, these constraints repeatedly appear on the monthly and quarterly radar reports presented to the enterprise. These constraints will most likely have imposed risk. Without a constant stream of reminders, the business will forget how these constraints have impacted the solution.

Additionally, building a system that perpetuates *false facts*, even if spun as *true facts*, is foolish. Make the future data solution smart! We are entering an exciting future where data and information solutions will become smarter and support knowledge and intelligence capabilities. Embrace the change and know its implications on your data engineering choices. DataOps needs to be adopted by data professionals as a critical approach to managing data in today's complex, data-driven world.

One size does not fit all and as such, building with data contracts in mind will force the development of data stores with the same logical data into the physical data architecture as fit-for-purpose parallel instantiations. Correctly building data solutions to be future-proof requires a vision, strategy, mission, and architectural approach to prevent the implementation from dying an untimely death due to the juggling needed to get the solution serviceable for the business.

Third-party vendors and cloud providers will produce well architected solutions that do not integrate, or worse yet, that foster architectural anti-patterns that must be avoided. As such, the data mesh and the cloud provider's data fabrics are only buzzwords until the concepts are fully understood and rationalized into your architecture and organization's objectives. Design data solutions consistently to the architecture you develop, develop use cases across the system, and test, regression test, and monitor them for continual service in order to maintain the trust established through data contracts.

Lastly, stay agile! Read! Learn! Be innovative! Once the big picture is grasped, the forward-looking perspective will grant you the foresight to look beyond the obstacles that will be encountered. You will be able to keep the data solution and its data fresh and current with a governed, agile architectural process.

In the next chapter, you will be presented with the architectural background challenges that build on this overview.

2

A Data Engineer's Journey –
Background Challenges

The purpose of this chapter is to explain the challenges you must face to navigate a successful journey as a data engineer. The intricate nature of managing vast data, evolving technologies, and ensuring efficient data pipelines are some of the hurdles that make data engineering a demanding field. In this chapter, we will explain why data engineering is hard, and provide a foundation for the development of principles to be discussed in later chapters. We will also provide insight when navigating these challenges. Finally, we will provide an overview of the data engineering approaches you can use to scope out the current and near-future technology landscape.

We will discuss three of the main data engineering challenges:

- Platform architectures change rapidly

- There is a high cost and impact on a solution's longevity from the strategy to buy rather than build

- The prolific evolving set of data repository patterns marketed to unsuspecting data engineers

> *"Imagine you need to fill a jar with some big rocks, pebbles, and sand. What do you put in the jar first?" (Marlow consulting, June 7, 2023,* https://getmarlow.com/article/big-rocks-pebbles-sand---how-to-manage-it-all-1563899084731x981090384055894000).

This metaphor is often cited when trying to prioritize what should go into your ordered list of priorities, with the "big rocks" being symbolic of the topmost priority, followed by "pebbles" (medium), and lastly "sand" (least priority). The big rock lesson is that if you don't put the rocks in the jar in order, they won't all fit later. We provide a list of the big rocks, such as data immutability, lineage tracking, data quality preservation, scalability, security and compliance, data discoverability and accessibility, and cost-efficiency, which are considered the major considerations, or focal points, to accommodate when navigating these challenges.

To help you understand these challenges, we will provide you with an overview of data engineering technological changes taking place today. You will also get an overview of common data engineering architecture patterns, such as data warehouse, data lake, data fabric, data mesh, in-memory data stores, and niche data aggregators. We will also discuss the following:

- Data Warehouse
- Data Lake
- Data fabric
- Data mesh
- In-memory data stores
- Niche data aggregators.

After reading this chapter, you will understand the background challenges within data engineering and be equipped to navigate them. You will be aware of some relevant aspects of data engineering history, as well as the different architectural patterns that have evolved over the years. By delving into the historical evolution of data engineering architectures, you will gain a nuanced understanding of present-day challenges and be better prepared to navigate them as you generate future-proof solutions. Lastly, you will understand the vital considerations that underpin the construction of a robust data engineering system. In this chapter, we will address three key categories of challenge. Within each are factors to be considered, as shown in *Figure 2.1*:

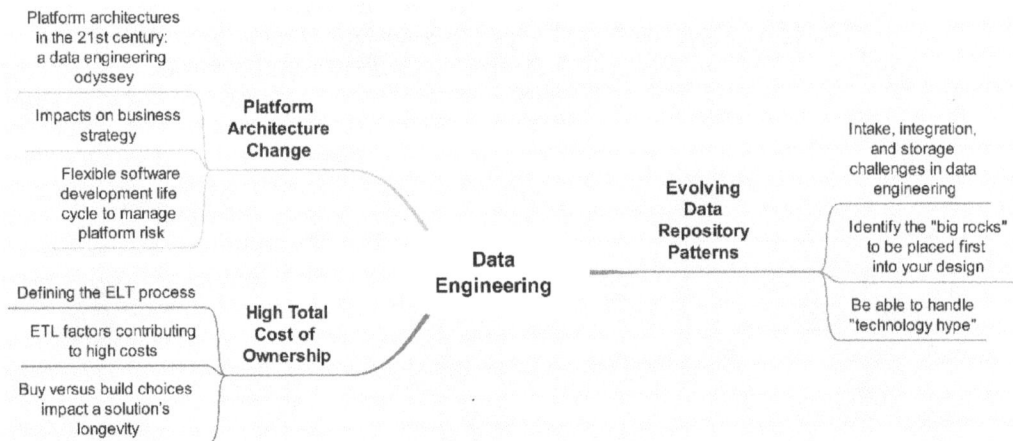

Figure 2.1 – High-level challenges mind map

The numerous data patterns are examples of the rapid rate of technical change in data engineering. However, as we'll discuss in the next section, change is pervasive throughout all aspects of data engineering.

Challenge #1 – platform architectures change rapidly

The dynamic landscape of platform architectures poses both opportunities and challenges for data engineers. As systems and technologies rapidly evolve, there's a pressing need for professionals in the field to stay abreast of these changes. Adapting to these shifts is not merely about understanding the newest tools or frameworks; it's about foreseeing potential bottlenecks, ensuring system compatibility, and optimizing processes to accommodate these evolutions. Moreover, these frequent changes compel data engineers to embrace a continuous learning mindset, ensuring that their skills and knowledge remain relevant in an ever-shifting landscape.

In this section, you will learn about systemic changes in platform architectures in the last two decades, such as the move from SQL to NoSQL, the rise of big data, and the migration to the cloud. Please refer to *Figure 2.2*:

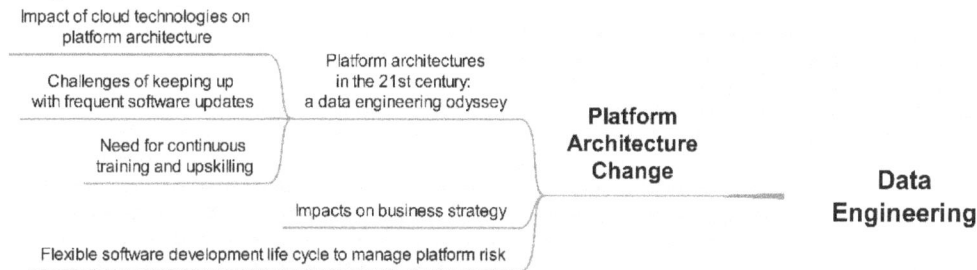

Figure 2.2 – Platform architecture mind map

We'll focus on how these changes impact business strategy and how you can manage platform risk through a flexible **software development life cycle** (**SDLC**).

Platform architectures in the 21st century

It is insightful to look back at how data engineering has evolved in the past 20 years. At the beginning of the new millennium **relational database management system** (**RDMS**) was the dominant paradigm. The start of the 21st century brought forth an unprecedented era of technological metamorphosis. As with previous shifts, those companies and individuals that adapted quickly thrived, while those resistant to change found themselves struggling. Through the lens of a data engineer, the evolution of platform architectures in this period was nothing short of revolutionary. In the coming sections, you will learn about these changes and how they affect your designs:

- **What effect does the dawning of big data have on your designs?** The new millennium kicked off with the promise of data, and lots of it. Data sources multiplied, pouring in from e-commerce sites, the **Internet of Things** (**IoT**) devices, social media, and more. Traditional RDBMS solutions, however adept in the late 20th century, began showing cracks under this onslaught. The sheer volume, variety, and velocity of data necessitated new architectures. Hadoop and its ecosystem, inspired by Google's **MapReduce**, symbolized this paradigm shift, offering distributed storage and parallel processing. However, with this shift, data engineering faced new challenges, such as cluster management, fault tolerance, and ensuring data consistency.

- **How do NoSQL versus relational schema decisions impact your designs?** The structured world of SQL met its versatile counterparts – that is, NoSQL databases. Whether it was the document-oriented **MongoDB**, wide-column stores such as **Cassandra**, or graph databases such as **Neo4j**, NoSQL architectures offered flexibility. But they also posed questions on data integrity, consistency, and transactional guarantees. The data engineering world was now split with structure and **atomicity, consistency, isolation, and durability** (**ACID**) guarantees on one side, and flexibility and scalability on the other.

- **How does the emergence of cloud providers influence your design?** The rise of cloud providers such as **AWS**, **Azure**, and **Google Cloud Platform** (**GCP**) changed the game altogether. Data storage became more economical, elastic compute resources reshaped processing paradigms, and **Platform-as-a-Service** (**PaaS**) offerings streamlined operations. This cloud movement, however, meant data engineering had to grapple with data transfer speeds, multi-cloud strategies, and cloud-native tool adaptations. Cloud computing makes it possible to deliver real-time predictive analytic solutions that scale. Cloud compute-optimized services provide cost-effective capabilities for decision support and analytics as opposed to static legacy reporting. These essential cloud advantages will need to be part of your future solutions.

- **How does the need for real-time streaming influence your designs?** Gone were the days when batch processing was the only answer. Platforms such as **Kafka** and **Spark Streaming** heralded the age of real-time analytics. Data engineering tasks grew in complexity – from ensuring low-latency data ingestion to managing stream processing at scale.

- **How do containerization and microservices influence your designs?** The monolithic architectures of yesteryears paved the way for microservices, and with them came **Docker** and **Kubernetes**. Data engineers now found themselves juggling data tasks between containers, ensuring seamless data flows across microservices, and managing container orchestration.

- **How do AI and machine learning integration influence your designs?** As machine learning models moved from labs to production, data engineering platforms had to evolve to support machine learning (ML) pipelines. Tools such as **TensorFlow** {https://packt-debp.link/ENz3e1} and **MLflow** {https://packt-debp.link/KJdU8O} added new dimensions to data tasks, from feature engineering to model deployment.

- **How does the need for data governance and quality influence your designs?** With the democratization of data came the essential need for governance and quality assurance. Data lineage, cataloging, and quality checks became paramount. Data engineers now juggled both the macro (architecture, flow) and the micro (data quality, lineage) aspects of data.

In essence, the 21st century has been a roller coaster for platform architectures, continuously challenging the conventions and demanding innovation. Data engineering, sitting at the crossroads of these shifts, has been an exciting field to observe, adapt, and innovate. As Stonebraker often emphasized, *One size doesn't fit all* {https://packt-debp.link/TzSXRT} and the journey of data engineering in this century is a testament to that mantra.

Impact of cloud technologies on platform architecture

The advent of cloud technologies brought about unprecedented flexibility and scalability in platform architectures. Cloud providers enabled organizations to offload infrastructure management, allowing data engineers to focus on design, optimization, and functionality. Elastic scaling, on-demand resources, and pay-as-you-go models revolutionized how businesses approached infrastructure investments. Moreover, managed services provided by cloud platforms – from databases to AI services – accelerated development and reduced time to market.

Challenges of keeping up with frequent software updates

The rapid evolution of software tools and platforms means that data engineers are constantly on their toes. Every update can bring improvements but also potential incompatibilities or new vulnerabilities. Staying updated requires a proactive approach, not only in installing patches but in understanding their implications. The balance between leveraging new features and maintaining system stability becomes a delicate dance.

The need for continuous training and upskilling

The rapid pace of change in the world of data engineering means professionals cannot afford to rest on their laurels. Continuous training and upskilling become imperative. Organizations need to invest in training programs, workshops, and certifications, ensuring their teams are equipped with the latest knowledge and best practices.

There are many training paths open to a professional. Expect 20% of one total career time to be allocated to upskill efforts such as the following:

- Mentoring or being mentored
- Participating/contributing to open source projects
- Coaching or being coached
- Being a member of a **center of excellence** (**CoE**) and participating as a trainer as part of a train-the-trainer program
- Reading journals
- Being a member of a vendor's user/product focus group

- Attending free online training

- Creating training materials

- Building a knowledge base of artifacts so others can learn from you, which organizes and solidifies your knowledge

- Creating a 5-year and 10-year career plan with skills needed at various stages (for yourself) or for others

- Attending company-paid conferences such as AWS Summit, Microsoft Ignite, IBM, Databricks' AI Summit, and more

- Obtaining certification path training (sometimes good but often expensive and requires continued updating)

- Speaking at conferences (requires self-assessment and lots of preparation and depth of knowledge)

- Writing a book (which is like a preparation for a conference speaking engagement but with even more preparation, research, and depth of knowledge)

Impacts on business strategy

In a data-driven age, platform architecture directly influences business strategy. The ability to quickly process and gain insights from data can offer competitive advantages. However, the choice of architecture also dictates costs, scalability potential, and adaptability to future changes. Businesses must align their strategy with their technological capabilities and be prepared to pivot as those capabilities evolve.

What are the business risks of not adapting to technological changes?

Failure to adapt to the evolving technological landscape can spell disaster for businesses. Risks include falling behind competitors, incurring higher costs due to outdated systems, being vulnerable to security threats, and not meeting customer expectations in an increasingly digital world.

A flexible software development life cycle to manage platform risk

A nimble **software development life cycle** (**SDLC**) is critical in the face of changing platform architectures. Adapting SDLC processes to be more flexible allows organizations to respond swiftly to technological changes, ensuring they harness the power of new tools without compromising system integrity. This section provides an overview of managing platform risk through leveraging agile methodologies in a flexible SDLC.

- **How do you adapt SDLC to agile methodologies?** Agile methodologies prioritize flexibility and responsiveness. By breaking down development into smaller, iterative cycles, data engineering teams can more effectively respond to changes, whether those are new platform features or shifting business requirements.

- **How do you address the rise of agile and scrum in managing change?** Agile and scrum have become the gold standards in managing change in the tech world. Their emphasis on collaboration, continuous feedback, and iterative progress aligns perfectly with the dynamic nature of data engineering.

- **What are the benefits of iterative development cycles in a changing landscape?** Iterative cycles allow for continuous improvement. Instead of waiting for a final product, teams can deploy functional modules, gather feedback, and make adjustments on the fly. This not only improves end-product quality but also reduces the risks associated with long development cycles.

- **How do you balance flexibility with stability in an SDLC?** While flexibility is essential, it cannot come at the cost of stability. Ensuring robust testing, maintaining thorough documentation, and adhering to best practices are all crucial in ensuring that the pursuit of agility doesn't compromise system integrity.

- **How do you address the challenges of frequent change to established processes?** Constant changes can be disruptive. Teams may face learning curves with new tools, struggle with migrating systems, or grapple with the loss of familiar processes. Managing change effectively requires clear communication, comprehensive training, and a culture that embraces adaptability.

- **How do you maintain documentation and knowledge transfer with a flexible process?** As processes evolve, maintaining clear documentation becomes critical. Proper documentation ensures that knowledge isn't lost during team transitions and that all members, new or old, have a clear understanding of the systems in place.

In the next section, you will learn how to address the next challenge: the high **total cost of ownership** (**TCO**).

Challenge #2 – Total cost of ownership (TCO) is high

Engineers are tasked with making it happen! But how they make it happen is subject to many constraints. The first is cost and the second is time. Issues such as the total cost of running and managing the solution over time and the feasibility of maintaining it operationally also come into focus. The TCO for a well-engineered data solution is affected by the **extract, transform, and load** (**ETL**)/**extract, load, and transform** (**ELT**) architecture and **buy versus build** tooling choices for selected adopted solutions and architecture patterns. Please refer to *Figure 2.3*:

Historical context: the origins of ETL

Defining the ELT process

Modern evolutions: ELT and cloud based ETL services

Hardware and infrastructure expenses

ETL factors contributing to high costs

Labor costs: ETL developers, data quality assurance, and maintenance

Consider the time-to-market of buy versus build

Advantages and disadvantages of off-the-shelf tools

Consider support and community available assistance when choosing buy versus build solutions

Caution - off-the-shelf tools often lack flexibility

High Total Cost of Ownership

If you build it - you gain full control over all capabilities and features

If you build it - then tailoring to specific organizational needs is possible

Be prepared for custom solutions downsides

Building custom solutions is alluring

Buy versus build choices impact a solution's longevity

Take into consideration maintenance overhead

Address the syndrome: not invented here (NIH)

What's the impact of buy versus build on a solution's longevity?

Beware of evolving data standards and formats

Future-proofing and scalability concerns

Be ready to adapt to organizational growth and changing requirements

Dependency on vendor for updates and patches

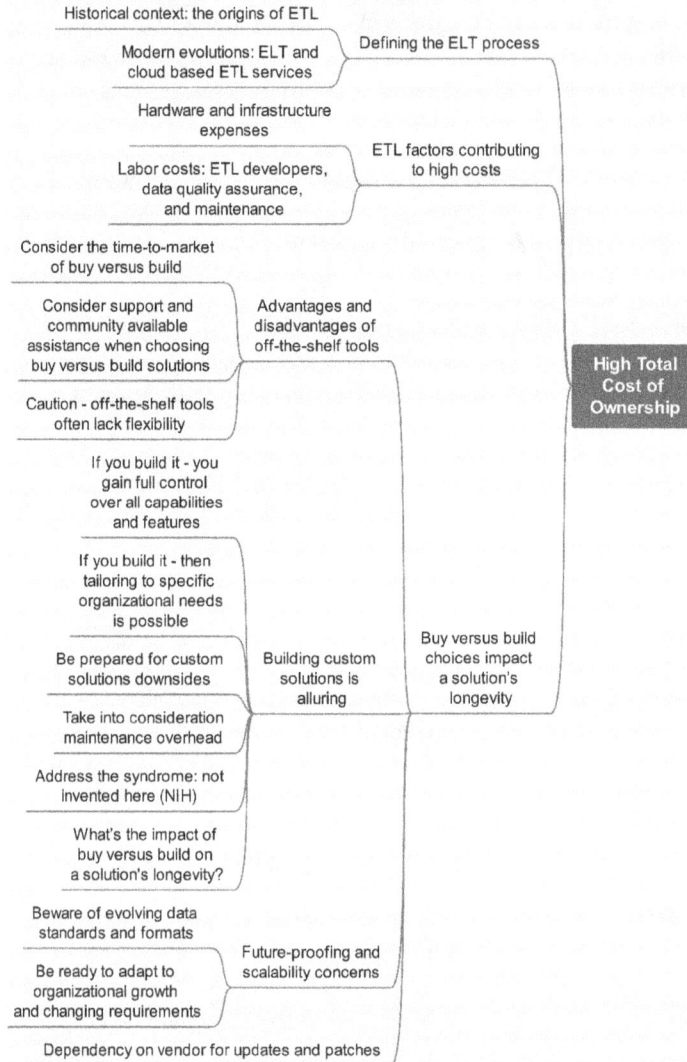

Figure 2.3 – TCO mind map

After reading this section, you will have a greater understanding of ETL/ELT: what it is, its origins and historical evolution, why its costs are so high, and the advantages of *build versus buy*. You will also learn why legacy master data management architectures are no longer in vogue.

ETL architecture costs are high!

ETL is a long-standing process blueprint that has historically been the first line of processing that affects incoming data for a data warehousing and **business intelligence** (**BI**) solution. Its application enables organizations to make sense of raw data. This, in turn, enables it to be transformed into information by various data pipeline transformations, with the final downstream insight generation stage made available to the consumer.

ETL – or its alternative, ELT – is an essential part of data engineering. Since being able to load data cannot be skipped, its high cost for processing and retention can be a major budgetary strain. This is made more evident when assessing the legacy **master data management** (**MDM**) offering costs that wrap data warehouses with governance, workflow, and rule engines to affect the desired result, which is establishing gold standard data in a warehouse. The cloud has changed all this and ripped the E, L, and T apart from the data warehouse and created what is advertised as a cheaper, more scalable set of performant alternatives in PaaS services.

Is it extract, transform, and load (ETL) or is it extract, load and transform (ELT)? The need to load data first and then transform it afterward leads to the ELT pattern for ingesting data. Raw data may be too raw, and it may not be normalizable until it is fully ingested and profiled for quality. Additionally, data being rendered fit for purpose requires it to be kept in as near raw form as possible… just not *too* raw. The shift in cloud service support for ELT is evident in **Apache Spark**, **Databricks**, **Alteryx Trifacta**, and other big data processing engines. Thus, ELT is now preferred to ETL.

Cloud providers such as Amazon and Microsoft have produced pipeline services such as **AWS Pipeline** {https://packt-debp.link/qqK7ZM} and **Azure Pipelines** {https://packt-debp.link/POYzEK} and used them to knit together ELT processing equivalents. What is evident when total cost of ownership for cloud pipeline ETL solutions are compared to legacy options is that you gain greater flexibility to implement logic using a cloud solution. There will be higher initial development and then run time cost since you are not buying a tool but building from more primitive PaaS services and finer grained third party tools. The third party ETL framework costs for a legacy provider's solution framework would justify many capabilities, but can these be integrated by an organization?

Many advanced features are left unused. This happens when human governance and workflows provided by the legacy ETL MDM tool are overwhelmed by the huge data volumes being encountered today. I've seen work queues for manual data intervention that are 10,000+ lines deep. They will get ignored and data errors will propagate downstream or worse yet get into the gold standard data of a data warehouse.

In the next section, we'll define the ELT process and each of its individual components.

Defining the ELT process

The benefits of ELT (or ETL) lie in its three underlying acronym identified processes:

- **Extract**: This is where we initiate the data journey, from the raw source to data storage. The essence of extraction lies in identifying and capturing the data that reflects the business operations needed. It is the first crucial step for analytical pursuits. Some argue that the scope of the extraction is to take it all versus taking just what is needed. The issue is that you don't know what you might need tomorrow. Your extraction process should take it all but store what is essential today in a set that is linked to the set that may be used tomorrow. Drawing the indices and maps to keep the dataset logically together but potentially separate is a key set of decisions that you should make as early as possible. Data is extracted from myriad sources and types, which might include transactional databases, flat files, external data sources, APIs, and so on. The data that's extracted is often a raw, unrefined resource awaiting its profiling and metamorphosis to a normalized loadable form.

- **Load**: In this leg of the data journey, data finds its home in data storage. It's stored in a fit for purpose form and meticulously organized, mapped, and indexed. It is made ready for data quality querying, profiling, wrangling, eventual downstream transformation, and further exploration. The structured repository that's created serves as a source for various analytical tools; however, it may not have all the metadata needed to make it actionable.

- **Transform**: Transformation is where the data is adjusted, normalized, and assigned semantic metadata. After being wrangled, profiled, mapped, restructured (even parallel stored), partitioned, and made ready in downstream processing, it's analyzed for quality. Data undergoes these processes to ensure consistency and accuracy, and that there's a format aligning the target data pipeline stage with the data lake or warehouse schema storage. This phase is essential to ensure that data is made fit for purpose and truthful over time. Iterative transforms can take place over loaded data, especially as corrections arrive, restatements are submitted, or additional extracted and loaded data is added to the transformed datasets presented to storage. Data transformation processes directly impact the quality of data pipelines in later stages since insights will be derived from what meaning is harvested from the loaded data and retained as metadata for dictionary reflection, analytics, and insight generation purposes.

In the next section, you will gain a deeper understanding of ETL by learning about its historical origins and modern evolutions.

Historical context – the origins of ETL

Database management systems have existed since the 1970s when organizations began to see the potential in leveraging data for analytical purposes. However, it wasn't until the late 1980s and early 1990s that the term **data warehousing** came into prominence, largely attributed to the seminal works of industry visionaries such as Bill Inmon and Ralph Kimball.

As the concept of data warehousing germinated, there arose a distinct necessity for a process to funnel data from operational systems into these analytical repositories. The inception of ETL was driven by the need to efficiently ferry data from operational systems into a data warehouse, ensuring it is neatly organized, accurate, and reflective of the business operations. The 1990s saw the proliferation of ETL as a critical component of data warehousing solutions. The term ETL began to be emblematically used in technical and business discussions as organizations embarked on building data warehouses. During this period, commercial ETL tools started to appear on the market, reflecting a maturing and formalization of ETL processes.

Modern evolutions: ELT and cloud based ETL services

The modern epoch has seen the cloud becoming a mainstay in the data management realm. Cloud-based services offer an allure of scalability, flexibility, and cost-effectiveness, traits that are quintessential in managing the modern data deluge. Platforms such as **AWS Glue**, **Azure Data Factory**, and **Google Cloud Dataflow** embody this cloud-centric narrative, offering robust, scalable, and fully managed ELT services. They promise a realm where data processing is no longer tethered to on-premises infrastructures but soars in the boundless expanse of the cloud, unhindered by the physical limitations of yesteryear designs.

However, the reality is more complicated as these services have high operational costs. According to Wakefield Research and Fivetran {https://packt-debp.link/ZajI4X}, a typical data engineering team will spend about $520,000 annually to build and manage custom ETL pipelines. Many factors lead to these high costs.

ETL factors contributing to high costs

The cost dynamics surrounding ETL processes are multifaceted and can significantly impact an organization's budget and resource allocation. Several factors contribute to the high costs associated with ETL processes, and understanding these factors is pivotal for optimizing expenses while ensuring robust data engineering practices. Let's look at a few factors that contribute to the aforementioned high costs.

Hardware and infrastructure expenses

The third party framework, hardware, and infrastructure required for ETL processes are often major expenses. ETL operations demand robust computational resources and efficient network infrastructures to manage data extraction, transformation, and loading. The requisite servers, storage, and network gear can be capital intensive. Many tools are not built for cloud deployments and come with unclear operational costs on top of being expensive. Moreover, as data volumes escalate, there's a corresponding surge in the need for additional hardware and infrastructural upgrades to handle growing workloads. It's also pertinent to factor in the costs of redundancy and disaster recovery solutions to ensure data integrity and availability.

Licensing costs for ETL tools

ETL tools, especially those from leading vendors, come with significant licensing fees. These tools offer a gamut of functionalities that simplify and accelerate the ETL process but at a cost. Licensing models vary, but many are structured on a per-user or per-core basis, which can quickly escalate with the scale of operations. Moreover, the costs are not just confined to the ETL tools themselves; additional costs can accrue from ancillary software licenses required for monitoring, scheduling, and orchestrating the ETL workflows.

Labor costs: ETL developers, data quality assurance, and maintenance

Labor is another major cost driver in ETL processes. ETL developers are specialized professionals whose expertise is critical for designing, implementing, and optimizing ETL workflows. Similarly, the role of data quality assurance personnel is crucial to ensure the accuracy and consistency of data being moved and transformed. Furthermore, ETL processes demand ongoing maintenance to accommodate evolving data schemas, business rules, and performance tuning. Maintenance also encompasses troubleshooting and resolving issues that arise during the ETL life cycle. The cumulative labor costs for ETL development, quality assurance, and maintenance form a significant chunk of the ETL cost structure.

By meticulously analyzing and addressing these cost/benefit factors, organizations can devise strategies to optimize their ETL expenditures. Whether through leveraging more cost-effective hardware, exploring open source ETL tools, or streamlining labor costs through automation and effective project management, there are avenues to navigate the cost landscape of ETL processes with fiscal prudence. There are over 300 ETL vendors. The following list may help you select from them the capabilities that you want:

- Ab Initio {https://packt-debp.link/UZgu5O}

- Airbyte {https://packt-debp.link/MHAl8f}

- Alooma {https://packt-debp.link/zft1xq}

- Apache Airflow {https://packt-debp.link/Aj59PY}

- Apache Hadoop {https://packt-debp.link/u6W9GO}

- Apache NiFi {https://packt-debp.link/UjDjQ9}

- AWS Data Pipelines {https://packt-debp.link/P3OjMO}

- AWS Database Migration Service {https://packt-debp.link/BwZ4vz}

- AWS Glue {https://packt-debp.link/v09M1X}

- BusinessObjects {https://packt-debp.link/fMQ2Z0}

- Census {https://packt-debp.link/lUQsaF}

- CloverDX {https://packt-debp.link/mDQPY3}

- Coupler.io {https://packt-debp.link/f6mN4K}

- Data Fetcher {https://packt-debp.link/t9xENa}

- Databricks Lakehouse Platform {https://packt-debp.link/RQs5GU}

- Dataddo {https://packt-debp.link/8QrFjR}

- dbt {https://packt-debp.link/no6Fd0}

- Designer Cloud {https://packt-debp.link/lO6wNZ}

- Fivetran {https://packt-debp.link/K6K1is}

- Google Cloud Dataflow {https://packt-debp.link/FYy9bK}

- Hevo {https://packt-debp.link/pfz7PY}

- IBM Infosphere Datastage {https://packt-debp.link/BgJ5hn}

- Informatica PowerCenter {https://packt-debp.link/XCIAk3}

- Integrate.io {https://packt-debp.link/551PJ0}

- Keboola {https://packt-debp.link/QVuoaY}

- Matillion Limited {https://packt-debp.link/jddYWI}

- Microsoft Azure Data Factory {https://packt-debp.link/cNpzdj}

- Microsoft SQL Server Integration Services (SSIS) {https://packt-debp.link/18IYLd}

- Mozart Data {https://packt-debp.link/J5GVsi}

- Oracle Data Integrator {https://packt-debp.link/qU09gw}

- Panoply {https://packt-debp.link/e4xwCg}

- Pentaho Data Integration (PDI) {https://packt-debp.link/bNqQFE}

- Pervasive Software {https://packt-debp.link/DMm3QG}

- Polytomic {https://packt-debp.link/9AXdTr}

- Portable {https://packt-debp.link/kRIPBD}

- Qlik Compose {https://packt-debp.link/G7i1qc}

- Rivery {https://packt-debp.link/CGszd1}

- SAP BusinessObjects Data Services {https://packt-debp.link/YqM1x7}

- Singer {https://packt-debp.link/ijslhp}

- Singular {https://packt-debp.link/fgux6J}

- Skyvia {https://packt-debp.link/KUskjY}

- Snaplogic {https://packt-debp.link/zYf6yF}

- StarfishETL {https://packt-debp.link/zkAVz9}

- Stitch {https://packt-debp.link/pP5WUg}

- Supermetrics {https://packt-debp.link/kgOCkJ}

- Sybase {https://packt-debp.link/KLg0IK}

- Talend Open Studio (TOS) {https://packt-debp.link/YdJMsN}

- Tray.io {https://packt-debp.link/UN2Uti}

- Weld {https://packt-debp.link/fzPxQX}

- Whatagraph {https://packt-debp.link/Gtwk7O}

- Workato {https://packt-debp.link/Q8L1gK}

- Xplenty {https://packt-debp.link/XPDLSt}

Identifying the capabilities that one or more of the preceding may afford your solution is a key takeaway from this chapter. You need to access the benefits of any tool for fitment into your solution since each of the preceding has brought something to market that was compelling to some of their audiences.

Buy versus build choices impact a solution's longevity

The conundrum of choosing between purchasing off-the-shelf tools and crafting custom solutions is foundational in data engineering endeavors. This pivotal decision influences not merely the immediate project goals but extends to its impact on the longevity and adaptability of the data solutions deployed.

Cloud vendors will assert that PaaS services are easily consumed and there are clear integration patterns for incorporating proprietary cloud designs. These patterns, if used, will reduce customer costs. **Software-as-a-Service** (**SaaS**) vendors will assert that solutions perform better in their hosted environments and that they will provide the data security, operations, monitoring, and other related services that are difficult to develop (but at a cost!). These are all taken care of, or are they? Try to pry open the black box of a successful SaaS vendor or get data to flow in and out of your cloud to that vendor in a cost-effective manner! Third party tool vendors assert that they have capabilities that are 80% ready for **general availability** (**GA**) – or worse, bundled into already cooked GA solutions that will never be used (probably a market test).

Worse yet, if you use them, they will probably not remain supported by the third party vendor in the future. There are advantages to using a buy versus build approach but if you are building toward a future-proof goal, then investing in a business solution is wise. You will want to make the case for building where the value is and buying where leverage is needed across a set of capabilities, enabling you to step up into the future. There are many implications to consider when selecting to build versus buy that go beyond just cost/flexibility. The business strategy must guide your decisions. You want to build for competitive advantage since your solution is a key investment and part of the business success strategy. You want to point to the business case and due diligence effort as outcomes of the business strategy required for these buy versus build decisions.

The bottom line is that if you buy any tooling or service, you will have to isolate, wrap, and prepare risk mitigation steps for that part of the solution. These may die on the vine and have to be cut out before the rot of trying to retrofit it over time takes hold. Without this reserve funding and related effort being maintained, you will end up with a vendor solution driving your architecture and lose leverage on cost negotiations and the architecture control of your solution.

Advantages and disadvantages of off-the-shelf tools

Let's explore third party **commercial off-the-shelf** (**COTS**) tools a bit more. Buying IT solution components is a necessity, so let's not think we can build *everything* ever! This approach offers advantages such as quicker time to market and standard support either directly from the vendor or through the community of users. You will want to put all decisions under a due diligence process that boils all decisions into a metric – try US dollars! Create the **total cost of ownership** (**TCO**) that it's modeled according to the limits and duration of the business's strategic plan. Itemize assertions, assumptions, and reasoning. Note that the open-book approach is the best practice! Unless you have an unquantified gut feel for a vendor's ability to deliver or the capabilities of a product that they offer, go for it! But beware that this subjective selection criteria will be discovered if there are issues – and there are always issues.

Part of the TCO is the ability to know what will be spent in the future, how much has to be put up in reserve, what capabilities can be used because the solution fits the architecture, and what cannot be used since that would violate the architecture, its principles, or the vision you are on a mission to achieve. The TCO of a solution also comes with a hidden cost: the risk reserve cost. This cost requires funds to be put into reserve in the eventuality that there is a need to stop using part or all of a purchased or custom built component. Not all TCOs include this cost; however, when faced with too much risk, you will want to have this reserve ready to be drained if needed. Insurance company technology project leaders are often challenged concerning their risk reserve estimates. These are based on the probability of a solution's success. This is ingrained in the insurance business leader's thinking for obvious reasons given the entire insurance business is based on risk.

The **return on investment** (**ROI**) of your solution has to be greater than the business TCO, including your solution, so a financial modeler is essential to bring onboard before going on a spending spree. If the architect and engineer are separated from finance at the table, you can bet that there will be an architect and engineer available to be blamed when the effort fails, or the cost overruns start happening.

Key to the ROI will be the **time to market** (**TTM**) for the solution since the revenue will not flow until market readiness is achieved. Buying solutions also involves integrating them and that effort has to be placed into the ROI as well as the spikes and sprints needed to provide agile **proof of concept** (**POC**) for that integration. These are small integrations, and they have to roll up to the larger systemic integration as the sprints progress and the integrated solution takes shape.

Consider the time to market (TTM) considerations of buy versus build

Achieving TTM as early as possible with third party COTS tools is usually considered the goal. Products are primed for immediate deployment with pre-configured features, significantly speeding up the TTM – that is the optimal outcome expressed by the vendor. It is often a critical advantage in competitive or time-sensitive business solutions. After all, there is always time to do it again, but if you miss the market, it remains missed!

Consider support and community assistance that's available when choosing buy versus build solutions

Commercial tools are often supported by in-house staff and an ecosystem of support provided by a community of users. Tools that come with community-based support capabilities, such as **Stack Overflow** {https://packt-debp.link/J3ZPXI} and **AnswerHub** {https://packt-debp.link/loNC0q} contain a treasure trove of help, troubleshooting, how-to resources, updates, code snippets, and best practices. Alternatives include **Askbot** (a free engine) {https://packt-debp.link/jw069h}, **Open Source Question and Answer** (**OSQA**) {https://packt-debp.link/1nFKPB}, **Rosetta Code** (multi-lingual algorithms) {https://packt-debp.link/ct3UOi}, **Shapado** {https://packt-debp.link/xZEYSl}, and various other internet forums. In many cases, you may be able to find a solution to an issue simply by searching such a community site. Even if a solution is not found via a direct answer, you may be able to find a similar topic and the contact information of a subject matter expert who does know the answer.

Caution! Off-the-shelf tools often lack flexibility

The use of COTS solutions comes with some risks. One such risk is the loss of flexibility in being able to adapt the solution to your needs. You will have to influence development via user groups, pay for customization (which means you may have a well-built solution after all anyway), and just add a lot of glue using a developed sub-solution or other COTS product. Many of these tools just lack the flexibility required for unique organizational capabilities. They will be embodying the one-size-fits-all approach that often does not align with your best principle if you adopt that from this book. Specific data workflows, potentially affecting solution efficacy, is just one example of inflexibility. Consider the excellent SaaS data warehouse solution **Snowflake**, which appears like a black box to cloud based PaaS services and even your own home grown distributed algorithms on Databricks. They are like kryptonite to each other's solution. Many cloud AI solutions in *Azure* and *AWS* depend on Databricks, but if you wish to or *must* leverage Snowflake, beware of the costs and data movement fees. This is where the TCO analysis should speak louder than the salesperson, and where POC results matter!

When proving a COTS integration, you are entering the agile development sphere of software engineering best practices. We won't get too much into the software best practices; however, the SDLC processes for development impact data engineering toolset development. Whether you have a lot of glue holding a solution together or clean layers (preferred), you will be developing and therefore building.

Building custom solutions is alluring!

Building custom tools is very appealing to some organizations that have experience in custom software development. That is less appealing to organizations that see IT as a support function to a core business. The ability to build just what you want to build is very appealing to either, but the costs are different. Think about the degree of enterprise control that is simply not possible using COTS solutions. Development also avoids vendor lock-in. Developers may also find building new tools more interesting and exciting than using off-the-shelf solutions. But the dark side of developed software has to do with *how* it is developed and *what* is developed. The *how* question is answered by the architecture processes (the organization's SDLC). The *what* question is answered by the conceptual, logical, and physical architecture of the solution. The engineer has to negotiate with the architects so that the developed solution has all the capabilities it needs and works, always works, and never fails!

In other words, the custom solution must implement comparable functionality to a COTS equivalent solution.

If you build IT, you gain full control over all capabilities and features

Custom-built solutions can afford complete control over the features, design, and underlying architecture, enabling a tailored solution to meet precise organizational demands. Whether they do that well has to do with your development and data teams' skills.

If you build IT, then tailoring it to specific organizational needs is possible

Since we assume the IT staff is well-trained, knows how to take direction from an architect, and is willing to work in an agile manner, with a software data quality mindset, then the bespoke nature of custom solution development allows for alignment with unique data processing requirements, regulatory compliance, or particular data handling and analytics needs. What is built can be directly aligned and interfaced with other components and the solution requires little glue to keep it together. The word *seamless* comes to mind and that fits the tailoring idea nicely.

Be prepared for custom solution downsides!

As you know, the best of plans can go awry during execution. The allure of custom solution development often causes organizations to overlook their downsides. You will be competing for other IT resources. You also have to consider that solutions need to be maintained over time. This means different things to one organization versus another. A reduction in the department's funding in next year's budget will inadvertently result in an entire support wing being lost. That tailored solution could be left with orphaned support, and worse yet still kept on the critical path to solution delivery. Through asset management controls, an IT manager has to know what is running at any given time and what value the service provides. It makes sense to have this type of robust asset management, but many organizations do not. Be especially aware if glue code is developed. This is especially necessary in the cloud, where glue code (integration script and configuration) is sprinkled like sugar all over the place. Don't be fooled and caught not knowing this fact and losing control of any of your developed assets; they are *not* in GitHub, I assure you!

Consider maintenance overhead!

At this point, I expect you are thinking about outsourcing your IT department and keeping just the data. Let me assure you that this is not the right way to go! Your people are inherently interested in quality solutions, but the outsourcer is only motivated by the terms of the outsourcing agreement. The luxury of software solution customization comes with the exigency of ongoing maintenance. The operational overhead of managing, troubleshooting, and upgrading custom systems is a significant consideration. All this will be boiled down in dollars and cents as part of the TCO model, and if it can't be easily distilled, *STOP!* If it is too complex to plan for the run-manage phase of the solution, it *SHOULD NOT BE BUILT*.

Maintenance for many COTS is 20% of the purchase price. Developed solutions take no less effort. Some will try to shave it – *a lot*. When that happens, be prepared to rewrite and re-solution since you have limited time before the technology melts down. On average, technology shifts every 7 years, and with that technology generation comes new items that force refactoring, re-working, re-platforming, and generally re-doing. It's expensive but it's our IT reality. Future-proof solutions allow well-architected capabilities to be upgraded with the least pain to the development budget.

Address the syndrome – not invented here (NIH), not used here

Some organization cultures grapple with the **not invented here** (**NIH**) syndrome, a mindset favoring internally developed solutions over externally developed or purchased ones, regardless of the efficacy or suitability of the available off-the-shelf options. This bias can lead to unnecessary expenditure of resources. As pointed out earlier, reinventing the wheel rather than leveraging existing, often well-supported, commercial solutions is essential in some cases.

Overcoming an organizational bias is a difficult task but should be handled as part of the architecture driven due diligence processes of your SDLC. If that is not possible you will want to escalate bias and subjectivity in design to the highest actionable authority and make your case for objectivity and clarity. Note that sometimes, you must yield to the imperfect decisions of an organization and pursue a course of action that is subjectively driven. After all, we can get anything working with enough time and money – it just may look a bit like a Rube Goldberg {https://packt-debp.link/wxR5DA} solution!

What's the impact of buy versus build on a solution's longevity?

Ultimately, both custom and off-the-shelf solutions pose risks for an organization. Off-the-shelf solutions contain risks due to vendor lock-in and vendor dependency. Vendors may go out of business or simply stop providing support or updates. However, custom solutions, if improperly managed, may also become dependent on a small number of engineers or even a single engineer for maintenance and updates, creating significant risks for the organization.

To mitigate this risk, the creation of documentation should be prioritized. Not just any documentation but the structure of the documentation must be operational. After all, it runs after deployment on day 1. It stays running due to people following the playbook or the **runbook**, something that those in legacy operational roles will remember. I've seen too many hunts through agile project history in **JIRA**, **Microsoft Teams**, or **Microsoft Azure Pipelines** to find the gold nugget of information required to identify and fix a problem in production. There is a lot that goes into a runbook, but make it right for your solution and include important items:

- Provisioning guide

- Configuration management (with physical architecture alignment)

- Development history

- Escalation processes

- Frequently asked questions and scenarios for prior issue resolution

- Code navigator (with repo links)

- Code review comments (for technical debt)

- Operational technical debt (what needs to be done that was allowed to slide in the past)

- Links to subject matter experts

- Links to the business owner

- Links to IT roles for maintenance

- Links to applicable tickets

- Budget and project metadata

- Links to observability and monitoring dashboards

- Links to security, privacy, and compliance audit outcomes

- Trend reports (data and software quality)

- Reliability, availability, scalability, and performance metrics

Additionally, organizations may overestimate the extent to which their needs are unique and may neglect standard features that would suffice. For example, an organization in the not-too-distant past developed an in-house analytics platform without accounting for the EU's **General Data Protection Regulation** (**GDPR**) requirements. When GDPR finally did take effect, the solution had to be scrapped. The engineer who built this solution was no longer on the team and there was limited documentation to be able to adjust the custom-built system. In contrast, various contracted commercial vendors long ago anticipated the need for GDPR compliance capability and still do provide viable solutions.

Sometimes, a custom solution is warranted if it is for a competitive advantage or to satisfy a need that does not exist in other organizations. This kind of innovative driver for a build-it approach was recently demonstrated by a retail analytics company that chose to build liquid data in a memory system as a core to their architecture. It was – and still is – highly successful, even if it did require a sacrifice. It was (and is) solely an on-premises hostable solution due to it requiring huge RAM provisioning with kernel modifications obviating **Infrastructure-as-a-Service** (**IaaS**) server hosting.

> **Caution!**
> Don't expect to create the next state of the art tool!

That being said, engineers often dream of creating a widely used tool. It is difficult enough to create a widely used tool when that is an explicit goal. As mentioned in *Chapter 1*, the market will always have more losers than winners and compete to develop the same capability. Thinking that a custom solution (primarily developed to solve internal problems) will become an industry standard is a deeply naïve thought. Unless broad applicability is the goal from the beginning this global use is not going to happen.

Consider **Thrift**, a tool created by Facebook for internal use to allow more efficient communication between services written in different languages. Although Thrift {https://packt-debp.link/kwGmpi} generated some interest and excitement when it was open sourced by Facebook, it failed to gain traction. Facebook may well have been justified in creating Thrift for internal use and it likely benefited from open sourcing it. But even with the immense resources of one of the largest tech companies in the world, it wasn't able to create a new standard.

This is not meant to discourage entrepreneurs – something we also are. Just pick your battles and see what your vison's implementation needs – don't compromise on the vision.

Future-proofing and scalability concerns

In this section, you will learn about specific techniques for increasing the longevity of your solutions. Specifically, we'll focus on maintaining awareness or evolving data standards and being proactive about adapting to organizational growth and changing requirements.

Beware of evolving data standards and formats

You must be conscious of evolving data standards and formats. These are in a state of flux, necessitating a clear architecture that is adaptable. Off-the-shelf tools might lag in being able to adapt to change caused by this evolution, while custom solutions could be built with future changes in mind. How this is to be done varies but the industry is converging on huge in-memory binary storage formats and *not* in parsable inefficient storage formats that run a big revenue check for the tool or cloud provider.

We built self-describing binary message formats 25 years ago to send binary data between machines with different architectures – taking into account big and little endian conversions. It was *fast*! And it was great. Even the TCP/IP stack did not exist at the time, so this hybrid between XML and serialized C structures (not even classes) was sent over many line protocol types: Ascii, X.25, and UDP/IP.

Why not build a competitive advantage again in the cloud or ask the cloud providers to open this grid like mesh for fabric computing? If they don't have it today – and we need it – why not build it with them? If not, be ready to handle the traffic as the herd heads in this direction.

Be ready to adapt to organizational growth and changing requirements

With the technical changes comes organizational growth. Growth is natural, and data solutions must scale and evolve to cater to increasing data demands and evolving organizational objectives. The scalability and extensibility of a data solution are fundamental to its longevity. Don't limit solutions to bottlenecks and checkpoints. Build that into your platform frameworks. You do not have ten different ways to implement a common pattern in your solution. It just builds technical debt.

Dependency on vendors for updates and patches

Reliance on vendors for product support is a double-edged sword. The discontinuation of products or support can leave organizations in a precarious situation, scrambling for alternatives and dealing with migration hassles:

- **Know the risks of vendors discontinuing support or products**: You can't know everything. Vendors will even use your contract to raise awareness of their value in the industry and leverage that to a successful private company exit. It's happened many times. After an IPO, sale, or other exit, the game is different, and the services are altered. You are left executing a contingency plan. 50% of the projects in production are affected by this event if buy-before-build is a principle of their architecture. You will have to migrate solutions to address tectonic shifts in vendor offerings that you have utilized.

- **The challenge of migrating to new solutions**: Migration is not merely resource-draining but fraught with risks such as potential data loss or integration issues. Especially when migrating from a discontinued or unsupported commercial tool, the challenges are manifold.

In summary, the decision to buy-versus-build a solution is a principle driven fork in the road with long-term ramifications on the adaptability, sustainability, and longevity of the data engineering ecosystem. A holistic analysis encompassing immediate project needs, future organizational growth, and the broader evolution of the data engineering landscape is imperative for making informed and beneficial choices. Let objectivity rule your due diligence and be transparent with any results. Reserve funding for risk mitigation and get financial modelers on your side as well as the architects according to at least a five-year rolling strategic plan. Any third party tool or built design requires analysis that should account for internal biases such as not invented here (NIH) syndrome, which could skew decision-making toward less optimal solutions.

Challenge #3 – Evolving data repository patterns – identifying big rocks for data engineers

As data engineering terrain continually expands, a deluge of data repository patterns has flooded the market, each promising a seamless pathway to robust data architecture. However, amid this deluge, it's imperative for data engineers, especially those embarking on this journey, to align their designs to the decided upon foundational principles. These principles will be applied to the solution as the big rocks to be handled first. These patterns must be effectively applied to the common challenges down to the smallest as the solution architecture is formulated. This section unravels these large items: data immutability, lineage tracking, data quality preservation, scalability, security and compliance, data discoverability and accessibility, and cost-efficiency (later chapters will depict the principles to be applied to them). Please refer to *Figure 2.4* and *Figure 2.5*:

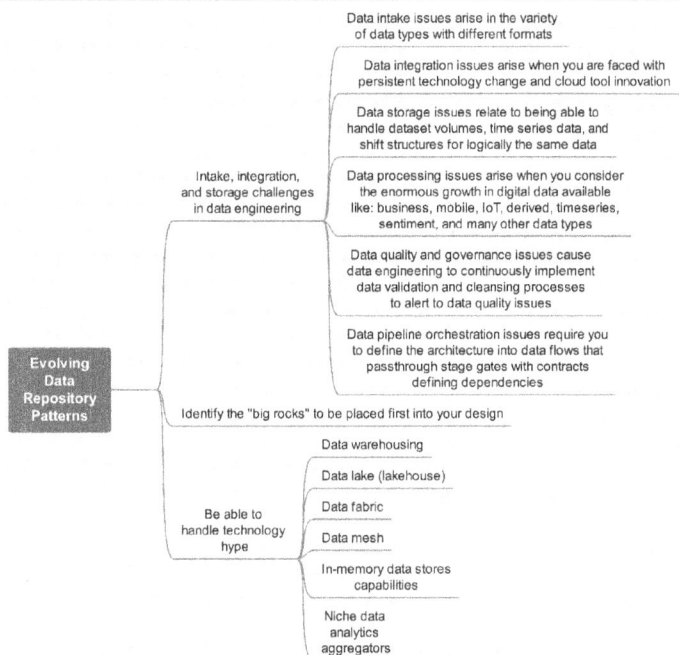

Figure 2.4 – Evolving data repository patterns mind map

Figure 2.5 – Evolving data repository patterns technology hype mind map

Casting light on what to prioritize amid the ever-evolving data repository frameworks is essential.

Intake, integration, and storage challenges in data engineering

Data engineering has numerous challenges. In this section, you will be made aware of these and learn the challenges facing data engineers. Being aware of these challenges is an essential part of your success.

Data intake issues arise in the variety of data types with different formats

ELT necessitates transforming data after it's loaded but before any further downstream processing or analysis. This is complicated by the need for real-time analysis. The following are some key takeaways:

- Many data source types

- Data quality (including domain correctness for contracted data such as trend correctness, timeliness, completeness, and not being overstated)

- Large volumes are handled given technical limitations

- Capability to handle real-time data ingestion, pipeline processing, analytics, and insight generation

Data integration issues arise when you are faced with persistent technology change and cloud tool innovation

The hyper-change cycle causes data systems to not lend themselves to integration with ease. Legacy systems, new on-premises systems, and cloud systems need to be knit together. However, tool integration patterns conflict with the service levels needed to effectively connect those systems. Having legacy systems glued together with modern systems brings in data sync and flow problems that can make a whole system fragile. Let's look at some key takeaways:

- Data structures are not uniform

- Data system technologies need a lot of technology glue to be integrated

- Data mappings need to preserve semantics, and these are complex

- Data governance, audit, compliance, privacy, and security optimized systems work against reliable, available, scale, and performance goals using current tooling and cloud architecture blueprints

Data storage issues relate to being able to handle dataset volumes, time series data, and shift structures for logically the same data

Data storage systems need to scale and be cost effective. They also need to be able to be distributed, expandable, performant under stress/scale, and affordable. Data locality and replication use cloud-based SaaS and PaaS services that conflict with important processing strategies like: "store data once, and process many times." The following are some key takeaways:

- Select a data storage strategy and optimize across key factors such as cost, scale, performance, availability, and integration for an optimized approach

- Scalability and performance will be issues when you're wrapping storage services for business consumption

- Optimized data partitioning, mapping, and indexing strategies can't solve all use cases and as such have to be grouped into zones

- Data security issues such as RBAC/ABAC versus group/user row-column-based access, entitlements, privacy, auditing, defense in depth response support, and support for zero trust are all to be considered when you're trying to glue a system together

Data processing issues arise when you consider the enormous growth in digital data available such as business, mobile, IoT, derived, time series, sentiment, and many other data types

Legacy processing blueprints have trouble scaling to handle the volume. Distributed processing such as **Hadoop**, **Spark**, and cloud microservices enable parallel processing on machine clusters. **Data quality management** (**DQM**) approaches are necessary to handle data in terms of gaps, errors, inconsistencies, overstatements, semantic misalignment, and other data profile related anomalies. A uniform data management strategy with governance is needed. The following are the key takeaways:

- Processing big data requires an architecture that delivers a contract enforcing data pipeline for the curation of insights from knowledge. Knowledge from information and normalized data that was ingested as profiled raw data is required.

- A distributed computing capability is necessary for processing.

- Complex data transformations, aggregations, measures, metrics, and trend stats that support contracts are necessary and need to be operable.

- Data pipelines need to be built to not be ad hoc cloud data flows and must make use of patterns that are repeatable and optimized to the architecture of the system.

Data quality and governance issues cause data engineering to continuously implement data validation and cleansing processes to alert to data quality issues

Anomaly detection for outliers, the ability to impute missing data and fit synthetic data to prior trend curves in the absence of real data, and profiling is needed to validate data and potentially correct it before downstream processing is affected. Regulatory compliance in a business data domain can vary and affect what can and can't be done with data, derived information, or even the presence of a data item's existence. Healthcare, finance, and government are particularly affected by the following regulations:

- The **Health Insurance Portability and Accountability Act** (**HIPAA**) of 1996

- **Gramm-Leach-Bliley Act** (**GLBA**) privacy

- **Children's Online Privacy Protection Rule** (**COPPA**)

- **Payment Card Industry Data Security Standard** (**PCI DSS**)

- **EU GDPR** and **UK GDPR**

- **California Privacy Rights Act (CPRA)**

- **California Consumer Privacy Act (CCPA)**

- **Virginia Consumer Data Protection Act (VCDPA)**

- **Colorado Privacy Act (CPA)**

- **Connecticut Data Privacy Act (CTDPA)**

- **Utah Consumer Privacy Act (UCPA)**

- New York State **Personal Privacy Protection Law (PPPL)**

- China's **Personal Information Protection Law (PIPL)** and OTDA

- **Russian Federal Law on Personal Data** (No. 152-FZ)

- Germany's **Bundesdatenschutzgesetz (BDSG)**

- Other privacy laws (Oklahoma, Texas, Kentucky, Hawaii, New Jersey, Rhode Island, North Carolina, New Hampshire, West Virginia, and Florida)

All these regulations affect data processing, security, and storage architecture blueprints. The following are some key takeaways:

- Data profiling, validation, cleansing, and normalization are essential.

- Data quality contract and compliance validation is a necessary capability.

- Data governance involves technology, processes, and people. Its implementation requires a framework to exist that includes solution architecture.

- Regulatory compliance demands privacy with auditing, like other security requirements form contacts that impact your solution architecture. Privacy, auditing, and security can't be handled later as afterthoughts.

Data pipeline orchestration issues require you to define the architecture as data flows that pass through stage gates with contracts defining dependencies

Making sure it all works according to contracts, services, and quality is an orchestration task and not just a monitoring or observability task. Building an architecture that supports flexible orchestration with minimal redevelopment as changes are required is a core goal. Cloud services will implement code solutions that do not lead to configurable orchestration and require code to be sprinkled through the implementation that is expensive to maintain and subject to more costly testing to ensure reliability. The following are the key takeaways:

- It is essential that you build solutions that are flexible, configurable, and orchestrated as complex data pipelines.

- Define contracts at each stage of the data pipeline and manage dependencies using configuration management. This approach supports the data factory management blueprint.

- Establish structured error handling patterns and ensure consistency both within and between pipelines. Avoid mixed approaches to prevent system fragility.

- Data lineage, version control of code and datasets, and structured deployment of data pipelines with rollback capabilities are required for smooth operations.

We now have an idea of the common challenges regarding data engineering but have yet to identify the big rocks. In the next section, we will do just that – identify the aspects of the solution that need to be part of the architecture of a future-proof data engineering design.

Identifying the big rocks to be placed first into your design

First, why do we call these big rocks? Picture the metaphor where if you were to place rocks into a jar to fill up as much space as possible, you begin with the biggest rocks and shake them down until they form the best fit. Then, you add smaller and smaller rocks while shaking them until there is no air left. The jar is full and maximally absent of unused volume. The same goes for our design. We need to handle the tough issues first and then the less important ones.

So, what are the big rocks? These are as follows:

- **Data immutability**: The principle where unaltered data remains in the system post-creation

- **Lineage tracking**: The blueprint of data's sojourn through data pipeline transformation stages

- **Data quality preservation**: The crusade for reliable, consistent, complete, accurate, and relevant data

- **Scalability**: The ability to gracefully dance to the tune of burgeoning data volumes

- **Security and compliance**: The sentinel standing guard over data's sanctity and legal edicts

- **Data discoverability and accessibility**: The signposts guiding toward data treasures

- **Cost-efficiency**: The fine art of balancing the scales between cost and value

It almost sounds like we're talking about religion here, but the issues are serious. Let's explain each of these in sequence:

- **What is data immutability?** Amid the throes of transforming raw data into glittering gold, the essence of data immutability stands as a steadfast guardian. It's the unseen bond keeping the data's integrity unscathed through the tumult of transformation. Ignoring this sentinel could plunge the data lake into a whirlpool of inconsistencies, spawning ghostly data replicas with every unauthorized alteration – a perilous path leading to corrupted insights and fallacious decisions. Data can't change after passing a data pipeline's stage gate. It can be subject to change within the zone via its data pipeline stage processing but once committed at the end of the zone's path, it is stamped as final. These immutable datasets are baselines that form checkpoints. When processing huge volumes of data in an environment where failure can and does occur, you have to start from those stable points when reprocessing. So, often, systems do this in an ad hoc manner and become fragile.

- **What is lineage tracking?** Lineage tracking, akin to an architect's blueprint, illuminates the path that's treaded by data from ingestion through its various stage-effected metamorphoses. It's the forensic lens peering into the data's past, unveiling the narrative of its transformation, a narrative indispensable for unearthing and rectifying errors. The absence of this lineage lore could morph the data lake into a labyrinth, with each error a menacing minotaur lurking in the shadows, awaiting the hapless data engineer. The key here is that data lineage has to show the data semantics at rest and as it is transformed. It also needs to show parallel lineage for data that is being curated and not yet committed to a release set. Only when designated as a delivery set can it be subject to archive or roll off. On a transactional level, lineage technologies such as **blockchain** have arisen to establish trust in the transactions underlying a controlled dataset. Blockchain technology makes it possible for you to deliver an immutable record of data lineage with a data item. The **Blockchain Distributed Ledger Technology** (**DLT**) records transactions in the distributed ledger so that the data's origin, ownership, and transformations that are applied over time are tracked. This can help ensure that data is not altered when delivered through reports or business intelligence tools. Blockchain technology also helps show where remediation should take place by following the data lineage path. A few financial services use cases follow from *"Tech Target's 10 blockchain use cases in finance that show value"* (https://www.techtarget.com/searchcio/feature/5-blockchain-use-cases-in-finance-that-show-value) by Mary K. Pratt:

  - Blockchain technology has a number of potential use cases in financial services like:

    - Faster, cheaper, and more secure scalable **financial services**

    - **Collateral management** (tracking assets used for loan collateral)

    - **Stablecoin** (crypto currencies)

    - **Tokenization of real-world assets** (enables assets to be tokenized while supporting fractional ownership)

    - **Virtual world transactions** and reward tokens (metaverse support)

    - **Crypto staking** (questionable crypto asset lock agreements to validate other agreements)

    - **Invoice factoring** (to borrow against overdue accounts)

    - **Supporting new types of B2B networks** (driven by need for better end to end customer experiences)

    - **Facilitate and track data flow in a financial institution** (inter department tracking; intra company tracking has been a challenge with many financial institutions pulling back)

    - **Paper currency replacement** (a long time away but crypto can replace digital currency payments)

- **What is data quality preservation?** The quest for data quality is a saga that ensures data's essence remains untarnished through the rigors of numerous data pipeline transformations. The **Data Quality Management (DQM)** processes are the unseen shields guarding against the onslaught of inaccuracies, ensuring the data's tale told to the analytics engine is devoid of fallacies. It even sounds like our engineering designer is on a quest! But seriously, quality matters! The specter of poor data quality is a harbinger of misguiding insights, leading enterprises astray in the data-driven odyssey. Data quality is to be managed and then reported across various types, such as **raw** and **normalized data**, **transformed data**, **master data**, and **consumable data**. We'll talk about where this data resides and the types of DQM needed to support the detection of raw data gaps, time series anomalies, trend anomalies, and gross/volume anomalies later in this chapter.

- **What is scalability?** Where data grows with every tick of the clock, scalability is the sorcery enabling the data architecture to expand gracefully, accommodating the growing horde of data. Neglecting scalability causes performance bottlenecks and escalating processing times, preventing the timely glean of insights, leaving organizations languishing in the wake of missed opportunities. As an example, think about retail analytics processing needs where sales volume processing has to accommodate peak seasonal transactions (that can be ten times the normal weekly transaction flow) as well as day to day transactions. Your solution capacity has to be able to automatically scale-up and scale-down when necessary. As this Arthurian diatribe unfolds, we are going to be seeing that cloud offerings provide the best toolsets for handling your data factory pipeline's scalability needs.

- **What is security and compliance?** Security and compliance is the guardian of data's sanctity and adherence to legal edicts. In a domain where data breaches are the harbingers of doom, ensuring fortress-like security and steadfast compliance is the knight's vow to safeguard the kingdom's honor and avert the wrath of regulatory dragons. OK, you get it – managing security is going to be a problem. You will want to keep your system components updated with a host of current security patches and have these notifications and the update history ready for the auditor's review. One good thing is that with complexity comes obscurity and with that some degree of implicit security. Your auditors will make you aware of potential vulnerabilities, and your responsibilities to address risk. If you do not have exposure to auditors yet, you will eventually, so be honest with your designs and self-audit your solutions. With some cloud provider services, risks are minimized if best practices are applied; when that happens, any potential risks become clearer. Also, you will see that security integration is going to be complex and loaded with potential failures as you attempt to secure data. Our suggestion is to test, test, test, get other perspectives, and test some more – otherwise, this fire-breathing dragon is going to ignite your world!

- **What are data discoverability and accessibility?** Data discoverability and accessibility are the twin lanterns guiding the way through the dark woods of data repositories, leading toward the treasure trove of insights awaiting discovery. They beckon a culture where data is not a cryptic codex but a lucid lexicon, ready to narrate tales of insights to those who seek. Implement your metadata and be able to search it like it were primary data via facets, keywords, and structural components of the architecture you implement.

- **What is cost efficiency?** Cost-efficiency – the prudent, sage counsels with the fine art of balancing the coffers while reaping the bounties of a robust data repository. It's the whisper in the ear urging caution against the lure of fanciful yet financially draining data frameworks, advocating a path where value reigns supreme. Alright! Enough of the drama. You do need to humor us now and take cost accounting seriously. You can't spin off a data factory per silo. And the cloud service will *never* provide logical data flow accounting for you. So, you have to build this accounting facility and cost reporting capability on a multi-tenant hosted system; otherwise, your cloud bills will be enormous.

- **What are the big rocks takeaways?** Navigating the tempestuous seas of proliferating data repository patterns demands a compass. Data engineers are called upon to manage the largest issues first. Over time, it could be that you discover more of these big rocks, but they will be perceived as obstacles to your engineering effort if they were not considered from the start of your architecture's design. They will be tossed at your design like a big rock being tossed at someone living in a glass house. So, be ready to handle them. Market-hyped data repository frameworks will arise and try to sucker you into a purchase deal.

Keep a keen eye out for these while building resilient, efficient, and insightful data architecture.

Being able to handle technology hype

Technology will always be in flux, and as a designer, you will be faced with exploring the cutting edge. However, you are also advised to beware of the sharp edges that you will encounter. One of these is the hype that media, vendors, convergences, and blogs point to as the next shiny object to pursue. Not all are valuable, but a few are worth mentioning as being very important.

In this section, you will get an overview of these important technologies related to your data engineering efforts. Here, we will talk about **data warehousing**, **data lake (lakehouse)**, **data fabric**, **real world implementations**, **data mesh**, **event driven pipelines**, **in-memory datastore capabilities**, and **niche data analytics aggregators**.

Data warehousing

Data warehousing has long been a staple in the realm of data management, providing an optimized environment for querying and reporting against structured data. Understanding data warehouse capabilities is crucial for making informed decisions in an enterprise data strategy.

What is the evolution of data warehousing?

Over the years, data warehouses have undergone significant transformations to cater to the ever-evolving needs of businesses and advancements in technology. Initially conceived as a centralized repository for storing and managing structured data, data warehouses have transitioned to support big data, real-time analytics, and integration with various data sources and platforms.

The key questions are as follows:

- **What are the historical context and origins of data warehousing?** Data warehousing originated in the 1970s and 1980s, primarily shaped by Bill Inmon and Ralph Kimball. Inmon, known as the father of data warehousing, defined a data warehouse as a centralized, consistent, and time-sensitive data repository for decision-making. He set the foundational principles of data warehousing. Kimball contributed through dimensional modeling, focusing on making data comprehensible and usable for business users, with an emphasis on data marts and dimensional models. Their combined efforts, blending Inmon's broad principles with Kimball's practical approach, have significantly influenced data warehousing, evolving it into a key element of enterprise data strategies. Today, data warehousing integrates various organizational data sources into one system for effective analysis and reporting, with their methodologies still guiding modern practices and technology in the field.

- **What are some modern data warehousing trends?** With the rise of big data and advanced analytics, there has been a push toward the adoption of real-time data warehousing. Organizations now demand instant insights from their data, leading to the popularity of real-time analytics and data streaming. Moreover, the proliferation of diverse data sources such as IoT devices, social media, and mobile apps has necessitated the ability of data warehouses to handle unstructured and semi-structured data seamlessly. Additionally, there's a trend of adopting data lakehouse architectures, which blend the best features of data lakes and data warehouses to offer enhanced flexibility and scalability.

Fundamental components of data warehouses?

At its core, a data warehouse comprises various components that ensure its smooth operation and efficient data management.

The following are some key components to consider:

- **Database**: The foundational storage system where all the data resides. This can be relational, columnar, or a hybrid system

- **ETL tools**: These facilitate the process of extracting, transforming, and loading data into the warehouse

- **Data marts**: Subset repositories created from the main data warehouse, tailored to specific business functions or departments

- **OLAP servers**: **Online analytical processing** (**OLAP**) servers enable multidimensional analysis, allowing users to view data from various perspectives

- **Metadata layer**: This holds information about the data source, transformations, and the overall structure of the data in the warehouse

Some important technological capabilities and choices need to be addressed. These are as follows:

- **What are the types of data warehouses: star and snowflake schemas?** Core to data warehousing are specific design schemas such as star and snowflake, which define the structure and relationships within the data. While the star schema keeps data denormalized with a central fact table, the snowflake schema normalizes data, reducing redundancy but adding complexity.

- **What ETL processes exist within data warehouse ecosystems?** The ETL process is the backbone of data movement into a data warehouse. It allows you to integrate data from various sources, transforming it into a consistent format and loading it into the warehouse for analytical use.

With a solid understanding of data warehouse capabilities, you will want to contrast strengths and weaknesses in the data warehousing approach. These are identified in the following sections.

Key features and strengths of data warehouses

Here are a few data warehouse strengths:

- Data warehouses come with a suite of **features tailored toward delivering robust analytical capabilities**. They support complex queries and BI.

- Data warehouses employ **sophisticated indexing and optimization** techniques to hasten query performance, enabling swift retrieval of data, even for complex queries.

- Data warehouses provide **seamless integration** with **business intelligence (BI)** tools and analytics platforms, which enhances data visualization, reporting, and analytical capabilities.

- Data warehouses are structured to ensure a **high level of data accuracy and reliability**, which is critical for making informed business decisions. Handling historical data and tracking changes over time is facilitated through mechanisms such as **slowly changing dimensions (SCDs)**, ensuring data consistency and meaningful trend analysis.

But there are also challenges, as evidenced in the next section.

Challenges and limitations of data warehouses

No system is without its challenges, and data warehouses are no exception:

- **Scalability and performance concerns**: As data volume grows, traditional data warehouses may face scalability and performance issues. Modern solutions such as cloud-based data warehouses attempt to mitigate this by providing scalable infrastructure.

- **Evolving requirements and the need for warehouse redesign**: Changing business requirements can necessitate a redesign of the data warehouse schema, which can be resource-intensive and time-consuming.

- **Integration issues with external data sources and systems**: Integration with external data sources and systems can pose challenges, especially when dealing with heterogeneous data formats and inconsistent data quality.

In this section, we have elucidated the capabilities, strengths, and challenges associated with data warehousing. Being armed with this knowledge, data engineers can better navigate the terrain of data repository patterns, ensuring patterns are selected that align with organizational needs and goals.

Data lake and lakehouse

In today's data-centric world, the capability to harness vast amounts of unstructured and structured data is imperative for business agility and innovation. Data lakes, or the evolved version known as lakehouse, offer a versatile solution for organizations to store, process, and analyze data in a unified environment.

Fundamentals of data lakes

The term data lake has been increasingly popular in the data management world, signifying a shift toward more flexible and scalable data storage and processing solutions. Here, we'll delve into the core principles behind data lakes and how they emerged as a response to the limitations encountered with traditional data warehousing.

Data lakes can be conceptualized as vast storage repositories that can hold an immense volume of data in its raw form, without requiring any immediate transformation or structuring. Some of the unique characteristics of data lakes are as follows:

- **Scale**: Capable of storing petabytes of data

- **Agility**: Allows for quick ingestion of new data sources

- **Diverse data handling**: Accommodates structured, semi-structured, and unstructured data

- **Low-cost storage**: Often implemented on commodity hardware or cloud storage, reducing overall costs

- **Evolvability**: Easy to update and change the structure as business needs evolve

Unlike traditional data warehouses, which usually require structured, cleaned data, data lakes excel in storing raw and unstructured data. This storage paradigm opens up possibilities for holding data in its original form, be it text, images, log files, or a mix of different data types, without a predefined schema.

With the rise in the variety and volume of data, the flexibility in data lakes became a critical advantage. Data lakes support a wide array of data formats and adopt a schema-on-read approach, contrasting with the schema-on-write approach of data warehouses. This flexibility facilitates the ingestion of data in its native format, deferring the imposition of the structure until it's read, which can be crucial for exploratory analytics and data discovery.

The emergence of the lakehouse paradigm

While data lakes answered the need for flexibility and scalability, businesses still yearned for the analytical prowess and structured nature of traditional data warehouses. The lakehouse paradigm emerged as a middle ground, offering a unified platform that capitalizes on the strengths of both data lakes and data warehouses. A few questions arise that we will attempt to answer here:

- **Why is combining the best of data lakes and data warehouses important?** The lakehouse paradigm emerges as an attempt to meld the advantages of data lakes and data warehouses into a unified platform. This convergence aims at bridging the gap between the flexible, cost-effective storage solutions of data lakes and the structured, performance-optimized analytic capabilities of data warehouses.

- **What benefits exist for unified analytics and BI over data lake storage?** This blend facilitates unified analytics and BI capabilities over a single data lake storage, easing the management overhead and providing a streamlined path from raw data to actionable insights. The synergy that's fostered between analytics and machine learning workloads within a lakehouse setup drives enhanced BI and decision-making processes.

Key advantages and features of data lakes

Data lakes offer a host of advantages, especially when integrated with modern cloud storage solutions and big data processing frameworks. Some questions arise when elaborating on the use of data lakes in your solution:

- **What scalability and cost-effectiveness gains are evident when using cloud storage solutions?** Using cloud storage solutions for data lakes significantly increases the level of scalability. As data volumes increase, cloud platforms can adjust storage resources, often without manual adjustments, reducing potential downtime. This elasticity allows organizations to pay only for storage that they use. The pricing model of many cloud providers shifts away from large initial capital expenditures to a more gradual operational expense model. It's also worth noting that with the inherent redundancies in some cloud storage platforms, there could be potential savings on aspects such as disaster recovery and data backup, though individual results may vary.

- **What integrations exist with cloud providers such as AWS S3, and Azure Data Lake Storage?** Integration with cloud providers such as **AWS S3** and **Azure Data Lake Storage** further enhances the scalability and cost-effectiveness of data lakes, providing vast storage capacities with the advantage of reduced costs compared to traditional on-premises solutions.

- **What economical storage options and pricing models exist for data lake storage?** Cloud-based data lakes offer economical storage options and pricing models that allow organizations to store and manage their data efficiently without a significant investment in infrastructure.

Support for diverse data processing tools and frameworks

Data lakes can support a variety of data processing tools and frameworks that accommodate different tasks, from batch processing to real-time analytics and beyond. This adaptability means businesses have the option to integrate with a spectrum of tools without being tied to a singular ecosystem. However, the breadth and depth of integration would largely depend on the specific data lake solution and its compatibility with other platforms. Two questions arise regarding data lakes:

- **What integrations exist with Hadoop, Spark, and other big data platforms?** Data lakes support integration with various data processing tools and frameworks, such as **Hadoop** and **Spark**. This integration facilitates the handling of big data workloads, enabling efficient data processing, and analytics.

- **What ability to leverage serverless data processing exists for data lake processing?** With the advent of serverless data processing, data lakes now offer the ability to process data without the need for managing server infrastructure, thus reducing operational overhead, and allowing organizations to focus more on deriving insights from their data.

In the next few sections, we will delve deeper into data fabric and how it contrasts and complements the evolving data lake and data warehouse ecosystems, steering toward a holistic data engineering strategy that aligns with the organizational objectives.

Data fabric

In a rapidly evolving data landscape, organizations continually search for robust frameworks to manage, process, and garner insights from their data seamlessly. The emergence of **data fabric** addresses this need, offering a comprehensive and integrated architecture for data management across the enterprise.

Defining data fabric

Data fabric is an architecture and set of data services that provide consistent capabilities across a choice of endpoints spanning on-premises and multiple cloud environments. It provides a unified, intelligent, and integrated end-to-end platform to support new and emerging data requirements. Again, the topic raises a few questions we will attempt to answer here:

- **What's the evolution and rationale behind data fabric architecture?** Historically, data was confined to siloed databases, often with limited interoperability, making holistic analysis a challenge. As organizations expanded and the volume and variety of data surged, a more interconnected approach became essential. The data fabric architecture evolved as a response to these challenges. Its rationale lies in providing a cohesive data infrastructure that can seamlessly connect various data sources and services. By doing so, it offers a streamlined platform for data ingestion, processing, analytics, and management, irrespective of where the data resides or its format. Data fabric integrates not just data but also APIs, services, and applications in a hybrid or multi-cloud environment. This holistic approach not only simplifies the data landscape but also aids in driving actionable insights from vast and varied datasets. The name data fabric is based on the metaphor of weaving diverse and varied data sources, technologies, and environments into a cohesive, integrated, and unified data management tapestry, analogous to how various threads are interlaced to create a sturdy and versatile fabric. This architecture aims to enable seamless data connectivity, accessibility, and flow across the intricately interconnected data landscape, much like the uninterrupted continuity observed in a well-crafted fabric.

- **Why is there increasing complexity in data ecosystems?** The rise in data sources, types, formats, and processing requirements brings about a complex landscape that necessitates an integrated, orchestrated approach for optimal data management.

- **Why is there a need for agility and real-time data accessibility?** Swift adaptation to evolving data needs is crucial for staying competitive. Driven by customer expectations, as well as the need for competitive advantages and rapid decision-making, real-time data accessibility is a cornerstone in achieving this agility, thus driving the shift toward a more interconnected data fabric architecture.

The data fabric pattern contains a few core components and technologies that you need to understand so that you are equipped to make decisions. In the next section, you will learn about this core.

Core components and technologies

Data fabric is made up of the following components and technologies:

- **Data ingestion layer**: Intended to manage the preliminary entry of data, this layer addresses both streaming and batch data from various sources, striving to maintain data quality and uniformity throughout the integration process

- **Persistence layer**: Aimed at storing and managing data, this layer seeks to establish a balance among data retrieval, durability, and query optimization, aligning storage methodologies with diverse data types and organizational goals

- **Data preparation and data delivery layer**: This layer is designed to clean, transform, and enrich data, with an aspiration to make it conducive to analytical processes

- **Orchestration and data ops**: Orchestrating data operations involves managing, monitoring, and automating data workflows, intending to ensure a reliable and optimized data pipeline execution while enabling timely data availability

- **Data processing**: With a focus on managing real-time and batch data, this component is tailored to ensure that data is processed, analyzed, and made available according to organizational needs, through careful selection and application of processing technologies

- **Data management and intelligence**: Intending to uphold data quality and ensure governance and compliance, this component aims to manage metadata and maintain data integrity and usability throughout its life cycle, across the data fabric

- **Augmented data catalog**: Aiming to automate data discovery, classification, and metadata management, this component aspires to enhance data visibility and access throughout the organization, though actual efficiency may be influenced by various factors

- **Knowledge graph**: Devised to provide a semantic layer that identifies relationships and entities within the data, this component seeks to illuminate data connections, enhancing data discovery and usage more contextually and insightfully

Two questions have to be considered when integrating the data fabric pattern into your solution:

- **What data integration and orchestration patterns and tools exist?** Data fabric excels in integrating disparate data sources and orchestrating data flows efficiently, leveraging technologies that facilitate seamless data ingestion, transformation, and dissemination.

- **What data governance and metadata management exist for data fabric designs?** Robust data governance and metadata management are pivotal to a reliable data fabric, ensuring data quality, consistency, and regulatory compliance while enhancing data discoverability and understanding.

So, you will want to consider using a data fabric design pattern in your solution because there are several benefits and strengths to consider.

Benefits and strengths of data fabric

The inherent strengths of data fabric are rooted in its holistic approach to data management. It promotes unified data access, regardless of the source or structure of data, thereby aiding in real-time decision-making. With integrated metadata and governance structures, data fabric ensures data quality and trustworthiness, both of which are vital for accurate analytics. Moreover, its adaptability allows businesses to stay agile in the face of evolving data landscapes, ensuring sustainability and competitiveness. Microsoft's Data Fabric {https://packt-debp.link/tl11jS} is a good example of the way Microsoft has taken up the data engineering challenges faced in the cloud, as well as the data mesh concepts. The capabilities that have been envisioned are worth studying as you formulate your architecture since they may lower the cost bar if you choose to leverage the offering. Here are some aspects to consider:

- **Does data fabric facilitate seamless data access and sharing?** One of the primary objectives of data fabric cloud offerings is to make it easier for the data engineer to build robust systems that also loosely cover the data mesh principles that break down data silos, ensuring seamless data access and sharing across the enterprise. By supporting SaaS-hosted data factory capabilities such as advanced metadata, lineage, and governance, data can be treated as a product. Unified analytics is provided as is a unified access point. Streamlined data-sharing processes, both internally and externally, foster collaboration and drive informed decision-making.

- **Data fabric provides for data virtualization and federation.** Data fabric employs data virtualization and federation to mitigate data silo issues, providing a unified view and access to data scattered across different repositories. **Data factory**, **data engineering**, **data warehousing**, **data science**, and **real-time analytics** are some capabilities the cloud provider's data fabrics advertise as being available. A significant challenge is the use of the fabric or building your own. As the cloud providers morph their data fabric offerings to address additional data mesh principles and the various best practices presented in this book, they will undercut the competition and even your platform efforts as you try to build what you need *now*. You need to be ready for this reality. Remember, when faced with a challenge, smart people will start gravitating to similar solutions. You can build anything with enough money and resources (like Microsoft), and you can see that they will jump the line before others. Being first to market is so very important regarding data-fabric-enhanced offerings. The challenge is *do you jump on the wagon and buy the cloud data fabric services, build your own data factory with an eye on the potential for your competitive advantage,* or *do both?*

- **Data fabric designs enhance data discoverability and usability.** With comprehensive metadata management and semantic modeling (or similar semantic representation) for data at rest, data fabric architectures, designs, and services improve data discoverability and usability, expediting the data-to-insight conversion process.

Supporting advanced analytics and artificial intelligence (AI) capabilities

Data fabric architectures are inherently designed to support advanced analytics and AI. They unify diverse data sources, ensuring data quality. They provide fertile ground for advanced analytical processes. The streamlined access to data, combined with integrated analytical tools, empowers organizations to delve deeper into insights, harnessing the power of AI and machine learning for generative, predictive, and prescriptive analytics. The following are some aspects to consider:

- **What data fabric integrated analytics workspace capabilities are available?** By providing integrated analytic workspaces, data fabric designs integrate advanced analytic products such as Microsoft Power BI and **Azure Analysis Services** (**AAS**). Workspaces and tool integrations accelerate the analytics life cycle, fostering a conducive environment for advanced analytics and machine learning. One challenge across the capabilities is to use the legacy tools with their capabilities with the newly rendered data that the data factory provides. It sounds easy: SQL is the answer. But it's not! If you wish to process billions by billions in a matrix in Power BI and then add a non-additive calculated measure, then good luck! Likewise, if you want to have sparse matrix support in **Apache Spark** (or **Databricks**), then good luck! Getting support for that from the vendor is just not going to happen. If you want to hack the Spark memory model, be my guest – it can be done. To get some machine learning pattern clustering algorithms to work as expected, you have to live with severe limitations or be willing to dive into the open source. Then, if you want the value to be added in Databricks or **Microsoft Synapse**, you will be faced with a vendor development wall.

- **Data fabric enables AI for data management tasks.** Utilizing AI and machine learning technologies, data fabric can automate numerous data management tasks, optimizing operational efficiencies and enhancing the ability to derive actionable insights. The data fabric architectures of the cloud providers are a very nice way to implement a beginner data factory. They force you to think of data as a product and quality as releasing data with a dictionary/catalog and providing governance. You can use the services to build *your* solution. However, they are not so prescriptive as to give you a fully working factory. You still need to understand the capabilities that they offer and make them fit *your* architecture, not *your* architecture to their blueprints. Knowing this helps with the challenges that will arise. Cloud providers *will change* their offerings and you will lose the advantage when they do. It's important to strategically design your data architecture while considering potential changes in cloud provider's services. Balancing your reliance between your architecture and the cloud provider's offerings can safeguard against disruptions from future changes in their services.

Real-world implementations

We will delve more deeply into the pitfalls and real-world challenges of implementing data fabric in later chapters. Suffice it to say that, like most new technologies, the hype may not live up to reality. Challenges include the complexity of managing multiple platforms and data sources, data governance and security, performance optimization, and vendor lock-in risk.

How does data fabric architecture compare to data lake architecture? While data lakes are primarily storage repositories capable of holding vast amounts of raw data in its native format until it's needed, data fabric is an architecture that extends beyond storage to provide a comprehensive data management framework. Data fabric facilitates seamless data integration, orchestration, and governance across diverse data environments, whereas data lakes are more focused on storing disparate data cost-effectively.

Moreover, data fabric addresses the challenges associated with data silos, data quality, and real-time data accessibility, which are often prevalent in environments solely relying on data lakes. It offers a more holistic approach to data management, ensuring not just the storage but also the usability, accessibility, and quality of data across the enterprise.

Data mesh

In the dynamic arena of data engineering, novel architectures and paradigms emerge to rectify the shortcomings of their predecessors. The data mesh paradigm is one such evolution that steers away from monolithic, centralized data architectures toward a more domain-oriented, decentralized data architecture. In this section, we will elucidate the principles and capabilities of data mesh, contrasting them with data lake and data fabric to provide a nuanced understanding of their differences and similarities.

What capabilities exist for the data mesh paradigm? Data mesh offers several capabilities that are rooted in its decentralized, domain-oriented approach:

- **Domain-specific data products**: It facilitates the creation of self-serve data infrastructure as domain-specific data products, promoting more accessible and reliable data

- **Decentralized data ownership**: By distributing data ownership, it encourages teams to be accountable for their data products, ensuring quality and reliability

- **Data discoverability and interoperability**: Its built-in mechanisms for data discoverability enable smoother cross-domain data interactions and usage

- **Scalability**: By eschewing centralized bottlenecks, data mesh scales with organizational growth, promoting agility and real-time data accessibility

- **Intrinsic data governance**: With clear domain ownership, data governance becomes inherent, ensuring consistency, quality, and compliance across data products

The emergence of data mesh

The data mesh architecture was conceived in response to the challenges posed by centralized data architectures. As organizations grew in complexity, the need for agility, scalability, and real-time data access became paramount. The limitations of centralized systems, including their rigidity, scaling challenges, and ETL bottlenecks, drove the search for a more adaptive, scalable solution. Data mesh, with its decentralized, domain-oriented architecture, emerged as the answer, championing the cause of democratizing data ownership, and promoting a more distributed approach to data governance and management. Since it emerged as a force, many companies have attempted to implement data mesh designs, and some have failed miserably.

There are challenges in implementing data mesh, and they have a lot to do with the **data domain ownership** principle and lack of a clear understanding of the need for semantic and lineage metadata via a discoverable catalog to support data as a product. Proper training helps with such problems, but with hype comes the rush of the crowd. Confusion will result if the person leading the charge to implement the data mesh lacks the necessary knowledge and experience. Here are some things to consider:

- **How are you to overcome the limitations of centralized data architecture?** Centralized data architectures, although orderly, often become bottlenecks in swiftly growing organizations. Their rigidity could impede the real-time data accessibility and agility that modern data-driven enterprises necessitate. Data mesh designs help with the challenges of integrating legacy data stores, but they will require one to form contracts about data and service those contracts in caches linked to legacy central architectures. You may want to look at the *In-memory data store capabilities* subsection of this chapter for more information.

- **How do you shift toward domain-oriented decentralized data architecture?** The shift to a domain-oriented decentralized data architecture such as data mesh embodies a paradigm shift that seeks to ameliorate the scalability and agility challenges plaguing centralized architectures. The challenge to overcome is in the architecture of your factory leveraging the capabilities and principles of the data mesh without drinking the ocean of Kool-Aid that the cloud providers are pouring out in front of you.

- **What are the data mesh principles?** The following points illustrate what will be later elaborated on in *Chapter 4*. They have been included here as an introduction:

  - **Serving over ingesting by promoting data as a product**: Data is treated as a product with clear ownership, promoting its quality and usability across domains. This contrasts with data lakes and data fabric, where data typically resides in centralized or federated storage solutions without clear domain-specific ownership. Data mesh focuses on serving data to consumers in a domain-oriented manner. It also decentralizes data ownership and architecture, focusing on serving data in a domain-oriented fashion, which encourages cross-domain data interoperability and usage. Lastly, it fosters discovering and using over-extracting and loading.

  - **Enhancing data discoverability and cataloging**: Data mesh prioritizes making data discoverable and usable, moving away from the cumbersome ETL processes often associated with centralized data architectures such as data warehouses and data lakes.

  - **Encouraging data consumption and utilization**: It fosters an environment where data consumption is straightforward, encouraging its utilization across various domains within the organization.

Event driven pipelines

Centralized data pipelines have traditionally been employed to push data through predefined paths, processing and transforming it in stages. The following are some questions to consider:

- **Is publishing events as streams of updates versus pushing data through centralized pipelines a best practice?** These pipelines, while streamlined, can sometimes introduce bottlenecks or points of failure, especially when you're handling vast amounts of data or integrating with multiple disparate sources. In a modern data factory, many paths and pipelines need to be coordinated. Hooking them together with script snippets here and there and not being able to easily release them as a unit as code change is going to get you a resounding operational failure! A better way to handle this challenge is with event-driven architecture. But even here, the cloud service configurations will work against you until they fully embrace data fabric capabilities.

- **What are the benefits of event-driven architecture?** Event-driven architectures facilitate real-time data sharing and reactions, bridging the gap between data producers and consumers seamlessly.

- **What are some real-world use cases of event-driven architectures?** Industries with real-time analytics requirements, such as finance or e-commerce, can significantly benefit from the event-driven architectures enabled by data mesh.

- **Why use centralized pipelines versus decentralized data dissemination?** Centralized pipelines, given their structured nature, offer a more predictable and controlled environment for data flow. They can be optimized for performance, ensuring efficient data transfer and processing. However, they may not be as agile or adaptable to rapidly changing data sources or requirements. On the other hand, decentralized data dissemination, as championed by the data mesh paradigm, offers flexibility and scalability, allowing data to flow more organically across domains, but it might introduce challenges in data consistency and governance.

- **What comparative analysis exists for pipeline processing?** Unlike the centralized pipelines in data lakes and data fabric, data mesh supports decentralized data dissemination, aligning well with modern microservices architecture.

- **What's the impact on data availability and consistency?** This decentralization improves data availability and consistency across domains, fostering a culture of data sharing and utilization.

Treating data as a product involves the need to create an ecosystem about the data. You will need to grok the scope of this concept and the implications as you dive into this topic in the next section.

The ecosystem of data products over a centralized data platform

What are the key elements of a data product? The key elements of a dataset that are considered regarding data products are as follows:

- **Purpose and relevance**: A data product should have a clear purpose and offer value to its intended audience

- **Data quality and integrity**: Ensure the accuracy, consistency, and reliability of data within the product

- **Metadata and documentation**: Comprehensive metadata and documentation to aid in data understanding and utilization

- **Access control and security**: Proper mechanisms to control who can access the data and how it's used

- **Maintainability and life cycle**: Ensure the data product remains relevant, updated, and properly maintained throughout its life cycle

- **Usability**: The data product should be user-friendly, ensuring it can be easily consumed by its intended audience

It's important to note the following regarding data products in a data mesh:

- They are **self-contained datasets** or data services with clear ownership, catering to specific domain needs

- They are **self-documenting** and discoverable to empower decentralized self-service

- They are **trustworthy** datasets with accuracy that are subject to **service-level agreements (SLAs)**.

Data product ownership and life cycle management are data mesh essentials. Ownership and life cycle management are fundamental to ensuring the longevity and relevance of data products, a stark contrast to the often ad hoc data management practices in data lakes.

How do you transition to a data product ecosystem? One challenge with legacy data systems is being able to transition them to the data mesh. You must begin at the beginning and do the following:

- **Assess existing data assets**: The journey starts with understanding current assets. However, discerning their future relevance isn't always straightforward.

- **Identify data domains**: Segmenting data into domains is conceptually sound, but real-world data often blurs these boundaries, making this a challenging step.

- **Empower domain teams**: While it's crucial to equip teams with tools and training, ensuring consistent standards across autonomous teams can be a daunting task.

- **Establish governance and best practices**: Decentralization doesn't mean chaos, but creating a balanced governance model that doesn't stifle innovation is a delicate act.

Transitioning to a data product ecosystem necessitates an organizational and cultural shift toward recognizing data as an asset, a journey starting from centralized architectures such as data fabric to a decentralized data mesh.

There will also be many technical considerations and best practices to adopt. Technical transitions include adopting new tools, frameworks, and practices that support decentralized data ownership, governance, and distribution, aligning with the overarching principles of data mesh.

In conclusion, data mesh emerges as a decentralized data architecture paradigm that substantially diverges from the centralized models epitomized by data lakes and data fabric. Its domain-oriented, product-centric approach promises to alleviate many of the scalability, agility, and data ownership challenges.

In-memory data store capabilities

In the modern digital realm, the significance of accessing data at an unparalleled speed has become paramount for organizations striving to attain a competitive edge. This is where in-memory data stores step into the fray. Let's delve into the nuances of in-memory data stores, unraveling their core design, types, and the blend of challenges and advantages they usher.

In-memory data stores defined

In-memory data stores are a revolutionary breed of data management systems, storing data in the system's main memory (such as RAM) to provide blistering data access and manipulation speeds, as opposed to traditional disk-based storage systems. RAM costs have steadily declined over the years, and who better to make use of huge in-memory instances and memory grid computing than cloud service providers?

What core design patterns are available? There are several interesting in-memory database providers. They fall into a few classes:

- Caching providers

- Memory-optimized databases

- In-memory graph databases

- In-memory grid computing

- SaaS providers

This field is rapidly expanding with offerings. You can see that there is a great need for speed and scalability with high availability across the business landscape. A few questions arise here:

- **Do you build for business competitive advantage or use a third party data store as a core of your architecture?** Business domain data in **consumer packaged goods** (**CPG**) retail domains have Circana's **Liquid Data** and Nielsen's **NIQ**. In finance, you have Bloomberg's vast in-memory data store. All are locked up behind business services, but they are good examples of where data as a product is already a GA product and is being serviced.

- **What are in-memory versus disk-based storage capabilities?** Traditional disk-based storage systems entail data being written to and read from disk, a process that can become a bottleneck in high-performance scenarios due to the mechanical nature of disk drives. In stark contrast, in-memory data stores leverage the system's main memory, ensuring minimal data access latency and significantly higher throughput. The challenge is persistence! How do you ensure data is highly available? Does this not involve a lot of replications? Yes. In today's offerings, that is key and a cost challenge.

- **What are the performance benefits of some use cases?** The performance perks of in-memory data stores are invaluable in use cases demanding real-time data processing, such as financial trading systems, real-time analytics, and high-frequency trading. These systems require a level of performance that only in-memory technology can feasibly deliver. Read performance is great within memory storage systems; however, write-through performance can lag due to the need to support high availability and potentially geo-redundant writes. You are challenged to build that distributed, in-sync, multi-master type of mechanism to keep a modern in-memory system accurate. *MySQL* has such a multi-mast capability, but it is problematic if it's not built and tested very, very carefully.

The types of in-memory data stores available

There are two main types of in-memory data stores:caching systems and in-memory databases. Caching systems are utilized to speed up data access for frequently read or recently read data by retaining this data in memory, hence substantially reducing the database load. The following are some examples of caching providers:

- Aerospike {https://packt-debp.link/EcNwrc}

- ArangoDB {https://packt-debp.link/x1OEEY}

- Amazon DynamoDB {https://packt-debp.link/8ZVUy7}

- Azure Redis Cache {https://packt-debp.link/1UWpur}

- Amazon ElastiCache for Redis {https://packt-debp.link/zaGq4O}

- Azure Cosmos DB {https://packt-debp.link/SU1D3I}

- Memcached {https://packt-debp.link/RLNhUI} (limited scope)

- Redis Enterprise {https://packt-debp.link/99BZHh}

- SingleStore {https://packt-debp.link/s68U2M} (MemSQL {https://packt-debp.link/GzoYeY}

In-memory databases such as SAP HANA {https://packt-debp.link/x9X9jl} and Oracle TimesTen {https://packt-debp.link/aP0DjK} shift the entire database in memory, facilitating instantaneous data access and transaction speeds, proving instrumental for applications necessitating real-time responses. The following are some examples:

- In-memory graph databases:

  - Neo4 (a graph database)

  - OrientDB {https://packt-debp.link/9iBk8C} (a distributed graph database)

- In-memory grid computing:

  - Gridgain {https://packt-debp.link/fXeIul} (also known as Apache Ignite {https://packt-debp.link/AAqeit}

  - HazelCast {https://packt-debp.link/6xQcHJ}

  - Cassandra {https://packt-debp.link/hXN310}

- SaaS providers:

  - Snowflake {https://packt-debp.link/p1e5gE}

  - Oracle TimesTen {https://packt-debp.link/nzEztv}

- Memory optimized databases:

  - Clickhouse {https://packt-debp.link/f6o6Xo}

  - Couchbase {https://packt-debp.link/Z4SF2q}

  - MongoDB {https://packt-debp.link/6Z6Opa}

  - RockSet {https://packt-debp.link/i12aRE}

  - StarRocks {https://packt-debp.link/JHNkFo} (open source)

  - SAP HANA {https://packt-debp.link/wxyNyz}

Advantages and challenges of in-memory data stores

The advent of in-memory data stores carries a set of compelling advantages and intrinsic challenges that are imperative for data engineers to understand, as listed here:

- **Performance and scalability**: In-memory data stores can offer enormous performance benefits and in-cloud memory-optimized instances are getting bigger and cheaper to afford

- **Real-time analytics capabilities**: In-memory data stores amplify the capacity for real-time analytics by allowing organizations to process and analyze data on the fly, thus enabling data-driven decision-making in real time

- **High throughput transaction processing**: The remarkable speed of in-memory data stores makes them an ideal fit for high-throughput transaction processing systems, which benefit from the reduced data access latency, ensuring seamless and efficient operations

What are the hardware costs and durability concerns for in-memory data stores? Harnessing in-memory data stores can be cost-prohibitive due to the need for extensive RAM, which is pricier compared to disk storage. Moreover, the burgeoning data sizes further strain the costs.

What is the data durability concern for in-memory data stores? In the realm of in-memory data stores, data persistence emerges as a concern since data resides in volatile memory. In the event of a system crash or failure, there's a looming risk of data loss unless persistent backup strategies are robustly engineered into the system.

In a nutshell, in-memory data stores are a potent tool in a data engineer's arsenal, poised to redefine the paradigms of data accessibility and real-time processing. However, they necessitate a meticulous evaluation of the associated costs and data durability strategies to ensure they align with the organizational needs and long-term data management vision.

Niche data analytics aggregators

In the contemporary era of data-driven decision-making, organizations are often faced with the challenge of efficiently aggregating and analyzing vast datasets from various sources. **Data analytics aggregators** provide a viable solution to this challenge, acting as a conduit for assimilating data and facilitating advanced analytics. This section unfolds the core functionalities, key variants, benefits, and challenges associated with data analytics aggregators.

Data aggregators provide key functionality for the assimilation and analysis of data in a neutral setting. By this, we mean that the data analytics value proposition makes it beneficial to share data in a single SaaS or PaaS manner out of the on-premises silo that has curated that data.

What are some examples of data analytics aggregators? The following companies offer such solutions:

- Data Axle {https://packt-debp.link/EEEia2}

- Experian {https://packt-debp.link/7tENQ3}

- LiveRamp (formerly known as Acxiom Corporation) {https://packt-debp.link/qhT0q8}

- Lotame {https://packt-debp.link/i7HEop}

- Neustar {https://packt-debp.link/Mn8GQf}

- Nielsen {https://packt-debp.link/E4D0XS}

- Oracle Data Marketplace {https://packt-debp.link/a8xtTH}

The key capabilities of these companies' products and services are illustrated by LiveRamp's offerings, which power collaboration and analytics, perform identity resolution and translation, and activate your data via a digital platform.

Data analytics aggregators specialize in collecting data from disparate sources, whether they be traditional databases, real-time data streams, or cloud-based storage solutions. For instance, consider a global retail corporation aiming to amalgamate sales data from various regions to deduce global sales trends.

What are some data aggregator core capabilities? We want you to be aware that data aggregators are arising in various industries to support the business processes that underlying data also supports. Some data aggregators offer a generic platform and then data services on top of the platform, which makes them well-suited to a data consumer's use cases. Not only is data aggregated but it is served in a way to facilitate downstream insight generation, such as in marketing/advertising. Here are a few of those capabilities:

- A **neutral data marketplace** for data as a product seller to enable their segment data, signal data, and data services for consumption across an ecosystem of third party platforms, publishers, agencies, brands, and even other data companies.

- A **cross-device identity** that enables an individual to be identified via **personally identifiable information** (**PII**), cookies, or mobile device IDs.

- A superior **match rates** capability that enables entities to be resolved to known entities. This helps sellers minimize data loss (when unmatched data is tossed).

- **Data portability** is where data can be used across many use cases, such as brand, agency, publisher, platform, and TV partners, to enable marketing activation and measurement by known and matched identities.

- **Transparency** through metadata cataloging, where metadata provides detailed descriptions and organizations of the data, fostering increased use, understanding, and trust.

- **Unique data** offerings to a large number of consumers through the platform's processes.

- **Customization** for both standard and custom segment creation.

- **Strategic growth** is enabled since once a segment is activated, the data marketplace facilities enable the expansion of that data as part of future data seller offerings.

- A **cutting edge** platform for uploading and distributing data neutrally.

- A **permission based** system enables data sellers to manage consumer access and entitlement to provided data. Policies are also provisioned for which use cases can use seller data (**rights management**).

- **Segment combination** capabilities allow buyers to overlap and combine first and third party data before activating. This capability boosts demand for a data domain owner's supplied data. Marketers can combine segments before processing or sending analytic results to a recipient.

What's important to take away is to select the right aggregator if one exists for your industry or select one that can be extended to your industry based on a core platform. Here, buy and integrate is compelling for your use cases. The value proposition is that there are a significant number of data analytics aggregators across business sectors. In healthcare, aggregators allow patient data to be analyzed for better clinical decision-making. In finance, they can be pivotal for real-time fraud detection by analyzing transaction data from various banks.

A key takeaway is that you may desire the usefulness of an aggregator's services, but pursue that use with eyes wide open. Your data may be sidelined with the competitor's data easily. It may highlight deficiencies in your data's coverage or scope. The positive side of using an aggregator is time to market, and the ability to have your data enhanced when aligned with broader datasets.

What are some of the data variants and technologies that are possible when selecting aggregators?
You may wish to select a cloud analytics database and leverage the finer-grained capabilities of the cloud provider. Unlike traditional databases designed for transactional processing, analytics databases such as **Google BigQuery** {https://packt-debp.link/Dg7T5C} and **Amazon Redshift** {https://packt-debp.link/L6jwtQ} are optimized for query processing and analysis, rendering them integral components of data analytics aggregators. Additionally, data discovery platforms such as **Tableau** and Microsoft's Power BI fall under this category, enabling users to discover, visualize, and share insights from aggregated data. They often work in tandem with analytics databases to provide a comprehensive analytics solution.

Some key benefits and challenges of using aggregators

Aggregators offer a comprehensive analytics solution. They offer enhanced analytics capabilities with data that is aligned across a domain or industry. Additionally, they provide improved data accessibility for consumers. Data analytics aggregators make data readily accessible for analysis. For instance, a marketing team can swiftly access and analyze customer data to tailor marketing strategies. Lastly, they provide integrated analytics and visualization toolsets. By integrating with platforms such as Tableau or Qlik, data analytics aggregators not only provide robust analytics capabilities but also intuitive visualization tools to represent insights coherently.

What data quality and governance are afforded by a data aggregator? An aggregator should be able to facilitate data accuracy and consistency. The accuracy and consistency of data are paramount. Implementing validation checks and consistency algorithms are some practices to ensure data quality within aggregators. Regarding governance, aggregators support robust data governance policy services and make it easier to adhere to industrial or legal compliance standards (such as GDPR). Data cataloging and metadata management for discovery are mature offerings with aggregators. These are critical for maintaining data integrity, compliance, and discovery information needs.

The transformational impact of using data analytics aggregators is evidenced by enhanced data accessibility, improved analytical capabilities, and strengthened data governance practices. As the data landscape continues to evolve, so will the functionalities and capabilities of data analytics aggregators, potentially heralding new methodologies in data aggregation and analysis. For those interested in delving deeper into this domain, resources such as *Advanced Analytics with Google BigQuery* and *Power BI Cookbook* are recommended.

Summary

In this chapter, we explained the challenges data engineers will face when crafting a future-proof data engineered solution. Some core challenges have been outlined that will be faced when managing vast data, evolving technologies, and ensuring efficient data pipelines:

- Platform architectures change rapidly based on the cloud provider's user demands, combined with shifting technology opportunities

- The total cost of your ownership (TCO) is high if you do not build a future-proof solution

- The data and system architecture that your design conforms to must be rational and handle important items first and not have them appear as obstacles later so that your future-proof goals are attainable

Data engineering remains a hard task, but you are going into the effort with your eyes open. A solid foundation has been laid for the principles to be discussed later. We provided an overview of data engineering approaches so that you can scope out the current and near future technology landscape. You will have many choices to make when selecting foundational technologies and architecture patterns for your solution. The vendor choices are numerous, and the effort required to get various selections funded and into production is not to be trivialized.

In the next chapter, the need to develop a mission and vision will be elaborated upon and you will see how these challenges can be addressed in the mission with a clear vision. The set of principles you will discover after that will explain the guidelines for architecture and design.

3

A Data Engineer's Journey – IT's Vision and Mission

In this chapter, you will learn about the importance of a formalized vision, mission, principles, and architecture patterns as you design your solution. You will learn how to formulate these key artifacts for your solution and observe a few successfully implemented examples.

The key takeaway will be a renewed understanding of the need for a clear vision and mission so that what is developed is aligned with the business's needs. If requirements or designs deviate from the end goals, you will be able to provide justifiable pushback and then realign these requirements before unclear expectations (either of you or the solution) are created.

We will explain exactly why this is important to you, the data engineer. Often, engineers are fighting people and processes rather than technical problems, which could have been eased with social technology engineering. *Figure 3.1* shows how we will elaborate on the vision, mission, and strategy in the following sections:

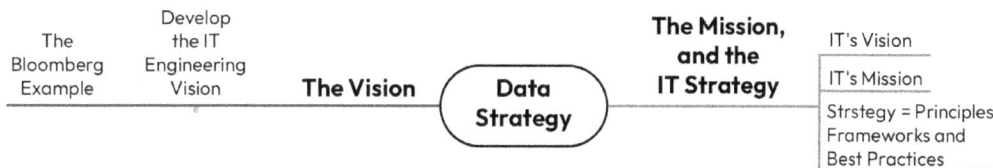

Figure 3.1 – Mission high-level mind map

We'll begin with the *vision* in the next section.

The vision

When a leader formulates a business strategy, a clear set of statements form in their mind. They begin with *"I see us…"*, after which the elaboration of that vision flows out describing the end state. It is like an artist describing in words what they intend to create. There is emotion, feeling, thought, and a clarity of purpose that comes from reflection on what will be! It is that vision that the leader uses to get up each morning and become excited. Progress toward the vision will be made that day… or not… but the leader will not be swayed in achieving the goal.

Every data journey should begin with a vision and a mission.

Unfortunately, the communication of that vision to the rest of the organization is often problematic. Everyone in the organization has to visualize the end state and drive their contribution to a strategy aligned with the vision. The vision statement is different from the vision itself. The statement is more like a flag that is waved to rally the troops. This flag is symbolic of the vision itself, which has many more words and detail, emotion, and effort than can ever be said in the catchy one liner that the vision statement portrays. Too much emphasis is placed on the vision statement rather than the vision itself. To keep focus on the organizational vision, everyone should write a vision statement that aligns with the overall vision and put it in their daily routine.

Even if a small section of a painting doesn't represent the beauty of the whole, the individual must see where they fit in and buy into that vision. Being emotionally drawn to its implementation is a key necessity. It is from this perspective that each participant in an organization sees themselves and knows why they are participating in its efforts. The vision aligns the emotional energy and provides the stiction when the going gets tough. It answers the keys questions *"Why am I here?", "Why am I doing this?", "Why am I enthusiastic?", "Why am I loyal to this company?"*

Your vision also provides the strength to backstep, sacrifice for others, and place yourself and your immediate needs subordinate to others and the whole of the enterprise. Have we not wanted to just give up at some point? It may have been too hard to please someone who we desperately needed to please… and that individual just wouldn't be pleased… yet we persevered and saw a way through because we were indomitable in our pursuit! Remember when an OKR was generated and given to us, but it felt more like we were handed a lead balloon! We had to reach down and see why it was important in the grand scheme of things (the vision). When our vision aligns, we see the way forward clearly.

Organizations agonize over the creation of the vision statement. However, before that, various leaders talk about the vision itself since that vision will drive overall strategy and the corporate mission. Each department must read the vision statement, and its associated business mission statement, and align it with the operational business strategy. The expectation is that everyone's objectives and corporate activities align with those directions. It is through the clarity of communications and shared participation in the vision, mission, and strategy that each participant gains clarity for their work. Asking ourselves: "What is my vision, mission, and strategy?" gives us clarity of purpose and drives us to achieve the goal. A wise IT leader is a representative of those they lead and must envision how each individual can grasp the vision and its mission statement and be able to move forward.

So, that all sounds good: job satisfaction, participation, direction, emotional energy, clarity of purpose, and so on. But we know circumstances get in the way and difficulties arise. One team member may leave, not buy into the vision, or not fully accept the mission. The standout individual may even be *right* from their perspective, or another may have a self-seeking agenda. Situations like these can erode the success of the entire organization. Before a person sees themselves in the bigger picture, they will need to let their guard down and be willing to adapt, change, and even look foolish. After all, someone might ask, "If it's so effective, why didn't anyone think of this new approach before now?"

Reviewing history reveals mistakes such as black spots on the canvas and is a retrospective review from today's lens.

We always clearly see the mistakes of the past when the evidence of those mistakes comes out in the present. It's always easier to see where we went wrong in hindsight. Visionary leadership can avoid pitfalls where others may not be able to conceive of them. Good leaders do not encumber their development teams with avoidable problems.

With group consensus, many mistakes can be avoided, and errors can be minimized.

A shared, group-formulated vision enables an organization to maintain its mission and strategy. When changes are needed, they are not jarring, but rather natural reactions to external forces that the mission can adapt to and for which the mid-term and long-term strategy has already conceived a playbook plan.

So, why is all this important for the data engineer? They must see themselves in the big picture vision, even if the corporate vision statement doesn't seem to apply to their work. The data engineer will want to reiterate their part and emotionally buy into a vision that enables them to engage fully, happily, and successfully. The data engineering architecture's potential must have a conceptual seat at the table, even if it's only represented by a more senior IT leader. Without commitment, loyalty to the mission will fade and disillusionment follow... bits and bytes are all that is left.

Data engineering is *hard*! You must be loyally committed, be ready to fight huge battles technically, and separate vendor and cloud provider marketware from real integrated software services. After all, you should always expect the 80/20 rule to be applied when delivering software. But often, we think we are getting a baked, generally available solution from a provider when it's sold to us. However, the problem is that the provider implemented 20% and made it look like 80% and that was sold to you with a smile! Do not buy into the technology hype cycle! Be wise in the creation of the mission, and create a strategy from the vision. These are very important stages of a company and its product development offerings.

You should be ready to set up a strategy to be refined yearly and always be reset to the next five years, with clear progress each year as part of the goal setting discussions for the next fiscal. But we have elaborated enough on the high-level need for a vision, mission, and strategy. Making up a personal vision, mission, and strategy for yourself should be a personal effort. Being brutally honest with yourself is essential.

"To thine own self be true!" - *William Shakespeare; Act 1, scene 3 of Hamlet* {https://packt-debp.link/inlwN7}

Why was I doing it wrong in the past so much so that I need to change to do it right now? Why don't I see it the way others see it? Why can't I align? What do I lack that does not enable me to get it? Apple Computer Pioneer David Sun says the following:

> **"Be a Sponge!"**
>
> "Every day when I wake up, I want to be a sponge," David Sun says. "I want to learn everything from everyone, from everything that I touch and see around me every morning. We should immerse ourselves in learning every day."

Setting up a safe communication environment for change is an essential part of the vision, mission, and strategy process. Be true to yourself; seek advisors who know more than you do and are willing to tell you passionately to your face how they think the vison, mission, and strategy should be formed. They will bring a piece of the overall direction and you will be satisfied with the knowledge you will gain, as well as friendship and open communication. Often, early misalignments in the vision and mission cause a huge number of issues to arise in strategy, architecture, delivery, and business operations.

> **"[Steve Jobs]'s dedication and attention to detail is beyond belief"**
>
> "He was also very, very consistent in his pursuit of excellence. He started to look at things from outside the box before that term became popular in the business world. He's a pretty focused person. He always gets projects done correctly. He always creates something that none of us realize we need at this moment, but he's already making it." - David Sun {https://packt-debp.link/93c3EZ}

By now, you have almost certainly registered our emphasis on the necessity of a business and personal vision, mission, and strategy. You may be wondering: why this emphasis for a data engineering best practices book? The reason is that this approach produces different engineering than one that brings a group of consultants together to work it out. The difference is that it produces a head turning shift in perspective for all participants. That shift keens focus on the end goal with confidence that it can be achieved. The many answers to questions that are raised during development will be clear as we form a solution that meets the business objective as we first produce the architecture. For now, we want to dive into the application of this in data engineering.

Data needs to be smart. This is not just an outcome of disconnected software processing but is engineered to preserve the data's context with the data itself. We want to preserve the connectedness of information and leverage available technology to that end. We are faced with technical and vendor limitations. We are forced to build more into the solution rather than buy off-the-shelf solutions or integrate a house of cards technology stack. The vision sets up a more complex mission (or missions) than some will view as expedient! The concept of data being smart is visionary. We intuit that making data smart is the right goal, yet it is not expedient to implement, nor is it risk free. Engineering designs needed to make data smart are not going to be a walk in the park! Likewise, it's not always safe for the IT leader to embrace a difficult mission aligned with a clear vision, but it is the right thing to do for the company.

A strategy-aligned mission can be created and supported by academic research that is not yet implemented in a customer facing system. It's great to be on the cutting edge – it attracts hungry graduates just itching to work on the latest and greatest new tech, and that can be exciting. But development projects can also go off the rails quickly if your project becomes just a technology playground for curious engineers. What can you do? The answer is to form the strategy anyway and find the right balance!

> **Don't let the mission's difficulty deter you!**
>
> Embrace the cutting edge, just don't get cut!

It is important that innovation and creativity not be dampened. In fact, you will want to foster it, but not at the expense of true delivery progress. 20% of an IT professional's workday has to be allocated to improvement; otherwise, in 5 years, that individual will be irrelevant as the next generational wave crests. When the next wave comes, you'll either be on top, in the middle, or getting swamped at the bottom.

That 20% rarely comes from an employer. It comes out of your own personal time, after the overtime, after the delivery, or in parallel with it, but it will come out. Alternatively, if the employer is gracious and values the staff, time during development is allocated to the spikes that are in addition to the core effort. Here are some things you should do:

- Look for ways to move forward.

- Mitigate the risks of being on the cutting edge but be courageous.

- Contribute to open source, use it, learn from it, and leverage it where possible. Go so far as to enhance it and contribute back to the community. Make sure you have complied with the copyrights and usage agreements.

- Read others' code, and learn from it and each other. Do not be afraid to refactor. A wise mentor once said to me *never get married to your code!*

- Think operationally, and not just functionally when designing.

Develop the IT engineering vision

This means you structure your error handling code first. You also design *testing first* as part of the design and you implement *shift-left testing* disciplines across the team. You will save that 40% waterfall test effort that the old Mythical Man-Month prescribed more than 40 years ago. As a story, years ago. when I worked in the Bloomberg Equities team, I asked, "How does the team test code?" The answer was *"Carefully!"* I then asked how to access the test environment and I was told there was none. And lastly, I was told, *"Bugs are personal!"* I was grossed out on many levels, but I persevered, becoming a human compiler and sweating each time code went into production. I'm not saying this is good at all, but it worked for the Bloomberg Equities team.

> **The developer must take charge of the delivery quality and NOT depend on defects getting identified or repaired by others**
> Mitigate the risk of falling on the cutting-edge sword!
> Build with a test-first design and shift-left-testing approaches.
> Think about operations and not just development when implementing DevOps!

IT leaders must build a safety net for developers to produce quality code that integrates and operates according to a clear framework. That framework should include the customized agile framework for the organization (not for each team). This minimizes the defect rate and time/cost to fix errors at integration time. To be effective, these errors cannot be caused by intrinsic management process omissions. Think about the complexity of today's cloud PaaS solutions, which require serverless componentry. The end-to-end integration testing poses a huge operational risk due to incorrect assumptions about cloud producers' PaaS operations. This is especially evident at the edge of the PaaS server's scale capability. Error handling/recovery should be built as part of *your* design, and it *must* be tested before integration.

The Bloomberg example

When you think about the development journey, you have to minimize the pain and maximize the gain. I apologize for the cliché but it's true! IT development can be very difficult and you don't want to change what works well too often. Building a solution that withstands the test of time (in other words, being future-proof) will require you to have a clear vision with a clear mission and be principle driven in your drive for success.

As an example, when Bloomberg began in the late 1980s, they implemented an architecture that was based on the IBM Mainframe and 3270 controller patterns. For those who study computer archaeology, let me expand on this legacy design pattern a bit. That mainframe controller pattern allowed dumb terminals to be controlled by a control unit in the back office. About thirty-two terminals could be connected to a cascade of controllers at a time. What was mimicked in the Bloomberg system was the controller that was rewritten on Unix servers and the dumb terminal was replaced with a PC running the turnkey Bloomberg PC-hosted application. The Microsoft Windows OS was not needed, just the ability to implement Bloomberg's new control protocol between the UNIX controller and PC. That controller drove the GUI. It was *not* a rich GUI originally, but it was very functional and fast. More than 1,200 functions were built, and today, many more exist – after entering four base-36 characters for a function name, you can get one of 1,679,615 potential function name combinations possible from 0000-to-ZZZZ. A key aspect of the vision was driven by the competitive financial information services industry! Data is key to this vision and protecting it from being stolen or used improperly was – and is still – essential. Having this data in a usable but protected form (that is, visual only) made its licensing enforceable through the technology implementation as all data was visually rendered and not supplied programmatically. This was an elegant solution to a key problem solved by technology, and it remains a competitive advantage for Bloomberg.

As a comical note, I remember back in the mid-1990s that a backhoe broke a communication line near the George Washington Bridge and three information providers lost service at the same time since they were clearly getting data from a fourth who was drastically affected by the lost line. It was clear who drove whose profit, even if the three were not licensed users or redistributors! You can imagine what took place after this event and the consolidation in the industry that followed.

Bloomberg was right to keep the data safe! So, raw data was not let out of its closed system. It was shielded by a customized Bloomberg Unix mid-tier control server that sent opcodes to a custom PC application running on the customized PC with a customized keyboard. The Bloomberg controller protocol drove thousands of screen functions. The app was braindead but processed the rich GUI control commands from the mid-tier server. It was smarter than an IBM 3270 control system, but it leveraged the same architecture pattern! However, the GUI rendered information was available to the user's screen, not the data. And that was fully encrypted and protected. Data, analytics, and news functions were and still are provided at blazing speeds with a delivery service level that exceeds expectations at scale and remains robust today. This is due to the nature of the financial services end user. Shall I say that they want it now, and it better be correct!

Data could not possibly be scraped by the nefarious competition with that architecture, nor was it made available via HTML, text, or in any discoverable way. Later, a data license API was developed for which one of the authors of this book was involved. However, it was priced high, regulated, throttled, and usage validated for fare use compliance. The closed data Bloomberg information system addressed the vision for speed and accuracy, as well as the competitive landscape. It is invaluable to its users. It implicitly addresses the usage agreement and is sold as one terminal per user, licensed by month for a cost with *no access to the raw data*. Trust is guaranteed and verified by the business strategy alignment and the implemented missions, one of which was the adoption of the architecture as a key mission directive. Data is, was, and always will be a competitive advantage. The overall mission was really simple: *"Don't screw it up!"* Mike Bloomberg still says this today when launching a project.

Providing unparalleled access to information with total coverage via a 100% SLA is still as invaluable today as it was when Bloomberg launched.

You could also add a corollary to the mission: make and keep a lot of money for the company, its investors, and its employees.

> **Note**
> Bloomberg implemented the solution with technology from the early 80s written in *Fortran*, wrapped it in *C*, and interfaced it with *C++*. Then, the *Java* service was wrapped for scale, and *JavaScript* enabled it for graphic rendering. All this happened while they always ensured data was separated from the user and the solution was scaled to support more than 400K users with no tolerance for error or delay. Note that the code still runs after 25 years, but good luck finding a Fortran coder!

I should also tell you about the speed of the Bloomberg distributed database: BBCOM. The original design put all market data in memory and because of hardware limits, such as cost and RAM availability, it had to be distributed across 80+ servers. Engineers rewrote the TCP/IP stack, threw out the seven slot sliding window protocol of TCP, and created a window of 1,600+ slots called *PCOMM* and ran it as /IP. They did this to ensure data could be supplied to the mid-tier controller faster on the IP network than what was commercially available. In *Chapter 12*, we will elaborate on data services, and you will appreciate the genius of the Bloomberg system even more. You will want to leverage technology to meet and exceed the business's goals so that an out-of-the-box solution is elegant, future-proof, and profitable. The lesson learned is that if you can envision it but it's hard, do it anyway! Can cloud providers commoditize this type of solution in a data mesh for all use cases? The answer is yes, but they haven't yet! You, the customer and IT leader, will have to push the envelope to keep functionality high, costs down, and scale up!

So, what makes a successful vision? One that considers the end state and is real, not pollyannaish. It addresses the goal as well as the competitive landscape that could impede the goal. Its corresponding vision statement is short and quick but cannot cover all the details; therefore, it is meant to be a flag for others to look to when the going gets tough.

> **The vision statement**
> The vision clearly sets the direction for the mission, and that mission provides guardrails for the rolling strategy.

The vision and its mission acknowledge, or directly address, potential obstacles to success, as well as provide clear sight into the future. It is the visionary's goal to set this kind of vison for their company. That is the very essence of the term visionary. The IT leader will want to be that sort of visionary.

Vision summary

A vision should answer the question, *"What is this strategy going to achieve?"* To do so, the vision must do the following:

- Focus on the future, hence why the focus in this book is on being *future-proof*

- Set an aspirational tone

- Be motivating to the organization

- Be inspiring for all who *want to achieve the vision*

- Be free of rhetoric and jargon

- Be authentic, relatable, and meaningful

Now that you understand the need for principles and vision, you will be exposed to the type of wording needed for the mission statement.

The mission and the IT strategy

The mission should also have a mission statement that can be used as a guide for the development of the business IT strategy and the set of department missions that will arise from the vision. *Figure 3.2* shows how we will elaborate on the mission and strategy in the following sections:

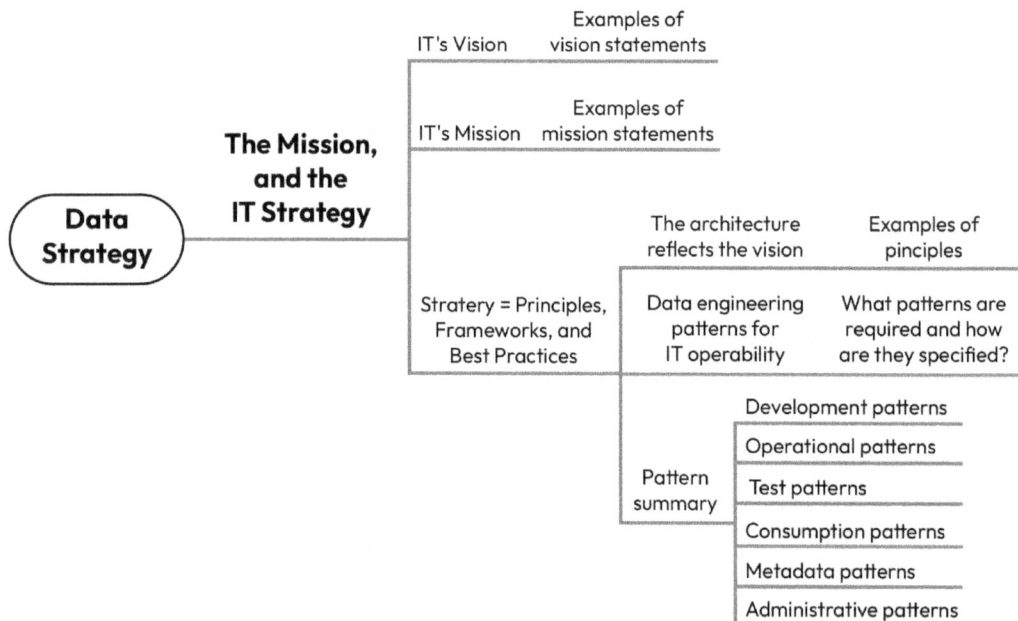

Figure 3.2 – Mission and IT strategy mind map

While vision is a painting of the future, the mission is more like a sculpture – it adds depth to this view. Walk around the sculpture and you will see many facets that are invisible from most other perspectives.

> **Everyone in an organization will see the mission from their own perspective and ask...**
> "What does this mean to me?"

The answer to this question is going to drive the development of the departmental missions:

- Finance will see how to arrange and monitor costs with a focus on **return on investment (ROI)**.

- **Human resources (HR)** will ask how many resources and how much training is needed to accomplish the overall mission.

- **Information Technology (IT)**, which is **not a cost center** but part of the profit center of the vision, will engage fully to ensure they have a competitive advantage regarding the company maximizing ROI.

- IT, which **is a cost center**, will focus on **total cost of ownership (TCO)** reduction.

- Each business silo leader will drill into the mission and what it means to the department's mission. Each leader will try to drive data engineering efforts toward their needs, and they are not going to make it easy for you to converge on a one-size-fits-all data architecture. A data engineer needs to leverage the overall mission statement to keep the silos aligned with the enterprise's goal, or at least point out when those goals diverge from each other.

What holds all in alignment will be the business strategy that was set up when the mission was communicated. The strategy considers the pragmatic needs of each group and weighs the cost of optimizing a solution for one group over another. This is where consensus is written down. It is hard and hard choices must be made; otherwise, the downstream fight comes for the architects of solutions to deal with when it should have been resolved by the business and IT strategy teams.

> **Tip**
> Identify when a strategy doesn't resolve key issues that the architects would have to resolve.

The mission and the strategy are interconnected; otherwise, one is misdirected, and the other one is made irrelevant. After all, running a successful business is part science, part art, and a whole lot of luck! The luck part is not as problematic when a mission and a strategy are aligned and you consider the perspectives of each key group within the organization.

IT's vision

The IT vision statement is subordinate and aligned with the company's business vision: It sets the engineering team's direction for the long term to help them obtain the company's strategy.

The clarity of a well formulated IT vision provides guardrails for the objective setting exercises of a company. In these sessions, business, product, and IT team members formulate and accept OKRs that can be enthusiastically and fully embraced. A proper IT vision fosters and demands a culture of inclusion. This makes integration possible and enables an open flow of communications.

The IT vision statement should be all of the following:

- **A mechanism for setting goals**: To illustrate the company values that will be applied

- **A statement of purpose**: To guide those ideating products/services or competitive challenges

- **An indicator of the measures of success**: This ensures that what is achieved is measurable

- **Brief**: It should be easy to remember and two or three sentences at most

- **Challenging**: It should be a bit vague since it has to cause employees to see how it applies to them, so this is necessary

- **Clear**: It should use understandable language and be concise

- **Oriented toward the future**: To set directions for future activity for which progress can be tracked

The IT vision points to the key values of an organization. It is important that these values appear in the principles that will be developed by the IT organization.

> **Note**
>
> We have avoided discussing the development of IT policy documents since these are compliance-oriented and focus on restraining the negative behaviors of an IT organization's employees. They rarely focus on the positive, building up aspects of the IT vision. Although required, they just need to be created and dealt with... but in the back of the IT leader's mind, not the foreground.

Examples of vision statements

Amazon

> *"To be Earth's most customer-centric company, where customers can find and discover anything they might want to buy online."* {https://packt-debp.link/IeR7in}

Bloomberg

> *"We believe profit and principles are not mutually exclusive. They reinforce one another. And doing the right thing - by our people, our customers, our communities, and our planet - is also the best thing for our business."* {https://packt-debp.link/vjcVhs}

Elsevier

> *"We help researchers and healthcare professionals advance science and improve health outcomes for the benefit of society."* {https://packt-debp.link/mvjiDn}

Microsoft

> *"…to help people and businesses throughout the world realize their full potential."*
> {https://packt-debp.link/Of6iO4}

IBM

> *"IBM's greatest invention is the IBMer. We believe that through the application of intelligence, reason, and science, we can improve business, society, and the human condition, bringing the power of an open hybrid cloud and AI strategy to life for our clients and partners around the world."*
> {https://packt-debp.link/cBgFNf}

Oracle

Note: Oracle's vision hasn't been published, but it's implied by its principles.

> *"…becoming a company that helps its customers to simplify their IT environment in such a way that they are adequately equipped to grow and thrive."*{https://packt-debp.link/KblKg5}

IT's mission

The purpose of the IT mission statement is to direct the IT organization on how to achieve the IT vision.

The overall business strategy and the aligned department strategies are the methods to achieve the overall mission that was derived from the vision. Goals are statements of what needs to be accomplished to implement the strategy, and methods define the streams of activity within the strategy.

IT leadership will take aspects of the business strategy such as revenue growth and enhanced customer retention and build a strategic IT plan to support that goal tactically and strategically via various methods, such as a lower customer service cost-to-user ratio. The method needs to be measured for success by increasing the **net promoter score** (**NPS**) by 15%. The IT strategy should support and align with the business strategy. Generally, an IT mission and strategy are used to make informed budgeting decisions based on empowered experts who set spending plans. This is critical for an IT cost centered group versus a profit centered IT group where the developed system is the product. IT Strategic planning contain the following elements:

- **Business portfolio value** by fiscal year and, if necessary, by quarter (if it's a public entity).

- **IT costs** that are fed to a decision support system for business unit analytics and product group analysis.

- **IT costs for infrastructure spending** including those from the cloud, PaaS, SaaS, IaaS versus on-premises, networking, and telecom.

- **IT costs for investments** including business-as-usual, maintenance, break-fix, and transformation.

- **IT costs** for professional services and customer specific solutions.

- **IT ROI tracking** by business initiative.

- **IT TCO** for various areas: application, services, and so on.

- **Metrics for customers** engagement, satisfaction, sentiment, returns, cart abandonment, visits, and so on.

- **Operational expense** versus capital expense variance.

- **Product** in terms of demand backlog, execution efficacy, and defect rates.

For the data engineer, the IT strategy's status reporting will not properly reflect the value being added by the tough choices being made when developing the data architecture aligned with the mission. For example, an elegant data architected engineering approach will positively affect many of the metrics listed here. It will be the task of the IT leader to create a scorecard showing how the architecture positively supports the IT strategic metrics. The C level needs to align the IT strategy with the business strategy; otherwise, when issues arise, the problem is always the architecture's fault! Blaming the architect has always been the easy path of a poor leader. Making sure that there are success metrics aligned with the recurring IT reporting process removes this excuse and replaces it with incremental feedback that is fair and clearly communicated. Faults will arise but knowing where to find the issue is empirically better than guessing.

Protecting the architecture, the architect, and the engineers should be an IT leader's goal. When I was coming up in my career, I heard a colleague moan about a manager they had that brazenly said, "*Never start a project without selecting someone to blame.*" Why was this guy not rooted out of the organization earlier? I don't know, but he didn't last that long from what I recall. As an IT leader, you want to preserve the assets and not play games with people because of the mistakes made by others, and having those mistakes be the cause of angst among staff. Engineer your success through correct and sometimes defensive positive success reporting. Do not begin with the solution definition followed by some way to pass the inevitable blame down as failures arise. Our suggestion is to do the following:

- **Architect the solution** so that it's aligned with strategy.

- **Report IT strategy aligned success** metrics in a way that the story may be absorbed by the C level.

- **Enable your architecture's success to be supported by empirical evidence** that is consistently aligned with the strategy.

- **Show positive value** and progress incrementally to provide evidence of continued success and alignment.

- **Identify impediments to progress** and ask for help early when problems arise. C-level types want to know, and they also want to support your solutions. Just don't expect to shovel problems upward, even if the clear cause is an upward problem. Illustrate the analysis of the options and point to a preferred type of help being requested.

- **Gather best practices**, seek the perspectives of technical experts, and collect successful patterns/stories for reuse in your architecture. However, do not act on all of them.

It could very well be that executive processes, resource shortages, erratic customers, product misdirection, or general development process faults (management/governance) have caused issues you have to tackle. It is still essential that the tracking of the strategic execution as you see it is communicated upward at a consistent pace. Your reports are not to be separated from the IT leader executing the mission that's aligned with the strategy.

Your IT mission is to execute the business mission and IT strategy with your mission as you develop the architecture with its implied best in class data engineering for the organization! The data engineer can't be just an order taker of impossible missions. The data engineer needs to be fully engaged early so that any design efforts are acceptable.

Examples of mission statements

Amazon

"We strive to offer our customers the lowest possible prices, the best available selection, and the utmost convenience." {https://packt-debp.link/XFzQBk}

Bloomberg

"We connect the world's decision makers to accurate information on the financial markets – and help them make faster, smarter decisions." {https://packt-debp.link/Z7kNMC}

Elsevier

"We are a global information analytics business that helps institutions and professionals advance healthcare, open science, and improve performance for the benefit of humanity. A world-leading provider of information solutions that enhance the performance of science, health, and technology professionals, empowering them to make better decisions, and deliver better care." {https://packt-debp.link/osy2aJ}

Microsoft

"…to empower every person and every organization on the planet to achieve more." {https://packt-debp.link/vutE0l}

IBM

"Our mission and purpose is to be a catalyst that makes the world work better. A catalyst is an agent of change. As catalysts, IBMers collaborate to release new energy. We forge partnerships and bring together powerful combinations of people and technology. We combine the methods of science with the dynamism of business to do things neither could do on its own. We solve difficult problems and amplify the impact of our solutions with speed and scale. Our work is focused on the most mission-critical systems on the planet: electrical grids, airlines, mobile networks, banks, transportation systems, and healthcare systems. These systems are more than just engines of economic growth. They are the systems that support modern society. In making them faster, more productive, and more secure, we don't just make business work better, we make the world work better." {https://packt-debp.link/DFU417}

Oracle

"We foster an inclusive environment that leverages the diverse backgrounds and perspectives of all our employees, suppliers, customers, and partners to drive a sustainable global competitive advantage." {https://packt-debp.link/DRfJjC}

IT mission summary

The mission should be clear in everyone's mind once the OKRs are created, and the business strategy made tangible in the end state architecture definition. Until then, it keeps everyone looking in the right direction. What's needed are principles that will be applied when making choices, a framework for how we are going to meet our objectives, and the best practices to scope the team's excessive desire to wander off the path toward the goal.

Principles, frameworks, and best practices

At this point, the IT strategy should be clear, and the corresponding mission embraced. So, what's next? It's time to think about how to corral the troops. It's beyond 2020 and we do Agile development. We just don't want to dive in. I can assure you that you will want to conform to a clear Agile manifesto for your company. Senior business leaders have not yet drunk the Agile Kool-Aid. They set a budget and expectations and then want to know when it's done.

> **Tip**
> We suggest that you build 20% slack into your waterfall execution plan and create the OKRs for your teams with the business but without that extra 20%.

Drive toward the implementation of the IT strategy using your Agile processes. You will need slack time to not blow out the strategic business expectations. To make sure your troops (Agile-enabled and trained) develop toward the goal, you need to be concerned about the following:

- **People**: Who is doing what and how aligned are they with the best practices?

- **Process**: Are the processes clear and aligned with the framework and is progress measurable?

- **Technology**: Are our technology choices aligned with our principles?

The architecture reflects the vision

When asked to define architecture in the past, we broke it down into the preceding three main divisions. From this, we can see that best practices, frameworks, and principles need to be roughly applied to people/skills, processes/methods, and technology/choices. The reality of development forces is not as rigidly segmented as the approach I have depicted but it works for ease of explaining where these fit into the architecture. This book will drill into a detailed architecture and framework for data engineering with best practices, but it will not elaborate on these areas in this section. However, for clarity and to align with the vision and mission discussed earlier, several examples of company principles have been included for you to digest.

The stage is now set for you to drill down into the data journey through our reference system. With this stage being set, you can share our perspective and appreciate that to future-proof a data solution, you must gain perspective from a metaphorical higher ground. When reviewing the examples of principles provided in this section, you will see why vendors see the technology landscape differently and will produce products and services that do not always play well in your environment sandbox. Supplied capabilities may not align. A lot of glue (that is, IT architecture and development effort) has to be used to create the solid future-proof data solution we desire.

Examples of principles

This section provides examples of company principles for entities that view IT as core to their product offerings and, as such, the IT principles and business principles align closely. Companies that treat IT as a cost center will have a set of drill down IT principles. What's clear is that the culture of the company will be evident in the values and principles made publicly available for all to view. Note that cutting edge companies are posting IT principles for AI. Since this is relevant to our best practices, I've included them for review. They will service a company well as it re-envisions data engineering to support the technology shift that's occurring.

Amazon's guiding principles

Amazon strives to be the Earth's most customer-centric company. This is stated on their *"Who we are"* page {https://packt-debp.link/3rRrrF}. They want to be the *"Earth's best employer"* and the *"Earth's safest place to work."* The principles can be identified:

- **Customer obsession** (rather than competitor focus)
- **Passion for invention**
- **Commitment to operational excellence**
- **Long-term thinking**

A takeaway from these principles is that the data engineering strategy, implementation patterns, and blueprints will be customer based to support innovation and the long term success of the organization. You can expect a lot to be built rather than bought when architecting engineering solutions at Amazon.

Amazon's leadership principles

Separate from the guiding principles, Amazon focuses on a specific set of leadership principles {https://packt-debp.link/lcrfNn}. The development engineering process capabilities are developed from these principles. A clear focus on collaborative discussion, consensus, and growth is evident. This impacts data engineering output in many ways. The reference to cost (frugality) stands out and with that, simple and innovative solutions are given room to be developed:

- **Customer obsession**: Leaders start with the customer and work backward.
- **Ownership**: Leaders think long-term and don't sacrifice long-term value for short-term results.
- **Invent and simplify**: Leaders expect and require innovation and invention from their teams.
- **Be right, a lot**: Leaders have strong judgment and good instincts.
- **Learn and be curious**: Leaders are never done learning and always seek to improve themselves.

- **Hire and develop the best**: Leaders raise the performance bar with every hire and promotion.

- **Insist on the highest standards**: Leaders have relentlessly high standards.

- **Think big**: Leaders create and communicate a bold direction that inspires results (thinking small is a self-fulfilling prophecy).

- **Bias for action**: Decisions and actions are reversible and do not need extensive study – speed matters, and there should be valueccalculated riskctaking.

- **Frugality**: Accomplish more with less to breed resourcefulness, self-sufficiency, and invention.

- **Earn trust**: Leaders listen attentively, speak candidly, and treat others respectfully.

- **Dive deep**: Leaders operate at all levels, stay connected to the details, audit frequently, and are skeptical when metrics and anecdotes differ.

- **Have a backbone; disagree and commit**: Leaders are obligated to respectfully challenge decisions when they disagree, even when doing so is uncomfortable or exhausting.

- **Deliver results**: Leaders focus on the key inputs for their business and deliver them with the right quality and in a timely fashion.

- **Strive to be Earth's best employer**: Leaders work every day to create a safer, more productive, higher performing, more diverse, and better work environment.

- **Success and scale bring broad responsibility**: Amazon leaders impact the world and remain humble and thoughtful about even the secondary effects of their actions. Local communities, planets, and future generations need us to be better every day.

So, as a data engineer at Amazon, you have a lot of room to succeed, but your thinking will remain driven by principles that channel your development direction.

Bloomberg's core values

Bloomberg is unique in that they offer values rather than principles. These are similar but not the same. With value statements, principles can be viewed as flexible, which can cause technology to shift more as a response {https://packt-debp.link/WhK71o}. When this happens, expect costly impacts. However, when hyper change is going to be constant and money is not an issue, the approach works well:

- **Diversity and inclusion** at Bloomberg innovation drives the business forward. It is understood that disruptive, breakthrough ideas come about when diverse teams look at challenges from different angles. The culture at Bloomberg values differences, fosters inclusion, and promotes collaboration. When every employee is empowered to impact the business - everyone wins.

- **Tech at Bloomberg** pushes the boundaries. Bloomberg looks decades ahead to create what clients will someday need. This is stated in other company information as their mission from day one, when the Bloomberg Terminal was built. Pioneering technology transformed the financial industry, where information was transmitted slowly and inefficiently before the platform launched. Technical innovation drives and infuses the product as the company is constantly imagining and investing in the future.

- **Philanthropy and service** is an integral part of Bloomberg. The company is committed to giving back to the cities in which it operates, using the time and talent of its employees, as well as resources. With the company's data insights, news, and innovation, the company looks to address unmet community needs and deepen engagement with colleagues, clients, and partners. Improving the lives of others around the world is a core value.

- **Sustainability**. Bloomberg hopes to transform the future by tackling sustainability from every angle. According to Bloomberg, businesses have a critical role to play in shaping a more sustainable future. Employees across Bloomberg are engaged in building a better world, delivering data and insights that empower the global financial community to take action on climate change by reducing carbon emissions and giving back to the community.

The legacy values of innovation, collaboration, customer service, and *"doing the right thing"* still apply at Bloomberg. It's clear that these principles will affect your data engineering solutions and how they are produced.

Elsevier's privacy principles

Elsevier, as a scientific and medical publisher among other business areas, is unique in that they have a strong focus on privacy, accountability, and value {https://packt-debp.link/f3sD0y}:

- **Value**

  "We collect and use personal information to facilitate efficiency and productivity in research, healthcare and education."

- **Transparency**

  "We tell users about the personal information we collect, including how and why we will use and share it."

- **Choice**

  "Users are given choice over the collection, use and sharing of their personal information."

- **Anonymization**

  "We depersonalize and aggregate personal information where individual identification is not necessary."

- **Accountability**

  "We are committed to acting as a responsible steward of personal information."

As a data engineer at Elsevier or a similar company, you will have an academic view of data and, with that, a sharp focus on gleaning value from data while not exposing privacy or security risks. At this point, you can contrast the differences between Bloomberg, Amazon, and Elsevier to see how different the solutions would be given these principles (or values).

Elsevier's responsible AI principles

Elsevier goes a step further than others and nails down ethical AI principles so that what their data engineers do with the knowledge gleaned from academic sources is not used irresponsibly, resulting in trust violations {https://packt-debp.link/VYagTl}:

"We consider the real-world impact of our solutions on people."

"We take action to prevent the creation or reinforcement of unfair bias."

"We can explain how our solutions work."

"We create accountability through human oversight."

"We respect privacy and champion robust data governance."

It's great to see a company explicitly state that it will not violate the academic and business trust that has been established in the company's prestigious history, even if that technology can unlock a lot of potential. Without AI constraints, expect innovation to sometimes cross the line and result in privacy/IP violations, embarrassment, subsequent government regulation, and corporate litigation.

Microsoft's core values

At Microsoft, the focus, regarding values, is like Bloomberg's {https://packt-debp.link/xqWy8h}:

- **Citizenship**

  "We work to be a responsible partner to those who place their trust in us, conducting business in a way that is inclusive, transparent, and be respectful of human rights."

- **Trustworthy Computing**

  "We build our Trusted Cloud on four foundational principles -Security, Privacy, Compliance and Transparency."

- **Innovation**

  "Using the power of AI and computing, we deliver technology innovation that inspires people of all ages and abilities, eliminate barriers, improve lives and strengthen communities."

- **Diversity and Inclusion**

  "Being inclusive is not just something we do - it's who we are. We celebrate diversity. Our continued success is a corollary of the unique skills, experiences, and backgrounds that our employees bring to the company."

- **Environment**

  "Discover how we lead the way in sustainability and use our technologies to minimize the impact of our operations and products."

What stands out is that trust, innovation, inclusion, and environmental consciousness drive design thinking. After touring Microsoft's data centers and plans in Redmond, I saw how they put these into action. The effort that's expended is just enormous. How the company adhered to these values is truly evident. Without these values, Microsoft would be so much less, but with them, the solutions they develop are on the right future-proof path to success.

Microsoft's responsible AI principles

Like Elsevier, Microsoft has identified several AI principles {https://packt-debp.link/0y9u20}:

- **Inclusive and respectful**

    "AI technology must be inclusive and respectful of human rights, enhance human capability, not replace it."

- **Transparent**

    "AI systems must make decisions about data used and consider any limitations to build trust; as such AI results should be transparent."

- **Accountable**

    "Microsoft is accountable for AI systems' impacts on society."

- **Fair**

    "Microsoft strives to ensure AI systems are free from discrimination and social bias."

- **Reliable**

    "AI systems are to be accurate and dependable."

- **Private and secure**

    "AI systems must preserve privacy and security of user data."

AI principles will evolve to include even more responsible AI features, but at least Microsoft has put a focus on this cutting edge area. The *"transparent"* principle is key to trust and the fair principle helps counter bias. The *"reliable"* principle is often missed, and it will drive difficult features of your data solution. That repeatable/reliability capability is not easy to obtain and lock into a solution's patterns!

IBM's core principles

IBM developed a number of principles {https://packt-debp.link/zUG9jc} that cut across the legacy technology behemoth's technology landscape. They are generic enough to cover the overall need and provide wide latitude for data systems development:

- **Alignment to values**

    "A company must be true to its values in all of its activities. These are: Dedication to every client's success; Innovation that matters, for our company and for the world; and Trust and personal responsibility in all relationships."

- **Cross-sector collaboration**

    "We work closely with the public and private sectors, including local, regional, and national governments, nonprofit organizations, universities, research organizations and school systems."

- **Solving problems**

    "by leveraging the full range of our company resources and then bringing solutions to scale."

- **Impact and measurement**

    "Whether we are taking on some of the unique and complex problems of the world's cities, helping to transform global health or working to prepare students for 21st-century careers, we endeavor to effect widespread, measurable and sustainable change."

IBM's principles can be viewed as softer than some of the others identified earlier. What stands out is the *alignment with values* principle. The trust and personal responsibility focus has implications for how data is made available to others both internally and externally. Expect solutions to be measured for success. They will also have an analytic focus. Enforced trust and transparency can also be expected.

IBM's principles of trust and transparency concerning AI

IBM, like others, views AI as a technology that needs to be constrained by principles {https://packt-debp.link/dc4UyZ}. Given the prior focus on trust, you can see the language inserted throughout IBM's AI principles:

- **The Purpose of AI is to Augment Human Intelligence**

    "AI should make all of us better, not just the elite few."

- **Data and Insights Belong to their Creator**

    "Clients' data is their data, and their insights are their insights."

- **New Technology, Including AI Systems, Must Be Transparent and Explainable**

    "We must be clear about who trains their AI systems, what data was used in training and, most importantly, what went into their algorithms' recommendations."

A takeaway from looking at IBM's AI principles is that the company does not want the AI to be confused with a **real** person. There has to be a focus on the *"augmentation of man"* rather than the *"replacement of man."* Being able to drill into the answers to the questions: *Why?* and *How?* an AI assertion is being made is akin to someone asking, *"How did you come that that conclusion?"* This self-reflection capability is a key to AI correctness, and it is often missed.

Oracle's core principles

Oracle's core principles focus on data system operations {https://packt-debp.link/zz79mc}. They are as follows:

- **Design For Operational Simplicity**

    "Keep each solution as simple as possible."

- **Strive For Hybrid Solutions**

  "Solutions must embrace a Hybrid Cloud model."

- **Take a SaaS First Approach**

  "Deliver the solution through a SaaS application wherever possible."

- **Ensure Use-Cases Drive Solution Design**

  "Solutions should address specific requirements and use cases as closely as possible."

- **Design for Business Continuity**

  "Consider high availability and disaster recovery requirements for business continuity."

- **Focus On Legal and Security Compliance**

  "Meet or Exceed Legal, Compliance and Security Requirements."

- **Automate**

  "Design Solutions and Services that can benefit from automation."

- **Maintain Separation of Environments**

  "Keep different environments, such as production, dev, test, user acceptance testing, and training separate from each other."

What is unique to Oracle as a data engineering solution company and vendor is the focus on security and use case driven solution designs. Users have to want, need, and ask for features before attention is given to developing solutions. Simplicity, automation, and integration (hybrid clouds, separation of environments, and so on) are key to the company's success.

Oracle's responsible AI

As a leader in data technology, Oracle also has stepped into the AI development field, but not without a set of responsible AI statements {https://packt-debp.link/pNfmgy}. Refer to the following Oracle paper:

In May 2023 Oracle produced a paper entitled

"Responsible AI: Oracle's Guide to Ethical Considerations in AI Development and Deployment", by Dr. Sanjay Basu, Senior Director of Cloud Engineering. Although the company has not published a specific set of responsible AI principles, it can only be rationalized that at the time of writing, Oracle Corp is following the guidance in this exhaustive work ((https://blogs.oracle.com/ai-and-datascience/post/is-responsible-ai-synonymous-with-ai-ethics).

In the paper Sanjay Basu, identified numerous non-profits and research institutes that are establishing ethical standards for how companies can use AI to protect consumers and employees, such as *"Partnership on Artificial Intelligence,"* the *"Institute for Human-Centered Artificial Intelligence,"* and the *"Responsible AI Initiative."* Oracle is aligning with the industry guidelines and creating awareness that ensures companies are using AI responsibly.

Principles summary

Principles and their underlying values are standards that drive downstream decisions. They are not arbitrary thoughts, nor are they 100% inclusive of all guidelines during the development of a solution; however, they are strong statements that will be the subject of future discussion if violated. They can also be pointed out when decisions are taking place to keep discussions on track. They evolve with the advent of new technology such as AI. They are forward looking growth enablers, just as corporate and IT policies are backstops against failure.

Next, we'll discuss how you can apply the principles that have been created for your organization or design efforts. You'll want to apply principles to the creation of various patterns that will define the framework of your solution's architecture. Consider a building analogy. Triangles and hexagons, when combined, will lead to the creation of round buildings, domes, and other circular shaped edifices. Squares and rectangular shapes will work well within an office building framework. Spheres can be stacked to fill a space or strung together like pearls. Such choices that are made early on drive the choices that are made during solution development. Having many divergent patterns at that same time leads to an unoptimized solution. The patterns you envision should form early on in the process. An important pattern is the operations pattern.

Data engineering patterns for IT operability

At this juncture, you will want to know, "How do I apply all this? I want to get started, and I don't want to reinvent the wheel and I don't want to end up driving into a pothole, or worse off a cliff." I've heard all these sayings and suffered a few of these outcomes myself.

What do we mean by data engineering for IT operability? It is being able to right-size data, metadata, and data access service levels. To accomplish this, you must be pragmatic regarding today's technology, speed, scale, and cost. In *Chapter 1*, the need for data zones was raised. Bronze, silver, and gold zones were minimally identified with the need for a raw landing zone and a consumption zone for analytics. Each zone should have clear interfaces for ingesting and egressing data and ensuring access patterns are documented and governed. What goes into this interface contract is key to how data underneath is organized.

- **Raw zone(s)**: These provide support for raw data ingestion:

  - They support data profiling and data normalization use cases, and they also reference data expansion at the syntactic level to preserve data quality. After all, you don't want bad, gap-filled, or redundant data to get into the data journey. You can also sanitize data that is in gross error and feed business operations teams' workflows to implement corrective steps when inputs need to be corrected promptly.

  - Supports **extract, load, and transform** (**ELT**) use cases to maintain the business value of the data.

  - Supports historical reprocessing and reinjection of ELT data because of systemic code changes or business rule changes in the subscribing downstream zones.

  - Implements a **change data capture** (**CDC**) outflow to the bronze zone. This CDC capability is very important since data has to be made fit for purpose in each downstream zone. Often, the same logical dataset needs to be rendered into different forms in those processing zones.

Keeping all logical data in sync is a challenge given the diversity of technology in use between the zones. A generic CDC pattern that can be leveraged at the zone boundaries will facilitate consistent data/metadata movement for all of your solutions. In the past, legacy RDBMS systems relied on transaction log replaying to keep a subscribing system up to date. That feature is not available in many cloud data systems, but the need to sync data systems still exists.

- **Bronze zone**: This zone implements a data lake:

  - Supports big data and some **online analytical processing** (**OLAP**) use cases, but these are performance limited. These use cases support data mining and enhanced transformations of the data and are rule based or AI assisted.

  - Supports **data quality management** (**DQM**) of the data. Here, trends are assessed and data is smoothed for downstream processing. Imputation, ascription, and other statistical approaches are applied to the datasets to make them right for the business domain, even though there may be gaps or errors.

  - Supports trend analytics to ensure data is semantically correct before being allowed into the data mart of the silver zone.

  - Implements a CDC outflow to the silver zone.

- **Silver zone**: This zone implements the cache enabled data lake and data mart functionality of the data warehouse from data originating from the bronze zone, as well as some from the gold zone:

  - Supports derived, imputed, and inferred data that is drawn from processing data in the bronze zone

  - Supports **online transaction processing** (**OLTP**) use cases and stores processed output in the gold zone

  - Implements a CDC outflow to the gold zone

  - Implements a CDC outflow to the subscribed consumption zones

- **Gold zone**: This zone implements a data warehouse as part of mastering the data and governing its correctness and long-term retention:

  - Supports on-line transaction processing (OLTP) use cases directly against information that has been mastered.

  - Implements a CDC outflow to the same datasets that are subscribed to the silver zone. This may look like a backflow but it's required for practical performance reasons.

  - Implements a CDC outflow to the subscribed consumption zones.

- **Consumption zone(s)**: These zones provide measures, calculated metrics, group rollups, custom trend analytics, and end user decision support:

  - Supports on-line analytic processing (OLTP) needs of the analytics user

  - Implements a Pub/Sub flavor of CDC outflow to external consuming users and feeds

In *Chapter 1*, it was pointed out that keeping all this data flow in sync can become a complexity nightmare without a clear framework and best practices to keep the system correct.

The architecture must build patterns that enforce the best practices for CDC, subscription, bulk data loading, data replay, and data reset. Best practices and governance must be implicit in the framework and best practices. This avoids the trials and errors that will arise.

> **Note**
>
> The cloud providers and third party vendors do *not* provide this framework!

They all provide unintegrated tools such as Kubernetes for the development of microservices, Kafka for queuing, Kinesis for scale, and so on, but none provide the necessary CDC at the cloud PaaS level! This is what is key to all zone syncing. Syncing zone data is required with today's cloud architectures.

Figure 3.3 shows what will be discussed in the following sections:

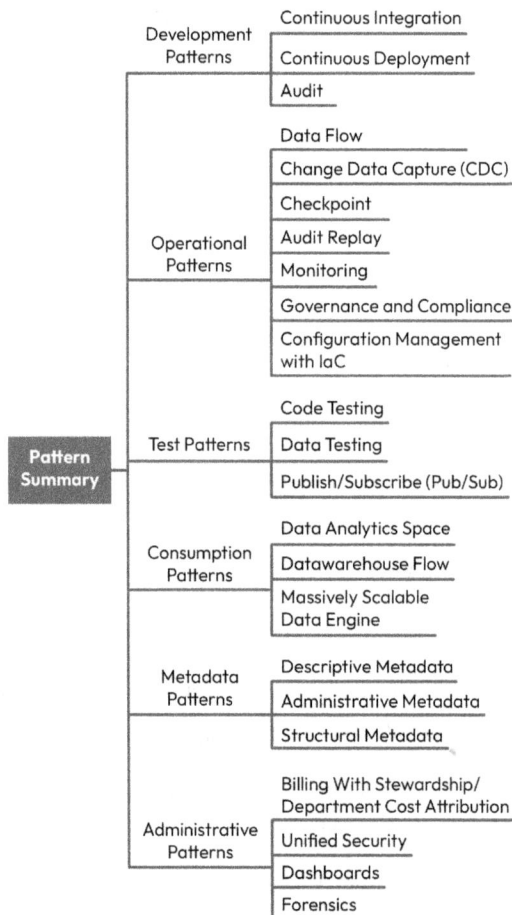

Figure 3.3 – Pattern summary mind map

Let's take a walk through the patterns that we've identified and the issues that need to be addressed when formulating your architecture, a prerequisite for the data engineering designs to follow.

What patterns are required and how are they specified?

As a data engineer, you will want to have a toolbox of *repeatable* – if not *operationally* reusable – components, at least component integration patterns that conform to framework blueprints. They can be created from a set of proven primitives and integration patterns that fit into an architected framework. You don't want to reinvent the wheel over and over; you know it's round and it works… so use it! The pattern is that the wheel needs an axel that, in turn, must be mounted on the cart.

Likewise, the data engineer will have to have tools organized so that they can be picked up and applied when ready. I've created a quick taxonomy of these patterns for consideration that I have used in the past. They begin with a test first, shift-left testing framework, then support for the operational SLA, and then business logic. It's sort of like eating your peas before chowing down on your steak and potatoes.

> *"The phrase "eat our peas" is not a historical metaphor. However, according to the context given by CBS, US President Obama was telling the nation that they had to buckle down and do what might hurt but would be good for them."*
>
> - *"What does "Eat our peas" mean - where does it come from?"* (https://english. stackexchange.com/questions/33935/what-does-eat-our-peas-mean-where-does-it-come-from)

You *must* optimize the easy patterns first so that when failures arise, they are all handled cleanly and do not inflict excess time and effort during the development debugging process, or the smooth operations required to push data through the implemented cloud pipelines. Since we are breaking the old straight through processing paradigms of input-processing-output into modern cloud computing, we have threads, queues, and PaaS and SaaS services that all need to work seamlessly and perform at scale!

As a simple example, you will want to implement data checkpoints to enable errors to be recoverable and not incur multi-hour reprocessing just because a small hiccup occurs. This approach is not supplied with cloud offerings; the paradigms are often missing and potentially built or bought with third party tool products. Cloud providers will not make any integration framework mandatory. They empower the third party providers. It is essential to study the *capabilities* of the third party vendor tools and develop a pattern inventory that will compose the data platform's framework that *you* desire. There will be *no single* framework that does not have to integrate, and many do not integrate well!

From prior experience, one of the architecture solutions in a retail data analytics setting required Azure cloud services, as well as a data platform for big data processing that scaled to support many billions of retail transactions per month. A data profiling capability was needed on a level not envisioned by either provider. The cloud vendor consultants listened and proposed solutions but ultimately neither could handle the load and complexity required at scale. A data profiling capability was needed and that drove the development of a new combined deterministic rule based and machine learning based data profiling ingest framework to get in front of the data quality issues before they became systemic data quality issues visible to the end user. Trifacta (now Alteryx) was selected to support this data profiling need. If garbage is allowed in, then garbage will come out; that is not good.

For this, you can select an **enterprise application integration** (**EAI**) framework – there are many to choose from, as defined by the well-known set {https://packt-debp.link/KnFDWt}. Note that the way they're ordered doesn't mean one is better than another:

- ActiveBatch by Redwood Software

- Azure Integration Services by Microsoft

- BizTalk by Microsoft

- Boomi AtomSphere Platform by Boomi

- Celigo Integration Platform by Celigo

- Ensemble (Legacy) by InterSystems

- Fiorano Hybrid Integration Platform by Fiorano

- Flowgear iPaaS platform by Flowgear

- Harmony by Jitterbit

- IBM App Connect by IBM

- IFTTT by IFTTT

- Informatica Intelligent Data Management Cloud by Informatica

- Interlok by RELX (Adaptris)

- Magic xpi Integration Platform by Magic Software

- MuleSoft Anypoint Platform by Salesforce (MuleSoft)

- OpenLegacy Hub by OpenLegacy

- Oracle Cloud Infrastructure Integration Services by Oracle

- Oracle Service Bus by Oracle

- SAP Extension Suite by SAP

- SAP Process Orchestration by SAP

- SEEBURGER BIS by SEEBURGER

- SSIS Integration Toolkit for Microsoft Dynamics 365 by KingswaySoft

- The SnapLogic Intelligent Integration Platform by SnapLogic

- TIBCO BusinessWorks by TIBCO Software

- TIBCO Cloud Integration by TIBCO Software

- UiPath Business Automation Platform by UiPath

- webMethods Integration Platform by Software AG

- webMethods.io Integration by Software AG

- Workato by Workato

- WSO2 API Manager

- Zapier by Zapier

- ZigiOps by ZigiWave

After digging into each, you'll see that some focus on transactions, others on security, message replay, and Pub/Sub, and yet others on the ease by which an API can be constructed. The list goes on regarding capabilities and capability categories. How can you possibly choose and then integrate these divergent and often competitive platforms with wildly divergent interaction patterns? Buy versus build is a great approach if you have stock in the needed glue manufacturer! You will find that what you need is different types of glue, something that can quickly produce a cloud version of Frankenstein's monster.

Your job should not be to perpetuate an integration meltdown but to converge solutions to correct patterns. While doing this, you must ensure you communicate the capabilities and the framework's fitment and show continuous processes as you integrate the features of the objectives in the development plan. It's a careful balance that can be destroyed by two anti-patterns:

- **A vendor can solve your integration problems**: If you are empowered and lead, your architecture and engineers can develop patterns conforming to the framework of the well communicated architecture.

- **A vendor has done this before and can be trusted**: None have built your system; otherwise, you could have bought it! If they said that they have, then they lied to you, and you can't recover from it… even if you can legally sue them for breach.

Pattern summary

Let's look at a few pattern categories that have value for our best practice solution:

- Development patterns

- Operational patterns

- Test patterns

- Consumption patterns

- Administrative patterns

Once you've grasped the impact of needing these pattern categories in your solution, you are ready to dive into a detailed explanation of each.

Development patterns

In regard to **Development Patterns** the following subsections point out the kinds of patterns needed in this category.

Continuous integration

Continuous integration (**CI**) is a way to apply code changes to a system as a stream of change with backward capability, with integration with **shift-left testing** (**SLT**) processes enabling regression testing of code with acceptable data comparative states. These rule-based tests preserve the integrity of data at rest; allow for flexible criteria for data correctness while applying rules and machine learning (ML) based assessments and judgments; and report on correctness as changes to code and data are released from repositories as changes.

Continuous deployment

Continuous deployment (**CD**) is a way to inject code and data success criteria into the systems at deployment in a backward compatible or transformative manner. It also allows the code bundle with the data state bundle to be captured in the audit system for glacial archive recovery.

Audit

An **audit** is a way to make sure that the audit trail is kept in the CI/CD process and can enable a data state to be reproduced after a stream of code changes has been applied over the years. It may be wise to create the audit record so that it's independent of the data source and use that as the 100% backward compatible form of data in downstream pipeline processing. This means much thought must be put into the syntactic and semantic format of that audit trail dataset pattern.

Operational patterns

In regard to **Operational patterns** the following subsections define the core set of patterns needed to run and manage your solution.

Data flow

Data flow is a way to manage the flow of data at a business transaction level and implement prioritized queuing, with scaling, and resubmission in support of application transaction processing across a logical dataset.

Change-Data-Capture (CDC)

Change-Data-Capture (**CDC**) is a way to implement a stream of changes with publishing capabilities to a downstream subscriber or set of subscribers. Modes include bulk refresh, streaming updates, and resetting to a past data state at a prior point in time.

Checkpoint

A **checkpoint** is a way to create a transaction bracket with *apply*, *rollback*, and *re-apply* capabilities all aligned with a business process flow.

Audit replay

An **audit replay** is a way to replay the audit and get to a checkpoint within the stream of code and data changes going back to the audit period: 7 years, 20 years, 50 years, and so on.

Monitoring

Monitoring is a way to inject synthetic transactions and trace them as business process data flows through the system. Often, too much is monitored in the hope that enough trace is available to find out what may have happened later when anomalies are detected. These telemetry collection techniques are wasteful! Only with machine learning-based approaches such as InsightFinder {https://packt-debp.link/BOcOcw} can head or tail be made of these huge logs. I would not attempt to build such a solution. You need to be able to increase the trace for a probe and be able to inject adjusted trace logic quickly at any stage when also injecting trace datasets to see what may be happening from end to end in the data flow's telemetry, but only after that trace has been turned on. If non-functional bugs are arising in the clouds' decentralized pipeline of PaaS and SaaS service calls, you need that excessive level of trace and will want big data analytics to be applied to the trace. If you find that you need this type of analytics, the solution is on the slippery slope of failure. Sloppy coding and a lack of an architected data framework will be the cause, and remediating this journey is not for the faint of heart.

Governance and compliance

Governance and compliance allow you to implement complex multi-account best practice capabilities for compliance. Regulated customers in various industries are challenged with finding timely and cost-effective solutions to managing resources while managing risk and providing custom business security controls that address service relationships. Delivering business products can be adversely affected by a prolonged **time-to-value** (**TTV**) for regulated services without a solid framework. IT compliance frameworks such as AWS Control Tower {https://packt-debp.link/YM4yZd} exist, as well as group controls by AWS services and control objectives from other compliance frameworks, such as NIST's {https://packt-debp.link/UKAeDz} 800-53 {https://packt-debp.link/xj6q0j}, CIS AWS Benchmark {https://packt-debp.link/neQ7Of}, or PCI SSC {https://packt-debp.link/hqxwFy}. These enable controls for meeting customers' compliance objectives in regulated industries. The following are some best practices:

- Adopt proactive approaches for continuously monitoring and testing various controls

- Adopt a defense in depth approach by enabling controls across all behavior categories

- Apply detective controls before preventive controls

- Automate the detection and remediation of non-compliant resources

- Create your organization's controls so that they extend cloud or third-party tool capabilities

- Test controls on non-production **organizational units** (**OUs**)

- Understand and assess workloads and OUs to apply the correct controls

- Fully understand the capabilities of tools such as AWS Control Tower's behavior and mechanism of control

- Measure progress toward implementing streamlined governance processes and enhancing the TTV

Configuration management with Infrastructure as Code

Configuration management with **Infrastructure as Code** (**IaC**) is a way to provision infrastructure from structured data that's maintained with metadata and conforms to the three-level architecture definition (conceptual, logical, and physical):

- This pattern may be confused with a single well-known vendor offering (ServiceNow {https://packt-debp.link/flUyAV}), but beware, although it's a great product, it can be expensive to integrate with **cloud configured items** (**CIs**). A single vendor silver bullet does not exist since the IaC will express itself with cloud specific tools: AWS Cloud Formation or **Azure Resource Management** (**ARM**) templates. You would need to coerce the CMDB output into the IaC of the cloud provider.

- Tools such as Terraform {https://packt-debp.link/3FiIQF}, and Pulumi {https://packt-debp.link/UvGaGB} offer some help but fall short of complete integration with the native cloud provider or security tooling. You will expend a lot of sweat, tears, and scripting to uniformly manage the cloud IaC provisioning without a CMDB. Even if a CMDB were selected, gluing all the IaC pieces together will have to be architected carefully before engineering "just does IT," and using the cloud native IaC tools. With complex and dynamic scalable cloud provisioning, many IaC templates would have to be redone as the cloud IT system agility morphs into its end state. This defeats the purpose of cloud IaC, which was to make this process easy. Without a structured IaC CMDB approach, the IaC will become a serious weakness and cause many problems.

- Security integration is a key CMDB and IaC concern. Without Hiera {https://packt-debp.link/hXbCMv} for Puppet {https://packt-debp.link/FYCwCh}, other integration keys and tokens are in the clear when using native cloud provided IaC templates.

- We want to avoid the tooling wars that will arise between Puppet, Ansible, or Chef {https://packt-debp.link/VmUK8F}. This may be a very hard discussion because the IT operations folks can be religiously fanatical and one-sided regarding even discussing changing tools. This is often why startups script IaC and avoid tooling (and use Terraform or Pulumi); the developers are the operations folks in the startup DevOps/agile development process. But with established organizations, the operations folks will (for compliance reasons) want the developers out of their infrastructure environments completely!

Test patterns

The following subsections point out the types of testing patterns that need to be part of your solution regarding test patterns.

Code testing

Code testing is a way to implement various regression tests for functional and non-functional requirements. These tests are for backend services as well as logic in the frontend. *Chapter 10* takes a much deeper dive into testing, and you will appreciate the focus on quality. Regarding code testing, we recommend that you look at the **model-view-controller** (**MVC**) pattern for basic test organization. Most data systems have a **model** for the data being serviced, a **view** (presentation) of that data, and a business logic **controller** for data services/transactions across the application. Each of the three: M, V, and C needs testing. The test driver application should also be an MVC patterned application that can be driven by test tools such as Selenium.

Software systems regression testing must also account for iterative: scale, stress, SLA, and performance testing as changes are made during development and after a system's release. These non-functional tests are very important for cloud solutions that exhibit end-to-end weaknesses due to complex PaaS service integrations when implemented as part of a complex business data flow. Thinks of all the functions, configurations, and potential bottlenecks that have to be debugged and verified to remain correct over time.

Data testing

Data testing is a way to evaluate the quality of data at rest and that data's trend correctness. Data tests should be associated with code tests and anomalies lead to resolution through the forensic analysis of trace. Data trend testing is a type of data profiling that's done after performing analytics. This means we must sometimes push the potentially incorrect curated dataset through the data to apply these data trend analytics tests. Then, we may have to backoff that dataset to the last known good checkpoint when a trend anomaly is detected. It's good to know that checkpoint processing exists so that such rollbacks are possible in an automated manner. Additionally, data at rest and in transit (lineage) needs to be tested so that we know the logic was deployed and running as expected. We must *never* take it for granted that somebody turned on the system since the trace does not indicate errors. This is even more evident in the cloud, where a code snippet may not be deployed or made active due to an operator error. There are two types of data testing:

- **At rest testing**: A way to assess the correctness of data at rest when a checkpoint processing step has been achieved.

- **In transit testing**: A way to assess the state of data as it undergoes its lineage transforms. This involves data deduplication, rollbacks from a checkpoint to a prior known good checkpoint, and the ability to apply data across multiple transform steps with the result being correct, as well as the lineage trace being available for forensic analysis. Sometimes, there are too many transforms, so being able to look back can become opaque without this capability.

Consumption patterns

Consumption patterns, or the ways users/data consumers will access curated information to harvest insights, require unique patterns due to performance and scale requirements. Let's take a closer look.

Publish/Subscribe (Pub/Sub)

Pub/Sub is a standard way to support the value-added publication of a final dataset with the capability to support distribution rights, billing, fair-use policy enforcement, security, and encryption all in stream mode. It also supports the bulk operations of a checkpointed publication, with all the benefits of streaming and the unit of transfer being a much larger grain.

Side capabilities include the ability to resend a dataset that is logically the same but pertinent only if it's updated if it were to be resent in the future; bulk refresh of the dataset that's been subscribed to; incremental updates at a checkpointed time – after code or data correction; audit/inventory of changes if data is resent by request or by originator necessity; and fan out data if multiple recipients are to receive data at the same for contractual reasons and industry fairness.

Data analytics space

Data analytics spaces are areas that you set up where you can publish datasets and are kept in sync with a master dataset in whole or part via adapters. This builds on the Pub/Sub pattern:

- Consuming users are not always able to wait to have huge sets of entitled data sent to them. Often, they require it to be ready when needed. For this, you must create consumable data mart formats to be able to consume external vendor feeds, something that can be done via programs such as LiveRamp {https://packt-debp.link/KSgatB}. The data formats vary, and cloud providers enable solutions such as Azure's PowerBI Premium with Azure Synapse for the creation of these analytic spaces.

- Consumers may wish to have blazing performance in their real-time analytics results. Here, REDIS Enterprise and Gridgain (Apache Ignite) can provide support.

- Others may wish for real-time in-memory performance with SQL access capabilities, something that can be provided by Clickhouse and StarRocks. Others may desire cloud-hosted solutions such as Snowflake and Rockset. However, the TCO must be considered when making these decisions.

It is very important to consider the analytics use cases when solutioning for analytics. Performance and scale for execution across a wide range of metrics, measures, and additive and non-additive calculations with and without summary rollups are essential considerations.

Data warehouse flow

Data warehouse flow is a cloud-native way to sync a warehouse to and from an analytics sandbox. This sandbox is an operational analytics space that's set up by the backend system to scope the datasets undergoing analysis. There is a need to keep the warehouse in sync in the following instances:

- **To and from an analytics space** (such as Microsoft Azure Synapse Link) to populate the analytics space from a data warehouse. With this comes the need to attribute costs to the consumer and not absorb those costs as part of the backend system's development run/manage costs.

- **To and from a big data lake or cached lake** (such as Databricks Delta Lake) to populate the big data areas from a data warehouse into the analytics space that supports the analyst's use cases, and again with cost attribution.

Keeping two data stores physically in sync across a set of data items in disparate technologies is no easy task, especially when these third-party data storage systems are closed and inflexible.

Massively scalable data engine

A **massively scalable data engine** is a way to process data using cloud-native PaaS or SaaS services to scale the execution of logic. This allows huge IoT-type arrivals in the order of billions of tweets a day to be handled; massive device collection; operational telemetry for a running system; or other exabyte-level collection goals.

An example includes the user of AWS Kinesis for web usage activity stream data collection. What lessons you learned when playing with the cloud services need to be put into a repeatable pattern of your framework. You cannot just code to these services without accounting for what happens at the edge of the service's functional and scale limits. When required, error flows must be executed, throttles applied, and scale rules configured, which can and does cause loss of data! Yes, it's a bug, but it's an operational one that's caused by the cloud service's quirky behavior that the coder will *not* be aware of until the PoC finishes, and then again with lots of overtime spent. This is a critical issue and is only found with trial and error. Once found, you can build a pattern to make it easier for the next guy. You don't want to have 10 types of wheels on your designed solution. Each developer/engineer will address these issues in different ways, and none completely until they put their heads together and discover the framework pattern of the architected solution.

Metadata patterns

Metadata is a way to describe the syntax and semantics of the data, as well as its lineage through the data factory and its data pipelines. This is essential to the current understanding and use of data. There are three significant metadata {https://packt-debp.link/1xV0UJ} types:

- Descriptive
- Administrative
- Structural

These will be discussed in the next few sections.

Descriptive metadata

Descriptive metadata enables semantic discovery, identification, and selection of various dataset information. Descriptive metadata can be formalized into semantic models (see *Structural metadata*) and object relationships can be enforced, or it can be loosely labeled based simply on a book's author, its title, or subject areas, for example. This type of metadata gets placed in the data dictionary or catalog for end user discovery purposes.

Administrative metadata

Administrative metadata captures data lineage through a data factory across many of the architected logical and physical components. The context of the data journey must be relatable to the business, as well as the IT professional, hence the need to capture metadata that references architecture entities at the three levels of the architecture. Also to be captured are the data contexts, which include entitlement and usage rights, as well as contracts for fair use. This metadata is essential when supporting adherence to an enterprise's data policies underlying auditing, governance, and compliance use cases.

Structural metadata

Structural metadata is needed for machine processing and will describe the context and semantic relationships among various objects of a dataset, such as the fact that a chapter is part of a section within a book. Of note are the implications for time series data metadata structure, which must be clear. Over time, data can lose its efficacy, or it will need to be reduced into summaries so that when you're looking back at older datasets, it renders a correct meaning through the backward lens. Structural metadata enables machine learning, algorithms, analytic algorithms, and big data software to not misuse the dataset that the metadata describes.

Administrative patterns

The administration of a solution is too important to be left to be implemented at the end. The administration capabilities involve features for billing. These support the practical aspects of running IT solutions in large organizations, such as the need to support departmental chargebacks. Likewise, there is a need to manage users and secure data assets while unified securing policies and enforcing administrative systems. Additionally, analyst as well as operational dashboards must be created to support management information systems' needs.

Billing with stewardship/department cost attribution

Billing with stewardship/department cost attribution enforces data rights, entitlement, fair use, abuse detection, and cost allocation reporting:

- One issue that's often found in cloud data factory solutions is that cloud scaling affects the throughput of the system, as well as adversely impacts its IT costs

- Scaling optimization with cost considered requires PaaS and SaaS shared (multi-tenant) processing of datasets that may be owned by competing businesses with separate budgets for IT operations, storage, and administration.

- Granular cost reporting allocation is necessary; otherwise, the small dataset owners are going to be paying for the big processing datasets unfairly

Careful cost management while attributing those costs to the data owner is essential. This is a practical aspect of your solution, and it is needed to manage costs. This is made difficult to optimize since some cloud services need to run in a multi-tenant shared mode without any built-in capabilities for usage reporting to support your goals for cost attribution.

Unified security

Unified security is a way to provide end-to-end security capabilities that preserve the context of the dataset from end to end.

Since each cloud provider and third-party vendor organizes these capabilities differently and sometimes incompatibly, the security pattern must be clear, published, and operationally administered to form a solid pattern for reuse. Note: Zero Trust {https://packt-debp.link/pHrHFR} is great! However, it can be a problem to implement in a big system with performance SLAs that have to be maintained. So, you need to make real compromises for the choice that's made. Also, note that you need to preserve the security of the data for a user associated with a role against data organized for per user or group level access to support row level access (refer to your knowledge about RDBMS row level entitlement versus cloud security model data entitlement differences).

Security and privacy are huge issues to address in the cloud, especially in hybrid computing environments. It is also a field that is shifting and responding to increasing threats sometimes caused by increasing cloud complexity. Forming the framework's capabilities and selecting tools will be essential. As an example, you will want to look at AWS Control Tower {https://packt-debp.link/1wftYL} and assess potential competitor tools so that you can select first the capabilities needed for your business. You must be aware that many tools lock you into a cloud provider's offerings. Also, be ready to build where buying is not possible. Let's consider some competitors:

- Microsoft Defender for Cloud provides cloud compliance and workload protection. It provides security and threat management in a hybrid environment to visibly prevent, detect, and respond to security issues.

- Google Cloud Platform Security Overview's {https://packt-debp.link/H9wjZX} capability is cloud workload protection. This can be deployed on a GCP infrastructure from IT experts in secure networking and software development.

- Wiz's capability is cloud compliance and workload protection. It is a **cloud-native application protection platform** (**CNAPP**) {https://packt-debp.link/j7yFJO} vulnerability management tool that combines the following with container and Kubernetes security support:

  - **Cloud security posture management** (**CSPM**) {https://packt-debp.link/ArEGdy}

  - **Kubernetes Security Posture Management** (**KSPM**) {https://packt-debp.link/VvkpQo}

  - **Cloud workload protection platform** (**CWPP**) {https://packt-debp.link/gYeQe4}

  - **IaC scanning** {https://packt-debp.link/Ap3Ktn}

  - **Cloud infrastructure entitlement management** (**CIEM**) {https://packt-debp.link/pENyny} and

  - **Data security posture management** (**DSPM**) {https://packt-debp.link/WgtLbS}

  - **Cloud service network security** (**CSNS**) {https://packt-debp.link/DA6a3m}

- Vanta's capability is cloud compliance and provides a trust layer on various services to provide security for data safety.

- Google Cloud Security Command Center's capability is cloud workload protection. This allows security teams to gather data for threat assessment proactively before businesses are adversely affected. It also provides insights into application risk for mitigation, as well as cloud resources and their general health.

- Drata's {https://packt-debp.link/wo7TTl} capability is to act as a cloud compliance security and compliance automation platform.

Dashboards

Dashboards allow you to control the state of the system and the state of the data as it procures through the data factory's {https://packt-debp.link/NlaUNc} data flows. Dashboards offer the following benefits:

- What you measure can be improved! Being able to visualize the state of the data flows in a data factory is very important to sustaining the agreed-upon SLA.

- Dashboard metrics also enable automated proactive responses to analytic anomalies before they impact quality and become observable to consumers.

- **Business process management** (**BPM**) techniques can be applied. Third-party tools such as Datadog {https://packt-debp.link/A8oizk}, Cloud Health {https://packt-debp.link/MG3Pmv}, and CloudCheckr {https://packt-debp.link/eGaQ4o}, as well as native PaaS cloud provider tooling, are part of the solution, but only part.

The best practice is that if you built it, it needs a steering wheel and dashboard; otherwise, the engine should *not* be started!

Forensics

Forensics allows developers and data **subject matter experts** (**SMEs**) to access data, audit the scope of access, and write data correctors into the system.

"What can go wrong will go wrong… and it will happen to me!" This is a motto I live by, and it has served me well over the years. Data will have errors and require correction. Never touch it directly. Build a corrector and register it as part of the data fix – just like code may be patched, so will data. Remember that it is audited and must be recorded as part of the system when corrected.

Additionally, being able to access the data outside the app is required for forensic analysis and before building corrector code or adjusting the code of the data pipeline. This access must be granted and removed for the forensic session. This capability will save the engineers lots of time when they have to probe for problem fixes.

Summary

In this chapter, you learned how important it is to have a clear vision and mission when starting your engineering journey to build a future-proof data-engineered solution. This need is best illustrated through some examples that we have been directly exposed to over the years, and which have been included in this chapter.

In addition to the vision and mission statement outcomes that are produced after formalizing the organization's thoughts around this topic, you have been exposed to the need to develop clear principles that will guide the teams as they develop an architecture. That architecture will form the framework for a data platform. These principles are going to be key guides when you're making decisions – and there will be many to make. This structured, methodical approach will help explain all the choices that are made as design challenges are presented or when technical issues arise.

Lastly, you were exposed to the need for a toolbox of repeatable, operationally reusable components constituting integration patterns. These patterns will crystallize into architectural blueprints later in the design process.

The principles that govern choices are so very important. They will be the subject of the next chapter as we build on the vision and mission that *you* will generate as part of the data engineering process.

Architecture Principles

Thus far we have written about the importance of establishing your organization's mission, vision, best practices, and patterns, and general corporate principles; however, there are specific principles related to the architecture when developing a future-proof data engineering solution that will have to be addressed as we move forward. These principles carry with them the underlying corporate and IT values that will help keep a solution effective in the future. Tools, vendors, and various cloud offerings will morph to embrace new machine intelligence capabilities.

In this chapter, you will be provided with the foundation for developing principles. You will also be provided with a number of principles that are relevant to the task of building a future-proof data factory. Lastly, you will be provided with an abundant number of references for additional reading needed to flesh out a common understanding of challenges that will be presented as you form an architecture that will withstand the test of time.

We begin with an overview of the technology landscape and the way businesses jumpstart their efforts with research firm findings. These non-academic sources are highly compensated for the information they provide, and that information often defines how the herd is moving. We use the term *herd* because those who wish to lead the field of their peers are those who have the qualifications, knowledge, and experience to advance cutting edge technologies without excess cost or outright failure.

Architecture principles overview

The principles to be outlined next will give professionals a guiding light to make wise choices now that define future success. Research firms can be a good source for guidance, such as the following:

- Gartner {https://packt-debp.link/lXFfHo}
- Forrester {https://packt-debp.link/gEVSOD}
- Bain & Company {https://packt-debp.link/pAnKim}
- Frost & Sullivan {https://packt-debp.link/JrUuR0}
- International Data Corporation (IDC) {https://packt-debp.link/D4PrBA}
- 451 Research (now owned by S&P) {https://packt-debp.link/KF6rkc}
- Economist Intelligence Unit (EIU) {https://packt-debp.link/zy1quO}

- Experts Exchange {https://packt-debp.link/GvVItH}

- Omdia (was Ovum) {https://packt-debp.link/ZhkiPT}

- Nucleus Research {https://packt-debp.link/Wsg8SG}

- GigaOM {https://packt-debp.link/q1GLgK}

- Lexis Nexis Research {https://packt-debp.link/30GcOM}

- Olive Tree Insights {https://packt-debp.link/WX88AP}

- Aberdeen Strategy & Research {https://packt-debp.link/NQJFAu}

- Outsell {https://packt-debp.link/mmlLot}

- CEB (now owned by Gartner) {https://packt-debp.link/3mfY5h}

All the aforementioned produce valuable current insights for professionals trying to navigate the rough waters of our current technology journey. Sometimes the future is charted by riding the various *hype cycles* through their ups and downs. But is there a better way? Can the elusive future be made clear and plainly communicated with minimal cost?

Every development journey, including the development of the data factory's data pipelines, begins with preparation and knowing what to pack. The choices are driven by principles. Following the herd will not get you to where you want to go; rather, it will get you only to where the herd is headed. These principles will become the constitution governing your architectural design choices. They will be anchors for conversations and settling disputes that *will* arise. Some of the most successful people, companies, and project efforts are those that have clear guiding directions for what they want and do not let babble (or babblers) get in the way of success.

In order to remain grounded as you make progress, you will want to develop the organization's IT principles and remain securely anchored in them as you develop the architecture and the data engineering designs that are derived from this architecture. Be practical about the principles you develop from what will appear in this chapter. Be careful with the wording of the principles because each carries huge implications for the downstream development effort. Begin with a primer, and do not assume everyone is on the same level of knowledge as you formalize *your* principles. Your constituents, stakeholders, and engineers may have been drinking a vendor's Kool-Aid for years and can't talk about technology if it threatens their limited knowledge. Expect them to be defensive! The following sections were written to provide you with the foundation that will be used to level-set others with the knowledge necessary to gain acceptance for your solution. Many links are provided for you to deep dive into the subject matter. If you remain open to innovation and closed to compromises to quality, fragile design, short-sightedness, and waste, clear principles will arise that withstand the test of time.

Architecture foundation

When developing a modern data-engineered solution, you will want to be level set on the foundational technologies required for success. There will be many themes and disrupting technology cycles to be dealt with. One such technical area is the concept of the **data lake** and the **data mesh**.

Data lake, mesh, and fabric

So, what are some examples of modern data organizations that need to be understood as you roll out a modern data engineering solution? Consider the data lake {https://packt-debp.link/PoLMQS} concept, which was coined in 2011 by James Dixon {https://packt-debp.link/oO3q9r}. A data lake is a system/repository of data stored in raw format (for example, object blobs/files). So, what has happened since 2011? The data lake became the dumping ground for everything! Unfortunately, over time, data lakes have led to a loss of value. A way of properly understanding data was also lost when the context was not captured in a time series-sensitive metadata store. All this can quickly become a **data swamp**.

What other problems can appear with data lakes? If a data lake holds poorly organized data or too much data without the correct metadata with its required data governance, important data quickly becomes non-discoverable. Information retrieval time, insight discovery, and analytical misdirections (including **machine learning** (**ML**) hallucinations) increased since semantics was put into disassociated algorithms sprinkled like sugar across cloud vendors' **platform-as-a-service** (**PaaS**) services.

New data being added over time loses applicability since its processing rules drift! The data journey is not journaled, and the software and data life cycle pollute the data lake with misinformation that – worse yet – *looks* credible. Data begins to lose its usefulness and starts to incur retention costs with no measurable **return on investment** (**ROI**). Third-party vendors and cloud providers will attempt to fix issues by making the data lake perform faster and be more scalable, with increased caching and data mapping, but can miss the mark on the true need. That is to *provide more relevant data over the long haul*.

Even the proper representation of the *time context* for data can become lost in the data lake, and a stagnant data swamp results; this must be avoided. The data lake is a great idea, but as with all ideas, it needs to be created with thought regarding how to provide important metadata, governance, normalizing transforms, and numerous semantic contexts to data so that the data lake remains useful over time. The need to move beyond the data lake and toward the data mesh or its superset, referred to by Microsoft as the *data fabric blueprint*, becomes evident in the work by Zhamak Dehghani {https://packt-debp.link/MJdwA6}.

> **Note**
>
> The data lake is no longer the centerpiece of the overall architecture. Patterns for the data mesh, enhanced **extract, transform, load** (**ETL**) pipelines, and analytic use cases for dataset class patterns need to be created.

According to Zhamak Dehghani, the data mesh has four pillars:

- *Domain ownership*
- *Data as a product*
- *Self-serve data platform*
- *Federated computational governance*

It's worth reading through Zhamak's work because it comes at a time when the concept of the data lake has taken hold and organizations have begun experiencing data swamp issues. What the **artificial intelligence (AI)** and analytics communities point out is that there is a pressing need to go even further with the data as a product and self-serve concepts and provide commoditized, profitable software and data services to implement federated and domain ownership aspects of the data mesh blueprint. Microsoft calls this the *data fabric* since it adds several cloud-specific services that make the difficult job easier with, of course, the inevitable cloud vendor lock-in that is required to gain those benefits. To sum it up, the data mesh concept is the future but only with a data services and software services framework that makes the implementation something other than a theoretical discussion. Microsoft architects will admit that this is an evolving area. Expect rapid change but remain anchored in the principles.

Data immutability

Data lake datasets should be isolated into zones of clear value: *raw, bronze, silver, gold*, and *consumable zones* have been suggested in an earlier chapter. Cloud technology vendors have not built solutions that allow all these to be placed in a single physical data lake; so, expect your data factory-built data flows to be needed and migration of similar and enriched datasets via lineage-tracked transforms to be created. This means that operational costs will be higher than when the monolithic data lake pattern was envisioned. This will be the case until data fabric offerings mature in the future.

Beginning with the rawest form of any normalized dataset that can be catalogable, one populates the Raw zone. For example, when building out a retail analytics system, it became clear that data being too raw is just not acceptable – it would not be analytically consumable and incur too much cost to reprocess if not first normalized. It is useless until that ingested data is first verified as not being garbage data. For example, you do not want to retain the exact same logical resubmission of a raw dataset due to a technical glitch on the incoming FTP server (aka same data with different timestamps) because it would double the rolled-up data totals. You also do not want retail gray market sales items in a **point-of-sale (POS)** end-of-day summary file to be included with qualified sales numbers. Likewise, you do not need to miss a POS summary dataset for a set of stores in a region due to a power outage in that supplying region because the total store summary metrics will be way off. Raw data needs to be raw but not too raw to make it useful.

Data lakes' immutable data is to be available for explorations. Raw data zone data should be locked down when accepted and normalized. It is subject to various anomaly detection profile checks and lineage transforms after arrival. The dataset will produce raw second-order profile data such as its *trends*. This is derived data that is subject to adjustment. Additionally, time series tagging and summarization data should create new data and not adjust the immutable raw data in the raw zone. So, does the raw zone only contain raw data? Let's say it contains data derived from primitive data, as well as the original primitive data itself, with semantic context in a form that can be useful for analytics. The first analytics use cases leveraging raw zone data are data profiling use cases.

Data lakes' immutable data is to be available for analytical usage. In the past, Trifacta {https://packt-debp.link/UJPS9u} was a very useful and scalable tool to create raw zone data for downstream uses, particularly for the curation of bronze zone data before entering the data pipelines feeding the silver and gold zones. Whenever the data pipeline results in a transform, the metadata lineage tracking capabilities of the data factory are applied to enable reporting and backward traceability for forensic discovery of errors. Data cataloging of the original data entity, its transform, and its result are all to be retained for analysis. Data is transformed, but immutable sources are retained with this approach.

Data lakes' source domain data will be discoverable. Since data, data context, and data pipeline lineage are all retained in the proposed blueprint architecture, the framework that implements the blueprint will need to have strong capabilities to discover data in the catalog and its recorded lineage. Data catalog reflection and lineage observability patterns are necessary to allow the raw data zone's content to be iterated on until quality metrics are attained. Preserving the quality of the data and its service components is essential as you put data into various zone-to-zone transformations.

Third party tool, cloud platform-as-a-service (PaaS), and framework integrations

Data lake tool selections and the cloud vendor PaaS must not contradict the architecture framework. Enabling data readiness for zone migration requires tool selections for **data quality management** (**DQM**) to be integrated carefully and fully support real-time iterative processing until a set quality metric is attained. The quality metric is a gating function that allows data to be propagated forward in the pipeline. Cloud vendors and third-party DQM providers will not easily integrate at this stage in the solutions designed by data factories, so be prepared for integration headaches. Set up the architecture that works for your organization and never compromise on data quality. Once bad data passes the zone-to-zone transfer gates, it is treated as truth, and that can have deleterious effects on analytics insights in downstream zone-hosted systems. At least if and when these are detected, the lineage and cataloging capabilities will enable datasets to be resubmitted for reprocessing and posting. You will have to decide how many of these stream aggregates form a reasonable release set supporting the organization's data release strategy.

> **Note**
>
> *The data mesh pillar*. Data as a product sounds good until it also must be released just as software is released, with versions, quality, history, time series, branches, and so on.

The data mesh blueprint requires domain-driven ownership of data. Various business teams comprise the domain, and they will take responsibility for data in their area. Any analytical data will be organized around these same business domains.

Data mesh principles

Data mesh defines domain-driven ownership of data. Just as software and systems are developed in domain-aligned team boundaries, so should the data be organized within the zone. Ownership of the distributed architecture, analytical data, and operational data belongs to the business domain team and not to a central IT team. Given our best practice of hosting data in zones, the domain owner will own a piece of each zone's data as cataloged in the data catalog with data lineage trails that are also owned by the domain team.

Data mesh defines data as a product. The data mesh blueprint also defines data, metadata, and analytical data as a product, and it should be curated and sold with it being a business product in mind. Consumers for the data mesh's datasets are not the programs and people for which the data may have been originally curated. A business domain team provides high-quality data, metadata, analytics, and contracts for service, just as with any other public service API. There is a lot more to be said about data mesh data as a product. But it comes down to the product's value and service for a given cost, which is maintained for a duration of time as a contract. The data has guarantees and warranties and even allows value to be added.

Data mesh data is fit for self-service. In traditional IT systems, the service wrapper for data enables it to meet the characteristics of *ready for self-service*. With the data mesh blueprint, the data can be clearly understood.

Data mesh defines a self-serve data platform. Self-service data and analytics platform thinking can be applied to the design of the data infrastructure. The data engineering platform team provides functionality, tools, and systems to build, execute, and maintain data products across all domains. A core data platform team will provide the patterns each business domain team needs to curate data products. With a data plane infrastructure, the data describing the data catalog, the data pipeline curated lineage, and the value-added analytic/measures/factors/metrics and installable value-added code segments, the **self-service data** concept may be realized. The issue will be what third-party tools exist that can be coerced into the patterns needed to implement this principle aligned with the overall framework.

The data mesh blueprint for domain-driven ownership of data, data as a product, and self-service data drives the need for **federated governance**. With the central governance group setting up practices for governance and the actual governance performed in a federated manner, the overall organizational rules and industry regulations can be adhered to. Compliance with federated governance standards is provided on an honor basis; however, verification via standard reporting is required to maintain data contracts.

Data mesh defines federated computational governance. Integration policies for federated data governance are created in a **guild** manner. Domain owners designate empowered members to the central guild. Any data standards and documentation sets are published and refreshed by the federated governance group (also known as the guild). Specific care is given to data: security, privacy, contracts, service levels, and legal aspects of the data mesh blueprint. Metadata management will include branches of data lineage for a dataset's pipelines, which make it possible to retain parallel branches of a dataset's lineage within and between zones. This aspect of data as a product will make it possible to curate data and its metadata iteratively till ready for release. Federated governance also makes it possible for data rights to be preserved by downstream consumers. All too often, a user changes the column name and resends that data after adjusting a small value. The rights to do this have to be identified in the governance metadata for data at rest – prohibiting such value-added adulteration of the curated dataset. Likewise, data redistribution rights and attribution of data ownership are to be clear in the federated governance policies in the metadata for a dataset.

Data mesh metadata

A data sets metadata shall be treated as first party data, since metadata is data! As such, it is linked to the data and linked to other metadata semantically for any data at rest (any pipeline stage) or its transforms defined, versioned, and grouped into a data lineage trail for any curated dataset. Parallel versions of data at rest metadata or lineage metadata can and do exist in a system. How many prior versions are to be retained is a governance issue associated with each federated domain owner. Value added datasets with at-rest and lineage metadata will be treated as first party data. Metadata is defined for data at rest and data in **pipeline transformation** (also known as **transitioning data**), and it exists for the system implementing the pipeline to give the forensic viewer the operational context of the system operating the transform at the time of the pipeline's operations on the dataset curated in the factory pipeline. It is essential that any value-added data or lineage operations also reflect the data mesh blueprints mechanism enabling value-dated contributions to the factory pipelines. Any class of data will be cataloged in a dictionary. The data mesh's data catalog is a logical data dictionary that enables reflection on all data and metadata of the solution. It is searchable, contains changesets, and can be operationalized for self-service analytics, data quality assessment, and forensic data analytics use cases. It may also be used to age out old data with its metadata. It is not good to retain expired datasets, datasets that have lost their worth, or when the multiplicity of dataset versions is too costly to retain. The data retention policies combined with federated governance capabilities drive data life cycle rules. The rules themselves are represented as metadata and retained with the datasets. The data catalog will also drive **disaster recovery** (**DR**) and archival activities of the data mesh. Since data is a product, you must consider how to stream changes to consumers and group these into intelligent release sets since data does not stand alone but always clumps together. It can be said that data has **gravity**, and the more that there is, the more it produces a **mass effect**, and this results in the creation of **implicit meaning** for datasets in a release set. Releasable datasets that are produced by one or more data pipelines establish the data contracts of the data mesh that are enforceable.

Data semantics in the data mesh

Tim Berners-Lee {https://packt-debp.link/DBIXta} raised awareness of the need for a **semantic web**. The follow-on discussions focused on this new awareness that required web documents to be rendered with a formal definition of the page's semantics. This all took place when he invented the concept of the World Wide Web in 1989. When at CERN {https://packt-debp.link/NgzAKQ}, he developed the semantic web concept by observing physicists meeting together for scientific discussions and discussing common interest areas. Over the years, the topic has risen over and over in different forms. Data needs to be self-explanatory, unambiguous, and clear.

> **Tim Berners-Lee**
>
> Berners-Lee stressed the decentralized form, allowing anything to link to anything. This form is mathematically a graph or web. It was designed to be global, of course.

In 1990, the HTML standard {https://packt-debp.link/BB77OC} was produced, and in 1994, Tim Berners-Lee founded the **Open Data Institute** (**ODI**) {https://packt-debp.link/kInTo5} in London. The ultimate ambition of the semantic web, as Berners-Lee sees it, is to *enable computers to better manipulate information on our behalf*. He proposes that in a *semantic web*, the word *semantic* indicates *a web of data that can be processed directly and indirectly by machines* {https://packt-debp.link/bC8ErQ}.

The brilliance of this insight must not be lost on those seeking to engineer future-proof systems. You cannot build intelligence about this sum of human intelligence that exists in the World Wide Web of data if it is misunderstood, if it is not linked in order to preserve context, or if it is full of gaps and omissions. A web of data without semantic representation is missing the mark. To date, the vision of the semantic web has yet to be delivered. We are still hindered by large, centralized repositories of information. We are sheltered by the website logic that business folks wrap around data to provide secure data services. To practically minded data and software engineers, the semantic web is a utopian dream.

The semantic web's basis is a new set of standards required to annotate web pages. Structured XML (or RDF) would have no effect on page rendering. These metadata structures would be readable by software to gather page meanings that could be visually, implicitly, and inferentially determined by a human reader. In all pragmatic engineering approaches, you tend to code to the requirement or simplify the solution. You do not build *more* than can satisfy immediate requirements. Machine-readable page semantics may not have been created for web pages, and as such, even if it were available, it would not be leveraged by many external entities. Even if that level of information were available, by publicly exposing it a business could be put at risk. You can see that the development of semantic web-enabled pages could be costly, may not be useful to a consumer, and even poses a risk.

The same argument existed for the development of the World Wide Web, yet it was built anyway. The key was that there is tangle, tacit, and direct value in collaboration. But there was a cost, and you had to *put it out there*! This does not mean you have to give all your data away; it only means that you need to be *smart* about the data.

Smart data is just that – data that can be shared and not subject to abuse just because it can't be contextualized or carry its meaning with it when consumed. It's just right to make sure what is curated in a released dataset has this correct meaning. Metadata is the key, but as pointed out, pragmatic software engineers will not take direction from a data engineer easily, nor will they in turn take direction from a *knowledge engineer*. When the data engineer takes up the mantra of the knowledge engineer, there will be a wider acceptance of the solution to structured metadata that Tim Berners-Lee envisioned.

The recent developments in generative artificial intelligence (AI), rather than the prior technology generations machine learning pattern matching, **deep learning** (**DL**), or predictive capabilities, bring the need for proper data engineering into focus and will drive semantic web concepts forward. Will it be the XML/RDF/OWL 2 {https://packt-debp.link/MvV5CH} standards of 2012? Maybe.

The development of operational ontologies as **knowledge bases** has been valuable in solving generative AI's hallucination problem {https://packt-debp.link/vjX44A} encountered with OpenAI, ChatGPT, and related generative solutions. We know that machine processing of data into information to gather key insights is a key capability in the future state of the IT industry. The technology will enable so much more for mankind's productivity; however, it is being hindered by current data organization, which was forced by pragmatic software engineering decisions of the past and further hindered by cloud providers solidifying those solutions as easily consumable platform services. This has been good and bad since dead-end architectures are used by those who are not thinking through the ramifications of their organization's architecture choices. Cloud providers are driven by a healthy desire to give the customer what they want today and do not always look out for the customer's future needs. This also produces a huge drag on the development success of the data engineer, because they are mandated to implement

IT in ways that just do not make sense for the future. Think about the complexity required for cloud pipeline implementation and the lack of clear observability from end to end with huge gaps in metadata representation offered today.

So, how should we proceed? The architect and data engineer see what is possible but can't possibly build every aspect of the solution to execute on the vision. When building the first production car in America, the Model T, you can read the following quote from Henry Ford:

> *"I will build a motor car for the great multitude. It will be large enough for the family, but small enough for the individual to run and care for. It will be constructed of the best materials, by the best men to be hired, after the simplest designs that modern engineering can devise. But it will be so low in price that no man making a good salary will be unable to own one – and enjoy with his family the blessing of hours of pleasure in God's great open spaces."*

The assembly line was invented by Ransom E. Olds when he rolled out the Oldsmobile Curved Dash in 1901, yet Ford's engineers made many efficient innovations for the Model T. Ford's vision required workers and engineers to assemble in a line and not have to craft each car from scratch.

So, the takeaways from the preceding Ford example focus on the general guidance, which is to follow his pattern of success:

- Build data curation pipelines in an assembly manner, buy the assimilable components, and integrate the whole

- Never let the vision and mission be compromised

- Don't sacrifice data quality!

Remember – Ford first developed Model A through Model S before releasing the Model T. Spend the time to build the support scaffolding, and then be ready to take it all down or reuse it on the next iteration of the pipeline data flow.

Okay, enough of the analogies! What are the data factory assembly components that we *must* have to make data pipelines work correctly in the data mesh? Let's explore these:

- **Knowledge graph** {https://packt-debp.link/M8lcwd} is the ontology conformed to the formal definition of the domain model

- **Knowledge base** {https://packt-debp.link/rOuDFy} is the operationalized ontology defined by the knowledge graph

- **Semantic validation** of ingested data leading to correct incorporation of data into the knowledge base

- **Data lineage** semantics of datasets as they transit data pipelines prior to release and consumption

- **Performant and scalable data store** cloud storage classes with in-memory down to glacial access contracts

- **Data contract validation** capabilities to *observe* data lineage and data quality of the data rather than just the correct software functionality of the solution

From the preceding list, you can take a more in-depth dive into the specific selection process for this core assembled solution components of the data pipeline assembly. What is very important is the focus on quality that is itself ill defined if not standardized and then measured.

So, what does it mean to have quality data? I go back to my applied math days – an item is correct if it is true, always true, and never wrong.

The ability to make sure ingested data is correct is to check it, build assertions, and through a lot of tests, if no errors are found, then it's good enough! Right? The answer is "No!" Data needs to be correct, and that means it needs to be associated with assertions that can be checked by the contact of the data as it is being loaded.

You need not throw tests at data; data needs to throw tests at itself and tell the program it has encountered a semantic error if data is being applied in such a way as to violate its context constraints. Imagine in a retail setting, having a shoe's *item ID* put into an attribute field of a kitchen appliance. It's lunacy!

Today, we often encode meaning within the *item ID* as alphanumeric characters. We abuse the pure intent of it being a numeric *item ID* by overloading it with some encoded semantic meaning. There are hundreds of constraints on the semantics of data (aka knowledge) within a domain, and this needs to be codified in the data's associated metadata. Semantic metadata is defined in the knowledge base to make it possible for data to be fit for purpose and right! Semantic data validation is required to define metadata for data at rest. This metadata should also be put under life cycle management. Semantic validation of knowledge base items is performed using **Shapes Constraint Language** (**SHACL**) {https://packt-debp.link/2Z9cj3} (or its competing standard, **Shape Expressions Language 2.1** (**ShEx**) {https://packt-debp.link/UFNZvJ}). Leveraging these tools and approaches is not for the faint of heart. It is the lack of clear, easy to use third-party tools in this area that leads software engineers to raise developmental red flags to the data engineer. Some common objections heard are the following:

- This is not taught in undergraduate schools!
- I only know RDBMS and SQL!
- How is this to work or perform in production?
- Bring in the architect – let's hang that individual!

My answer to the faint of heart is: *Suck it up, buttercup* {https://packt-debp.link/f9CZmY}! It is the right thing to do, yet making the case and bringing others along will be a difficult journey, and you will need some solid vendor help from sources such as these:

- TopQuadrant {https://packt-debp.link/xjFYBF}
- Stardog {https://packt-debp.link/gxG2os}
- Ontotext {https://packt-debp.link/yKIsHK}
- Cambridge Semantics {https://packt-debp.link/rToggK}

The preceding vendor tools may be used to assist in the development of the model, the semantic validation, or the operationalization of the knowledge base vision.

This all has a goal of making data correct, and the one who makes this vision a reality for your data architecture is a winner. If the architecture is right, the subsequent data engineering will be as well. Note that there are alternatives to the use of model-based semantics:

- Neo4j {https://packt-debp.link/IeVzwA}

- JanusGraph {https://packt-debp.link/F2tw8U}

- TigerGraph {https://packt-debp.link/sHzkVP}

- Microsoft Graph {https://packt-debp.link/BPp1Kq}

After incorporating data into the knowledge base, you must make it perform. This is where, even in the consumption zone, data performance needs will outpace the technology available with on-premises tools or cloud solutions. Data within the zone will need to be performant. For this, you will need to replicate the data and keep it in sync when it needs to be kept in parallel physical forms, even though it is logically the same. Some examples of performing in-memory data stores follow:

- Snowflake {https://packt-debp.link/jTIKqT}

- ClickHouse {https://packt-debp.link/vqn4WW}

- StarRocks {https://packt-debp.link/tuRbdO}

- Rockset {https://packt-debp.link/7XgW9h}

And in-memory caching solutions exist, such as the following:

- Redis {https://packt-debp.link/bXS5ds}

- Gridgain {https://packt-debp.link/Y9Babe}

- SingleStore (formerly MemSQL) {https://packt-debp.link/ZwmBt9}

All offer solutions for in-memory data storage. Each may be integrated, but consider not allowing non-persistent third-party databases to take on a master data role unless parallel-hosted persistent storage engines are also leveraged.

It will be clear at this point that there are many choices to be made as the technology field matures and the data mesh capabilities become more commonly understood as essential. For this, you will need a clear logical and physical architecture to be produced and kept up to date at all times. The logical architecture and any engineering design choices reflected in the physical architecture need to focus on the following:

- The data with validation checking

- The factory as various datasets are curated, resulting in various transformations, and data lineage traceability

- The data flows that need to be monitored via *observability* capabilities guaranteed to operate within contracted tolerances

What becomes particularly difficult in the data architecture's definition will be the handling of time series data.

A logical and physical architecture for handling time series data as well as data freshness are business considerations to be addressed upfront in any design. The architecture and capabilities of data stores that effectively support time series data vary. You will need to look at the requirements for scale and performance before selecting one third-party product or another. I recommend you look at the **Time Series Benchmark Suite (TSBS)** {https://packt-debp.link/7Una8J} that currently supports many of these time series-capable databases:

- Akumuli {https://packt-debp.link/QtZRVQ}

- Cassandra {https://packt-debp.link/WSUvep}

- ClickHouse {https://packt-debp.link/E4xMkJ}

- CrateDB {https://packt-debp.link/QeMePV}

- Druid {https://packt-debp.link/41QVqY}

- Elasticsearch {https://packt-debp.link/nDiCwJ}

- Graphite {https://packt-debp.link/88E7jW}

- InfluxDB {https://packt-debp.link/ts5TO3}

    - InfluxDB 3.0 {https://packt-debp.link/yMbo8u} (not open source)

- KDB+ {https://packt-debp.link/OtFAtb}

- MongoDB {https://packt-debp.link/LVI2kQ}

- QuestDB {https://packt-debp.link/Dp3UX4}

- SiriDB {https://packt-debp.link/0AJo4H}

- TimescaleDB {https://packt-debp.link/y1H4Nt}

- Timestream {https://packt-debp.link/Yhc5vi}

- Victoria Metrics {https://packt-debp.link/fRzW1s}

Not yet officially supported with a TSBS benchmark are Apache Druid {https://packt-debp.link/41QVqY}, Graphite {https://packt-debp.link/88E7jW}, Elasticsearch {https://packt-debp.link/nDiCwJ}, and InfluxDB 3.0 {https://packt-debp.link/yMbo8u} (not InfluxDB OSS) time series data-enabled databases. Note that the TSBS benchmark may be extended for these or other time series databases. You may wish to take time to evaluate the performance results in the examples published by InfluxDB {https://packt-debp.link/Ekqclu} in its comparisons on the GitHub site to view the outcomes, or you may wish to review the footnotes for these products that contain links to sites with valuable information as you come up to speed on the competitive capabilities of each. You'll also want to look at how these products are best integrated by your chosen cloud provider where security and integration blueprints are available to facilitate solution development.

Remember that this is a hyper-competitive subject area, and you should expect vast changes in capabilities such as scale and performance as the field matures. Each vendor will seek an advantage over the other, and some of these areas will focus on the ease of cloud and analytic tool integration.

Data mesh, security, and tech stack considerations

When getting into the details of data mesh engineering, we need to step back and understand that the data mesh pattern is somewhat vague in regard to providing prescriptive integration directions. Consider security integration, for example. Should you implement a **zero trust** design?

> **Data security (zero trust) adds complexity to the data engineering design**
>
> Zero-trust security designs provide many rewards in regard to software solution simplicity.

Additionally, foundational capabilities such as **test-first design** (**TFD**), data profiling, metadata design, audit, and ML anomaly detection reduce costs and lead to a future-proof solution. When building out the physical architecture, there is a lot of room for innovation. You should study examples of data mesh implementation that exist. There are several different ways to implement a data mesh architecture. Here is a selection of typical tech stacks that we have gathered:

- Amazon Web Services Simple Storage Service (AWS S3) and Athena {https://packt-debp.link/SWBKqs}
- Azure Synapse (Azure Fabric) {https://packt-debp.link/lI0tmc}
- Databricks {https://packt-debp.link/v0cmbU}
- dbt/Snowflake {https://packt-debp.link/3Ty7F7}
- Google Cloud BigQuery {https://packt-debp.link/pwulCh}
- MinIO and Trino {https://packt-debp.link/BNxAHg}
- Starburst Enterprise {https://packt-debp.link/77sMny}

Some excellent guidance exists with the aforementioned integrations; however, as you can see, some cloud vendors have yet to formulate unified cloud PaaS offerings to simplify what the MinIO and Trino integration stack and the Starburst recommendations direct you to implement. The engineer must look at the big issues first and then fill in the details. There are many niche problems that can be cause for concern, but you can't become stuck on thorny issues when juggling boulders!

This puts big issues into the crosshairs of our target architecture. You want to identify and resolve these first. The biggest issues that arise are not simple technical ones; but rather compound issues made worse by ineffective and fragile integrations. The following areas will be addressed: security, test first design (TFD), data profiling, metadata design, audit, and machine learning (ML)-based anomaly detection.

Security

Federated domain ownership is a primary principle of the data mesh architecture pattern. How do you enforce security as a contract as you get the identification, authentication, authorization, access, nonrepudiation, and rights management features correct? These need to be easily configured, provisioned, run, and administered in a managed way in order for data and metadata to flow through the data factory solution. You must also consider the effect on time series instantiations where the security context will change over time.

Third-party and cloud tools vendor offerings will conflict with the data mesh pattern's goals, but they are necessary because, on day one, the engineered solution design can't reinvent all aspects of the system. There is a lot of potential for fragile integration. The amount of glue holding together **identity and access management (IAM)** for **role-or attribute-based control (RBAC/ABAC)** compared to group/ user level row-level access and entitlement in the data store will be significant. Compromises must be made, and they require cloud provider and third-party security vendor tool integration.

Plan this carefully and completely. Provide for ethical hacking tests, regression testing and lockdown drills with testing scenarios, security response playbooks, release/change procedures, and **incident response (IR)**/escalation processes. Think about the creation of the runbook for IT operations of the DevOps/ DataOps solution first, and what gets engineered will be robustly handled. Even when there are failures, they will be managed and still look like success to the business owner.

There are numerous data mesh security hurdles to overcome. Federated domain ownership/governance will make the goal of obtaining 100% security effectiveness challenging. Making sure that security policies are uniformly implemented according to their contracts with policies that back up those contracts can be an integration headache. This is especially true when trying to get a data factory's pipelines performant and scalable in the cloud.

It is very important that there is a clear line of sight to the logical security fence around the datasets being protected. Setting the boundaries for data defense by establishing a zero-trust approach that's affordable and scalable in the cloud is a goal. The organization's processes enforcing global policies must be controlled with feedback provided to the **chief information security officer (CISO)** and clearly made identifiable in the organization's logical security dashboard.

If you lose trust, it is hard to get it back. This is especially the case in the federated and distributed data mesh architecture. Clear proactive and reactive security testing along with **incident response (IR)** scenario results give the security **incident response team (IRT)** something to work with when battling issues. There will always be issues to fight even if, on day one, the system is certified as secure, and there will always be new evolving threats to contend with. Trust in your process enables an adaptive response to new threats.

As part of a zero-trust implementation and least-privilege access to datasets, industrial, government and commonly accepted cloud standard blueprints require audit capabilities. Identity and access controls will be audible, and that is a very important aspect of security implementation. Making sure data security is correct for a given user is vital. Security can be maintained by role- or attribute-based security design approaches but must be consistent even when a federated data domain owner thinks they are special and wants to be treated as such.

This is one key reason the core team must govern and enforce the architecture that trumps optimal data engineering designs. Additionally, data rights (to do *something* with the consumed data), fair use (the amount of data that can be accessed in a time period), entitlement (ability to see that data exists versus the data content versus its metadata), and ownership/lineage (who is the domain owner and what value-added enrichment was performed) in the federated governance model add facets to the security architecture of the data mesh. Think about security as a graph problem, and what will come out will be a solution that can be flexible and extensible over time and be a source of consistency in the solutions' architecture.

To provide a certification that a data mesh dataset is correct, of high quality, and of timely value so that it can be considered ready for consumption is a contract. The contract needs to be set and enforced. Part of that contract is the clear line of sight to the dataset's raw origin and the rights that were set when the raw data was acquired.

Often data is purchased with contracted strings attached that define these rights. Even in an enterprise such rights exist and need to be enforced. And what you can enforce you can also have audited. This assures the domain owner and end user that rights are being preserved. You can go so far as to put all this into a blockchain for 100% auditability, but tools to track and validate lineage with metadata are still maturing. The goal is to certify data as if it were a precious jewel that requires expert ratification of genuine value. Once this thought syncs into the reader's mind, you can see that the data that the data factory curates is to be treated with a level of correctness and auditable genuineness so that data trust is always preserved.

It is recommended that the architect and data engineer look at the tools to be integrated, the conflicting security models of the cloud and third-party vendors and the organization's priorities, and create a unified security architecture for your data mesh.

> **Note**
>
> There is *no silver bullet solution* today that can be purchased and plugged in – it is a custom solution with trusted data at the center. Do not be oversold into thinking a difficult integration can be handled by the CISO – it is an architecture issue and implication required for the data mesh, and it is fundamentally a shared solution.

Test first design (TFD)

We develop software in an agile manner, but we do not yet release data in an agile manner. This is a common observation. When did you last check a dataset into GitHub and not have it bounced right back out? Even when zipping it up, you just can't do this! You need to think of data as being versioned in the data mesh, yet there are few tools to add that kind of versioning control to curated datasets. It is essential that there be a change set mentality and a new form of release processes for data in the mesh.

When you have implemented metadata quality checking, made datasets self-describing, and applied formal data quality checks throughout the various stages of the developed data pipeline, you are thinking about engineering a robust system with contract validation in mind. You do this for code released into the system via CI/CD pipelines, but what about the data tests so that the data may be released with quality?

At various levels of data quality testing, we need to leave enough of a trace to forensically fix anomalies when they occur. We will need to retain that trace with data lineage metadata. We need to do this for the data and not just for the software code affecting the state of the data.

You will also need to shift the data testing efforts as far left as possible in the development process, with the CI/CD quality checking steps aligned with the process. The effectiveness of the data pipeline execution needs to be verified to remain effective. A key goal is to not allow bad data to propagate. Knowing what is bad is a subjective measure transformed into an objective measure with a chain of quality checks with auditable results.

Data profiling

Garbage data in leads to garbage data out.

That has never changed. You need to assess the data's profile before it is used and considered valuable.

What does it mean to *profile* data? The answers lie in the need for data to be semantically correct (has data arrived and is it of the right class?). We will want to know if it is complete (has a partial receipt been received due to an error?) in that enough data arrived to process it as a cohesive dataset. We will also want to know if data is overstated (has a file arrived twice because of a transmission glitch?). Does the arriving data break the expected trend (it could be that arriving data does have increased volume but is not 100x the norm; maybe somebody had a really sticky keyboard)? Do the data's metrics and measures (what we've always referred to as **second-order data**) break trends or pass the test of being credible?

You will want to gauge data anomalies carefully by creating analytic probes, made operational as data profiling steps running within the data pipeline. Data quality is not always myopically observable after a single data pipeline processing step is complete, but only after being rolled up into a metric or having a calculated measure applied will anomalies be discoverable. Additionally, machine learning techniques for pattern identification, matching, and common feature clustering (followed by classification) allow anomaly thresholds to be exceeded and issues identified that are not obvious.

For example, in a prior engaged retail analytic setting, these machine learning techniques identified the occurrence of gray market retail items in retail store sales volumes that should not have been reported as true sales. This was detected by seeing a pattern of sudden rise in sales in a store within a particular region. The ability to detect anomalies and provide enough signals to enable you to forensically identify the root cause of anomalies is essential to preserving the credibility of any derived consumer-driven insight.

Does the **third-order data** (the *insight*) generated hold true during backtesting? You will want to perform backtesting of any insight using past data as part of data quality testing. This allows the truth of the insight to have been discovered in the past if such a test existed when prior data was received. Backtesting an insight's hypothesis adds credibility to the insight and solidifies it as a truth!

Data profiles need to be created and all data testing automated to define the trustworthiness of a dataset. Software development frameworks do not satisfy the need alone. Data profiling frameworks do! So, a data profiling framework is needed in the solution to support answers to the questions posed in this section. What frameworks exist for data profiling? You should look at Alteryx's Trifacta {https://packt-debp.link/UJPS9u}, Datameer {https://packt-debp.link/cMTXQS}, DataFlux {https://packt-debp.link/jUXuaI} (but you will most probably be working with SAS since it is a subset of SAS now), Hightouch {https://packt-debp.link/tA0Zm5}, and AWS Glue {https://packt-debp.link/mRLiTl}.

Data profiling is a goal. The tooling choices needed to support that goal may be part of the solution. Even if homegrown or chosen from over-the-counter solutions and then integrated, data profiling is essential to gauge the shape of the data to have it semantically fit the need and be of high enough quality for consumption.

Metadata design

Metadata patterns and solutions can be endlessly discussed and are often implemented badly. Refer to prior sections where various vendors and offerings are mentioned. See how cloud provider solutions are beginning to focus on the need for structural metadata to explain the chain of Cloud PaaS dependencies of the customer-implemented data services their pipelines require (for example, Microsoft Azure Purview {https://packt-debp.link/OET4Vj}).

Today, in many data processing systems, you can't easily trace the data journey from end to end. You can't afford **master data management** (**MDM**) systems of the past, and they don't fit the cloud model anyway. The risk of a full-on data quality meltdown exists at many levels. This is prevented with the clarity that a solid metadata blueprint provides. Structural metadata (cloud data transform operations with the data organization definition), metadata for data at rest (semantics, shapes, and contexts including time), and data lineage metadata (enrichment and transformations) are all required in the data mesh metadata solution.

Audit

We've mentored this thought to many teams in the past and often say the following:

It is not good enough to get the job done; it's essential that any individual following you do IT as well if not better than you!

Going from *hero to zero* is not just a cliché, but rather a very distinct possibility if you do not build that mountain of support!

Build a mountain of support for your success, and do not climb the flagpole with your banner to obtain your career heights!

We often begin prep talks as follows:

It is not good enough to just get it working; IT must work, IT always has to work, and IT can never fail! If you just get IT working, you've done one-third of your job!

So, what does this have to do with the topic of auditing and the data mesh pattern? The answer is, please leave breadcrumbs! It is essential that there be an audit trail containing the essence of *what your software does* and the effect on the data pipeline's released output after each stage of the data pipeline journey.

What the pipeline code does to the dataset's state has to make sense when looking backward at the audit trail. This data journey audit trail leads to its hookup into the data mesh (or fabric if in the cloud), which then needs to be discoverable. What your IT solution does to the state of your data is what the customer will see, use, and enrich. Change is a stream, but quality is discrete and assigned at the stage boundaries. Even if the stages are small, changes from one release state to the next must be explainable by those functioning as quality assurance team members. This will be needed when the analytics user wants to use that data, which leads to impactful insight.

Imagine presenting to a company's board a revenue forecast that is based on erroneous summary retail data! The analyst will often be the critical downstream user to be serviced; any change better be explainable, to everyone! With data being self-service, this is even more an issue for obvious reasons.

On another note, debug levels in your software quality analysis are analogous to the data trace levels of your data quality audit reports. Think about **data forensic** use cases when building data quality and assessment tools to clarify the features needed for the designed audit trail. In this manner, you only collect good trace and not piles of useless clutter. Sifting through it for insights can be fruitless.

Tools such as InsightFinder {https://packt-debp.link/AvNA70} exist (for assessing IT operations patterns), but ML algorithm-driven tools can miss important and necessary details. Changing data pipeline code after the fact to collect more audit trace and waiting for the off chance that a previously observed anomaly reoccurs, and doing this literally through spiral iterations, leads to many lost cycles, frustration, and bitterness directed toward IT developers. It is proof that an architect did not get involved early enough to stop the erroneous thinking in advance.

The cost of building a safety net for trace assessment may be high, but without this safety assurance environment, the ability to add new quality checks becomes difficult. If insights gleaned from trace assessment are missing, the feedback loop to improving quality checking is vastly slowed down. Here, we leave you with one of our favorite thoughts; we call it the first principle of paranoid programming:

What can go wrong will go wrong, and it's going to happen to me! So, I better just be ready to handle IT!

Turning a difficult software or data quality issue into a series of opportunities makes it possible for you to build a ratchet toward success rather than a house of cards doomed to many failures.

ML-based anomaly detection

The human capital cost of curating a complex or vast dataset can be expensive. Leveraging ML minimizes this manual effort.

Using ML-based anomaly detection reduces quality costs and leads to a future-proof solution!

Any quality application of ML or DL approaches to address data quality anomaly detection needs to be of high efficacy. Obtaining trusted repeatable results is the key. When data science meets data engineering, you want a repeatable quality assessment framework for ML results, with real-time observability approaches to be integrated to know the following:

- When any retraining is required
- When drift is occurring
- When precision/recall within a category (a logical segment of data) goes awry

There are many advanced ML quality metrics {https://packt-debp.link/F1a3Id}, and many are applicable to a tuned-up data engineering and data science team. Listed next are a few of these metrics:

- Area Under Curve
- Classification Accuracy
- Concordant Ratio and Discordant Ratio
- Confusion Matrix
- Cross Validation
- F1 Score
- Gain Chart and Lift Chart
- Gini Coefficient

- Kolomogorov Smirnov Chart

- Logarithmic Loss

- Mean Absolute Error

- Mean Squared Error

- R-Squared/Adjusted R-Squared

- **Root Mean Squared Error** (**RMSE**)

- Root Mean Squared Logarithmic Error

In the past, when building data engineering processes for data scientists, it was like trying to tell an artist that their art is not logical. It is where art meets science and where there is much contention. Often, this is due to mismatches between the language of the following:

- Statistics, computational linguistics, and mathematics

- Computer, software, and data engineering

Math is the common language of science and engineering, so bring it all down to the numbers, the proofs, and the algorithms.

You will want to build a workbench of processes so that what works in an inherently fuzzy processing area works well over time and can be understood by all mindsets. Any stochastic approach can go off track if assumptions are not codified, built as contracts, and tested to exist before applying a model. Once the contract is verified, it needs to be monitored for operational compliance to make sure drift does not occur, and retraining and then re-contracting take place within the ML workbench.

With hundreds of parameterized ML models, this can become an area for systemic meltdown without a framework. With the addition of any new dataset or new instance data aligned with an existing approved dataset, this meltdown can happen. No AI solution can remain effective without an engineering discipline to minimize risk to the data domain owner.

What are the key foundational takeaways?

In this overview section, a basic level set was provided leading to the generation of principles to be itemized in the next section. Each area has key takeaways that need to be kept in mind as the principles are elaborated upon so that they can be made part of your organization's data engineering approach:

- **Data lake, mesh, and fabric**: Know the architecture differences between data lake, data mesh, and data fabric and think of data as a curated factory product subject to all the life cycle needs of a product.

- **Data immutability**: Establish data zones such as Raw, Bronze, Silver, Gold, and Consumable to separate classes of data as being fit for purpose within each zone. Understand that the justification for this approach is a pragmatic one, given current technology limitations and vendor offerings that often are at odds with your integration goals.

- **Third-party tool, cloud PaaS, and framework integrations**: Focus on DQM in your data factory architecture and data engineering designs. Data as a product needs to trump niche third-party vendor offerings that may be difficult to integrate or force the necessary architecture principles to be violated.

- **Data mesh principles**: Grok {https://packt-debp.link/Y84Gs8} the data mesh pattern fully:

  - Domain-driven ownership of data

  - Data as a product

  - Self-serve data platform

  - Federated computational governance

  Build out the architecture knowing the conceptual, logical, and physical architecture implications for the data engineering effort.

- **Data mesh metadata**: Metadata is a key enabler of the data mesh pattern. It is formed in three classes:

  - Structural data

  - Data at rest (semantics)

  - Data in transit (lineage)

  Metadata needs to be discoverable via a data catalog and operationalized to be effective. It also must overcome and bridge cloud and third-party vendor gaps/inconsistencies in this key capability.

- **Data semantics in the data mesh**: Metadata explains data at rest semantically but the *how* is subject to debate. The need for a semantic web was raised by Tim Berners-Lee long ago, but that vision was not realized. The data mesh needs data semantics to enable quality and federated domain curation of data through the mesh. The data engineer's platform must address the need for comprehensive metadata capabilities.

  You need to weigh the effort to create a formal domain model creation (in OWL 2/RDF, for instance) as proposed by Stardog, Ontotext, and Cambridge Semantics in their approach versus the use of **labeled property graphs** (**LPGs**) proposed by Neo4j, JanusGraph, TigerGraph, and Microsoft Graph. This will be required as part of the logical and physical architecture definition of any modern data platform.

- **Data mesh, security, and tech stack considerations**: Security is the first capability of the dataset contract for service that is needed to preserve the quality and integrity of any curated dataset. It must be well architected for proper and effective fit and service of the data platform. It must not become the Rube Goldberg {https://packt-debp.link/pWgI9k} implementation task of the data engineer:

  - **Test First Design** drives the CI/CD pipelines need to be applied to data quality and not just software quality.

  - **Data profiling** is needed to transform data after it is validated, normalized, and proven to be of high enough quality to pass pipeline gating checks before propagating to the next stage of the factory.

- **Metadata design** is key to the data mesh's implementation of the federate and domain ownership principles.

- **Audit capabilities** are to be built for compliance, and forensic and governance needs to preserve the contract and validate the level of trust you can give to a curated data set.

- **ML-based anomaly detection** reduces the cost of maintaining the data quality contract, which leads to a future-proof solution.

In this overview section, you were given a walk-through of our thinking regarding what is needed before generating your principles. The key takeaways will be in line with your vision as the principles are elaborated upon in the next section.

Architecture principles in depth

The following principles will help guide your decisions when architecting and engineering solutions are built into a modern data platform.

Principle #1 – Data lake as a centerpiece? No, implement the data journey!

This may sound shocking and really an anti-follow-the-herd mentality, but it is true! Thinking that the data lake was envisioned as a source for all data that can be miraculously understood and repurposed over time leading to great insights is naïve. It can become a data swamp and a costly liability without semantics, context, time series structures, and a clear metadata pattern with governance principles aligned with the data mesh and operational data fabric capabilities.

Data needs to be curated in the factory from raw form to consumable form, and it needs structure and life cycle along its assembled journey through various zones (such as a number of logical data lakes) until ready for consumption. Data needs to be released like a product. You cannot just wave a magic wand over the data to get it to this state of readiness for the consumer. That data journey is worth noting, logging, preserving, contracting, securing, controlling with entitlements, and assigning value-added rights for long term sustainability.

Principle #2 – A data lake's immutable data is to remain explorable

A logical dataset should first undergo its contracted life cycle's curation process steps (as defined as assembly instructions) along with quality metrics within a zone. Logical data is then ready for propagation to the next stage in the data journey or made ready for placement into the downstream zone. Datasets will remain explorable and cataloged at the zone boundaries since that is where contracts are enforced. Datasets will need to have lineage recorded along the journey and then have the lineage condensed (or rolled up) once the dataset is readied for release. It is important that data catalogs have the capability to separate temporary lineage trails from final trails that must be retained with the released dataset. Many tools today do not have these parallel metadata trails available nor a way to roll up the trails to the release set. This is even more complex in a cloud environment where many platform as a service (PaaS) offerings are utilized along the data journey. Cloud metadata services provide tracking services (such as Azure Purview), but these need to be augmented to correctly implement this principle. *Once metadata for the data journey is available, the data should be made explorable.* Without the data and its metadata being linked together and made available for search systems, you cannot effectively identify a dataset's worth, its context, or fitness for use when being leveraged by a data engineer or consumer. Datasets and their metadata and search indices will need to be locked down and readiness state be a primary search facet when looking for data across the entire data factory platform-enabled system. It is far too easy to lose track of a widget on a physical factory floor, and this is true of a dataset element in whatever state it may be, across the entirety of the data factory.

Principle #3 – A data lake's immutable data remains available for analytics

After all, code is code, and it has bugs that need to be addressed in time.

Errors will exist in data, metadata, second-order and third-order data/metadata, and various quality processing steps. Data states will be affected every time a code snippet is adjusted, and with that, the history of the change and its effect on prior processed data is subject to auditing and downstream explanation. Analytics snippets are applied throughout the data factory's data pipelines. These are part of the shift-left testing and TFD methods that must be core to the software design approaches applied throughout the **software development life cycle** (**SDLC**). The analytics steps will need to be self-contained and cataloged. Also, the analytics output needs to be clearly cataloged as metadata for primary, second-order, and third-order data of the pipeline.

Data quality analytics, trend assessments, and consumer analytics will indicate that there may be errors or deviations to the contracted dataset's state. What do you do with corrective algorithms? In a prior system, corrective algorithms piled up change on change for a curated dataset so high that the truth of the original dataset was lost. How data corrective steps are applied must not change the immutable source but be applied as data lenses on the source data so that if they were removed – the original could still be leveraged. If necessary, all value added lenses should be rebuilt when gross errors are encountered.

Some data lenses are temporary in that they will apply only for a limited release or duration of time to correct a gross gap, understatement, or overstatement of a dataset's raw availability. These also need to be cataloged and auto-removed when no longer pertinent (such as for the next data release).

Data analytics is not an art but a science and when engineered into a data factory will result in the need to explain the state or quality of curated datasets. This will require change when the explanation is: *That's an error!* When errors arise, new data corrective steps are added that need cataloging and life cycle management as data lenses rather than changes to the immutable raw or curated data.

Principle #4 – A data lake's sources are discoverable

A dataset's source will need to be clear when leveraging the data catalog's search capabilities with the data lineage trail, and all steps, stages, and corrections evident. This gets very complex and can look like a tree's roots seen from a bird's eye view. There are lots of branches, merges, algorithm processing nodes, and ultimately, way down deep, you will find the row and column of the source. When tracking data lineage metadata, it is important that the journey be backward discoverable. This is not easy since a processing step may have implemented a nested update such as *Change a column on a table using a transform after a join that first merged a lookup table with a conditional*. What caused a particular row/column change was just lost. Many metadata tracking systems just report the query and leave it to the forensic analysts to figure out if the transform was effective. Worse yet, when the update join operation was performed, it may be that the lookup table itself was being updated in another parallel processing step at just the time the join was being made.

This really happened in a prior experience, and the audit trace looked like a perfectly executed operation when it was really producing a gross error since the joined lookup table was being updated at that exact moment. You could have fixed this with a global system table lock but that would have slowed the entire factory down. The update was being performed in an optimized service that was out of the development team's control. A snapshot of the table followed by a quality check was implemented to fix the problem, and then the snapshot persisted for backward traceability once the pipeline was completed.

It is important that you do not forget the need to assess a data source's original source when looking backward from the end state. Being able to get past a data pipeline processing *black-box stage* by making sure they have a *gray box* level of inspection trace will enable the state of the original data source to be discoverable.

Principle #5 – A data lake's tooling should be consistent with the architecture

Many third-party tools and cloud services are over-marketed to data engineering management. These tools and services are often only 80% production ready, even though they have been launched as generally available. They come with the need for excessive training since they exhibit peculiar error-handling characteristics that will appear only at the edge of performance, scale, or functional use! A **proof of concept** (**POC**) may never expose these characteristics. Only a data engineer with a bulletproof quality standard with clear methodology will be able to smoke out these bugs (and they are *not* features) yet a developer must code around the handling of these anomalies. Sometimes the workarounds do not warrant the use of the tool, and this can come to light too late in the project and force re-solutioning of the pipeline. This is not refactoring. It's a disaster of architecture and design caused by believing in a vendor's marketware rather than proven software.

The need to assess tools and services correctly with an objective due diligence process, with a clear demonstrable winner with the optimum fit for the architecture, must precede the POC to vet vendor fit. Once deemed fit, a POC should go forward to vet all integration needs, and then finally the iterative agile development of the solution.

All too often, the industry herd moves to the next shiny thing that is offered by a start-up vendor. The result is a fragile mess of a solution that exposes the business to risk. The architecture is implemented based on principles and required capabilities. It stands above the niche vendor's capabilities. It should be used to guide and align any selected vendor capabilities and reject any misaligned capabilities and marketware.

Principle #6 – A data mesh defines data to be governed by domain-driven ownership

We are suggesting that datasets in the data mesh be curated by data pipeline processes with contracts set by domain owners. This will mean that datasets stand alone and do not cross domain boundaries. This is not always the case since data can be adjusted with downstream value-added additions. Ownership becomes shared between the primary domain owner and the value-added owner. The contracts for subsequent downstream consumption are then a mixture between the primary owner and a chain of value-added owners.

With ownership comes rights for entitlements, fair use, and commoditization with or without redistribution rights for the consumable dataset. You can even go so far as to put all contracts and lineage into a blockchain, but that capability has yet to mature into easily integrateable third-party offerings. However, it is still required to maintain contracts and make them discoverable in a data catalog along with the metadata for any data at rest with lineage as well as all value-added adjustments.

The domain owner's data change life cycle and potential for contract default exists, and as such, trust can be eroded over time. Who has not used data thinking it would be refreshed yearly only to find out that it was only partially refreshed due to the business failure of the domain owner? Departments in a large organization come and go, ownership changes, funding stops, or management is shaken up. The data consumer may never be notified of contract breaches before they impact production. It is required that contracts be subject to compliance tests and predictive alerting enabled before failure arises. This way, the consumer has time to react. If data is domain owned and a product, it must be treated as such.

The total cost for the domain owner must consider the intangible revenue and costs to the contracted consumers. The proper data-engineered solution must provide that dashboard to the financial accountants, who often do not know that the domain owner's business data obligations impact consumers. The changes in a domain owner's funding or viability have ripple effects on the business. Centralizing domain ownership into a core IT group is often the solution for orphaned domain owners' data products. Over time, the data contract changes since a steward has replaced the domain ownership. Consumers need to see this in the changed contract when the architecture refreshes yearly and tech debt is re-assessed. Additional data debt remediation costs are incurred implicitly, which need to be added to the strategic plan's rolling 5- (or 10-) year total cost estimate as the future state horizon moves out in time.

There are many practical data engineering effects on domain ownership of data in the data mesh. These need to be discussed and planned for since what is a great idea on day one must meet the practicalities of the long-term run/managed solution. That solution is developed in the context of a fast paced dynamic business environment.

Principle #7 – A data mesh defines the data and derives insights as a product

A dataset (with cataloged metadata), its derived data, its quality analytics process steps, its transforms, its corrections/adjustments, and various versions as it appears in the data pipeline's zones are considered products in the data mesh. As a product, a dataset comes with all the implications of it being a product as defined by Zhamak Dehghani in her work on the data mesh concept while at Starburst {https://packt-debp.link/77sMny}. What you also have to know is that any directly observed insights as well as implicit insights/inferred insights are *also* products. Imagine acting on a dataset's insight today only to have it change tomorrow or next month because of the dataset's restatement, a software bug correction fixed in reprocessing, or a trend break that was standard for years. It could spell financial disaster for a consumer. Heads will roll if the impact is not assessed and the change provisions of the data contract are not communicated.

We can't begin to tell you how much time was spent in the past explaining data changes and why a data change was made going forward or, worse yet, in the past. Products can be recalled and so can data. You must be ready for reality. Data and insights do not just get profitably produced but can be recalled and restated. You must handle the ugly parts of treating data as a product and not just build anticipation for the revenue that the thought evokes. There are real costs to maintaining data contracts, and some of these are practical corrective costs for restatements, parallel version maintenance, differential sets (for **change data capture** (**CDC**) log replays), communications of data inventories in stock, quality metrics, and change logs (dataset inventories that could even be snapshots of the data catalog at the time of distribution/consumption).

We caution the reader to look at all aspects of data being a product and not skimp on any capability that puts the data contract at risk or the dataset's quality into an unknown state.

Principle #8 – A data mesh defines data, information, and insights to be self-service

We have observed that data being self-service is a great concept but rarely do analysts want to spend an enormous time curating it from its raw state. Consider the bricklayer working with clay and sand and cement rather than bricks and mortar when building a wall. It would be insane not to build with some degree of prebuilt material. Items must fit together seamlessly rather than having to be manufactured on demand at the time of assembly to become an insight. Data, therefore, must have edges or facets that align with other data. You can't have a year be defined as a calendar year when the data is aligned with hundreds of different company corporate years. Retail calendars have to align for weekly, quarterly, and yearly data to be analyzed comparably. Datasets subject to be used as self-service need to be faceted, subject to contracts, and fully discoverable in the data catalog with lineage explaining contracts. Only trusted clear data can be put out as self-service. Any difficult analysis that requires internal data facets to be exposed must be wrapped as second- and third-order data that then becomes subject to self-service.

Once any complex data is too hard to explain, it can't be made available self-service! So, to solve this, you can provide open services and analytics to put into code and scripts to interpret complex data. These analytic wrappers and pre-canned services help solve the complexity problem while leading to data maintenance issues. Analytic code changes in Microsoft Power BI or a big data notebook can be made incorrectly, leading to unchecked data abuse. Financial information providers are acutely aware of data abuse issues in analytics. The financial numbers should tell a story but even a small gap in a time series can lead to analytics errors.

For self-service data mesh goals to be truly effective, self-service data must be correct, complete, timely, high quality, and at all times aligned with the data contract. But what if the data has gaps or is missing key fields? Can it be augmented with synthetic data, and if so, how can that be good regarding maintaining the data contract?

It all depends on how the data contract is written. If data must be 100% complete and the raw data is not, does this practical and real lack of some data points hold up the assembly of the final product? The answer is, "No!" Can the contract enable the factory pipeline to produce data points in a dataset that are fiction just to maintain a trend or fill a gap or subject points that are redundant? The answer is, "Yes!" The data's use case is the important goal. The purity of the dataset as a whole has to meet the terms of the data contract to apply to the entire dataset, not just individual data points in the set.

Principle #9 – A data mesh implements a federated governance processing system

Federated governance has an implication. There is a central governance function, and some governance can be delegated to others. It is a fine balance that must be maintained between the two major divisions of data governance: central versus distributed. The effect is an ability to change and adapt as data and contracts are brought online in the data mesh. The enterprise's organization must be made ready to handle this model of governance. Also, the organization must promote the federated model with standards from the highest level (also known as the core architecture) and incentivize compliance.

Martin Fowler {https://packt-debp.link/o4dhRZ} points out that centralized golden datasets are no longer pertinent. You have to comply with the data mesh federated principles and established architecture while maintaining data contracts within the data mesh.

Principles drive how the following subject areas impact your designs:

- Data quality standards
- Data contracts
- Security and entitlement
- Audit and regulations
- Modeling and cataloging features
- Self-service and metadata governed
- Metrics and measures captured
- Code steps and transforms
- Error detection and correction

These subject areas then drive the need for a federated governance model in order to keep your solution relevant in the future.

Principle #10 – Metadata is associated with datasets and is relevant to the business

Metadata defines the data and how it became the end-state dataset used by downstream consumers, whether that be internal users, automated processing steps, big data repositories, or **business intelligence** (**BI**)-tooled end-user analysts. Metadata facets are exposed in data catalogs and subject to time series organization to provide filtering capabilities of the data in the data mesh, or data fabric if hosted in the cloud. Metadata is important since it represents the data and can be used to certify data in the mesh as meeting the data-contracted requirements of the business. It establishes the trust required to enable the data itself to be self-serviced.

Capturing metadata and retaining auditable versions of it in the past enables past data to be verifiable and understood in a prior context. It also provides change traceability that adds to current data credibility.

Metadata provides for the creation of facets that may be discovered through dictionary search capabilities. Metadata lineage capabilities aid in the forensic analysis of data quality issues. Domain ownership of key datasets and any value-added business enrichment may be discovered from metadata lineage. Shortening the time to assess errors, finding owners, and determining entitlement issues is an accelerator for the business. Exposing data facets created from metadata also enables you to implement facet intersection and, as such, obtain new insights. If you were to be given the goal to *do more with less*, you would want to leverage metadata.

Principle #11 – Dataset lineage and at-rest metadata is subject to life cycle governance

Metadata, whether it is semantically definitive of the domain owner's truth or reflective of the data journey as lineage, will need to be tracked as if it were a core dataset. This form of second-order data being put into change control and treated as a product makes sense since it feeds the searchable data catalog of the architecture. **Life cycle governance** is needed for all datasets, including the metadata defining pipeline curated datasets. A key addition is the linkage between the curated dataset and its metadata instance as part of the named branch noted as third-order metadata.

You can envision multiple instances of datasets with linked metadata within a pipeline, with the final instance being the change set that is allowed to be released and, as such, propagated to the downstream zones. What determines the acceptability of the releasable instance of data with its metadata is the quality metric. Governance rules check for those contract conditions to be met, and if not, direct action is taken to make them acceptable. These include reprocessing, gap filling, synthetic data creation, data removal, trend smoothing or fitting, factorization, and other statistical enrichment processes.

Principle #12 – Datasets and metadata require cataloging and discovery services

The data catalog is not a static service. It is a searchable up-to-date reflection of the state of data within the data mesh. It maintains the data facets that make the data self-serviceable and the contacts that make it trustworthy.

Discovery and visualization will be key to the data mesh since over time, data loses pertinence. A consumer will want to dial in data at various levels. Often, data must be timely, or it needs to be up to date and correct for some other fixed point in time desired by the analyst. The ability to dial in the zoom when looking for insights must be a key self-service capability of the discovery tool of the data engineering solution.

Principle #13 – Semantic metadata guarantees correct business understanding at all stages in the data journey

It's all about the semantics when bringing data together for analysis. Then, it's all about the quality of the combined dataset before the assessment of an insight's hypothesis for truth. You know that you can't combine apples with oranges and get a valid resulting fruit. They don't cross-pollinate, although you can do this for plumcot, which is a plum and apricot hybrid. The details of what can and can't be combined are in the dataset's semantic at-rest metadata. Preserving the domain owner's understanding of the curated dataset enables the fair use, entitlements, and rights of the consumer.

Clear semantics may be preserved in the creation of a model in an OWL 2/RDF model knowledge graph that supports forward and backward inferencing. As an alternative, you may create an **labeled property graph** (**LPG**) knowledge base, but this is inherently unsuitable for reasoning purposes. For advanced knowledge graph capabilities in an LPG, you need a lot of code to glue the semantics together, but direct inferencing capabilities may have to be given up. A knowledge graph with instance data forms a knowledge base. With a formal knowledge base, you have a very strong and enforceable representation of the domain owner's data in the mesh.

Principle #14 – Data big rock architecture choices (time series, correction processing, security, privacy, and so on) are to be handled in the design early

With this principle, you are being asked to handle architecture choices early and set up a framework for data engineering design success:

- Time series data

- Data corrections

- Data entitlement, data rights, and data privacy

As stated earlier, data is time series sensitive. This was noted as a key issue requiring a solution in a data mesh. Often, data loses value based on age, and this affects how it is to be represented over time. In the simple case, it can be partitioned in a relational system and rolled off after the audit period expires, but that removes it from use unless rolled up as summary info when aged out. How this affects its metadata lineage is also important. Data with its associated metadata together have to be rolled off (or partitioned off) when the primary data ages out. Modern knowledge graphs often lack partitioning capability making this possible and, as such, must be cloned or custom copied over into special archival semantic models created to support a compressed form of historical data.

Likewise, current data corrections, historical restatements, and the effect on downstream data pipeline consumers must be clear in the design of the data mesh. This can't be left to an afterthought without the data mesh, losing the trust it was built to preserve. We've built data systems that preserve data if the assessed change is less than 10% but create a new set if the change exceeds 10% deviation because of correction before becoming a restatement. That percentage should be a dialed-in number and not fixed as per the domain owner's contract, but once set, it should not change for a given contract. A correction is a small change, and a restatement requires full downstream reprocessing.

Entitlement to view the existence of a dataset, the contents of a dataset, or reflect on (inspect) the dataset's metadata must be separate types of entitlement. Additionally, you will want to add enhanced entitlements for fair use (how often data can be accessed), redistribution (ability to send data to another), value add (to enrich data), and resale (to redistribute with profit). Basically, if you can protect the datasets with a constraint, it can be an entitlement. Once you protect data with entitlements, you want to define the type of entitlement structure in your design. This is either via roles (RBAC) or attributes (ABAC), but you must deal with legacy group/ID-based IAM security of third-party products and convoluted cloud security services. These tools and services implement the latest security system *du jour* and do not integrate well with RBAC/ABAC– and you still have an OKR to implement a zero-trust secure system. Our suggestion is to generate the security architecture and keep to it even when faced with many integration obstacles.

Principle #15 – Implement foundational capabilities in the architecture framework first

For this principle the foundational principles componnents include: zero trust data security, test first design, data profiling, metadata design, auditing, and machine learning anomaly detection

The architecture of a future-proof data engineered modern solution will need to address a number of key areas identified in the principles discussed in the various sections of this chapter. These are the following:

- Security
- Test first design (TFD)
- Data profiling
- Metadata design
- Audit
- Machine learning (ML) based anomaly detection

Do not dive into the details of the data mesh implementation without scoping out necessary logical components first because you will not be able to fit them into the solution later if you do not handle them here.

Summary

One needs to grasp the concept of the data mesh and its core principles but apply practical data fabric DataOps features to make the data mesh implementation work as expected.

Data being a product makes sense, but it needs to be fabricated, produced, and curated to the point that data-derived information is ready for self-service, which can be leveraged for knowledge pop-out insights, leading to wise decisions. This progression in the past was forced through big systems software development, but today and going forward, it will involve automated intelligence brought about by the data engineer.

Data needs to stand alone and be smart as well as enabled for insight harvesting. Data needs to be smarter than it is today! The wording of that implies a personification of data. This is a data engineer's goal. Engineered data is no longer hindered by today's big data approaches. Data that is of the highest quality, fully understood, linkable, transparently curated, and contract enforced (to maintain trust and truth) is data enabled for consumption. The principles described need to sink into your data engineering approach. The result will be the creation of systems that lead to the discovery of insights that become data, information, and newer insights in a rapid progression. Intelligence is driven by facts that are supported by information organized by knowledge. So, when we engineer systems to support artificial intelligence, we must not forget that the future-proofed data organization must support similar structures and processes we use to bring about human intelligence. In the next chapter, we'll elaborate on a principle-based conceptual architecture that you will develop based on what you have learned in this chapter.

Architecture Framework – Conceptual Architecture Best Practices

The need for an architecture framework will be elaborated upon in this chapter. The framework is documented as a three-level *conceptual*, *logical*, and *physical* architecture, which is needed to clearly communicate the intended state of an IT solution. These levels consist of blueprints. There will be a clear focus on the conceptual architecture level in the chapter and, with it, the best practices for its definition as a current state artifact, as well as important considerations regarding best practices for its content. You may ask: *What's in a conceptual architecture?* The answer is: the document that contains a human-readable conceptual rendering of the solution that is used to level-set the business on the IT solutions' capabilities and promote dialogue. High-level capabilities are part of the entities depicted in the conceptual architecture. These are conceptually knitted together with dataflows. The capabilities coalesce into platform blueprints and are also defined at the conceptual level, and they look like flat, stacked layers of blocks that even a child could put together. The intersection forms blocks, which are the *interfaces*. A single-page representative diagram best illustrates the conceptual architecture level of the solution, and from that, you are able to peel the onion, so to say, in order to enable the business stakeholder to be able to dive ever deeper into the details. A goal is to not overwhelm the audience with the details, but in effect produce an *ah ha!* moment. This way, the key messages are communicated; however, the solution architecture will not be complete until being fully fleshed out at the *logical* and *physical* architecture levels.

Conceptual architecture overview

The goal of this chapter is to do the following:

- Define the contents of the conceptual architecture and illustrate it

- Provide the best practices for that illustration and how that could be rendered in a way to make it effective for stakeholder communications

- Provide best practices for future-proofing the architecture, with an emphasis on the dataflows and blueprints for a data factory/data pipeline approach to curated data delivery

The three goals for this chapter will align with subsequent chapter elaborations for the logical and physical architectures of a reference solution.

Best practice organization

In this chapter, you will observe the following structure:

- Best practice composition of the conceptual architecture:

    - What is the structure of a conceptual architecture?

    - Why is there a need for data classes and data processing zones in the conceptual architecture?

- Learning the definition of data classes in data processing zones and how datasets are moved between zones:

    - The *bronze class* is used for data profiling, data normalization, and data cleansing use cases for big-data-lake processing.

    - The *silver class* is used for data pipeline processing or big-data-lake, **online transaction processing (OLTP)**, and some **online analytical processing (OLAP)** operations.

    - The *gold class* is used for enterprise-grade mastered data and for OLTP and OLAP operations.

- How to implement data processing zones:

    - The *raw zone* is used for data ingestion to the data factory. In this zone, raw data arrives and is normalized. Note: This often leads to a data lake implementation.

    - *Transformation (bronze, silver, and gold) zones* are used for data warehouse and data lake cloud storage. Note: This often leads to a delta lake or cloud data warehouse implementation.

    - The *consumption zone* is used for high-speed access and data analysis where consumer use cases and dataset deliveries take place. Note: This often leads to a **business intelligence (BI)** in-memory or data mart OLAP implementation.

- Learning best practices regarding how to tie the conceptual architecture to the logical architecture.

With this structure, a data engineer will be able to communicate effectively to those at the highest levels of an organization without the details yet also be able to drill into details if challenged and still relate back to the high level for justification.

How does the conceptual architecture align with the logical architecture and physical architecture?

The software (DevOps), data (DataOps), test (TestOps), **machine learning operations (MLOps)**, and infrastructure (operations) teams can build/deploy a solution the architecture defines. This can happen only during and after the logical and physical architectures have been formulated. Additionally, special attention will be given in this chapter to the dataflows that implement the various data pipelines of the data factory solution. This must be very clear later at the logical level of the architecture but communicated to the business at the conceptual level so that data contracts and **service level agreements (SLAs)** may be formulated with the business owner and data consumer. Fully configured dataflows appear only in the physical architecture, but this detailed level of the architecture adds huge complexity to the total definition and can easily be confusing to non-technical stakeholders and data domain owners. The details are still required to get an IT solution provisioned, deployed, scalable, performant, and supported. If the business stakeholders and data domain owners get lost in detail, they will not support or fund your efforts, so this is where the architecture framework is applied. Low-level details will blur the great message you are trying to deliver to the business; therefore, high-level architecture at the conceptual level will be required to communicate upward and outward in the organization.

At this point, you may want to point out that there are lots of enterprise architecture frameworks to pick from … why not leverage one of these?

- The Open Group Architecture Framework (TOGAF 10) {https://packt-debp.link/qWK0FS}

- Zachman Framework {https://packt-debp.link/vhPR63}

- Federal Enterprise Architecture Framework (FEA) {https://packt-debp.link/QCU6tF}

- Others include C4ISR {https://packt-debp.link/AKHvam}, DoDAF {https://packt-debp.link/SfP0qn}, FEAF, Gartner Enterprise Architecture Process Model {https://packt-debp.link/dprb7Q} … and even more

This approach we have chosen for you is a pragmatic one. It is similar to the **Gartner Enterprise Architecture Process Model**, in that we realize that without a governing mandate such as the US Federal Government organizations have to pick an enterprise framework that works without the cost and formality that a more rigid framework or methodology requires. We are rooted in the decision to select a pragmatic approach that is not exactly the same as any approach listed previously and more aligned with the **4+1 Architectural View Model** {https://packt-debp.link/SEvxb6} shown as follows:

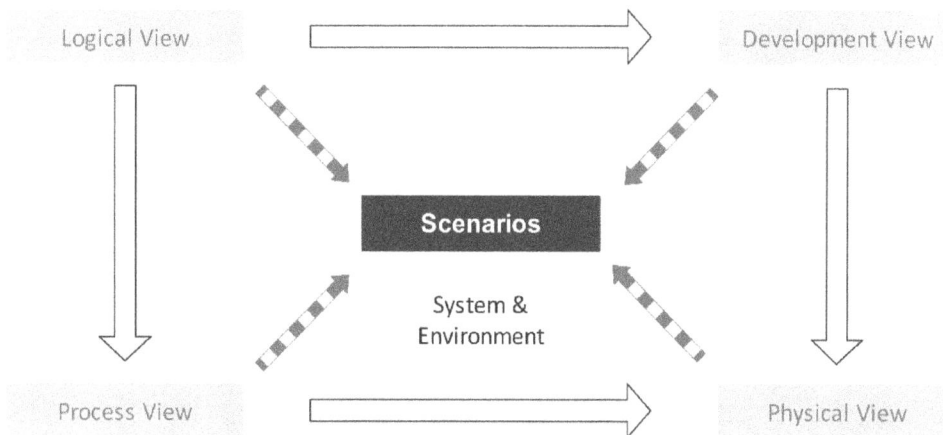

Figure 5.1 – 4+1 Architectural View Model

Along with the *4+1 Architecture View Model*, an architecture has to be represented with another set of facets that describe the people, process, and technology aspects of the organization's methodology. This is needed to set up governing processes. These processes enable the architecture to be kept evergreen.

Just as an evergreen tree stays green during the winter months in the north, so should the architecture remain green, living, and fresh in various business climates. Those who make their living in the IT field have gone through their winters of discontent, where the architecture is on the defensive due to various technical and business disruptions. At these times, the architecture future-proofing effort will be greatly appreciated. When the architecture is clear, it can be communicated to various audiences with their different levels of technical expertise and depth of interest. When it is fuzzy, it leaves room for errors in engineering. Where it is clear, it can also be maintained and remain solid into the future. With architectural clarity comes the ability to have great discussions and rational thinking based on the current state and future to-be states of any IT solution. Creating the **architecture round table** is not just an Arthurian {https://packt-debp.link/U6dmtf} dream. It can be a reality for you if you choose.

The key data engineering role has to be established, and it must have a seat at the table when architecture choices and, especially, capabilities are being evaluated and then turned into IT blueprints. Critical dataflow blueprints will appear in your conceptual architecture, and these will define the flow of datasets through the zones that need to be established. Business and other non-technical stakeholders can grasp the intent of the architecture's solution, and the engineers can build it to be reliable, available, scalable, and performant. In this chapter, you will learn how to generate a conceptual architecture for your data solutions and observe how it may be used to effectively communicate.

Conceptual architecture best practices

At this point in your reading, you will want to know: What's an example of a conceptual architecture, and can I see a picture? This picture is worth a thousand words, but note that the thousand words are still mandatory. In *Figure 5.2*, you can find the *Conceptual Architecture Reference Model*.

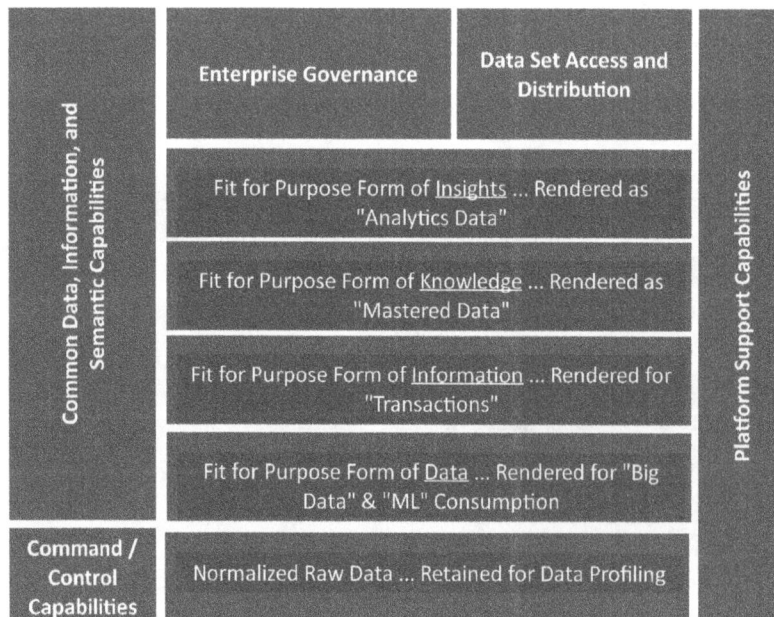

Figure 5.2 - Conceptual architecture reference model

We will use this reference model throughout the book as we flesh out the best practices. It leaves a lot of room for discussion, and it brings out many points that we will make clear in this chapter as well as the chapters that follow. The clear logical tie into the conceptual model has to be evident. If not, you will hear that critique from the domain stakeholders. Our advice is to listen to them and adjust, except where the principles, vision, and mission are violated. Likewise, for the logical architecture, **why** and **how**, types of questions will be covered in the physical architecture level as that also has to be tied into the conceptual level.

Conceptual architecture description

Before we dive into the best practices, you will need to know the following:

- What goes into conceptual architecture and its reference model?
- How are choices made to future-proof the architecture?
- Lastly, you need to understand some key trends: **data gravity** and **smart data**.

There are many references to read concerning the first question: **What goes into the conceptual architecture and its reference model?** The answers may cause you confusion! You will not want to get into dataflows at this conceptual level of the architecture. You will want to define all high-level capabilities. Beware of defining all of them as I have in the past: the list can be hundreds of lines long, so roll them up into categories of capabilities if the list gets large. Identify capabilities that are shared or function as utilities for other capabilities. Examples are monitoring, security, PaaS cloud capabilities, cloud **Information Technology Infrastructure Library** (**ITIL**) process support (even if that term is considered by cloud providers to be an anti-pattern for them – learn from it), and custom dashboards.

These shared capabilities often appear as columns to the left or right of your one-page reference model. Up the center from bottom to top (or top to bottom if you think that way), you place the capabilities in some categorized order that illustrates the concept the solution architecture is trying to convey. In our case, we group them into zones. This is what you should take away when the logical layer is examined since it has to show how the concepts are implemented. The logical architecture will illustrate the dataflow between services built to support the high-level capabilities. At the conceptual level, there is *no* dataflow defined yet; but it does *imply* a degree of functional flow and maybe implicitly stated data movement. One should also depict any grouped interactions of capabilities (in the Raw Zone of our **conceptual architecture reference model** you can observe the types of raw data that are being normalized; for example: email, FTP, or unstructured data), but you will want to stay at the capability definition level at this conceptual level. Just show how the capabilities roughly interact with the core of the architecture. Avoid all arrows and connectors in the conceptual architecture definition. If you need them, use a light pointer when presenting the diagram. If you specify the interaction in overlays, you'll get more questions than can be answered by the one-page diagram or the definition sheet (glossary) of each of the terms used to define the capabilities.

The conceptual architecture involves a number of areas that we usually like to break down into clear parts. These are the 5 areas:

1. Core platform
2. Semantic information

3.   Governance and data access

4.   Platform support

5.   Command and control

The five sections will be illustrated and elaborated on as we go over all the capabilities in the following sections as we explain the architecture and the glossary of definitions needed to navigate the concepts.

Conceptual architecture glossary

In the tables of the sections to follow, you will find a glossary of terms used in the architecture. The area and description columns will be of great interest. The glossary is divided into divisions and subdivisions in order to enable you to focus on the capabilities and features that each area will require. They form major blocks in just about any conceptual architecture you will envision. *Figure 5.3* is a simple visualization of the major areas:

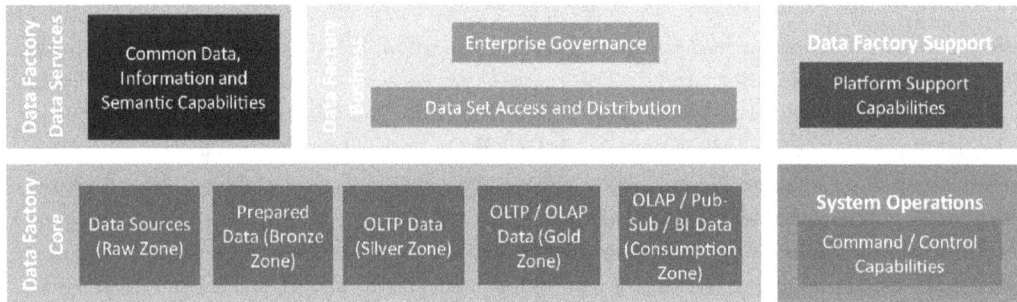

Figure 5.3 - Conceptual architecture block diagram

The glossary's major divisions and their subdivisions follow:

- Data factory core

  - OLAP/pub-sub/BI Data (consumption zone)

  - OLTP/OLAP data (gold zone)

  - OLTP data (silver zone)

  - Prepared data (bronze zone)

  - Data sources (raw zone)

- Data factory business

  - Enterprise governance

  - Dataset access and distribution

- Data factory support

  - Platform support capabilities

- Data factory data services

  - Common data, information, and semantic capabilities

- System operations

  - Command and control capabilities

Each of these will be elaborated upon in the following sections.

Data factory core

The core platform capabilities define the flow from raw data to consumable data through zones of activity. This is shown in the following diagram:

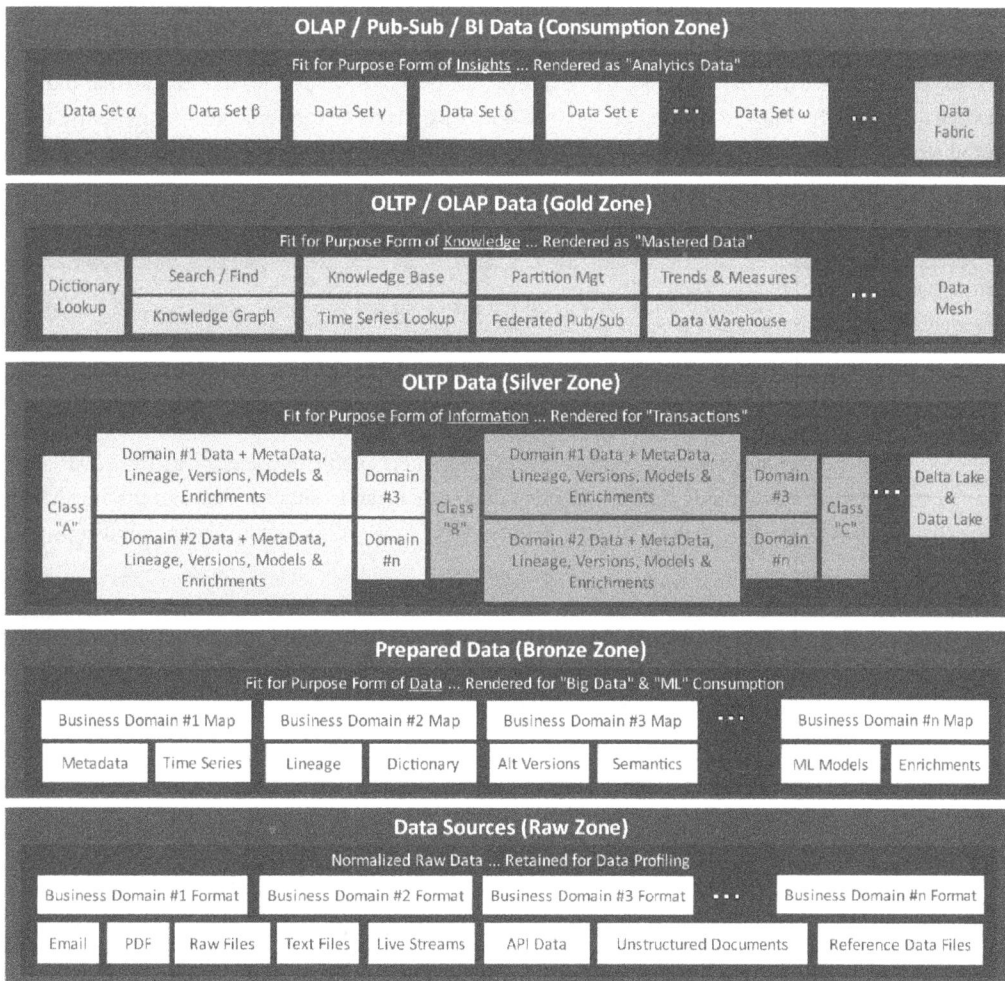

Figure 5.4 - Conceptual architecture core platform

The core of your data factory architecture should consist of the processing zones that implement the data journey from raw data to a consumable released set of productized data for user consumption.

> **Note**
>
> A user could be an algorithm, data scientist, analyst, or other value-added reseller. After all, data is going to be a product in the future!

OLAP/Pub-Sub/BI data (Consumption zone)

We'll begin with a discussion of the final data product. It may be a bit backward to discuss the data product's end state first but it makes sense to approach the problem this way. This is because what you will be producing for the consumer has to meet very high quality standards. This is possible with a data factory approach. Knowing where we are heading helps you grasp the reasons why we are being so formal in the design definition of the data factory's processing steps that we are suggesting you adopt. It is the overall quality of the data product that matters. The data consumer does not care all that much about what goes into the sausage-making factory but does care that they can understand how to use the final product unambiguously. As an engineer, you care very much about the preceding steps. So, you will have to be very methodical in the design approach taken:

Area	Description
Fit-for-purpose form of insights, rendered as analytics data	Published and subscribed data is rendered for analytic processing, leading to insight harvesting.
Dataset α	One of a number of datasets (Greek alphabet used to contrast with class datasets containing business domain data) organized to support data as a product life cycle: version controls, releases, time series, dimensions, and relationships.
Dataset β	
Dataset γ	
Dataset δ	
Dataset ϵ	
Dataset ω	
Data fabric	Data mesh data that is placed into cloud provider-specific fabric services. Unification of security is key. Refer to Azure Fabric Warehouse and SQL endpoints that now support column-level and row-level security {https://packt-debp.link/iZ7oV5}.

Table 5.1 - OLAP/Pub-Sub/BI (Consumption zone) data areas

Datasets consisting of clearly defined data classes comprise release sets of the consumption zone. Cloud providers are supporting data fabrics that are advertised to implement a superset of data mesh capabilities. It will be your designs that leverage these capabilities into features for your cataloged data classes that the consumer will be able to access.

OLTP/OLAP data (Gold zone)

Datasets that are hosted in the gold zone constitute the enterprise's mastered data:

Area	Description
Fit-for-purpose form of knowledge, rendered as mastered data	Business domain data and metadata forming enterprise master data fit to be re-rendered for other purposes.
Data mesh	Data conforming to the data mesh principles, particularly supporting business domain ownership, and self-service goals leading to data as a product through semantics.
Data warehouse	Data is physically organized to perform and support relational normalization, particularly supporting current state tool capabilities and services.
Dictionary lookup	Human- and machine-readable data lookup support using lineage and semantic metadata to discover meaning.
Federated publish/subscribe (pub/sub)	Support for change data capture (CDC) subscription to business domain data and downstream re-renderings as changes are made by a data domain owner.
Knowledge base	Enables data to be semantically organized with instances of class data appearing in a graph.
Knowledge graph	The abstract relationships of domain data class within and between any other data in a graph model lead to complete clarity in defining domain semantics.
Partition management	The practical ability to manage data partitions for scale, performance, volume, recovery, and auditing use cases.
Search/find	Enhanced discovery services over any dataset and its metadata, including semantic search, inferencing, and reasoning.
Time series lookup	The capability to support time series data as first-party data and not versioned data. With this, you have the ability to record trends, measures, stats, metrics, and derived data.
Trends and measures	Statistics of versions and time series data, and the abstract form of measures, metrics, and trend calculations that will be applied across the time series data.

Table 5.2 - OLTP / OLAP (Gold zone) data areas

Mastered data is still not ready for end-user consumption, but it is available to the data factory's pipeline processing steps. This mastered data can be aligned with other datasets in order to produce consumable data.

OLTP Data data (Silver zone)

Data classes in the silver zone are being assigned, aligned, classified, categorized, and enriched with data semantics. This way, data class meaning is clear and understandable to all. Mastering this data comes next, but making sure the data is smart first is essential:

Area	Description
Fit-for-purpose form of information, rendered for transactions	Domain data pivoted into classes that lead to correct semantic organization.
Class A domain #1 data + metadata, lineage, versions, models, and enrichments	
Class A domain #2 data + metadata, lineage, versions, models, and enrichments	
Class A domain #3 data + metadata, lineage, versions, models, and enrichments	Classes of data are formed from raw data. Current technology supports the creation of graph data that is highly relational. Preserving relationships facilitates join speed, which would be an impediment to performance in large relational database systems.
Class A domain #n data + metadata, lineage, versions, models, and enrichments	
Class B domain #1 data + metadata, lineage, versions, models, and enrichments	Abstract classes support the generic relationship, and instances of the class support the data itself. The metadata is defined by the class model that is formalized in the ontology and walks are defined by taxonomies.
Class B domain #2 data + metadata, lineage, versions, models, and enrichments	*Special note*: Implementation choices are LPG versus RDF/OWL in special in-memory graph databases with class linkage
Class B domain #3 data + metadata, lineage, versions, models, and enrichments	to traditional dataset storage.
Class B domain #n data + metadata, lineage, versions, models, and enrichments	
Class C, and so on	
Delta lake and data lake	Capabilities are big data storage, distributed map/reduce processing for job segmentation, and a high degree of caching (also known as delta lake) supporting lakehouse concepts.

Table 5.3 – OLAP (Silver zone) data areas

Silver zone data classes fit in the data lake and lakehouse and can be consumed by the data factory's processing to form mastered data. The danger is that giving the data scientist direct access to this data can appear to be beneficial, but since it is subject to data drift and errors, it can lead to the propagation of errors into the insights that the consumer wants to obtain. This is why subsequent processing steps are required, then hosting in the gold and consumable zones.

Prepared Data data (Bronze zone)

Bronze data is characterized as normalized data that has been ingested and made complete, correct, and aligned for syntactic and rough semantic correctness:

Area	Description
Fit-for-purpose form of data, rendered for big data and ML consumption	Domain data is mapped for optimal access patterns in the lakehouse.
Alternate versions	Alternative versions of data are renderable from a core dataset and kept in sync within the zone for performance considerations.
Business domain #1 map	Normalized datasets are mapped, and the data and its maps are made available for use cases run in the zone supporting enrichment, metadata extraction, and business domain validation.
Business domain #2 map	
Business domain #3 map	
Business domain #n map	
Dictionary	The raw and normalized datasets are categorized, cataloged, and placed under dictionary process controls.
Enrichments	Raw data is enriched to support the clean normalized business domain data and subsequently enforceable contracts for downstream use.
Lineage	Data lineage metadata is first formulated and stored setting up the conceptual wrapper for all subsequent metadata to be retained.
Metadata	Semantic metadata is harvested from raw and normalized datasets, then categorized in preparation for data category-defined pipeline enrichment leading to a dictionary entry.
ML models	Developed models and parameterized instances to those modeled are retained with change lineage so that data lineage can identify the exact model and parameters that may have produced downstream effects (because they often have effects that must result in new models being created).
Semantics	Teasing out semantic implications from any raw dataset is needed so that what is normalized is correct, complete, and specific.
Time series	Data that is semantically the same and 100% aligned with other instances of data with the same format will be a type of data requiring time series storage – as such, that will be flagged for special pipeline treatment. Unlike versioned data, time series data will have multiple versions that form the current dataset.

Table 5.4 – OLAP Prepared (Bronze zone) data areas

Bronze zone data is data that is correct and appropriately tagged before undergoing more scrutiny in the silver zone where semantics is assigned. Data lineage is important in this bronze zone's processing because some raw values are distilled out and others magnified due to the nature of the data cleansing operations that take place. Traceability to the data source's raw zone is established in bronze zone processing. The focus will be on data lineage, enrichment, and normalization of the raw zone's data.

Data sources (raw zone)

In the raw zone, data is first ingested and retained in its original form. It is essential to preserve raw data and the form in which it is received in order to provide data lineage traceability. That data needs to be viewed from the perspective of the downstream zone's processing, which is subject to change over time. Raw data processing software changes can then be applied after committing those changes without having to create compensating transactions. Rerunning processing software from the last known good data state helps a lot since change happens, and a lot of software change takes place with raw data processing:

Area	Description
Normalized raw data, retained for data profiling	Domain data is stored in the format is it received.
API data	Data can be acquired via a gateway API from any source.
Business domain #1 format	Business data can be of any type, but the value is subject to processing and normalization. Processing in this zone is limited to making sure it is ingested syntactically correctly, stored, and auditable. Service levels of the business domain owner will be applied outward to the supplier and enforced in the processing rules of this zone.
Business domain #2 format	
Business domain #3 format	
Business domain #n format	
Email	Support for email data arrival, with audit, and header retention as metadata.
Live streams	Supports live stream of data arriving as pub/sub streams. Particularly supported are segmentations of streams into datasets for processing and retention.
PDF	Support for unstructured text within a structured container, PDF being one of many types of structured containers: HTML, PDF, SGML, XML, and so on.
Raw files	Support for raw delimited and undelimited files of various template formats and parsable forms.
Reference data files	Data file reference data – often indexed to some other file for pseudo-relational flat file transmission to another system. Included in this are **relational database management systems (RDBMSs)** or SQL data dump files.

Area	Description
Text files	Structured or delimited test files: Unicode, ANSI, or other encodings fully specified but subject to downstream normalization.
Unstructured documents	Fully unstructured free-form text in some human-readable language but subject to downstream normalization into a core language before translation and localization changes are applied at the final domain owner's discretion.

Table 5.5 – Data Sources (Raw zone) data areas

Data in the raw zone sometimes produces alternative versions of the original raw data that are still considered raw data. An example is the creation of a graph of concepts contained within an unstructured text article. Both forms, unstructured and structured, are semantically equal but the derived structure form is semantically equivalent to the unstructured form. Building the raw zone with this in mind is important to the design. Likewise, time series data is a special case. Here, the same data for different times is being collected and annotated. The cardinality of the time series instance data can be vast for the dataset, each with the same semantics!

Data factory business

Businesses require data to be governed, and that is a practical reality. **Master data management** (**MDM**) system vendors have built entire ecosystems to govern data in large enterprises. This is required because datasets, and sometimes even columns in a dataset, are owned by different siloed groups. Most company cultures still suffer from this issue, and it's not going to change quickly. The ownership has to remain with the domain owner, but the attribution of that has to be retained in the dataset's metadata and enforced with enterprise data governance.

Enterprise governance

The solution you design will need to have governance and access controls. Under the banner of governance, you will need to tie the enterprise to the solutions architecture for the following:

- Mission

- Vision

- Strategy

- Policies

- Assortments

- Competition

- Markets

- Clients

The access and security of the system needs to have these capabilities (see *Figure 5.5*):

- API gateway

- Confidential compute

- Data fabric or data mesh exploration

- Delta data exploration

- Bulk FTP services

- Rights and fair use management

- Third-party data aggregator (integration

Enterprise Governance				Data Set Access and Distribution			
Mission	Vision on themes	Goals & Strategy	Policies	3rd Party Aggregator	API Gateway	Confidential Compute	Data Fabrc Expoloration
Markets	Client (profile)	Assortment	Competition	Data Set Bulk FTP	Rights & Fair Use Mgt	Data Mesh Exploration	Delta / Data Exploration

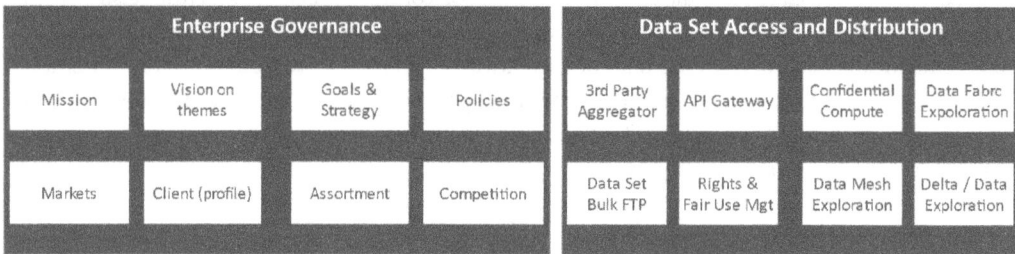

Figure 5.5 - Conceptual architecture governance and data access

Many assigned data attributes drive governance rules, and processing steps are defined by the domain owners. Workflows, issues triage, and resolution all are systems built to support the governance of data at any zone processing stage. It is important to know that data rights are assigned and enforced, security is assigned and audited, and change workflows are implemented and applied for data, metadata, derived data, and data attributes (security, privacy, and so on) set. Who has permission to change any of these is part of the enterprise data governance capability. It is a key part of your data engineering design:

Area	Description
Assortment	The process to treat data as a product where datasets are subject to a variety of product bundlings as offered to customers. A range of data products are held in data factory inventory in order to meet customer needs and preferences. Specific bundles are supported, and with that, pricing, accounting, auditing, and cost attribution (within the factory) are all part of the data factory's assorting processes.
Client (profile)	Client profile data is retained for marketing, sales and usage cost, opportunity mining, and auditability.
Competition	Competitive data is assessed for efficacy measures as data is compared and risks identified for the domain owner to maintain competitive pressure.
Goals and strategy	Strategic goal metrics, objectives, and observations are retained to provide dashboard input to the business domain owners and enterprise leaders regarding the strategic efficacy of the conceptual architecture's goals.

Area	Description
Markets	Market support will involve support for sales within the markets the domain data owner is operating in. In a data-driven organization, Excel spreadsheets and out-of-band (OOB) business models are replaced with processes that enforce accountability to objectives during development and business as usual (BAU)operations.
Mission	Dashboard compliance with the mission. How is the organization doing overall with drill down to deviations, including tech debt, business debt, and mission debt that will be addressed in the future?
Vision on themes	Does the company see things the same way? Does data in the systems support the vision when the data factory was launched? Are we driving toward the vision's implementation?

Table 5.6 - Enterprise governance areas

Beyond enterprise governance, you will want to set up access and processes and controls for data distribution. We'll discuss this in the next section.

Dataset access and distribution

Released datasets need channels for distribution and data service. There are various mechanisms for data consumption. Each requires security, privacy, and rights management features to be understood and then preserved with the correct granularity at the dataset level:

Area	Description
Third-party aggregator	Sometimes the best way to distribute data is to allow a third party to handle the complex issues of fair use, contracts, scale, performance, entitlement, and confidential computing. Data collaboration platforms such as LiveRamp come to mind, as well as their competitors: Terminus ABM Platform, HubSpot Marketing Hub, ActiveCampaign, Demandbase One, CallRail, AdRoll, and Looker.
API gateway	A custom API gateway for data distribution similar to Xignite for finance. 21 companies have contributed their investment banking APIs (mostly). APIs allow a company to circumvent traditional **financial service providers** (**FSPs**). Here are some companies identified as contributors: 1) *Workflow APIs*: Advisor Software, ChartIQ, EdgeLab, Planwise, QuantConnect, Tradier 2) *Analytics APIs*: Autochartist, TipRanks, Insight 360, PsychSignal 3) *Data APIs*: Estimize, Estimates APIs, Nasdaq, Streamdata.io, StockTwits, SRLabs, Yodlee, Xignite 4) *Incubators*: Draper University, ValueStream, Fintech Sandbox, and Level39

Area	Description
Confidential compute	Ability to bring algorithms to a shared data area, preserving the trust of all parties since access to data is prohibited except the algorithm and its output that must treat the data in the box as black-box data since it cannot be used except through the algorithm in a trust compute area.
Data fabric exploration	Being able to launch data into a cloud, or SaaS, provider's shared area and to have it treated as first-party data in a shared or limited access data fabric.
Data mesh exploration	Being able to launch data into a provider's shared data mesh and have it treated as first party data.
Dataset access and distribution	Any customized third party raw dataset transfer system.
Dataset bulk FTP	Any customized third party raw dataset transfer system that flattens deliveries and data product releases into sets/brackets that are subject to life cycle and auditing and scheduled releases.
Delta/data exploration	Being able to launch data into a provider's shared delta lake infrastructure and have it treated as first-party data.
Rights and fair use	Preservation of dataset brackets for release management purposes. Also, the ability to track data usage at a granular level while preserving entitlements including rights for access, to know a fact exists, or for redistribution.

Table 5.7 – Dataset access and distribution areas

A particularly thorny and often forgotten capability for dataset distribution is the preservation of digital rights. Without this, data gets locked into silos. Implementing the correct rights management policies and enforcing them with each dataset within the factory and external to the factory is going to be a challenge. Your solution will be made possible with your cloud provider's service support. Some providers support policy-driven data controls better than others, so there are design variations that you need to accommodate based on your cloud of choice.

Data factory support

The platform needs to be supported with a number of services, such as the following (see *Figure 5.6*):

- Cloud service (**infrastructure as code (IaC)**) management
- Configuration management
- PaaS monitoring
- DataOps (for data mesh and/or data fabric)

- Data privacy compliance

- InfoSec (SecOps)

- Credential and key management

- Release management

- Third party tool SLA management

- Authorization, encryption, and SecOps auditing

- Usage metering

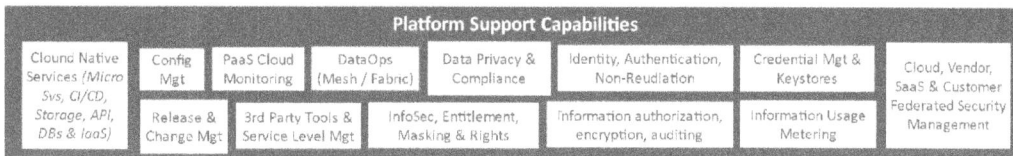

Figure 5.6 - Conceptual architecture platform support

Data factory support capabilities from the cloud provider will help you design solutions aligned with best practices that are also embraced by the provider. Some have a very different vision than you will want to adopt, and none share a common vision. You can expect to have to make many accommodations.

Platform support capabilities

You will want your cloud platform to support key areas like **service level management** (**SLM**), federated security, cloud storage, configuration management, release management, security key management, data privacy, and various monitoring capabilities:

Area	Description
Third-party tools and SLM	Capability to manage service levels and report on deviations before they are incurred and become violations.
Cloud, vendor, SaaS, and customer-federated security management	Centralized security command and control to address key **Open Worldwide Application Security Project** (**OWASP**) requirements. The top 10 are: 1) Broken access control; 2) Cryptographic failures; 3) Injection; 4) Insecure design; 5) Security misconfiguration; 6) Vulnerable and outdated components; 7) Identification and authentication failures; 8) Software and data integrity failures; 9) Security logging and monitoring failures; 10) Server-side request forgery.

Area	Description
Cloud-native services (microservices, CI/CD, storage, API, DBs, and IaaS)	Microservices frameworks for Kubernetes with orchestration: Amazon Elastic Container Service (ECS); AWS Elastic Kubernetes Service (EKS); AWS Fargate; Azure Kubernetes Service (AKS); Azure Container Instances; Azure managed Openshift service; Cloud Foundry; Digital Ocean Kubernetes service; Docker Swarm; Google Cloud Run; Google Container Engine (GKE); HashiCorp Nomad; Linode Kubernetes Engine; Mesos; Mirantis Kubernetes Engine (formerly Docker Enterprise); OpenShift; Portainer; Rancher; Rafay; and Red Hat OpenShift Online. CI/CD frameworks: AWS CI/CD (AWS CodeCommit + AWS CodeBuild + AWS CodeDeploy + AWS CodePipeline); Azure (Azure DevOps + Azure Pipelines + [a selected source control system] + Azure Key Vault + Databricks); Bamboo; Buddy; Buildbot; CircleCI; Codeship; GitLab; GoCD; Google (CloudBuild + Artifact Registry + Cloud Source Repositories); Jenkins; Nevercode; Semaphore; Spinnaker; TeamCity; Travis CI; and Wercker.
Configuration management	The capability for the management of configurations is essential to tie the physical architecture to the logical and conceptual architecture. It is useful in determining the state of the cloud infrastructure at any time, especially when forensically assessing issues that arise in data and code. Lastly, it should define what can be, should be, and should not be the current and future state of the architecture as the system operates in primary, backup, or **high availability** (**HA**) modes (especially during transitions between the modes).
Credential management and keystores	Vaults are necessary to abstract keys from code and to enable them for use in an obfuscated manner to preserve the integrity of the system.
Data privacy and compliance	Privacy, as with security and entitlement, also needs to be implemented and **personally identifiable information** (**PII**) clearly tracked through the data pipeline as metadata and policy enforced before rights are given to a consumer. Add also to this the dashboarding required to support auditing internally and externally applied.
DataOps (mesh/fabric)	Choose capabilities from some of the top DataOps tools: Composable.ai; K2View; RightData; and Tengu. DataOps capabilities: 1) Agilely orchestrating data pipelines for ML and analytics; 2) Easily connecting key datasets to consumers; 3) Simplified pipeline processes for continuous insight generation; and 4) Data source: access, **internet of things** (**IoT**) streams, data automation (pub/sub), big data/lakehouse/data warehouse access, BI, ML.
Identity, authentication, non-repudiation	In a federated manner, support identification, authorization, authentication, access, non-repudiation, and entitlements.

Area	Description
Information authorization, encryption, auditing	Implements information authorization, delivery encryption, and auditing.
Information usage metering	Capability to meter usage and provide usage tracking dashboards.
InfoSec, entitlement, masking, and rights	Implements information security, internal entitlements, publisher rights tracking, and user masking (if needed to implement security at the edge).
PaaS cloud monitoring	Cloud provider's PaaS monitoring is integrated into the business operations dashboard.
Release and change management	Data release and change set management are similar to that applied to the software that operates on the datasets (they are *not* the same).

Table 5.8 - Platform support capabilities areas

What you will find is that your cloud provider is more than willing to help you build your data engineering solution. Know that the help has to be vetted against your vision and mission and not aligned exclusively with their market-driven goals. Your future-proof solution should withstand the changes that the market forces drive the cloud provider to adapt to.

Data factory data services

The creation of data factory services enables your vision to be implemented as part of the mission. Metadata lineage, data quality, data-at-rest metadata, dictionary, and catalogs that you envision will allow the consumer to understand data as a product and leverage it to its fullest extent.

Common data, information, and semantic capabilities

The information semantics capabilities are delivered and then retained across a number of key repositories (see *Figure 5.7*):

- Metadata repository
- Big data repository
- Dictionary
- Workflows
- Job flows
- Systems quality (tests) repository
- Data quality (tests) repository

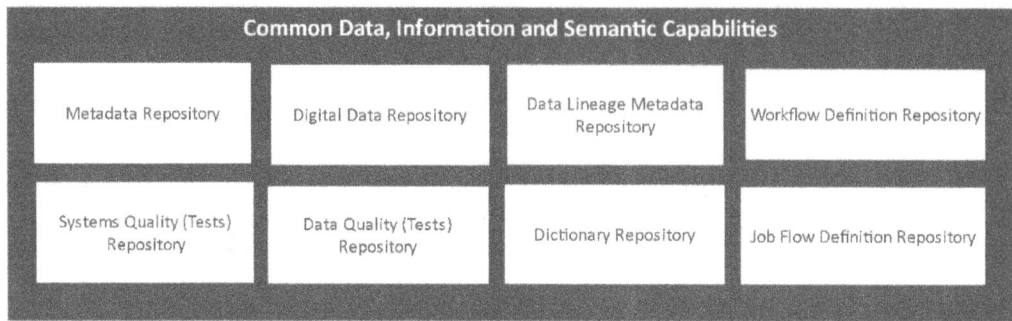

Figure 5.7 - Conceptual architecture semantic information

You will see that there is a great focus on semantic metadata and its discovery on the part of the data consumer. Knowing what is available, its quality, its coverage, its lineage, and its domain semantics are essential to provide trust in the insight the consumer will render based on the data you expose:

Area	Description
Data lineage metadata repository	Data lineage metadata needs to track changes to data as the pipeline executes. First to be tracked is the pipeline definition in place at the time data was processed. Then change data capture (CDC) effects at each stage of a tracked dataset as it undergoes assembly in the data factory have meaning. Lastly, being able to collapse parallel executed streams of metadata once a release set is flagged, so as not to add lots of dissociated lineages to the repository. Data changes (dataset and its metadata) must be retained to make it discoverable. You have to hold one or more data lineages and their output active and in parallel (multiple swim lanes will be tracked) until a data release set can be committed to downstream processing.
Data quality (tests) repository	Data tests and test results are data lineage metadata that needs to be tracked with primary data as data is curated in the pipeline.
Dictionary repository	Dictionary services are required to enable the discovery of data, metadata, release sets, datasets, and data lineage in swim lanes.
Metadata repository	The repository of choice varies. OpenMetadata is a good start to evaluate capabilities, and further options include Alation Data Catalog; Alex Data Marketplace; Alteryx; ASG Enterprise Data Intelligence; Azure Data Catalog and Purview; Collibra Platform; erwin EDGE Portfolio; IBM InfoSphere Information Server; Immuta; Infogix Data360 Govern; Informatica Metadata Manager; Lumada Data Integration and Analytics; MANTA Platform; Octopai Platform; OpenMetadata; Oracle Enterprise Metadata Management; OvalEdge; PoolParty; SAP PowerDesigner; Smartlogic Semaphore; and TopBraid EDG-Metadata Management.

Table 5.9 - Common data, information, and semantic capabilities areas

No data or metadata is usable unless the system producing it is trustworthy. The data operations part of the solution has to be solid.

System operations

Lastly, command and control capabilities give you a view of your operational system with these capabilities (see *Figure 5.8*):

- DevOps monitoring

- Insight mining (from logs)

- SLM (dashboard)

- Security dashboard

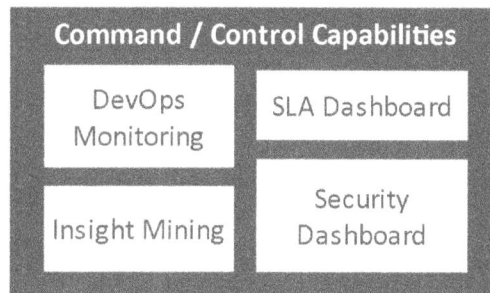

Figure 5.8 - Conceptual architecture command and control

Systems operations are going to be addressed in later chapters where DataOps is a focus. There are particular conceptual requirements that are key to DataOps. These are job flows (scalable technology-driven processing steps) versus workflows (human-in-the-loop processes), system quality, and final data repository (for audit and compliance).

Command/control capabilities

Along with the required capabilities comes the need to implement command and control for your solution. You only fly when the preflight checklist is executed by the pilot, and you should not service customers without operational discipline. This comes with the dashboards you have designed so that dataflows optimally through the factory:

Area	Description
Job flow definition repository	Job flow definitions specify the processing steps affecting constituting the data pipeline. This is needed to break up big processing into executable parallel running chunks (threads, distributed jobs, SaaS API calls, and so on). These need to be tracked in a dashboard and associated with the SLM of the data pipeline step they were initiated from.

Area	Description
Systems quality (tests) repository	Implement DevOps testing capabilities from these tools: Ansible (task automation); Appium; Bamboo; CruiseControl; Docker (build, ship, and run distributed applications); Functionize; GitLab (test repo); HashiCorp's Terraform (infrastructure management); HashiCorp's Vagrant; Jenkins; JMeter; Kobiton; Nagios (monitoring); PagerDuty (**incident management (IM**); Puppet (configuration management); SaltStack (remote execution); Selenium; Snort; SoapUI; and Stackify Retrace. Also, implement systemic testing from one or more of the framework types that follow: 1) Linear automation (record-and-playback driven) 2) Modular-based testing (application segmented into isolated testable units, functions, or sections and driven results an abstraction layer) 3) Library architecture testing (similar to modular based testing except testing is grouped by a common objective) 4) Data-driven (tests running from scripted orchestration for the same code using different datasets) 5) Keyword-driven (feature-driven testing orchestrated from a labeled keyword in a table aligned with a sequence in a use case) 6) Hybrid testing (any combination of the other frameworks for your organization).
Workflow definition repository	Pipelines should themselves be data-driven and not hardcoded into the configurations of the CI/CD pipeline or homegrown code. Abstract workflows are themselves needed for metadata data lineage capture, and they are important to be defined and captured.
Digital data repository	A data catalog must have an inventory of products in final form for shipping as well as a backlog for items in pipeline progress before becoming ready for shipping. Once a set of releases is ready, the shipment is made along with a neat packing list for the customer. So, a customer gets a group of release sets, and each release set contains datasets, all neatly bundled into a shipment.

Table 5.10 - Command/control capabilities areas

At this point, you have read the area descriptions for your architecture and are ready to assess the key issues and formulate the conceptual architecture that you will communicate to all as design and development proceed.

What are the data architecture's key issues identified in the conceptual architecture?

There will be some clear messages the conceptual architecture will convey. At least the stakeholders will have a place to put their finger on and ask: *What's that?* Expect questions to arise such as: *How are choices made to future-proof the architecture?* These have to be backed by due diligence, experience, external reference, and **subject matter expert** (**SME**) confirmation. Once the concept is clear, expect the challenge. Be ready to anticipate the need to justify everything … after all, you are communicating foundational thinking and want all to see the solution as you see it and before they do … they will see a solution from their own perspective. So, let's look at a few key concepts that this architecture requires: the need for data to stand alone and be *smart data* {https://packt-debp.link/WbVvOt} and a practical feature of big data – it has **data gravity**.

In order to generate data as a product, you will want to know where data clusters are, when they form, and what meaning can be gleaned from that clustering. This means the data has to be smart. The days of building IT solutions with a build-it-and-they-will-come mentality are far gone. We build to where there is value! And a data product that has value will encompass the boundary of a cluster of information, which better be smart enough to have value added rapidly. If data is to be built upon with value-added enrichment in a rapid cycle, then this acceleration leads to the cluster being considered to have data gravity. We'll begin by making you aware of data gravity.

What is data gravity?

What is data gravity? Data attracts data! It therefore has *gravity*! When datasets grow, they attract various customers, their applications, and services. Thus, an ecosystem develops around the center of that dataset that is subject to value enrichment and even more data gravity. You may ask, when did this start to occur? The answer is when IoT and other data sources started pouring data into the cloud and on-prem solutions because storage was cheap. Data volumes ballooned and not always equally in each domain. Digital Realty {https://packt-debp.link/OUxWbh} (a global provider of multi-tenant data center capacity) measures this industry occurrence in the **Data Gravity Index** (**DGx**) report. They maintain a unique perspective on technology infrastructure and how it should be built, deployed, and operated. The report is their effort to quantify and predict data gravity globally. The report's goal is to help in the infrastructure decision-making process leading to effective solution operations. In the report, data is refined and correlated with a country's **gross domestic product** (**GDP**) growth. The DGx results indicate the degree of global economic centricity around data. The report was first issued in 2020 when it helped customers shift strategy to address hybrid IT challenges (during the global pandemic), which highlighted this data gravity reality. The DGx continues to provide excellent insights and the important need to provide secure data exchange services about data as the global economy begins to shift from digital toward a data-powered environment. Data gravity drives a lot of the need for the creation of zones within your future-proof data solution. Even within the enterprise, I can assure you that data gravity and technical tool, cloud, and cost limitations will force data to cluster into zones that will exhibit the data gravity effect. If implemented well these zones can be extended, and patterns developed for internal service, parallelization, transformation, publication, movement, and subscription (with security) will be extensible to the external consumer. These are the trends that are notably influencing data gravity.

Digitally enabled interactions are increasing the digitization of enterprise workflows augmented by data and **artificial intelligence** (**AI**). 70% of the new value created over the next 10 years will be based on digitally enabled platform models (from *World Economic Forum, Shaping the Future of Digital Economy and New Value Creation, June 2022*). This increases enterprise data exchange volumes globally as explained in the following points.

- Data localization is expanding legal and regulatory policy requiring local data storage. 78% of IT leaders will maintain local copies of customer and transaction data for compliance (from *Digital Realty, "Global Data Insights Survey", April 2022*). This increases the number of enterprise locations for data aggregation.

- Mergers and acquisitions (M&A) are driving globalization, which in turn is driving corporate M&A to achieve scale. 63% of organizations are preparing for an M&A deal in 2023 (from *Deloitte, Could M&A Activity be a Springboard for Controllership Transformation?, December 2022*). This increases the number of data sources participating in data exchange.

- Cyber/physical security integrations of physical and digital systems to prevent intrusion of IoT systems are forcing data to be clustered. $12.6 trillion is spent on IoT, which could enable up to $12.6 trillion in value globally by 2030 (from *McKinsey, "The Internet of Things: Catching up to an accelerating opportunity", November 2021*). This increases data creation and exchange types and volumes.

- *Enterprise data stewardship* is increasing in the enterprise and fast becoming the world's data steward. By 2025, 80% of data worldwide will reside in enterprises (from *IDC #US44413318, Data Age 2025, "The Digitization of the World from Edge to Core", November 2018*). This increases the volume of data that needs to be aggregated and then retained.

In April 2022, the *"Digital Realty Global Data Insights Survey"* garnered responses from 7,295 C-level executives, business, and technology leaders, representing large multinational enterprises across 23 countries and 9 industries, with revenues ranging from $100 million to more than $1 billion. In the survey, the following was observed:

- A data-first strategy will prevail. 75% plan to use data to improve customer experience and build new digital products.

- Data distribution is going to increase over time. 72% plan to add new business locations in the next 2 years.

- Data is localized to be closer to where it is geographically used and in the language of the user. 78% maintain local copies of customer data for business and compliance purposes.

- Data latency matters for consumer use cases that depend on vast amounts of locally accessible data. 77% identified data latency-specific performance requirements.

Multinational companies are best positioned to address the unrealized $100 trillion value opportunities that rapid digitalization adoption has made possible. Regarding data gravity, data classes and their volumes are growing with a shift in how data is created, processed, stored, and exchanged. This fact is compounded by the need to locally store data (aka localization) driven by regulation and sovereignty needs, which add to the variety of data gravity challenges for large multinational companies. With complex enterprise systems and many millions of consumers or application endpoints, these companies are feeling data gravity pressures. When a data-first strategy is enacted by multinationals, they profit from a shift to data-driven workflows that address data gravity where it is observed to be the highest. This occurs in major population centers such as New York with 33.6 EB (where **EB = exabytes**), London with 13.9 EB, and Tokyo with 23.6 EB of data. Capturing business opportunities where data is located accelerates efforts to place data where data gravity is evident. This is an important business goal {https://packt-debp.link/axLdWO}. It's interesting to note the aggregate company profile (in 2022) for all known large multinational companies is as follows {https://packt-debp.link/GqT0C7}:

- 13+ countries with a business presence (this is an average per enterprise)
- 19K **business units** (**BUs**) in 53 metros
- 36+ **points of presence** (**PoPs**)
- 7K+ data center PoPs
- 100M+ employees
- 11M+ applications
- 57K+ SaaS applications
- $3T+ Annual IT and network spend
- $18B+ Annual IaaS spend
- $8B+ Annual PaaS spend
- $40B+ Annual SaaS spend
- $7B+ Annual colocation spend

As you can observe, the scope of the $3T spend total and the enormity of the number of applications (11M+) in 7k+ data centers is a clear ringing bell to address observed data gravity issues in your platform. The design choices will enable parallel instances of processing or consumer-facing data zones to be created and not give up on the enterprise's mission, vision, or IT strategy goals.

Why is smart data a foundation?

Why is smart data a foundation for our best practices? You will first want to know how to best process highly connected data that is semantically defined in metadata and structured in a way to facilitate a high degree of connectivity (think about lots of relational joins if one were to implement a solution without a graph design, but I'll leave that to the logical and physical architecture discussion). This smart data is linked with the domain owner's data and snaps in figuratively upon arrival. There should not be software-defined implications, processing, or massaging required to have smart data aligned with other smart data. It is this ability to fit easily that makes the data smart. The graph structure can be leveraged to minimize the cost of joining tabular data, and the graph nature of smart data naturally fits with human thinking in that we naturally think about what goes with what, form the connection, and leverage that in our future thoughts as a known hypothesis and potentially a truth to be used in our logic. Likewise, the machine of the data factory should be able to grow knowledge in a way that is not hindered by human modeling and be able to absorb smart data even if the processing logic does not yet exist to leverage it. When the insight becomes clear, the logic can be added and that reinforces the correctness that the smart data was already brought in to reinforce. It's when the system has an ah-ha moment that the smart data way of organizing data will be very valuable. A backtesting hypothesis is now possible. You can see implications in your empirically collected data and derived information that inferencing reveals.

That all sounds like it would fit into our data factory vision nicely, but there will be many caveats and nuances and even more nay-sayers as one tries to integrate what is obviously needed into the conceptual architecture. The merger of data engineering with knowledge engineering comes with the data engineer upping the game on semantic technologies. These include graph database blueprints being formulated and the adoption of a stance on **labeled property graphs** (**LPGs**) compared to RDF/OWL structures for your knowledge base.

The knowledge base that will host smart data should be accessible from the bronze and silver zones of our architecture but retained in the gold zone. It will be slow and must be isolated for ease of use with interfaces created for mere mortals to use correctly. It should perform as well as possible because it is an in-memory technology solution. We have used the following:

- RDF/OWL graphs technology such as Ontotext GraphDB; PoolParty and Stardog (semantic graph engine)

- LPGs such as Neo4j, TigerGraph, JanusGraph, AWS Neptune (BlazeDB)

- Hybrid LPG and RDF/OWL graph engines such as Cambridge Semantics' Anzo and AWS Neptune

- … as well as others

At this point, you are probably rolling your eyes. You may have thought that a relational database vendor should have put all this into a data pipeline offering already and just sold it to me! It's all very *complex*! Well, the field is maturing and data and knowledge engineers (the new DataOps discipline) are not yet at the table when software, infrastructure, and test engineering (DevOps) is leading architecture discussions. The need to communicate the value of having data be smart, self-contained, and used to drive data pipeline processing steps in the data factory is just another type of inversion of control pattern that software engineers can grok. They just have to rethink (with you) how that can work and still preserve end-user quality expectations. This requires the code and configurations of the data fabric (also known as the cloud superset of data mesh design pattern) to be directly driven from the data's state and not from top-down programming or the current sprinkling of deployed scripts all over the cloud PaaS service offerings. It's more like cloud configuration management on steroids when conceptualizing how smart data can affect the correct step-by-step defined states of a future system. You will not want to compromise the concepts presented here, and the cloud providers will say that they are listening to the customer base, yet many are hearing through their legacy perspectives. They and the customer have to look beyond the past and even the current offerings and grasp the future to be bold enough to envision how data should be driving processing and not processing the data. State changes are to be event-driven data events (a new stream of true facts) driving rich new insights.

You will want to know how to best communicate the need for a data factory that implements the smart data vision. You will need to know what goes into the conceptual architecture.

Best practice composition of the conceptual architecture

What is the structure of a conceptual architecture? The answer is that whatever you need to convey the concepts necessary to build out the logical architecture has to be in the conceptual architecture.

The need to create a reference architecture as a diagram and explain the terms is fundamental to being able to discuss the concepts in written and presentation form. What is missing is the capabilities inventory. These capabilities combine the architecture concepts into business values that can become statements that the domain owners and business stakeholders clearly grasp. These are defined in the use cases that will be used to demonstrate the concepts of the solution. What will need some explanation to all audiences will be the threads of activity, maybe even through the creation of illustrative dataflows that underpin the use cases. Understanding the effects on each stakeholder's values will transfer ownership to the stakeholder in an emotional way as they begin to see the positive impact the solution has on their interests. What they did not tell you in school is that as an IT engineer (data or software), you are going to have to get into the heads of your audience to get them to see what you see … even when they potentially will not know what the heck you're talking about (if you only use your technology lingo). Translate *your* language to theirs, and your concepts are then communicable. So, back to the techno lingo; the translation I'll leave to you.

We also need to drill into the structure of the data classes in the data zones. Since zone data is fit for consumers' purposes, for data in each zone you will see a data class structure or the information and knowledge that will be different from that for big data, and likewise for analytics usage. The metadata about each class will be cataloged. Some of the metadata will be semantic and other metadata will be descriptive, some structural, and finally, some format (syntactic) based. The data class itself has interfaces that enable it to be drilled into for reflection purposes and for discovery. You will want to know that the data zones are retaining similar (if not the same) logical data yet have the same conceptual meaning. You will also want to develop CDC processing methods to keep logical copies in sync with their peer datasets.

Why is there a need for data classes and data processing zones in the conceptual architecture?

I reference the need for data classes specifically to address the smart data processing goals mentioned earlier. Without getting into the details of semantic knowledge engineering, I'll sum it up to say that data in modeled classes is instantiated for those classes to populate a graph data store, but for practical purposes, the data store can't be accessed without addressing scale and performance objectives of the overall solution. Conceptually, we will *not* want to architect the impractical but have in mind logical and physical reference implementations without defining them at the conceptual level. In the case of data classing, we want to think about the capability to support pre-query analysis of the consumers' data request and then map out how best to satisfy the request and iteratively execute second-level processing to get it all combined, and lastly to render it in a performant manner.

Performance and scale requirements do affect how the conceptual architecture is defined and its capabilities. There will be various classes to be processed and they serve different needs, even when classes of data are logically the same as other classes, and they may be hosted in different zones.

What data classes are to be curated within processing zones?

The need for the creation of processing zones was introduced in previous chapters. One size does not fit all, and that is evident when you realize that to support this goal, your data has to be replicated across many zones and then kept logically in sync as it is transformed between zones. Your solution has to make use of change data capture (CDC) capabilities, which means that if not created as a repeatable, operable design pattern feature, you will have many different implementations created by your developers. This will cause your solution to be unmanageable and costly to develop. The metadata capability supports the need for smart data to drive data pipeline step processing. Data zones are conceptual and logical boundaries for a class of data contracts that must be enforced so that use cases are not created with improper expectations of the data within the zone. The base data classes in data processing zones define what data can go into the zone and how datasets are moved between zones via bulk transfer (full release), and a CDC stream of **create, read, update and delete (CRUD)** transactions:

- The *bronze class* is used for data profiling, data normalization, and data cleansing use cases for big-data-lake processing.

- The *silver class* is used for data pipeline processing or for big data lake, OLTP, and some OLAP operations.

- The *gold class* is used for enterprise-grade mastered data and for OLTP and OLAP operations.

After data is adjusted, then kept in sync with downstream zones and cataloged as metadata and lineage, one can enable that data for application query processing in any zone it has transited.

It sounds like a complex query executor and optimized query processing capability is now needed. It's like you want to rip a legacy RDBMS apart and build it new in the cloud with the concepts of semantic metadata, data zones, and data classes supported. That is exactly what is being proposed! The prize goes to the cloud or third-party vendor that can make this easy and least costly to implement. This is why you need another feature of the conceptual architecture, and it's vendor due diligence!

Vendor due diligence is needed to reduce the total cost of capital needed to bring capabilities to market

Once the conceptual architecture capabilities are identified, you will have an issue when looking for vendors to cut the time and effort to develop the solution. The *buy versus build* discussion will take place during the logical architecture discussion, but for that to be effective and not contentious, the capabilities of these vendor products will have to be a superset of your solution's capabilities. Sometimes, they may be a subset requiring additional software integration (also known as glue) necessary to bring the solution to market.

It is essential that you only pay for what you use when bringing on third-party software of services and not be sold into a great deal with capabilities that will not integrate (I've been there and done that!). It sucks the funding out of the pot for other more important capabilities to be developed. Worse yet, the purchased tool may come with a conceptual architecture that conflicts with your solution's vision, mission, or strategy. Care must be taken to identify and communicate to the often-biased business stakeholders that their pet vendor will not be allowed to market to IT groups, nor will they be allowed to sell what will not and must not be used in the solution. All these issues come out in the vendor due diligence process, and this is why we have principles and capabilities derived from the architecture to hold up to all and answer the question: **Why?** The objective nature of vendor due diligence must be preserved. Adding scoring and weighting for each feature of a product supporting overall capabilities' needs is essential. Reporting the due diligence results objectively via a Kiviat chart {https://packt-debp.link/SpXf5O} multi-dimensional charting graphic helps. If you wish to render due diligence results, do so with transparency and objectively bring in the subject matter experts (SMEs) to support the findings. Make the case for a purchase, purchase with feature augmentation, or remove the vendor from consideration.

In the reference architecture, you see the glossary has many vendors listed. Use these to help form your vendor due diligence but begin with capabilities and the features of a product that support or detract from the capability desired.

What capabilities are necessary for the conceptual reference architecture?

The conceptual architecture will be represented in various ways as you need to communicate with stakeholders in the language that they feel most comfortable with. That is not just English, but rather pictures, diagrams, flows, and concept lists for many. One such list is the list of solution capabilities your design will deliver. The subheadings under this section are the capabilities you will want to closely consider for implementation in your data engineering designs.

Advanced analytics capability

Advanced analytics requires you to design software as a platform that boasts a suite of advanced and innovative algorithms to help users solve all types of intractable problems. The technology has solutions to aid in driving business impact – from data mining, text mining, predictive analytics, data visualization, and ML.

Analytics workbench capability

The analytics workbench capability we envision has business Intelligence (BI), insights, **decision support systems** (**DSSs**), and analytics features. These appear at the end of the data factory pipeline's output for user consumption. An analytics workbench is going to be where the data, information, and knowledge are turned into insights. The analytics workbench has the capability to aggregate data and visualize it. This is detailed in *Chapters 13* and *15*.

API, data lake, ML, and end-user analytics capability

APIs are going to be where your data lake and data services are visible to a data consumer. These consumers may be your machine learning infrastructure, or end user analytics consumer. The data lake is going to form the core of your data at rest, so that processing through the zones of the factory is efficient. It is the central location in which to store all your data, regardless of its source or format. The data can be structured or unstructured. One can use a variety of storage and processing tools to transform the data. Typically, in the past, that was performed with the Hadoop family of tools. Today, that is replaced with Spark processing or a microservices framework such as Kubernetes. This way, you can extract value quickly and make informed decisions based on data. Look for tools such as Databricks, Delta Lake, and Open Data Lake for performant services.

Application usage log consolidation and data mining capability

Application usage log consolidation and data mining need to be supported. Mining logs for usage patterns and forensic evaluation of data or programming problems is a constant effort. This capability should be supported, as are other use cases for various log types such as application user logs, system/application usage logs, security logs, access logs, and so on.

Business intelligence capability

Business intelligence (BI) tooling choices vary widely and are very diverse. If the solution is to produce generic datasets, then BI choices need to support integration across those datasets. Due diligence selection processes will choose the best product for your analytics use cases that have to be addressed. Tools such as SAS, SPSS, Tableau, MicroStrategy, Alteryx, Microsoft Power BI Premium, Azure Analysis Services, Cognos, and so on are all choices to consider. There are many more, and the key to success is to select tooling the analytics and business will accept. The business will get married to the tool, so beware if you ever want to change the selection once the choice settles into the organization.

Business intelligence, factory ELT analytics, and end-user reporting capability

BI, factory ELT analytics, and end-user reporting require third-party tool integration of an advanced data analytics platform so that rapid analytics and insight generation can take place *within* the factory. Often, you think of these tools being used at the end for end-user consumer-based analytics. These tools have a place within the factory as datasets transition from zone to zone and pass through quality gates. Also, these capabilities should be as self-service as possible for internal and external consumption.

BI tooling should be built with the intention of serving the needs of a business analyst looking for a self-service solution. Analyst services also include ML and **generative AI** (**GenAI**) use cases where **large language model** (**LLM**) integration is required.

Calibration and factorization services capability

Calibration and factorization services are statistical management processes and systems that are needed to level the data values when a resulting data trend does not make sense. This can be caused by many factors. Data may be missing, it can be overstated, or it can be caused by gross errors (gaps), late data, or internal processing errors. It is not good enough to detect data errors; it is required to create data patches and adjust data to be right to annotate that this has happened in metadata and be able to reverse a data fix/patch later after processing a data restatement or data correction as part of the long-term repair.

Change-data-capture (CDC), data movement capability

Change-data-capture (CDC) and data movement capabilities are defined as a way to capture low-level changes in a data store (also known as transaction log mining) and stream them out to a subscriber so that the subscriber can capture the change stream, act on the change, and if necessary propagate the change event downstream to other subscribers. A CRUD stream is processable by a subscribing receiver who can keep the subscribed system in sync with the source. The issue also requires a clean starting point, so bulk resync is a necessary part of this capability. This capability is needed for zone-to-zone data transfers. Note that there will be different ways to implement this capability's features in order to obtain an optimized result. All receivers should conform to the standard API formats and tooling that the architecture requires. Deviations will be costly in time and effort if diverse ad hoc mechanisms are implemented.

Classification exceptions management capability

Classification exceptions management enables judgments to be made on the system's activity. Telemetry needs to be collected and used to remediate fixes and control what is correct in a dataset and then tweak the data if necessary. No manual adjustments should be allowed; all changes need to be scripted, automated, tested, and rolled out as code.

Cloud accounting, security audit, and cloud security capability

Auditing capabilities are essential for establishing and verifying the trustworthiness of your solution. Auditing processes are tools to give organizations insight into cost savings, security opportunities and operational efficiencies not otherwise available in the cloud platform's PaaS services. Especially, given how many accounts a large organization opens across cloud providers, management of these can be a challenge without a comprehensive approach with integrated tools.

Cloud and external file movement (bulk datasets) capability

External file movement (especially for bulk datasets) requires third-party tooling or network integration. Consider tools such as the following:

- Ipswitch makes networking safe to share data, assuring secure data sharing and high-performance network infrastructures.

- MOVEit offers secure file transfer and full management visibility with email and web browser clients. A flexible deployment model includes cloud and on-premises offerings. MOVEit ensures that secure file transfer between users, systems, customers, and partners takes place. Advanced management capabilities give you visibility and control over data transfer activities through one secure system. You can choose an optimal deployment model to fit security and infrastructure operation needs. MOVEit's features offer guaranteed delivery and non-repudiation. MOVEit is also used for business-critical large secure file transfers of sensitive data. It is often used for regulatory compliance.

- Other cloud provider PaaS services such as Microsoft's *"SSH File Transfer Protocol (SFTP) support for Azure Blob Storage"* {https://packt-debp.link/dC66Ry} exist for transfer but not all have the capabilities you will need without customer development.

Cloud backup capability

The Azure Backup Service backs up data to the Microsoft Azure cloud. You can back up on-premises machines and workloads and Azure **Virtual Machines** (**VMs**). Having a stable backup and **disaster recovery** (**DR**) process is going to be required. Cloud providers insist that high availability (HA) designs are the way to implement disaster recovery (DR) and HA, but you will always want to know you have a business contingency plan well thought out, and that includes the data, metadata, catalogs, and the software repo at the time of backup creation. All have to be in lockstep with each other for the backup set to be valid. Also, make sure there is a way to restore backups and run a fire drill to be sure you have not missed something vital.

Cloud configuration management database (CMDB) capability

Cloud **configuration management databases** (**CMDBs**), such as Snow Inventory, and blueprints allow one to discover devices, software, and cloud services on all platforms, audit software installations, and track usage. Snow Inventory discovers all assets, including hardware configuration, software deployments, and usage. This insight gives stakeholders such as a software asset manager, IT asset manager, service desk, security operations, or IT operational team an overview and the detailed information needed to make critical business decisions. Also important to capture are licenses and keys to enable cloud infrastructure to be re-instantiated in the case of a cloud account crash (or failure due to a nefarious actor's activity).

Cloud disaster recovery (DR) capability

There is going to be an operational need for off-site and off-cloud provider backups. As an example, Microsoft provides infrastructure, platform, and software services via its Azure Cloud Services Portal. Azure Site Recovery (ASR) is one of those services, and it is a cost-effective, self-supported SaaS offering that provides recovery options for on-premises virtual and physical X86 workloads. In June 2018, Azure-to-Azure Recovery was also made generally available to provide recovery options for Azure IaaS workloads through the same interface. If you want fully managed or assisted support, you can also engage Microsoft partners who are certified in ASR. These are the features that this capability should provide:

- **Primary support approaches**: Self-service.
- **Primary workloads supported**: Physical and virtual x86.
- **Regional recovery presence**: As with Microsoft, more than 60 regions around the world, including sovereign clouds such as Germany and China, and government clouds.
- **Customer complexity**: Experienced support for all server images and additional non-image hosted volumes. Integration also for application-specific replication and recovery with some additional scripting.
- **Recommended use**: If you have a requirement to use a hybrid cloud strategy based on a primary cloud provider such as Azure, you will want to explore ASR self-support.

Cloud file and internal file movement (bulk datasets) capability

Bulk file movement and bulk data transfers can be performed offline or over a network connection. Choose a solution depending on these criteria:

- **Data size**: The size of the data intended for transfer will drive your choices.
- **Transfer frequency**: One-time versus periodic data transfers will require SLAs to be set and managed.
- **Network**: The bandwidth available for data transfers in your environment has to be balanced against retries, data size, and any network **quality of service** (**QoS**) constraints (such as rate limits).

Cornerstone is a **managed file transfer** (**MFT**) solution offering secure, automated file transfer, collaboration, and storage. Cornerstone Managed File Transfer provides high availability transfers. It also integrates with existing authentication systems to provide security and compliance. Cornerstone is a package that deploys quickly. You will want to study the features of the tool so that you can implement this capability effectively for your solution.

Cloud orchestration capability

When setting up your cloud infrastructure, you will want to have the capability to bring up the environment from a blank account and, in an automated manner, unfold all service configurations, software deployments, cloud processing configurations, and observability frameworks. To do this, you need tools! Here are a few to consider as you decompose their features and integrate one or more for your solution's orchestration capability:

- Terraform is a tool for building, changing, and versioning infrastructure safely and efficiently. Terraform can manage existing and popular **service providers** (**SPs**) as well as custom in-house solutions. Configuration files describe to Terraform the components needed to run a single application or your entire data center. Terraform generates an execution plan describing what it will do to reach the desired state and then executes it to build the described infrastructure. As the configuration changes, Terraform can determine what changed and create incremental execution plans that can be applied. The infrastructure Terraform can manage includes low-level components such as compute instances, storage, and networking, as well as high-level components such as DNS entries, SaaS features, and so on.

- Pulumi is an alternative to Terraform and it is used by many start-ups. This tool is also worth your study as you harvest features for your solution's orchestration approach.

- Other choices appear in the various cloud provider offerings, such as AWS CloudFormation, Azure Resource Manager (ARM) templates, Azure Bicep, and Google Cloud deployment templates.

Cloud orchestration and cloud deployment capability

Being able to coordinate all solution activities in a smooth and cost-efficient manner is essential. To do this, you need the following features:

- **Application release automation**: Get full control and visibility over your software delivery process.

- **Automated provisioning**: Automate provisioning across your heterogeneous IT infrastructure.

- **Code management**: Manage IaC using your favorite **version control systems** (**VCSs**) to better enable **continuous delivery** (**CD**).

- **Discovery and insights**: Quickly discover resources that need automated management, and drive change with confidence.

- **Infrastructure automation**: Define and continually enforce IT configurations no matter where your infrastructure lives.

- **Infrastructure CD**: Test, promote, and deliver infrastructure code across your infrastructure confidently.

- **Node management**: Define your infrastructure as code and save yourself the manual work of classifying and managing nodes.

- **Orchestration**: Orchestrate change with control, visibility, and automated intelligence.

- **Role-based access control (RBAC)**: Assign permissions to teams in line with company and regulatory policies.

- **Support, services, and training**: Get access to world-class technical education, training, and support to get up and running faster.

- **Task management**: Make changes or remediate urgent problems alongside your model-driven automation management.

- **Visualization and reporting**: Gain insight into your infrastructure, audit changes, and get rich reporting in a full-featured graphical console.

Cloud orchestration and cloud security key services capability

When implementing data-at-rest security protection designs, note that there will be performance impacts on your decisions. Ultimately, security could well mean the design is just too slow for your needs. You must implement **proofs of concept** (**POCs**) and be prepared to make trade-offs and create walled gardens.

Cloud orchestration, cloud security, security auditing dashboard, and security compliance capability

Regarding cloud orchestration, cloud security, security auditing dashboards, and security compliance, services are *not* consistent across cloud provider offerings. Check your cloud provider's version support list for compatibility with your needs before proceeding to leverage the offering in your solution. Even then, prove the functionality advertised works as expected first via POC (aka don't trust the cloud provider's literature). As an example, look at Azure's **System and Organization Controls 1** (**SOC 1**), **SOC 2**, and **SOC 3** coverage list:

- **Azure Trust Center SOC 1 and SOC 2 covered services (as of 2019-09-03)**:

    - Azure, Azure Government, and Azure Germany detailed list

    - Cloud App Security

    - Dynamics 365 and Dynamics 365 US Government detailed list

    - Graph

    - Intune

    - Microsoft Flow cloud service either as a standalone service or included in an Office 365 or Dynamics 365 branded plan or suite

    - Office 365, Office 365 US Government, and Office 365 US Government Defense detailed list; Yammer has achieved a SOC 1 Type 1 report

    - Office 365 Germany

- PowerApps cloud service either as a standalone service or included in an Office 365 or Dynamics 365 branded plan or suite

- Power BI cloud service either as a standalone service or included in an Office 365 branded plan or suite

- Stream

- Azure DevOps Services

- **Azure Trust Center SOC 3 covered services (as of 2019-09-03)**:

  - Azure, Azure Government, and Azure Germany detailed list

  - Cloud App Security

  - Graph

  - Intune

  - Microsoft Flow cloud service either as a standalone service or included in an Office 365 or Dynamics 365 branded plan or suite

  - PowerApps cloud service either as a standalone service or included in an Office 365 or Dynamics 365 branded plan or suite

  - Power BI

  - Stream

Cloud service monitoring capability

Our recommendation is that you store and analyze all operational telemetry in a centralized, fully managed, scalable data store that is optimized for performance and cost. Test various hypotheses and reveal hidden patterns using the advanced analytic engine, interactive query language, and built-in ML constructs. Integrate with DevOps issue management, **IT service management** (**ITSM**), and **security information and event management** (**SIEM**) tools.

Also, monitor applications to get everything needed regarding availability, performance, and usage of web applications and APIs, whether hosted on Azure or on-premises. Azure Monitor, for example, supports popular languages and frameworks, such as .NET, Java, and Node.js, and it integrates with Azure DevOps processes and your DevOps tooling such as Jira and PagerDuty. You can also track live metrics, streams, request and response times, and events with cloud provider tools such as Azure Monitor.

Command and control systems administration and configuration (rules configuration) capability

Command and control systems administration and configuration (also known as rules configuration) services are needed to manage dashboards, SLAs, and data contracts at the logical and physical levels. This way, they are given a form that makes sense for the business but still provides meaning to technology staff.

Custom data load management capability

Custom data loaders and derived data generators (hierarchies, taxonomies, and so on) support your import processing methods. Semantic alignment begins with data preparation at raw data ingest time for your datasets.

Note: Flat files often have a one-level hierarchy, but many complex datasets come with normalized forms of implied hierarchy and cross relationships. These require cross-referential lookup tables to exist and be ingested first. All this needs to be defined at raw data load time. Call it a template feature for raw data storage that has to be configured and processed to correctly ingest raw data.

Custom dictionary management capability

A custom dictionary manager allows for human-readable additions to the data dictionary supporting raw dataset errors and omissions. The dictionary also identifies when additional dependent reference datasets are required to ingest, process, or transfer data at zone boundaries.

Customer information management capability

Customer information management (**CIM**) provides an ability to enable customer self-service across a select set of data classes, sources, or entitled artifacts. This is where entitlements and governing policies are set and managed. Some of the details are complex rule driven and some are stochastic based on fair use limits and traffic patterns, where a nefarious user could drain a system of all data quickly if restrictive policies were not enforced.

Data factory data quality management (DQM) for self-service big data analytics capability

Data factory **data quality management** (**DQM**) for self-service big data analytics integrates data from various sources, sizes, or varieties. One can use a DQM set of features to assure users of the highest quality data delivery.

One can empower internal business staff (for example, IT staff, DataOps, data owner, and so on) rather than relying solely on IT developers for big data insights. Smart data features are key to this self-service analytics capability.

Data factory rules execution engine capability

A rules execution engine usually is a third-party tool or rules engine choice. It can be built internally based on tools such as Drools, or you can explore more expensive third-party offerings if you wish; however, they are not used much except when embedded within other system components. Human-in-the-loop workflows also use rules engines to drive next-step processing, so you'll encounter them there if your system needs this support.

Data lookup services capability

Data lookup services enable data to be found, hopefully from a catalog with dictionary lookup features. If search criteria are described by a consumer and that description results in a metadata match, then that underlying data can be found. This feature is needed as part of any enterprise search functionality so that data may be found with high accuracy.

Data loss prevention services capability

Data loss prevention (**DLP**) services include third-party tools such as McAfee Data Loss Prevention (Enterprise) and others. Being able to detect data leaks is a SecOps feature that has to be supported at the highest level so that what is normal usage can be separated from abnormal usage and feed DLP detection logic.

Data pipeline provisioning, administration, and monitoring services capability

Data pipeline provisioning, administration, and monitoring services provide blueprints for data pipeline definitions in your solution's platform. The low-level physical provisioning definitions have to make sense at the logical configuration level and conceptual level, else it all is a pile of stuff that can't be managed easily. This capability provides end-to-end transparency in your design.

Data tools (end-user applications) capability

Purchased or custom-developed data tools allow your data to be accessed. The tools ease data consumer integration and understanding. When developing custom data tooling, it is a best practice to provide sample code that shows how to leverage the power of the data through its API or tool so that all can be integrated with ease. The examples provided help consumers understand the technical ways needed to access your system's data.

Deep raw dataset analysis capability

Deep raw dataset analysis facilitates data profiling tasks. This often requires third-party tools or custom-developed ML and analytics code to be created to assess raw data for anomalies. This is discussed a lot more in *Chapter 13*.

Deep raw dataset trend analysis capability

Deep raw dataset trend analysis supports DQM, the data mesh's data as a product principle, and various IT zone processing use cases (normalization, enrichment, and so on). Often these data trends are generated using customer code or via integrated third-party products such as Datameer.

Dictionary management capability

Dictionary management supports dataset versioning, time series data, metadata, and searchable data to maintain the integrity of data facts across the entire solution. The dictionary is part of the catalog that also contains models and pointers to where artifacts reside in the solution.

DR backup capability

Operations need off-site, off-Microsoft, off-AWS, and off-GCP backup capabilities. An example is the Iron Cloud Server backup service, which protects data in a resilient, secure environment to keep businesses up and running with predictable costs. You can manage backup and recovery processes for critical data and systems. The solution is deployed in an on-site environment. Iron Cloud Server Backup stores copies on a local target and in the cloud or a secondary location. Optional on-site hardware and cloud failover allow you to maintain your **recovery time objective** (**RTO**) target.

Dynamic access method generation to data capability

Dynamic access method generation (the automated creation of SQL, skeletons/stubs, interface code, API access methods, and more) to your data is a necessary capability. This also includes access methods for various data classes in internal data zones so that permissions can be granted to selected customers who may access internal data as part of their data contracts.

End-user analytics via recommender engine capability

Recommendation engines provide consumer services that seek to predict the correct data needed or the correct preferences that a user would require given prior knowledge of the user's profile or prior data access patterns. This is a capability greatly facilitated by GenAI capabilities in the analyst workbench (see *Chapter 15*).

End-user analytics and factory analytics capability

End-user analytics and factory analytics capabilities consist of features from tools you choose to integrate along with languages such as the R suite of BI analytics software algorithms. An execution environment wrapped by an engine such as SAS Viya for operationalization will provide necessary analytics features.

Tools such as Revolution R Enterprise are distributions of CRAN R, created by Revolution Analytics, Inc. (now Microsoft R); while CRAN R is at the cutting edge of R development and research, Revolution R is designed for commercial use, where stability and performance are needed as well as access to new functionality. Revolution R Enterprise has a less frequent release cycle than CRAN R. New updates are produced only when significant updates have accumulated in CRAN R and they have been proven stable by the R community. Rebases of Revolution R Enterprise occur on the last stable point release of CRAN R. We mention all this because you will want to be able to code analytics processing into your data factory pipeline, and these tools help with that effort.

End-user analytics, factory analytics, ML via data science laboratory capability

End-user analytics, factory analytics, and machine learning via a data science laboratory are all part of end user data consumption. You will want to build features for a **user acceptance testing** (**UAT**) environment that leverages data carved from the production environment and made usable by business operations, IT, and those developing statistical methodology. This is needed to preserve overall data quality. All this is also needed to provide for data readiness of release datasets prior to user consumption.

As part of the data factory's processing, various analytic models can be built and then applied in an orchestrated manner via **user acceptance tests** (**UAT**). These were created by the data scientist in the analytics workbench and run in a designated test environment.

End user reporting via BI and analytics reporting tool capability

End user reporting via BI tools such as Tableau reporting, provides analytics reporting capabilities. This capability has a focused goal: to help users see and understand data. Organizations in various industries empower users with data that is presentable in visually appealing ways. Some ad hoc refinements and decision support features, along with data visualization, enable business insights to be harvested from these visualized datasets.

Entitlement management capability

Entitlement management for all datasets with their metadata is an architecture capability that needs your attention. These objects are subject to security, scale, and performance contracts. Expect integration issues with your cloud provider's supported RBAC or **access-based access control** (**ABAC**) versus user/group security controls for low level row/column security.

Entity classification capability

Classification management capabilities enable judgments to be applied to raw data as they are being processed. Exceptions are resolved when loading and matching, as well as inputs to machine learning algorithms gathered and leveraged when retuning ML models used in profiling data.

Extract, load, and transform (ELT) semi-structured mapping services capability

Extract, load, and transform (**ELT**) of semi-structured data requires mapping services to extract tabular data from unstructured text.

Extract, transform, and load (ETL) utility capability

Extract, transform, and load (**ETL**) utility capabilities are needed for ingesting data at various stages in the data pipeline. This way, data is stored as transformed since the prior zone retains its raw state from the perspective of the downstream zones' processing.

Factory data access object (DAO) capability

A **data access object** (**DAO**) provides the capability to access objects with an API to provide logical business value. An example program can be provided with the DAO to provide simplified access to data stored in persistent database storage.

Factory ELT real-time data ingest capability

Factory ELT real-time data ingest and real-time processing deals with data stream processing for captured real-time data. The goal is to implement low-latency processing and keep up with the rate of arriving data. Many real-time processing solutions need a message ingestion store to act as a buffer for arriving data (as messages) and to support parallel scale-out adjustments so that processing can proceed in a reliable manner. Data delivery and other message queuing approaches are expected features of any real-time data processing.

Factory message flow capability

Factory message flow services use tools such as Apache Kafka, which is a pub/sub messaging system where messages are immediately written to the filesystem and replicated within a cluster to prevent data loss. Messages are not deleted when they are read but retained within a configurable SLA. A single cluster serves as a central data backbone that can be elastically expanded without downtime. Apache Kafka is a distributed, high-throughput message bus that decouples data producers from consumers. Messages are organized into topics, topics are split into partitions, and partitions are replicated across nodes – called brokers – in the cluster. Compared to Flume, Kafka offers better scalability and message durability.

Kafka comes in two flavors: the classic producer/consumer model, and the new Kafka Connect, which provides configurable connectors (sources/sinks) to external data stores. Kafka can also be used for event processing and integration between components of large software systems such as your data factory. Data spikes and back-pressure (fast producer, slow consumer) are handled out of the box. In addition, Kafka ships with Kafka Streams, which can be used for simple stream processing without the need for a separate cluster as with Apache Spark or Apache Flink. See this article for more details: *"Big Data Ingestion: Flume, Kafka, and NiFi"* {https://packt-debp.link/KqI5Ct}.

Factory execution environments parallel computation engine capability

For factory execution environments and parallel computation engines (such as Spark with streaming, which is an extension of Spark), you will need to implement parallel processing features as part of your solution. Spark Streaming extends Spark for doing large-scale stream processing and is capable of scaling to hundreds of nodes with second-level scale latency. Cloud service PaaS facilities also exist for implementing logic that scales, such as with Azure Functions.

Factory infrastructure file storage distributed filesystem capability

With a redundant factory infrastructure distributed filesystem (such as **Hadoop Distributed File System** (**HDFS**), even though it is less popular than in the past), you have a distinct advantage since it provides high availability and performance:

- Cost-effective storage solution for businesses.

- Uses commodity direct attached storage and shares the cost of the network and computers.

- Is a highly scalable storage platform, because it can store and distribute very large datasets across hundreds of inexpensive servers that operate simultaneously.

- Can use Hadoop to derive valuable insights from data sources such as social media, email conversations, or clickstream data (flexible).

- Maps data quickly, wherever it is located on a cluster.

- When data is sent to an individual node, that data is also replicated to other nodes in the cluster, meaning that, in the event of failure, another copy of the data is available for use.

Alternatives to HDFS are available since Hadoop development and use activity is declining. These are the following:

- Amazon Simple Storage Service (S3) is AWS's cloud storage service

- Amazon EFS is AWS's NFS implementation

- CephFS is another high availability distributed POSIX filesystem for objects

- Cloudera and EMC/Greenplum's Isolon is a distributed storage system

- GlusterFS is also a high availability distributed filesystem that can scale up to petabytes

- Google Big Table is GCP's cloud storage

- IBM Spectrum Scale is a highly performant distributed filesystem

- Lustre is a parallel distributed filesystem with a focus on throughput/low latency

- Microsoft's Azure Blob Storage is like Amazon's S3 service

- Microsoft's HDInsight for Azure is Azure's HDFS implementation

- Quantcast is a distributed filesystem based on Reed-Solomon erasure coding

- ... others, such as legacy NFS, Disco, Filemap, and now defunct Map-R

Factory infrastructure, factory file storage … distributed filesystem manager capability

Cloudera Manager is an end-to-end management application for Cloudera's distribution of Apache Hadoop. It gives you a cluster-wide, real-time view of nodes and services running, provides a single, central place to enact configuration changes across the cluster, and incorporates a full range of reporting and diagnostic tools to help optimize cluster performance and utilization. What you want in a distributed file manager is illustrated by this product, and you need the management capability with any solution that you select or build.

Factory infrastructure, factory … SQL over NoSQL distributed database capability

Your data factory infrastructure will need an SQL access method even over any NoSQL distributed database selected. With the Hortonworks Data Platform (HDP) now defunct (as of 2023), Azure HDInsight's HDP software equivalent provides some years of needed runway for any legacy systems. Cloudera (Cloudera Distributed Hadoop, or CDH), Cassandra, DataStax, Redis Enterprise, and ClickHouse provide other alternatives, but they also come with vendor lock-in that you will want to consider.

Factory infrastructure, SQL over NoSQL database (data warehouse and data mart) capability

A data warehouse is often required for high transaction processing environments. These don't just append arriving data but conduct transactions before updating zone data.

This is where your data models call for table based key-value pairs and consistency checking that a warehouse makes possible over NoSQL environments. Data warehouses are highly responsive and always online. To achieve low latency and HA, instances of these applications need to be deployed in data centers or clouds that are close to their users. Applications need to respond in real time to large changes in usage at peak hours, store ever increasing volumes of data, and make this data available to users in milliseconds.

A great **NoSQL DB** example is Azure Cosmos DB, which is Microsoft's globally distributed, multi-model database service. Cosmos DB 1 and Cosmos DB 2 enable you to elastically scale throughput and storage across regions. You can elastically scale throughput and storage, using APIs such as SQL, MongoDB, Cassandra, Tables, or Gremlin. Cosmos DB provides solid SLAs for throughput, latency, availability, and consistency. Alternatives to Cosmos DB include ClickHouse, StarRocks, Rockset, Redis Enterprise, GridGain, MS Power BI Premium, and MS Azure Advanced Analytics Services (AAS) to implement your data mart.

Factory infrastructures enterprise data warehouse capability

An **enterprise data warehouse** (**EDW**) is fed from a variety of enterprise applications. Naturally, each application's data has its own schema. The data thus needs to be transformed to conform to the EDW's own predefined schema.

Factory job control capability

Job control systems are meant to replace the Unix cron scheduler across the enterprise for scheduling business workflows and obtaining proper business service-level metrics.

Factory process flow, microservices – real-time data input capability

Real-time data input with lambda-based microservices architecture is a pattern to implement for system performance. Lambda queuing had its origin in cell phone traffic coordination when cell technology was invented. Queues could clog up if the proper load and number of servants were not mathematically modeled in advance. The basic flow was the lambda rate into those models. **Jacksonian queuing theory** is alive and well, and the approach still could be used to model and plan for complex queuing solutions in order to keep dataflowing through a data factory.

Factory rules execution engine capability

The use of a rules engine in today's cloud systems is controversial. Microsoft takes the view that rules engines are not necessary, and one can see why. It's easier to code a step than specify a rule for code to later execute. But if you want a non-technical person to specify rules, then an engine to execute those rules makes sense. We'll take the position that factory rules should be codified where possible. Apache Spark Streaming is a high-capacity execution environment, where rules are codified and not specified.

Key features of a factory execution environment are the following:

- Easy to implement in order to increase ELT developer productivity (for example by 50%).

- Effective in that is able to port the majority of coded existing ELT processes with little to no changes.

- Economical so as to optimize ELT performance by implementing an efficient computation fabric.

- Scalable so that it can handle an increasing number of nodes or executors.

- Enable operational views of ETL processes in real time for service level management (SLM).

- Run in a cloud-agnostic manner so that a dataflow can run on or across cloud providers as well as any in-house clusters/local machines.

- Not depend on any one distributed file system type like HDFS.

- Remain execution fabric agnostic so that it can run with AWS Glue, Spark Batch, Spark Streaming, MapReduce (MR), Tez, Flink, and so on.

- Support multiple sources and targets to enable extensibility.

Creating a meta-model to document your factory's processing is clearly advised.

Financial report management services capability

The solution has to be able to support **financial report management** (**FinOps**). An example is NetSuite reporting for financial reporting purposes. There is much that could be said about FinOps, and we will do so in *Chapter 7*.

Full life cycle API management capability

Full life cycle API management provides an abstraction layer over your data services that takes generically modeled data elements and/or features and prepares them in a more specific way for a targeted developer or application.

Graph database capability

A graph database provides a data structure intended to hold data without constricting it to a predefined dimensional model. Highly dimensional data is best fit into a graph. Data may be stored in a semantic manner, preserving relationships without having to define these relationships explicitly. RDF/OWL2 modeling is possible using a graph organization of data, or a data architect could implement topic-modeled graphs that do not require a formal modeling specification (but lack inferencing capability). The advantages are that the graph eliminates nested joins for highly dimensional datasets and enables entity connections with formal relationships to other entities.

> **Note**
> Semantic metadata for data at rest is most formally modeled in a knowledge graph that contains a domain ontology; with instance data, that could be retained as a graph in a knowledge base.

Hierarchy and dictionary maintenance/reporting services capability

A toolset is used to create taxonomies through the ontology (for example, two-dimensional directed walks through the semantic graph stored as a knowledge graph), therefore enabling the walk to become a facet often used in a search facility. Knowing what a data item is semantically remains a dictionary task, and that definition needs to be cataloged and reported via reporting services. As data changes are made, subscribers should be notified so that impact assessments can be made. This is often a forgotten feature of legacy systems. It is still essential for future systems.

Identity and access management (internal user access, internal user entitlement, internal app entitlement) capability

Identity and access management (**IAM**) enables internal user access, internal user entitlement, internal app entitlement, and so on. All these need to be driven by policies and managed by a console with a well-defined process. Expect difficulty in integrating various security solutions. Go with federated approaches with **SAML** standards where possible to offload integration complexity underpinning your logical security architecture. Conceptually, security is easy, but in the details comes complexity!

In Memory NoSQL cache capability

In-memory NoSQL cache-like solutions (for example, Redis) deliver a vast number of operations per second at sub-millisecond latency. Redis Enterprise technology is the commercial version of open source Redis with enhancements, supporting all of its commands, data structures, and modules. The commercial version adds management, extensibility, stability, high performance, and resilience via multiple deployment options and topologies. A geo-distributed active-active architecture with built-in secondary indexing and support for very large data sizes using **solid state disk** (**SSD**) data offloading helps make the product a compelling choice. Alternatives also exist, such as the following:

- ClickHouse
- StarRocks
- Rockset
- Redis Enterprise
- GridGain
- MS Power BI Premium
- MS AAS

Input dataset contract management capability

Contract management (aka data contracts) for datasets is an essential capability to be implemented and reported on. Outputs feed into the enterprise accounting and sales systems since data should be considered a product. This is a huge competitive advantage if implemented correctly. It provides part of the **total cost of ownership** (**TCO**) and **return on investment** (**ROI**) models for the business data owner.

Insight generation capability

Insight generation is the analytics process using tools, applications, and methodologies needed to tease insights from enterprise knowledge, information, and normalized raw data. It is a key capability of what we propose you design, which is the analyst workbench.

Instance resource monitoring capability

There are many monitoring tools available. An early example is **Zabbix** with its enterprise level software capabilities designed for real-time monitoring of system metrics collected from thousands of servers, virtual machines, and network devices. Zabbix is an open source solution and comes at no cost except for tuning and integration. You can read about more alternatives in *Chapter 11*.

Job control management capability

Job control management implies that you require a work order job control blueprint within your architecture. This way changes to datasets could appear in bulk or micro-batch stream forms and be handled effectively. The key is that all activities are choreographed and auditable via job controls.

Log consolidation and mining capability

The acronym ELK stands for Elasticsearch, Logstash, and Kibana. The ELK Stack is a collection of these three open source products maintained by Elastic. The introduction and subsequent addition of Beats later turned the stack into a four-legged project and led to a renaming of the stack as the Elastic Stack:

- Elasticsearch is an open source, full-text search, and analysis engine, based on the Apache Lucene search engine.

- Logstash is a log aggregator that collects data from various input sources, executes different transformations and enhancements, and then ships the data to various supported output destinations.

- Kibana is a visualization layer that works on top of Elasticsearch, providing users with the ability to analyze and visualize the data.

- Beats are lightweight agents that are installed on edge hosts to collect different types of data for forwarding into the stack.

Together, these different components are most commonly used for monitoring, troubleshooting, and securing IT environments (though there are many more use cases for the ELK Stack, such as BI and web analytics). Cloud providers have created similar alternatives that integrate PaaS services only they can integrate, so ELK has lost a lot of followers in the recent past. **OpenTelemetry**, elaborated on in *Chapter 11* and *Chapter 16*, offers you some options and explains what will be needed with monitoring and observability in the cloud.

ML, deep learning, and GenAI services capability

ML, **deep learning** (**DL**), and GenAI services are going to be part of any modern data solution. Algorithms and statistical models perform tasks without using explicit instructions, relying on patterns and inference instead. Algorithms build mathematical models based on sample data, known as training data, in order to make predictions or decisions without being explicitly programmed to perform the task. Read a lot more about this in *Chapter 16*, where we will discuss the necessary processes to implement **Machine Learning Operations** (**MLOps**).

It is important to be able to gauge a model's efficacy/quality/performance using standard measures and replace those models when they fail to meet expectations. Separating science from art and making it all work in a production environment with quality is one of your solution's goals. You will be building MLOps processes for your analytics workbench with policies and procedures for bringing tuned models online.

ML engine execution environment capability

ML model creation and execution environment requires the development of a data science laboratory and an analytics workbench. To know more about this, read *Chapters 15* and *16*.

Metadata maintenance capability

Metadata maintenance involves the management of metadata driven from a formal distribution set/release set/dataset taxonomy, which makes the metadata searchable and administrable within a given context. This also includes any associated time series data within the logical dataset that must be retained to make the release set have meaning. In these systems, swim lanes (see *Chapter 13*) are supported where parallel data production pipelines can be managed till one or more complete to delivery.

Metadata management capability

Metadata may be manually assigned, such as tags, or generated from third-party data added when data is ingested and aligned with existing semantic graph data. It can also become a fully standalone graph created for new classes of data. Either way, data is profiled and assigned metadata if not already supplied at ingest time.

Network operations center capability

ITIL-like services for the cloud will be needed in your cloud solution. An example is the comprehensive response management, event triage, and reporting available from Logicworks. All of these services implement the *"Cloud Operations Manifesto"* as follows:

- **End-user experience is a key responsibility**: Uptime and high performance should be constantly measured and optimized.

- **Infrastructure is ephemeral, immutable, and utilitarian**: Don't fix it; replace it.

- **Infrastructure is software**: It's programmable, modular, and ideally deployed as part of the application. Everything can change quickly, is tested frequently, and is ready to scale.

- **Make data useful**: Log everything centrally, ingest logs into a BI tool, and get data that benefits the business.

- **Let automation do the boring stuff**: Patching, backups, key rotation, and so on, should be automated.

- **Security first**: Think about security at the beginning of the architecture process, not as a last-minute add-on.

- **Understand the difference between security and compliance**: Security reduces operational risk. Compliance reduces business risk. You need both.

- **Anticipate failure**: Expect everything to fail, plan for failure, and deliberately break things to test failover.

- **Have a runbook**: No matter how sophisticated your technology, you still need a response plan for issues, especially after-hours.

- **Don't be afraid to fail quickly**: Not every cloud technology is going to be right for you, but you won't know unless you stay current and test.

NoSQL read-optimized database capability

NoSQL read-optimized databases are tuned for fast read operations. You may have issues with writing performance with this type of database. They are used when required to address evident performance/cost issues with current often slow NoSQL distributed systems. You can contrast Hive with HBASE access performance as an example over HDFS. Likewise, you can see Cassandra's read versus write (poorer) performance.

Parallel batch computation engine capability

A parallel batch computation engine such as Apache Hadoop's software library is a framework that allows for the distributed processing of large datasets across clusters of nodes using simple programming models. It is designed to scale up from single servers to thousands of machines, each offering local computation and storage. Processing capabilities for query and analysis are primarily delivered using programs and utilities that leverage MapReduce and Spark. Other key technologies include HDFS and YARN (a framework for job scheduling and cluster resource management). But note that Hadoop has fallen out of favor, replaced by Databricks (aka a commercial Apache Spark system with lots of platform enhancements). Another alternative is Airflow. You will want to pick one that is monitorable and administrable, especially when errors arise. A clear feature winner is going to be one that provides the best forensic tooling because you'll need that feature!

Problem management and escalation processes capability

Problem management and escalation capabilities are often supported in integrated problem resolution and ticketing systems. Some come with more advanced features and support integration as well as **root cause analysis** (**RCA**), such as Remedy. You will want to build data issue management into your processes.

Quality management capability

Quality management is a process for establishing the quality of software, human processes, and datasets in your data factory. It is comprehensive and used to enforce and verify contract compliance (customer facing or consumer facing data).

Report management services capability

Report management services (quality reporting) are used for quality control purposes. They usually provide internal reporting for distribution dataset preparation destined for consumers.

Rule engine management of parameters for pipeline processing capability

Rule engine management enables high level parameters to be set for allowable additional data pipeline customizations to be made by the business data owners.

Sales orchestration capability

Sales orchestration is often performed with a third party integrated solution such as Salesforce, SAP, Microsoft Dynamics, and so on. You will want to assess your business data requirements before building integrations. Some of these integrations will consume a lot of effort.

Scrum Agile user experience capability

A scrum-developed agile **user experience** (**UX**) brings agile software development together with the data product with interactive design. Agile teams require detailed data understanding to envision a product. This means allotting time and budget for the UX's full development process, including research and testing.

Security audit capability

Standard audit and compliance verification is going to be needed at some point in the solution's lifetime. Here are some examples:

- Azure traffic analytics analyzes network security group (NSG) flow logs across Azure regions and equips you with the information needed to optimize workload performance, secure an application and its data, provide for any audit reports required, and stay compliant.

- SOC 2 (System and Organization Controls - SOC2) provides customers with assurance regarding the controls in place that protect a system or its data. Testing is based on defined principles and criteria published by the American Institute of Certified Public Accountants (AICPA). Reports cover common criteria/security but can also cover availability, processing integrity, confidentiality, and privacy.

As a backgrounder: SOC Type 1 reports focus on your organization's controls relevant to a user entity's financial reporting. SOC Type 1 examinations are performed in accordance with *SSAE 18*, resulting in clearer, more detailed information regarding your control environment. Given their limited scope, SOC Type 1 reports are best suited for organizations that must instill confidence in their controls and safeguards over their customers' financial data. Such organizations include providers of financial transaction services and various technology services, such as the following:

- Data center services

- Cloud computing and network monitoring services

- SaaS

- Payroll and medical claims processing

- Lending services

The difference between SOC Type 1 and SOC Type 2 reports lies in the time period upon which they focus. SOC Type 1 reports address the suitability of your control design and implementation at a specific point in time. In contrast, a SOC Type 2 report concentrates on control design and operating effectiveness over a period of time. A SOC Type 2 report, therefore, enables the user auditor to assess the risk of material misstatement of financial statement assertions.

Security compliance certification capability

An ISO 27001 certification demonstrates conformity of a company's **information security management system** (**ISMS**) with the documented standards. Obtaining an ISO 27001 certification exemplifies the maturity of a company's information security environment as well as the ability to meet contractual obligations and gain a competitive advantage in your industry.

Security provisioning and administration capability

Security provisioning and administration services capabilities provide for the security of the data supplier and data consumers according to the enterprise's security policies, operations, and governance models.

Usage analytics (source data profile, incoming file reporting) capability

Usage analytics (for example, source data profile and incoming file reporting) are needed for gathering critical incoming data statistics to source data contracts.

UX and content distribution network capability

A UX **content distribution network (CDN)** is a highly distributed platform of servers that helps minimize delays in loading web page content by reducing the physical distance between the server and the user. This helps users around the world view the same high-quality content without slow loading times. Content delivery can efficiently supply web content to users. CDNs store cached content on edge servers in POP locations that are close to end users, to minimize latency.

Workflow management capability

Workflow management adjusts workflow definitions that underpin a data pipeline's processing step. The data pipeline itself can also be a workflow and managed by the workflow manager.

Capabilities summary

It is understood that there is a lot to grasp in the aforelisted capabilities, but they will be essential as you put the solution together in the logical architecture where concrete choices are made. What is critically needed is to implement the capabilities identified that are related to data movement in and between data zones, which will be elaborated upon in the next section.

Data processing zones – ingestion, transformation, and consumption

A data pipeline flows data through processing stages, with one stage progressing to completion and then forwarding to the next stage. Along the way, metadata for each dataset and its derivatives is created. All data with metadata should remain discoverable after a stage completes. Both will appear in the release set. We are, after all, treating data as if it were a product, so it has to have a product-like life cycle. This forms a factory-like assembly plant method of curation. The dataset will logically exist in multiple zones, so conceptually, we will need to be clear on the rules and conventions of base data classes in each zone. We will also want to be clear that these datasets will appear in metadata lineage as flavors of the same dataset that gets curated into a release set and finally made accessible to consumers via contracts for service and quality. Contracts for internal curation consumers, analysts, and pipeline code in the zones also exist and are enforced. The contract will determine the features of the capabilities selected in the logical and physical architecture, and these have to be very clear.

In the *"Conceptual Refence Architecture Glossary"*, the zones and data classes are identified. Data is normalized, mapped, pivoted, flattened, and cached for consumption in different zones. Metadata ties the datasets together, unlike what cloud providers often tout where PaaS metadata services keep code fragments, and their execution invocation results together as metadata. Code should be orchestrated and data effects logged into metadata tracking services to support software CI/CD release processes across the cloud PaaS offerings. They are complex and brought about by the need to weave a chain of PaaS service calls together to implement business use cases. Dataset data lineage tracking in the cloud with customer dataflows was in the past an afterthought, but that must not be the case when building a future-proof data-engineered solution. You will want to leverage the cloud PaaS metadata services (such as Azure Purview, Dictionary, and Databricks Unity Catalog) but also add your necessary integration.

At this juncture, you see that there are a lot of moving parts in the data factory, and data transfers, transformations, pub/sub, CI/CD, and CDC approaches have to be created. To keep all working well you need to support standard ways to implement the zone data transfer with metadata tracking at the dataset/release set/delivery set level. Think of a standard release management approach that will work for your business. You will come to the conclusion that by creating a delivery, your organization releases software that affects the data released as a product. The data state defines the release. The delivery set is what you are contracted by the consumer to deliver, cleanly and completely with superior service. The following question arises: *How should we implement data processing zones?* Let's explore this further:

- The *raw zone* is used for data ingestion to the data factory. In this zone, raw data arrives and is normalized … Note: This often leads to a data lake implementation. Raw data is just that, but not too raw! Sometimes we want to retain *too much* and forget to get rid of previously received versions after accepting something that can pass the gating criteria for acceptance. Fix all this in the raw zone data and retain the contracted number of clean versions in as raw a form as possible so that transformations that take place in the downstream bronze zone are back traceable to a supplier's raw data. The adage *If you touch it, you own it* applies. So, make sure any changes to incoming raw data are explainable and the stewardship remains with the supplier.

- *Transformation (bronze, silver, and gold) zones* are used for data warehouse and/or data lake cloud storage. Note: This often leads to a delta lake or cloud data warehouse implementation. Datasets pivot in form from their raw but normalized mapped formats in the bronze zone to the data classes needed in the silver zone and the data mesh structure of the gold zone.

- *Consumption zone* data is used for high-speed access and data analysis and is where the consumer use cases are executed and release sets are delivered. Note: This often leads to a BI in-memory or data mart OLAP implementation. The format of the data will be directed by the cloud's data fabric services.

Conceptual to logical architecture mapping

What we need to take away after pondering the conceptual architecture as we move toward the development of a logical architecture is the answer to the question *Why?*:

- *Why* did you choose that product? *Answer:* It mapped to the capability and had to be integrated.

- *Why* does this data migration toolset have to move data generically? *Answer:* It will be put to use across multiple zone boundaries to move data – it has to be built commonly and operate well.

- *Why* this product over another? *Answer:* Look at the due diligence and the product's capabilities. The decision, if objectively produced versus subjectively dictated, should be evident.

There will always be challenges to the architecture. Just be ready to be clear and transparent. The reference diagram, the due diligence, and the capabilities go a long way toward building confidence with the business and IT groups as the solution is developed.

Summary

In summary, this chapter brought you on a journey where you were able to discover what goes into conceptual architecture and conceptual reference architecture. You also discovered key aspects of the architecture driving the best practices that will be evident in the book. By forming the conceptual reference architecture diagram, you created an architecture that could be communicated. Via a due diligence process, you made important objectively obtained vendor and tool decisions, and you learned that by itemizing capabilities needed in the solution, you justified the drivers for the logical architecture.

Additionally, new terminology was explained and a glossary was produced for clarity. Data class, data movement, and data formats in zones were identified. ELT is better than ETL (refer back to *Chapter 2*, where this was discussed), so we can get back to the raw data of the system that is being transformed multiple times in order to get it curated into final form. The data factory approach is used to prepare datasets into release sets for inclusion into a distribution set for customer release supporting the data as a product principle. Very important here is the concept that data has to be rendered as fit for purpose and support the *one-size-does-not-fit-all* approach to data organization. Lastly, a well-thought-out and communicated architecture provides the best chance of your implementing the future-proof vision for a best-in-class data engineering effort. This includes an emphasis on the dataflows and blueprints for a data factory/data pipeline approach to curated data delivery.

In the next chapter, you will see how to develop a logical architecture that aligns with the conceptual architecture.

Architecture Framework – Logical Architecture Best Practices

The need for a more detailed architecture framework will be elaborated upon in this chapter as the **logical architecture**. As stated in the prior chapter, the framework is documented as a three-level *conceptual*, *logical*, and *physical* architecture, and in this chapter, you will be exposed to the need for a formal way to depict the logical components of the system design that will flow together to implement the capabilities identified at the conceptual level of the design. This will enable you to clearly communicate the to-be state of the IT solution to implementors and vendors.

It also justifies the choices made at the conceptual level where design choices, vendors, and tools were selected after much due diligence. The due diligence process will continue when formulating this logical layer of the architecture. This layer produces an implementable blueprint that is essential to be able to design a solution. It is formulated in collaboration with the engineering staff before moving forward with various proofs of concept (POCs) and full-scale agile implementation. There will be a primary focus on the logical level of the architecture in this chapter, which will include various best practice definitions.

Logical architecture overview

In the process of developing the logical architecture, there will be important integration considerations and risks to address regarding the technologies selected, the implementation processes, or the integration patterns needed. The best practices for them will be identified as key process methods for developing subsequent engineering designs. A logical reference architecture will be presented with a focus on the current state of the art in regard to implementing the concepts previously presented for the creation of data pipelines.

As with the question posed in the prior chapter, you may ask, *What's in the logical architecture?*

The answer is: The level of communication with the engineers who implement features aligned with the capabilities presented in the conceptual architecture.

In feature-based development, you will find that if features are to be turned on and off, be subject to structured feature releases with their data, and have all that subject to service-level monitoring, these features should be implemented from one end to the other across the various data pipelines of the architecture. These observable features are to be experienced by the data consumer, and they are supported by end-to-end threads of activity as dataflows originating from raw data to end data publication. This is all orchestrated as the data pipeline (or runnable instances of generic data pipelines). Through the logical components depicted in the logical architecture, this flow will become clear.

> **Note**
> The flow of data is critical to being able to organize various software components' activity and the data lineage tacked data states as data is curated through dataflows!

The data engineer is an important participant in the development of the logical architecture in order to represent the principles agreed upon when the high-level capabilities were created. These are made effective through features created in the logical component definitions of the data factory. The platform concepts depicted in the conceptual architecture are mapped to various software and data components/services defined with component features. As the glossary was to capabilities, so here the logical components are to their features. You will see the word *feature* used many times in this chapter and in the notes.

The processing of logical components defines dataflows of the factory, but the details will have to wait till the physical architecture is strawmanned and then completed when the solution is made generally available. Some of the user facing features will depend on repeatable patterns that flesh out blueprints that were presented at the conceptual level. The picture of a logical architecture should form in your mind now. Picture it as a diagram that reads from left to right as if data were flowing through a set of processes. This is what will be elaborated upon in this chapter.

In the flow, you can think of data going though zones, similar to how a boat traverses locks on a canal. These zone transitions are prominent in the logical layer that we are presenting as a best practice. The data zones needed to segment raw data from consumable data clearly are identified through the stages of the data factory.

A single-page representative diagram best illustrates this logical architecture level of the data engineering solution, similar to the one-page definition of the conceptual architecture. From this, you have to dive into the details and peel the onion to get to the next layer of the physical architecture for a complete understanding of the current state solution. Therefore, all layers should be kept up to date and in sync with each other at all times for ease of communication as changes (inevitably) are made over time.

The goals of this chapter are the following:

- Define the contents of the logical architecture and illustrate them.

- Provide best practices for that logical illustration and how it can be rendered to enable effective stakeholder communications.

- Provide best practices for future-proofing the logical architecture with an emphasis on the features of a data factory, component tool selections, and cautions regarding anti-patterns detracting from best practices.

The three goals for this chapter will align with subsequent chapter elaboration regarding the physical architecture of a reference solution.

Organizing best practices

You will observe that we have adopted the following structure for the logical layer:

- Best practice composition of the logical architecture.

- What is the structure of a logical architecture (through an example)?

- What capabilities and logical features are needed in the logical architecture?

- Learning about features needed in the data processing zones and how datasets should be propagated between zones with a special focus on the types of use cases each data type will be subjected to:

  - Bronze data is used for data profiling, data normalization, and data cleansing use cases for big-data-lake processing.

  - Silver data is used for data pipeline processing or for big-data-lake **online transaction processing** (**OLTP**) and some **online analytical processing** (**OLAP**) operations.

  - Gold data is used for enterprise-grade mastered data and for OLTP and OLAP operations.

- How to implement data processing zones:

  - The *raw zone* is used for data ingestion to the data factory. In this zone, raw data arrives and is normalized. Note that this often leads to a data lake implementation.

  - *Transformation (bronze, silver, and gold) zones* are used for data warehouse and data lake cloud storage. Note that this often leads to a delta lake or cloud data warehouse implementation.

  - The *consumption zone* is used for high-speed access and data analysis where consumer use cases and dataset deliveries take place. Note that this often leads to a **business intelligence** (**BI**) in-memory or data mart OLAP implementation.

- Learning best practices regarding how to tie the conceptual architecture to the logical architecture.

With this structure, a data engineer will be able to communicate effectively to those at the highest levels of an organization without getting too much into the details, yet also be able to drill into details if challenged. Being able to still relate back to the high-level concepts for justification and low level details is an essential part of the logical layer of the architecture. It ties the physical implementation together with the concepts driving implementation.

How does the logical architecture align with the conceptual and physical architecture?

Development teams may only build and then deploy a solution once the architecture is defined to an acceptable level. This happens when the logical and physical architecture has been proven to be effective.

Why? Because sometimes it's *not!* Even the best of plans sometimes go awry, and changes are required. Technology can fail to work as expected or have integration anomalies, among other issues. Keeping the focus on what's important will guide you as to when a proof of technology is needed before a plan to implement that technology.

Special attention will be given in this chapter to dataflows that implement the various data pipelines of the data factory. Data contracts and **service level agreements** (**SLAs**) are important features when formulating the architecture layer. These non-functional but very important features of a solution can't be left till the end. They form a framework for your logic that will enable your team to develop robust solutions. How these features will support the capabilities of the architecture has to be made crystal clear. These will be subject to enterprise development processes such as *release management*, *change management*, *entitlement*, *service management*, and *data contract management* because they support the data products being curated.

The details are going to be important for your developers, but note that the fully configured dataflows will appear only in the physical architecture. Detailed choices for third party products and their integrations must be solid in the design and specified at this logical level. Selections have to be absent of scale, performance, reliability, or integration error; so, expect many POCs. These are required to vet out unacceptable risks (and there will be a lot of risk). The logical architecture will remain refinable as a work in progress until the physical architecture is complete. However, it is *not* as flexible since much subsequent design work is based on it being correct.

You are sometimes aware that you do not know how something works but can get answers; other times, you are not aware that you do not know how something really works and are clueless. Lastly, there are times when the integration just fails to live up to the hype, and workarounds are desperately needed to save the architecture from failure. Lastly, capabilities and their associated features may even be removed from the **minimum viable product** (**MVP**) if POCs fail to resolve some identified risk. That being said, you have to begin at the beginning.

Let's look at an example of a *logical architecture* in real life that illustrates many of the concepts we wish to highlight as essential in a *future-proof* data engineering solution. After all, it was Albert Einstein who said: *"The only source of knowledge is experience."* And we *do* want to learn from those who experienced the pains of putting together a great logical architecture.

Case study – accelerating innovation at JetBlue using Databricks

Data and **artificial intelligence** (**AI**) technology are critical in real-time proactive decision-making. Leveraging legacy data architecture platforms will impact future business outcomes. In the past, JetBlue data was served through a multi-cloud data warehouse that resulted in a lack of flexibility for advanced designs, latency changes, and cost scalability {https://packt-debp.link/VTQXrf}. Some issues the new logical architecture at JetBlue addressed are the following:

- High latency of over 10 minutes cost the organization millions of US dollars.

- A complex architecture with multiple stages, across multiple platforms with a lot of bulk data movement was inefficient for real-time streaming.

- The platform's high **total cost of ownership** (**TCO**) was evident. Many vendor data platforms were present in order to manage the dataflows, along with their high operating costs.

- Scaling up the legacy data architecture when trying to process exabytes of data generated by many flights was just not possible.

JetBlue adapted Databricks's **medallion architecture** {https://packt-debp.link/S21L4y} to their needs. But what does this medallion architecture pattern/blueprint define? In essence, it is a blueprint used to logically organize data in a lakehouse (as pointed out by Databricks), with the goal of incrementally (or progressively) improving the structure and quality of data as it flows through each layer of the architecture (from bronze to silver to gold data). Medallion architectures are sometimes also referred to as **multi-hop architectures**, among other names. This approach is clearly in line with the best practices set forth in this book.

Multi-hop designs are pragmatic! They cover the need for a broad formalized process similar to what is diagramed in *Figure 6.1* but with more ridged zones. It is worth studying many architectures to learn from their success as well as minor failures that later turned into successes through agile development. Refer to the following diagram and follow the components, observations, and especially the features noted afterward:

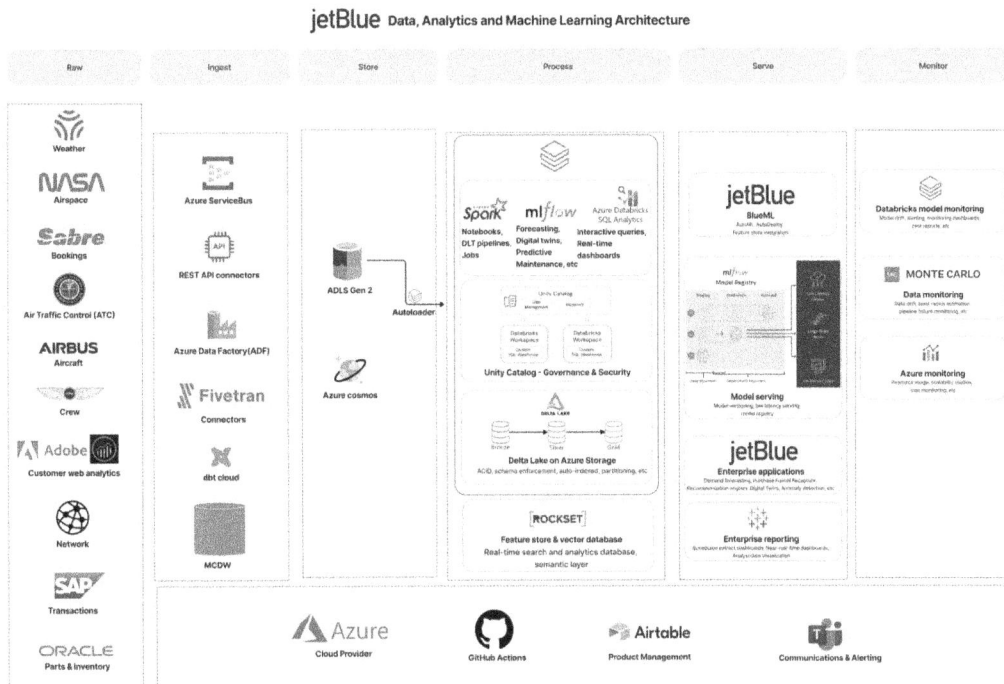

Figure 6.1 – JetBlue's data analytics and machine learning (ML) medallion
logical architecture {https://packt-debp.link/9eKqW2}

Looking at the inventory of components in the logical dataflow, we see the following list (Note: We have omitted the monitor (at the right side of the diagram) and support features (at the bottom) since they are mainly related to the software processing rather than the data engineering features of the system):

- Raw (raw zone data):

    - Numerous input sources of raw data are consumed, such as Weather, NASA (Airspace), Sabre (Bookings), Air Traffic Control (ATC), AirBus (Aircraft), Crew, Adobe (Customer Web Analytics), Network, SAP (Transactions), and Oracle (Parts & Inventory).

- Ingest (bronze zone data):

    - Ingest of normalized, but still relatively raw data is provided via Azure Service Bus, REST API (connectors), **Azure Data Factory** (**ADF**), Fivetran (Connectors), DBT Cloud, and MCDW (data warehouse).

    - Store data will retain all transformed bronze data via **Azure Data Lake Storage** (**ADLS**) Gen 2, Databricks Autoloader, and Azure Cosmos DB.

- Process (silver zone data):

    - Notebooks, pipelines, and jobs are used to internally process data via Databricks Spark with notebooks, **Delta Live Table** (**DLT**) pipelines, and Databricks jobs.

    - Forecasting, digital twins, and predictive maintenance are services provided by MLflow, custom software, and so on.

    - Interactive queries and real-time dashboards are big data services from Azure Databricks SQL Analytics.

- Process (gold zone data):

    - Metadata governance and security features are provided via Databricks Unity Catalog.

    - ACID transactions, relational schemas, auto-indexing, and partitioning are usually data warehouse features or delta lake features provided by Azure Delta Lake.

    - The feature store for real-time search, analytics, and semantics layer is a very fast scalable data cache, in this case provided via Rockset's in-memory SaaS DB.

- Service (consumption zone data):

    - Transformations are provided by AutoML, AutoDeploy, and feature store integration from BlueML custom applications.

    - Transformations are additionally affected through models that are operationally contracted by MLflow's Model Registry, serving, and versioning.

- Analytics applications are created for many business domains through customized tooling: custom enterprise applications (such as demand forecasting, purchase funnel recapture, recommendation engine, digital twins, anomaly detection, and so on).

- Dashboards for the consumer, DataOps, DevOps, and Ops are provided via custom enterprise reporting (such as scheduled extract dashboards, real-time dashboards, and analyst visualizations).

A primary takeaway from the preceding real-world example is that the *Process* stage appears to cause silver and gold datasets to undergo **data munging** {https://packt-debp.link/X6bpE6} in order to curate consumable datasets from the relatively raw bronze datasets. This is so that applications/consumers can analyze data. Data factory best practices should be applied so that datasets of different classes *do not ever* mix with each other. Without clear separation of this in the *Process* stage, complications can arise and cause two classes of data to be comingled during analytics, leading to incorrect results.

This does not take away from the excellent solution that was designed by the architects at JetBlue IT and depicted here. It was built to serve their business needs, and they did not have to treat data or its downstream enriched data as a data product, as per the principle we have adopted for our architecture. The data mesh principle has been applied where data is considered a product. In the logical architecture, we see that the choices made are *not* theoretical but grounded in business reality; practical choices had to be made. We also see that in the JetBlue logical architecture, features are often identified under the component delivered. This happens when the tool itself is the center of attention and not the feature set of the tool.

How those features are going to be brought into the dataflow and acted upon by third party tools, technologies, or Azure cloud services is less clear. You have to specify just what is needed to get the message across to the engineers and owners/stakeholders of the solution; so, you have to be willing to compromise on the level of detail evident in the logical architecture. Some architects use enterprise architecture tools at this point in the architecture design process, such as the following:

- BOC Group's ADOIT {https://packt-debp.link/qYHNVV}

- Ardoq {https://packt-debp.link/2TGV6w}

- Avolution's Abacus {https://packt-debp.link/zM6aND}

- Bizzdesign's Enterprise Studio {https://packt-debp.link/AiyFx6}

- EA Essential (open source) {https://packt-debp.link/KD81Bq}

- Mega's HOPEX {https://packt-debp.link/oCX2XT}

- LeanIX's Enterprise Architecture Management (EAM) {https://packt-debp.link/rCY3na}

- OrbusInfinity {https://packt-debp.link/sKz6Xx}

- Planview's Planview Portfolios {https://packt-debp.link/oKIQoI} (formerly Planview Enterprise One)

- Software AG's Alfabet Enterprise Architecture Management {https://packt-debp.link/y6TZaC}

- Sparx Systems Enterprise Architect {https://packt-debp.link/xIp3vJ}

Using one of these tools helps provide the drill-down mappings that will make the physical architecture definition accurate and easily maintainable as changes are made. They also take away the vendor tool focus. A tool also helps with the level of detail needed for bigger solutions without cluttering up the high-level, single-page logical diagram. If you have a budget for it, then you will find that funding spent on a tool is money well spent. What you will also need to know is that *somebody* has to be designated as the trainer and enforcer of architecture modeling processes, which can be onerous with some of these tools.

As you probably already know, architects can get enamored with their illustrations and especially their tools; so, get everybody on the same page before the selection of the tool becomes a battle not worth fighting! Focus on capabilities, and in the next section, we'll elaborate on this.

Detailed capabilities of the ingestion zones

The ingestion zones are comprised of various instances of conceptual raw zones defined in the conceptual architecture. Capabilities will be mapped to features in this chapter of the book. ETL pipelines and bronze data standards will be discussed because the total output of the ingestion zone will be the bronze data used in the downstream transformation zones.

ETL data pipelines

There are features of the logical architecture that need to be assessed for your solution. Integrations are very important to dataflows in a pipeline factory. Any vendor tool-driven approach with the features needed to facilitate a robust integration blueprint can speed up development and shorten development delivery timelines. You may then ask the following:

- *What are some data integration best practices?*

- *So, what do we recommend for the electronic data interchange (EDI) features needed for your solution?*

Let's get answers to these questions.

Some data integration best practices

Data integration is the process of combining data from multiple sources. It involves many data extractions from various systems, applications, databases, files, pipelines, and so on. These can be bulk, streaming, event-based, or supported by pub/sub gateways. After interfacing, the data is logically loaded to a destination such as a data warehouse, data lake, or another system. It is best to look at some features of important tools offered in this space. The following are some excellent third party tools to look at:

- Airbyte

- Alteryx

- Dell Boomi

- Informatica

- Integrate.io

- Jitterbit

- Matillion

- MuleSoft

- SnapLogic

- Stitch

- Workato

These tools are examined in more detail next since each has features that need to be considered. The important capabilities and features are highlighted for your attention. We have not listed the cloud providers' specific offerings (such as Azure's Data Factory). These are tightly integrated into the cloud architecture that was selected earlier in the process, and they are *not* portable. It is our goal to elaborate on the capabilities of alternative tools and allow you to decide whether a specific cloud service offering satisfies your solution's need or if a portable third party tool is necessary.

Airbyte {https://packt-debp.link/PzoOMo} has a number of features that stand out. In particular, the ELT workflow-oriented modular architecture has task automation for you to develop triggers to process specific events. This makes it possible for you to provide data orchestration, monitoring, and event scheduling. There are more than 300 built-in integrations to leverage, as well as a custom **connector development kit** (**CDK**) for integrating other features of your solution with the product. It is an open source integration platform, and as such, you can leverage the development might of the industry over time. Since it runs in the cloud as IaaS as well as on-premises, it can be an end-to-end part of your hybrid or multi-cloud solution. There are destination-placed SQL data transformations to complete your ETL flows. And the high code implementation provides great flexibility for you, the data engineer, to create exactly what you envision is required for your solution.

Alteryx {https://packt-debp.link/wzsIEa} is a no-code/low-code interface to support structured, semi-structured, and unstructured data. There is an open API for connector development. An advanced AI and ML capability enables data to be analyzed, integrated with profiles making downstream data integration easier. Blended data tooling and data analytics are available across many sources. Statistical and predictive modeling features make anomaly detection easier. The streamlined data preparation features enable the product to scale as your data grows in volume, depth, or breadth. Data transformations are customizable and support coded/scripted plugins. Analysis of data is made easier with this tool. The framework of the product enables you to implement custom data governance and security processes, assuring downstream usage remains in compliance.

Dell Boomi {https://packt-debp.link/D0myii} features more than 100 pre-built connectors, with a connector SDK available to enable custom-developed integrations. There is a no-code/low-code user interface/composability enabling reusable building blocks and automatability. The tool is built as a PaaS operation, and there is an API manager that integrates applications and/or microservices you wish to build and then call. The embedded AI capabilities make it possible for you to build fast and easily manage integrations. The service scales as needed to handle varying workloads. There is an enterprise-grade security capability, and the service is reliability built to service customers. The mission to implement democratized supporting roles is fulfilled. Also, **role-based access control** (**RBAC**) and team-based development is a central feature. Life cycle management is provided via built-in versioning with governance for various integrations, workflows, APIs, and data models.

Informatica {https://packt-debp.link/3Gk0jg} is a large provider of **master data management** (**MDM**) tooling with a special focus on ETL/ELT features. There are more than 1,000 built-in integrations. Security features (due to its market share and reputation for delivery) for data encryption/user authentication are very mature. Data profiling, data quality/data cleansing, data masking, and data governance are prominent features. The solution is scalable and performant. There is also a consumption based pricing model that can be attractive at entry but can be cost-prohibitive at scale.

Integrate.io {https://packt-debp.link/dLrRKl} provides a no-code data pipeline platform with ETL and ELT capabilities. It's easy to use, intuitive, and has many built-in connectors to make it possible for you to build comprehensive data pipelines in less time than coding them. Its purpose is to create a single source of insights leveraging its powerful features. With a no-code or low-code user interface editor to data pipelines, you can generate a solution quickly. Also, a native REST API connector exists for any more advanced features that you wish to implement to extract data from various other tools. The dynamically created clusters support your flexible scale needs. The built-in transformations include more than 220 options. The free data observability monitoring feature makes it possible for you to receive alerts in order to improve reliability. The reverse ETL features make mastered data usable for downstream use cases. Lastly, advanced security capabilities support accreditations/certifications such as **System and Organization Controls 2** (**SOC 2**).

Jitterbit {https://packt-debp.link/qG2SUm} offers a complete integration platform. Features enable businesses to connect and automate various processes. Data mapping and transformation is a supported feature. Workflow automation and real-time data synchronization bring your data to the state it needs to be in when consumed downstream in your factory. The API management features enable the framework to be integrated fully into a larger system. A low-code interface to create and then manage integrations and workflows with a drag-and-drop user interface also exists for you to customize the end-to-end solution built with Jitterbit. There are more than 400 integrations used to automate workflows along with other third party integrations. The solution is built as a PaaS solution that connects various SaaS to legacy applications. A no-code drag-and-drop application builder user interface exists to get started. The enterprise security supports your security governance integration for **General Data Protection Regulation** (**GDPR**), ISO 27001, SOC 1, SOC 2, and **Health Insurance Portability and Accountability Act** (**HIPAA**) privacy/compliance.

Matillion {https://packt-debp.link/nRoHNE} is primarily an ETL framework tool with a drag-and-drop user interface. There are more than 100 connectors available. As a cloud-based ETL platform with a focus on support for cloud data warehouses, insight extraction is made possible. There is support for large datasets using data transformation, integration, and observability features. API support exists for job monitoring, configuration edits, and other graphical user interface equivalent actions. The REST API connector is used to integrate with many data sources and destinations. Lastly, the advanced high-code and low-code tool features have been shown valuable to data engineering efforts.

MuleSoft {https://packt-debp.link/GD5tHz} is also an ETL framework used to streamline connectivity to enterprise applications. It also has more than 100 built-in connectors with templates from the Anypoint Exchange marketplace. API development support is available for REST, **Simple Object Access Protocol** (**SOAP**), and event-driven API designs. PaaS cloud-hosted services are core to the framework. Cloud connectors/services are enabled using existing **Open API Specification** (**OAS**) or **Web Services Description Language** (**WSDL**) specifications. MuleSoft is a high-code all-in-one data integration platform for data engineers.

SnapLogic {https://packt-debp.link/rhBmSS} has extensive ETL/ELT capabilities for on-prem data sources. The reverse ETL capabilities enable you to move cleaned, optimized warehouse data to other third party tools. A drag-and-drop user interface is available for configuring flows, and it has pre-built connectors. SnapLogic is also a cloud-based integration platform, so it works well for hybrid-hosted solutions you may envision.

Stitch {https://packt-debp.link/0AH5Oo} has more than 140 built-in connectors and offers low-code integration. There is built-in automation and lightweight simple data ingestion. ELT support exists for sources with a focus on loading data to data lakes and warehouses. ELT volume based pricing exists, which can be helpful when starting but can impact your ability to scale data as you grow. A small company/volume support is the focused market.

Workato {https://packt-debp.link/9PWs7f} features a no-code user interface with more than 1,000 integrations. The trigger/action automation activates unique workflows by *recipe-based integrations* that define the *trigger* and the *action* to be taken, and it has a visual configuration management interface for provisioning the framework.

The features of these tools, frameworks, and services need to be evaluated for your solution. You want to have as much help as possible when integrating your solution's ETL/ELT features. You want high-code support for complex logic, but often you will require a low-code solution for common ETL/ELT logic. You want to select tools that are compatible with your scale, performance, and cost needs and not be hindered by the tool's operational features when the functional capabilities are superior. Both functional and non-functional capabilities have to be aligned with your solution's criteria in order to be successful. Also, the connector or integration features of your other solution tools are not to be overlooked since these must all fit your overall needs. Lastly, consider reverse ETL needs as you move data between zones or populate digital twin datasets in integrated tools.

Recommendations for the EDI features needed for your solution

The answer to the question raised about recommendations will be based on your desire to buy versus build, your budget, and the scope of your integration pain. Some have only a few data integrations of high complexity, and others have many data integrations of low complexity. Either will add to your integration pain. The best practice is to figure out the total cost for a short list of tools and approaches. Just make sure that the capabilities you desire from the conceptual architecture are present in the overall solution's features with the tool(s) of choice, and then add the result to your logical architecture.

Significant features from the preceding EDI tools list are the following:

- **Reverse ETL** features enable **change data capture** (**CDC**) capability to be implemented more easily than with tools that support this feature. A CDC capability is a core need of the solution when moving datasets between your data zones. Then, prepare them within release sets before combining them into a distribution set for posting to the consumption zone. This is a lot of movement and synchronization. Standard patterns, testing, and discipline are needed so that this is 100% rock solid.

- **No-code/low-code** features for developing and maintaining user-directed dataflows when the analysts become more programming savvy, and they want control of their data analytics spaces. This gives them tools to make that possible. It is a wise choice; it removes you from their perceived thought as being the bottleneck to their success.

- **API support** for sophisticated data engineers who have to support complex data pipelines and configurations in a DevOps/DataOps IT environment.

- **Connector/cloud connector/extensible connector** support is required for any tool, else you get the power and speed to market without the flexibility to add and cover your total needed capability with features that could be limited.

- **Security** support that is feature rich/flexible enough to support your governance model. These integration tools are limited in their security flexibility, and they will cause you to change your integrated security design. If the tool chosen does not provide for ease of security integration, especially in the cloud with on-prem sources, think hard as to whether the benefits outweigh the pain.

- **Reliability, availability, scalability, and performance (RASP)** concerns will arise with any of the tools. How the features of the tool support your needed capabilities in this regard is important. Put your total integration of the tool as a logical architecture into the TCO, not just the tool's costs. The **high availability** (**HA**) footprint for these tools can double or triple your costs immediately!

- **Compliance** auditing and standards verification are important for some businesses that are regulated. The EDI tool selected has to offer features that map to your business needs. Any great tool should support some minimum audit features for basic forensic analysis of faults and data quality. Some tools are very weak in this regard and have to be augmented aggressively with your value-added integration.

A deciding factor for the decision to use one tool or another could be the tool's cost or the tool vendor's risk. They are not insignificant factors in your decision. Note: Cloud providers are implementing feature subsets of these tools with the corresponding cloud provider service lock-in. So, beware of cheaper alternatives that really hide risk. Perform due diligence across tools, features offered, and vendor profiles. Afterward, map the result to your features needed while identifying tools' features that *will not* and *must not* be used.

Without this guidance, the vendor will grow their usage footprint across your IT landscape and take control of *your* architecture. It's what they do! They are in it for profit. Beware now: once a vendor gets a foot in the door, they will leverage that foothold to gain more of your IT landscape and consume your IT budget! Set the vendor and their tools at limited use and stick to your decisions.

Once data is ingested and you have a clear idea as to how it will be processed at the zone boundaries of the architecture, you are set to define bronze data standards. In the next section, we will dive into bronze data standards since the logical architecture best practices we present depend on the quality of the normalized bronze data to be preserved throughout the solution.

Bronze standard datasets

In the raw zone, bronze data standards are first going to be defined and applied. These drive the data transformations, normalization, enrichment, mapping, metadata cataloging, and final placement activities of datasets into the data lake.

Be aware that you will need to make certain decisions about implementation. Firstly, you will have to decide on an initial data format for bronze zone-hosted data; for example, serialized data in Parquet formatted files or Avro files. Then, you will want to apply performance-enhancing techniques such as using Databricks Delta tables for frequently accessed data. Also, you have to look at the many options available for data serialization so that data can be accessed as objects. You can expect your data organization to change over time as input volumes grow. Be ready to reorganize your bronze data while preserving the interfaces that the downstream data processing code needs to use in order to remain stable.

Events and activities will be directed as orchestrated event queue-driven activities across a number of pipelines. These pipelines execute in parallel and require a great degree of coordination in order to squeeze as much performance out of the infrastructure with as little cost as possible. From past experience, in the following sections are a few observations and lessons learned that will serve as a background to our best practices.

How does big data processing complexity affect my data pipeline development?

When building a vast data processing factory in Hadoop, we encountered the need to optimize job scheduling and track low level job completion to high-level job invocation. It sounds like an easy task, no? Well, it was not!

The monitoring tool of choice was Pepperdata {https://packt-debp.link/CGR1U3}, (later HDP Insight, then back to Pepperdata – but that is a longer story). Although Pepperdata was a superior tool for monitoring low level **Hortonworks Data Platform** (**HDP**) cluster jobs, the job tracking lacked the capability to provide a view of low level jobs in the context of the high level jobs from which they originated. When looking at the dashboard, it was like a pile of stuff with uncoordinated meaning. No action to correct problems was evident since the mapping was absent. Confusion was the result. What was needed was clarity in the dashboard that could only be driven from segmenting the raw data of low jobs to their originating high level invocations. This was needed to meet the data SLAs set for the end to end curation and delivery of data releases to the analytic teams.

Note that the low level jobs could run for days with thousands of sub-jobs any of which could fail and cause the high level job to abend. This was a big problem since all that time and running cost was tossed when a failure occurred. As the jobs' code grew organically over the years, what took 5 minutes 5 years ago could take 50 minutes today. This was due to the ingestion of data as that data grew over time.

You may ask – *Was the best practice for implementing checkpoints implemented?*

Confirming checkpoint implementation

The answer to that was *no* – checkpointing was not implemented, and worse yet, the poor design pattern existed across hundreds of high-level jobs with literally 950+ instance runs per week, with 1.5 billion raw records to process each week. The final overarching issue was that the entire data factory had to be complete 1 week after the end of the month. Processing data incrementally throughout the month as it arrived was essential, and the platform ran on-premises, so you could forget about auto-scaling. If the schedule was missed, end-user data delivery was stopped.

As with most businesses, the customers would demand their subscription dollars back as per their contracts as a penalty. The processing was consuming 3,500+ cores across many server instances running HDP 2.6.5. So, to make a long story short, it was discovered that the failure of sub-jobs was caused by **Java Virtual Machine (JVM)** config changes that looked to be right, but because the data in the jobs processing scope had grown so large, without proper data stress testing they sometimes failed. It was like trying to figure out what dial among hundreds was causing sub-job failures. And there were a lot of dials to turn! A cascade through the system resulted after the high-level jobs abended after three high-level retry attempts. It was a huge waste of CPU cycles, resulting in tossed work till jobs *finally* run to completion. Trying to find out the root cause would go a long way toward fixing the system's sporadic instability, but the fix(es) had to be applied differently across hundreds of job types and then almost a thousand instances of these job types. Memory usage due to data volume increases was a chief contributor to the problem.

As a data engineer and computer scientist, you have to have a hypothesis, and then you test to make sure you rule out all errors and zero in on a conclusive affirmation of your hypothesis, then take action. Pepperdata consulting showed the team how to best use their product and then proposed some new real-time dashboards be built by their consulting unit to help identify when failures at low levels occurred. Enabling the capture of detailed logs, followed by dynamic turning up of log levels and finally rerunning sub-jobs till all errors were removed helped. Hadoop configuration complexity and unclear knowledge of the HDP platform's tuning correctness, followed by dwindling platform support (Hortonworks HDP Hadoop was legacy for sure since it is a retired product) created huge issues that needed to be addressed.

The Pepperdata tool helped demystify the black box that the complex distributed big data engine presented. When a solution is not future-proof, you will eventually have to justify why it's time to admit failure and move to a new solution that is future-proof. Not knowing what is causing you pain today will only become more difficult to fix over time as tooling and platforms melt down under your watch. You have to apply a key best practice…

Track the tech debt for logical architecture choices!

Make sure you have subject matter experts (SMEs) in data and software on call when you see that logical components are entering their sunsetting phase. They will all begin to melt down under you over time!

Know your configuration options and how they impact your solution!

Configurations are just as important as code! Make sure that they are controlled, checked in with code, and tested prior to code execution. A best practice is to have a configuration management process, governance, and an operational API for use in all aspects of service management. Configuration quality will affect data quality and data pipeline throughput. Build configuration assumption tests at the start of a pipeline's processing and for each step if at all possible; do *not* think that the configurations are just going to work for you auto-magically.

Some DevOps engineers will increase your JVM memory limits, garbage collector or its settings, or lastly, the distributed engine's parallel processing or retry limits and *not tell you*. All these affect your contract for data delivery within the factory. Also, make sure you have subject matter experts in data and software engineering on call when you see that logical components (or third party tooling) are entering that sunsetting phase and they begin to melt down over time.

In the next section, we will elaborate on the capabilities of the transformation zones.

Detailed capabilities of the transformation zones

The transformation zones are comprised of various instances of conceptual bronze, silver, and gold zones defined in the conceptual architecture. Capabilities will continue to be mapped to features in this chapter of the book. Data quality features, data lake and data warehouse access, and silver data and gold data standards will be discussed because the total output of the transformation zone will be the consumable data distribution sets used in the downstream consumption zones.

Before data can propagate downstream, it needs to be normalized and transformed. The illusion that you can take plain old raw data, retain it, and then process it in place without this transformative step is naïve at best. Raw data is garbage before QA processes are applied and the corrective steps are taken to adjust, transform, enrich it for mastering, and finally roll it off when no longer valuable due to time sensitivity. Additionally, do we have to mention the technical data debt that would build up over time as data standards, formats, and processing software change? Imagine if the simplest of quality metrics and adjustments were not retained along with corresponding raw data. The bloat would surely kill your budget, and you would have built a data swamp rather than a data lake. To fix this, you need to add features for data quality since the operational contracts discussed in the prior section also enforce quality contracts.

Data quality features

Data quality is a subjective as well as an objective goal. It is subjective in that a consumer may know information context and semantics that the system does not at the time of processing or during rendering its output in a deliverable set of data. It is what the IT solution implementors did not know that will cause the data to be subjectively wrong.

This brings to mind an experience from my past where store point-of-sale (POS) data was presented to a retail data aggregator that looked good, and it was even presented correctly; however, a small but still significant overstatement appeared for an item's sold quantity that produced a deviation from prior monthly data. What looked like good data was, in fact, hiding some offending data. Normally, correct POS data was received across thousands of such items sold throughout the US and EU. But hidden in the gross numbers was this small blip at a single retail location in France that sold more than 200 of the items in question. This was just not possible! Was it a data receipt error?

After calling the retailer out on the anomaly detected, the retailer admitted to receiving a large quantity of the offending items from a particular questionable source. After digging further, the source turned out to be a product counterfeiter from who-knows-where. The remaining gray market items had to be removed from inventory, destroyed, and a retransmission of the raw dataset (with real product sales) provided before our month-end close. What picked up the anomaly was the data trend check that required a human to fully investigate and resolve. No existing IT-based QA check picked this error up. The only indicator we had was a data trend deviation from the prior month. Data was received correctly – check! It just was wrong from the business/legal perspective! The lesson learned was that data trend anomalies are just as important as explicit IT data validation checks. It is in the checking of data metadata and statistics that some of the best data QA indicators arise.

These data anomalies still require forensic resolution, some detective work, and final data correction and restatement to correctly resolve anomalies. It is most important in a data factory that the data and its trends pass the business credibility test. But the following question arises.

Do subjective QA checks scale?

The answer to the question is: No!

Human-in-the-loop interactions must be replaced by statistical and AI-/ML-based anomaly detection, data wrangling, profiling, and even recommendations for corrective action. We've seen **master data management** (**MDM**) solutions melt down with tens if not tens of thousands of human data remediation workflow tasks. The IT implementor may have thought they solved the data QA problems presented by the business when they built to their requirements, but quality testing is a bit of an art of the statistician as well as a science for the engineer. Building quality QA does not mean you can fix the issue with just a few manual tweaks. Not so!

Trend anomalies are a simple class of the overall set of anomalies that can be detected. Consider calendar-driven item sale volume deviations from month to month that affect data trends or during one holiday season to the next year. A similar trend gets set within a trend. What about explainable skewed data due to real-world events, such as there is a spike in flashlight sales in a given region due to a blackout that lasted 3 days? False positives and false negatives have to be resolved as well in a much smarter way for systems to scale and continue performing at scale. You must be able to address objective and subjective data quality needs of the future.

You must apply automated QA and statistical anomaly detection to detect and then fix data quality issues

Important features of any data quality solution will most likely include tool selections. This will involve some detailed thought. Here are a number of features for consideration when selecting data quality features as part of your solution and/or tool selection for your designs:

- **Artificial Intelligence (AI)** and **Machine Learning (ML)** enable quality automation to scale QA processes with the still required human judgments.

- **Clustering** involves the merging of similar values using ML intelligent techniques, minimizing duplicates, and so on.

- **Data cleansing** involves removing duplicate records, addressing missing values, and rectifying inaccurate but correctable data points.

- **Data discovery** enables data to be found if buried in other contexts.

- **Data enrichment** involves adding alternative data to supplement data from external sources to enrich the primary dataset.

- **Data faceting** enables you to navigate and analyze by faceted filters.

- **Data governance** integration makes the inevitable human corrections or machine learning required judgments to effect actionable changes to current and future datasets.

- **Data masking** is a way to create a synthetic but realistic version of data. Sensitive data is thereby protected but the resulting data is still usable in your solution.

- **Data monitoring** tracks data quality metrics across datasets. It supports anomaly detection alerts and metadata lineage recording as well as data quality stats for transformations.

- **Data profiling** provides data-harvested insights related to quality, detected anomalies, and a general understanding of statistical data patterns.

- **Data quality management (DQM)** provides process support and dashboarding of the data solution's quality.

- **Data trust score** is a single rolled-up metric to indicate the overall data quality, trust, and confidence.

- **Data validation** defines specific rules, standards, and patterns that data must adhere to before being labeled as valid.

- **Infinite undo** and **redo** are features to reset a QA flow as a result of later QA stages proving or disproving the presence of previously detected or assumed errors.

- **Monitoring** and **reporting** deliver a continuous monitoring/reporting feature using standard metrics such as data accuracy, completeness, missing columns, and gross bulk receipt errors, with proactive data issues management being the goal.

- **Reconciliation** enables the matching of entities in a dataset with some other entity, possibly in another dataset.

- **Record de-duplication** involves the removal of redundant data.

- **Relationship mapping** includes features such as **address and entity** validation with the ability to extract, validate, and map entities together.

- **Scale and performance** are features of the Data Quality Management (DQM) solution to handle QA processes at scale for the data volumes expected.

From the preceding list, you will no doubt want to ask: *How do you get a head start on the development of these features?* There are some great tools to list here. Look at the ones that cover aspects of these features you require. Not all of them are in a single tool. Significant vendors have been separated from the lesser-known vendors, but both sets are worthy of consideration.

Here are some examples of significant data quality tool products:

- Ataccama ONE {https://packt-debp.link/syv4jM}

- Datiris Profiler {https://packt-debp.link/lGYntX}

- IBM InfoSphere Information Server {https://packt-debp.link/edd8rf}

- Microsoft Data Quality Services (DQS) {https://packt-debp.link/1Uq2Yu}

- Oracle Enterprise Data Quality {https://packt-debp.link/HMRAJm}

- PiLog Intelligence Data Quality Management {https://packt-debp.link/nD3tYq}

- SAS Data Quality {https://packt-debp.link/eoWBxO}

- Syniti Knowledge Platform {https://packt-debp.link/I1gyHn}

Other data quality tools include the following:

- Collibra Data Intelligence Cloud {https://packt-debp.link/ibiDKY}

- Datactics Self-Service Data Quality Platform {https://packt-debp.link/ZQF8qF}

- DataRobot AI Platform {https://packt-debp.link/G6NI11}

- Experian Aperture Data Studio {https://packt-debp.link/4tpOOW}

- Informatica Data Quality (IDQ) {https://packt-debp.link/izlrla}

- Precisely Data360 (Trillium) {https://packt-debp.link/hE64uu}

- SAP Data Services {https://packt-debp.link/TO0gT3}

- Pitney Bowes's Spectrum Technology Platform {https://packt-debp.link/NSDQGy}

- Talend Data Fabric {https://packt-debp.link/PM2pJa}

- TIBCO EBX {https://packt-debp.link/qo1yZt}

As you can see, there are a lot of vendors vying for an opportunity to show you how well they address your data quality needs. Please read the links in this section to the various vendor sites in order to generate your due diligence. Some enterprises require top-tier vendor tools and others can put together a credible solution from lesser-known niche players, yet others can start small and later grow to use larger vendors' tools. Just plan your QA features carefully, and if you do, your solution will have a built-in safety net against failure on your road to success. Even when a system begins to be stressed, that safety net will buy you the time needed to scale your data processing till you're ready to enhance the solution as the depth and breadth of your domain's data coverage expands.

Data lake house and warehouse

When talking about scalability in our logical architecture, the data lake is a natural fit to cover that basic need. It is natural to think of cloud storage as an infinite repository of cloud file storage, with Apache Spark processing over that storage. This is supported by tooling from Databricks that wraps the native cloud provider services supplied (by Microsoft Azure, Amazon AWS, and Google GCP) to form great solutions. With Delta Lake caching features, the solution is almost ready for online transaction processing, but hold on – think it through first. There are real latency costs for starting up a transaction in a system underpinned by Apache Spark. That delay is in many seconds, even if the system is warmed up beforehand.

How much will you be willing to keep that system hot when it should be scaled back after a big query runs?

Just because you can do it doesn't mean you should! Implement the right technology for the task. For data pipeline micro-bulk processing, the data can satisfy many transforms, but the data has to be laid out in an optimized mapped format in order to facilitate the execution of optimized queries. It then has to be replicated for OLAP and then kept in sync with potentially a third replica for AI use cases. Expect the data lake to exist as numerous data lake instances in any given zone when building out your solution.

Also, consider the security and departmental chargeback requirements for processing data optimally in a multi-tenant data lake. With or without built-in features such as chargeback activity-based accounting, the costs have to be separated for TCO and **return on investment** (**ROI**) accountability. In a real enterprise, these charges have to be sent back to the domain owner (also known as department chargeback). Small shops have maybe one or two domain owners. Larger organizations need to optimize costs and have many domain owners running jobs. These are on your platform, and they need clear separation. **Cloud service providers** (**CSPs**) rarely provide features for cost accountability in multi-tenant implementations. This is left to you as an implementation detail. It is not! It's an architecture detail and must be addressed in the logical architecture. This is where you do not design a solution without some thought of the practical physical architecture implications; otherwise, key features will be missed. There will be equal unhappiness with your IT architect, software, or data engineer if big misses occur when architecting a multi-tenant solution.

Implement high-speed in-memory caches and data marts in the transformation zone

Do not provide direct access to downstream gold-zone data warehouses. In addition to the previous data lake issues, you should consider whether any data warehouse should exist in the transformation or bronze zones. In past experience, the need for a high-speed data mart (drawn from silver or gold zone data) or an in-memory data store (drawn from the same sources) could be selected for this data warehouse purpose. Direct access to the silver or gold zone data of the data warehouse would slow downstream processes. For AI/data profiling, the read-only silver zone caches of the gold zone data will be scanned *MANY* times. Reference data, sources for matching metadata, profiled data, and so on all are examples of the classes of data that will need to be accessed. Also, any created data maps to core data in the data lake's dataset will need caching in a delta lake or other caches that you create as in-memory DB storage tables (also known as Redis, MemSQL, and so on).

After data profiling, data wrangling, data quality, and data transformations take place; data is adjusted, and the adjustments are sent as a CDC queue-formatted message as topics. The CDC message topics can be subscribed to and applied to the next zone's storage in that zone's data lake, data warehouse (or both), or other glued-together EDI integration requiring the application of that set of changes! The reason we are diving into this detail is that you want a few ways of implementing CDC in your system. Be aware that CDC is necessary for maintaining the logical correctness of datasets at zone transitions. If a chosen tool in your solution does not support CDC, you will need a custom workaround to provide it. You will want to govern and then police what the data engineers and IT software engineers use when gluing the system together. They should use the blueprinted architecture patterns and not deviate from them; otherwise, the system generates a lot of tech debt and baggage maintenance that will cause operational DataOps heartaches.

Cloud providers have continued to innovate in this area, and AWS has announced zero-ETL {https://packt-debp.link/QK1oUH} capabilities for its Amazon Aurora PostgreSQL, Amazon DynamoDB, and Amazon **Relational Database Service** (**RDS**) for MySQL integrations with Amazon Redshift. AWS customers are now able to analyze data without custom reverse-ETL integration code. This is possible since the PaaS services have been modified to capture log activity and transform that into a CDC stream that is applied at the destination with cloud adjustments to the target core service.

Gold and silver standard datasets

When you think of gold data, you think of data as being unique, rare, and *priceless*! That is how it should be viewed. It is the **single source of truth** (**SSOT**) for your system. From it flows all types of caching and versioning. It is the end product of the ingest and transformation states of various pipelines. It is relational, partitioned, and sometimes *huge*! Gold data will be used to reverse ETL the loading of consumable data stores of the consumption zones, and it will be reverse ETL'd and leveraged in the silver zone for use at any stage of the data pipeline of your factory either as an in-memory cache or as a data mart.

Gold data conceptually contains classes of data with semantic metadata representing the meaning of data at rest. The gold zone will be fenced off and tightly governed, audited, and protected with fair use rights that have to be preserved downstream. Contracts are assigned to the datasets to preserve all types of service agreements, reuse rights and entitlements associated with data, and any derivatives of that gold data. What makes gold data particularly important to an enterprise is that it closely maps to the legacy concept of it being *master data*. But it is not always master data, yet it carries the same data contracts that master data should since it resides in the gold zone.

Gold data is accessed as classes but may be stored relationally. This is important because semantics requires data classes to be instantiated. The instantiated classes, with properties, relationships, and contexts, define how a fact relates to another in a structured way. You may look at **Resource Description Framework** (**RDF**)/**Web Ontology Language** (**OWL**) {https://packt-debp.link/ggKDsj} models to define the classes and an operational knowledge base such as Stardog, Ontotext's GraphDB, or PoolParty {https://packt-debp.link/HFY1iU} to formalize the semantics with tools such as TopQuadrant to create the model.

Or you can create a custom semantics engine in software and leverage property graphs such as TigerGraph {https://packt-debp.link/pD2vTs}, Azure Cosmos DB v2 {https://packt-debp.link/4E031d}, Neo4j {https://packt-debp.link/71GBjq}, or JanusGraph {https://packt-debp.link/XkUvw6}.

You can also use dictionary tools such as Manta {https://packt-debp.link/4RoR74} or OpenMetadata {https://packt-debp.link/opcUwS}. Expect a lot of human maintenance with the use of the dictionary tools.

Silver data is data that is being prepared for transformation and reverse ETL'd into the gold zone, or it is data that is reverse ETL'd out of some gold dataset for caching in some other data pipeline as read only data. It will be used to facilitate processing in that data pipeline. Since data is cached or placed into a data mart format, you will be servicing that data with semi-offline OLAP use cases. We want the silver data to carry with it the contract that was set in the gold zone from which it originated if populated and/or synced from gold data.

Why all this formality about gold and silver data differentiation and zone-to-zone data movement?

We have found all too often that data can be used when it should not be used. If you do not build high walls, the data seems to escape, leak out, and get used where it should not be used. Imagine if someone's trading pattern history were exposed. As individual transactions, it may be useless, but if all those transactions were taken into consideration at the same time, a pattern would emerge, and if that were discovered, then what a compliance mess would arise! Just in setting entitlements for a data table, even fine-grained **row-level security** (**RLS**) and **column-level security** (**CLS**) will not protect the overall semantics of the data. That comes with an understanding of the data's metadata, rights to use that data in some context versus another, and an ability to add value to data. Even if a use case is entitled to read a given dataset row, that entitlement does not make it allowable to be combined with other rows. Dynamic entitlements and usage rights validation can get very complex and are subject to activation only at query time and then only in real time. Try to do that with RLS!

When building the logical architecture for moving data between zones, you will want to consider using scalable pub/sub features of cloud message infrastructure such as Amazon Kinesis, AWS Kafka, and AWS **Simple Notification Service** (**SNS**) {https://packt-debp.link/bQb6cH}, and maybe AWS **Simple Queue Service** (**SQS**) {https://packt-debp.link/TLl0zN} or the Microsoft Azure or Google GCP equivalents.

What you will want to do is understand and test completely and fully the queuing facility to determine its scalability, performance, and recoverability features. You will also have to address the question, *What size is the pub/sub micro batch?* You'll also want to learn how the cloud or third party queuing services work properly and then improperly. This is when queues run over or how they address dequeue latency. You will encounter these issues when you first get started dequeuing messages or experience recurring read latency after any dormancy between bursts of write activity on the queue. These tools and cloud services often exhibit odd behavior that needs to be worked around in your design.

As an example, SNS and SQS can be combined to create a more robust and reliable system. Together, messages can be delivered that require fast event notification and also have that event (or data) persist in an SQS queue so that it can be picked up for processing later in **First in First Out** (**FIFO**) message order.

Use queues for CDC updates, commands, and events

When implementing pipelines, we rarely send data in queues, but events, commands, and CDC updates are all part of the activities that should be sent. The transmission of large data in queues may simplify coding, but it will clog up the system and raise your costs. Even SQS queues run out of queue resources; we've been there and seen that happen! It's truly a sad day when it occurs, especially if you don't have a bulk resync feature between the source and target of your architecture. When a data engineer has to then tweak, copy, patch, and fix data, your architecture has failed to account for real-world cloud events, even the ones that the cloud provider imposes upon you at 2 a.m. when they cycle services, instances, and availability zones. Beware! Build to bolster the weakest link in your features dependency list. It will appear in the scalable high speed queueing under huge volumes! Make it all robust, bulletproof, fully operable, and observable. Then you will sleep well at night!

Additionally, we've had many discussions with software engineers who want to exclusively use microservices with networks of service dependencies rather than store-and-forward persistent queues for providing RESTful-driven state changes across an enterprise system. The answer is it's a great idea except when huge volumes come into play and you have to drop back to an activity-/command-driven message queue for notifying those microservices of pending bulk data operations.

Implement streaming data as micro-bulk operations

Architect streaming data changes via queued updates, but for large data volumes, implement these as micro-bulk operations with queued commands to apply bulk operations external to the queue.

Maybe we took a bit of a dive into the plumbing, so let's pop back up to the higher level and see the big picture again. It can be a maintenance hassle to manage each queue and data transfer in separate code snippets of the cloud. Even with the delays caused by auto-scalable queuing, you will have hundreds of active pipelines that are in flux. The last thing you want is to have each operate differently and uniquely report their activities in various logs. You will want to handle all these as a common architecture pattern blueprint with base code that is shared along with the configurations that will be driving specific logic for each data pipeline and physical queue.

In the logical architecture, you want to specify these patterns and be clear with the data and software engineers what the solution's selection is to be and then mandate that it be used. If necessary, enable developers to augment/add value, and make sure updates are backwardly compatible. But *do not* depart from the decision and use of fully customized queue mechanisms. We can't begin to tell you how this scenario has played out across the years and in many projects.

Detailed capabilities of the consumption zones

The consumption zones are comprised of various access methods and caches that were made available as distribution sets. All or part of the bronze, silver, and gold zone data, as defined in the conceptual architecture, is the target for this consumable data. Capabilities will continue to be mapped to features in this section of the book. A data analytics focus will be presented for BI, OLAP, and other big data access patterns across the data lake, data warehouse, and data mart technology spectrum. You may already anticipate the implementation as being a mix of in-memory/multi-dimensional storage for any distributed and consumable bronze data, silver data, or gold data that was discussed as the total output of the transformation zones. These were the consumable data distribution sets output by the transformation zones.

The type of synchronization needed to populate analytics spaces with mostly gold datasets (but not limited to just these sets) after discovering them in the dictionary/catalog will be a daunting task. It is assumed that the consumer is able to clearly understand the dataset after reflecting on its semantics defined as metadata at rest along with its data lineage. Every analyst, data scientist, or big data engineer wants all the data to be immediately accessible, in memory, and ready to be used to rock and roll for OLAP, BI, and ML use cases. Well, the truth is, data engineering is hard and the costs can be high if too much is opened up for high-speed access with the wrong type of use case. The architect and data engineer have to give the consumer access to data but not absorb the costs for supporting use cases or consumer requests that are exorbitant. It's prudent to give the consumer a cost estimate before launching instances, spinning up OLAP spaces, or granting access to expensive Power BI premium service instances. Some of the cloud services of the data pipeline will be multi-tenant capable and it is not OK to spin up areas that can't be shared when cloud or on-prem departmental costs matter.

Building this cost estimator feature is not going to be an easy task

The cloud services, third party tools, and even your homegrown integration designs may not provide billing transparency (do you have to wonder why?). Build a cost estimator feature into everything, and you will save the consumer from asking for something that they are not willing to pay for.

Now, we need to spin up the cloud service plumbing, the spaces necessary to populate single or multi-tenant analytics areas, and grant access to existing data zones for the requesting consumer.

Building an analytics data manager is not an easy task

Logically, building an analytics data manager for data quality makes sense. Don't think that one comes out of the box from a cloud provider or a third party tool. It would be nice to see an analytics data space manager feature. To make this a reality in the future, you have to build one. For now, the cloud providers do not offer a solution. In fact, many cloud service technology implementations today are going to contend with your needs. Logically, we are providing data to the consumer; sometimes, we have to populate it (with or without real-time pub/sub sync) as follows:

- Into a local cache

- Via vendor-supplied cloud database sync capability into a data mart

- Via custom yet facilitated OLAP space-populated mechanism (Azure Power BI to Azure Synapse direct query)

- Via custom yet facilitated big data bridge mechanism (Databricks/Stardog endpoint {https://packt-debp.link/vAVl6j} connectors)

- Via custom queue-based sync to implement CDC

After spinning up an analytics space, and after granting access to various cataloged datasets noted as part of the delivery set, zone data-hosted datasets will then logically appear as part of the analyst's space. All necessary metadata is provided for datasets copied or exposed to the consumer, along with drill-down features for full disclosure and understanding. Contract terms have to be hooked to service monitoring next. In the past, it was good enough to statically monitor system services, but the data contract we just committed to when populating the analytics space needs to be enforced and then monitored till that space is decommissioned. The data contract monitoring feature has to be built in a dynamic manner.

Now that the analytics space is ready, its features need to be drilled into. This will happen in the next section.

Data analytics

The analysis of data is a big topic, and as a data engineer, you will want to know how to service a few primary patterns for data analytics use cases:

- **Metadata discovery** leverages the bronze, silver, and gold data directly since it has dictionaries (aka data catalogs) for lineage and structural metadata at rest but *through* the consumption zone accesses grants.

- **Big data** leverages a clone of gold data as the silver data for highly distributed job segmentation purposes *into* the consumption zone.

- **Machine Learning (ML)** is built on the use case class for big data but includes in-memory caching (aka Delta Lake) as the silver data for highly distributed jobs *in* the consumption zone.

- **Business Intelligence (BI)** leverages bronze data, silver data, and gold data synced *into* the consumption zone for in-memory multidimensional analytics speed.

- **Online Analytics Processing (OLAP)**, as with BI data, is synced *into* the consumption zone.

- **Graph/spatial analysis** leverages gold data reverse-ETL rendered as a **labeled property graph** (**LPG**) *into* the consumption zone.

- **Semantic analysis (aka knowledge graph analytics with inferencing)** leverages gold data directly for at-rest semantic metadata, which is an RDF/OWL model of the business domain owner's data that will be accessed *through* the consumption zone.

- **Knowledge base analytics (advanced AI = ML with semantic analysis)**, as with the knowledge graph, leverages gold data-at-rest semantic metadata and instances of those semantic classes directly but with access *through* the consumption zone.

- **Security analytics** leverages all datasets in any location supporting the enterprise's **defense-in-depth** (**DiD**) and zero-trust assessment, penetration, vulnerability footprint, or forensic drill-down use cases. Access is provided *through* the consumption zone.

Each type of use case has a keyword emphasized: *into* or *through*. This is significant because we are faced with cloud realities and the technical limitations of technology. Sure, the analyst wants it *all* now in memory and ready to go, but that is not always possible with exabytes of data defined to be in scope for the analytics scenario! Sometimes you just have to allow read only direct access to a dataset in other zones, and at other times, a read/writable cloned snapshot has to be created for performance reasons (with eventual transaction writeback). Often, the rate of data change in a zone would slow analytics to a crawl even if only read access were permitted. Other times, data remains unstable for long periods; therefore, a data mart clone is required – with the option to bring the source dataset to a consistent state.

You will see from the previous list that each set of use cases could be explained with a book, but for brevity's sake, we'll dive into the data engineering needed to hook up the consumption zone's analytic spaces supporting these use cases. We will dive first into why and how silver data is leveraged.

Accessing silver standard datasets from the consumption zone

Silver data consists of datasets in the pipeline that are being transformed but not ready yet for finalization as gold data. Silver data may also be reverse-ETL synchronized with existing gold data (with or without pub/sub) to provide additional enrichment or transformation of the dataset. A cloned silver dataset could also be used in the process of enriching yet another silver dataset as it journeys toward its state of release readiness.

To sum it up, silver datasets are composed of the following:

- Data that is progressing toward being releasable and finally ETLs into some downstream gold dataset.

- A read-only dataset or portions of a dataset that were cloned from the gold data. If cloned data is changed or used to enrich other datasets, those changes have to be transactionally queued and then merged into the originating gold datasets affected.

One size does not fit all, so derived data is created for republication via ETL

You may ask, *Why not transact directly on the gold data when writing back rather than queuing?* After all, isn't the data warehouse capable of providing ACID transactions. Software engineers will want to transact writes directly on the gold data, but multiply that by billions of rows and you can envision how gold zone performance will slow to a crawl. The best practice answer will address these concerns:

- Scale (too many *writes*)

- Speed (data warehouse *writes* are slow)

- Isolation (so that performance bottlenecks do not have a negative impact)

You do not want to affect subsequent downstream analytics spaces of the consumption zone. To do that, you have to reduce the *write* transactional load on the gold zone:

Separate load from the bottleneck and optimize gold data write transactions as bulk operations via correct queue management.

The analyst, data scientist, and knowledge engineers are key consumption zone users of the datasets impacted by any poor performance of the solutions gold zone.

Technical limitations of memory speed and communications bandwidth exacerbated by *huge* data sizing all are factors affecting where and how the consumption zone is to be hosted. This affects the logical architecture as a hard, practical reality.

Trade-offs between public cloud, on-premises, and multi-cloud

Trade-offs between the public cloud, on-premises-based system hosting, and multi-cloud strategy must be considered within the architecture. The data and software engineers will work with the architect to pragmatically decide whether a solution is best suited for on-premise (your enterprise data center), cloud (IaaS and PaaS), or **service provider** (**SP**; SaaS). There will be many factors that should be boiled down to an optimal solution reflecting total time and total money over some run-manage time period denoted by the strategic plan (a 5-year budget, usually).

The best practice here is to get this issue in front of a financial modeler. Many times, we've had to take off the architect or engineering hat and put on a financial analyst hat in order to make a case for a clearly perceived but not yet proven architecture. Let's see some factors to put into the model.

Cost of ingest or egress for cloud data

Some cloud providers charge a lot for data input and little for data egress. Others think the reverse is true. These charges apply for SQL query results and queues, and they hit SaaS integrations especially hard where the SaaS provider is not in the same cloud, the same region, or the same availability zone or is in some other region even when hosted in the same cloud.

Cost of a dedicated network line to the point of service

The cost for direct connect has many options. Fractional line, fractional usage, adaptive usage, pay-as-you-go, or a 100% dedicated line with guaranteed **quality of service** (**QOS**). You get what you pay for! And you don't know what you need till you need it. Be prepared to wait up to 6 months if you choose wrongly with these options. Plan your costs well and factor in lots of lead time for testing. When arranging connectivity, also consider connected DNS issues between any on-premises users, local servers, cloud IaaS instances, and PaaS end service endpoint addresses when configuring name resolution. Lots of debugging is needed when integrating cloud and local DNS configurations. This is especially true with security integrations, **single sign-on** (**SSO**), Microsoft Active directories/cloud directories, and security federations.

Cost of provisioning

The provisioning costs of your solution must not be left as an afterthought. Pick an approach and make sure it works for your cloud and on-premises needs. Tools such as Ansible {https://packt-debp.link/4RmXiM}, Puppet {https://packt-debp.link/qOi3DE}, or Chef {https://packt-debp.link/cs246n} for server configurations, and Terraform {https://packt-debp.link/fQ6YEV} or Pulumi {https://packt-debp.link/DBzYN1} for cloud provisioning are great. When deciding whether the cloud's native provisioning tool features will work for you, please note that they will, but at a cost. It is advisable to wrap them if you must. You will have to depend on them when using a newly released cloud service. Examples are AWS CloudFormation or **Azure Resource Manager** (**ARM**), Azure Bicep {https://packt-debp.link/wd2a9B}, and Azure Blueprints {https://packt-debp.link/6maY7u} (or whatever the flavor is at Microsoft today; you see that we've been singed!).

Cost of monitoring and observability

You will be faced with trying to make head or tail out of the log trace, service signals, and cloud scale activities from various cloud services, your configurations, deployed code, and script snippets sprinkled about the cloud. It can devolve quickly into chaos on the first operational anomaly. We've seen micro-service designed systems melt down during a large scaleup, which produces downstream demands and a cascade of service failures due to bottlenecks, made worse by a high number of failed retry attempts, making the loading worse across the system. All this with a lot of wasted cycles. You will be faced with preventing meltdowns! You have to create operational firebreaks and safety nets for out-of-control scaling, with proper performance management even throttling at times. You may think that the engineers tested everything, but in their zeal, they made it possible for a DevOps or DataOps staff member to adjust just that one setting that was *not* tested in combination with that *other* setting that may cause runaway stress. Use tools to implement the vision of a bulletproof system and even if the system is not, you will have built features to recover before errors become visible to the data factory as a whole. The ability to think operationally is so important when formulating the logical architecture. Consider the features of products such as the following:

- Amazon CloudWatch {https://packt-debp.link/f5WC84}

- AppDynamics {https://packt-debp.link/vdteCZ}

- Azure Monitor {https://packt-debp.link/wJK7Kf}

- Datadog {https://packt-debp.link/wYQfpo}

- Dynatrace {https://packt-debp.link/W3qOwA}

- Elastic Observability {https://packt-debp.link/ufNUP8}

- IBM Instana Observability {https://packt-debp.link/gCVRYg}

- New Relic {https://packt-debp.link/ZLFIAK}

- ManageEngine Applications Manager {https://packt-debp.link/ICXF6f}

- SolarWinds Server and Application Monitor {https://packt-debp.link/3fL17G}

- Sumo Logic SaaS Log Analytics Platform {https://packt-debp.link/fXLDf5}

Other vendors include the following:

- Alibaba's Application Real-Time Monitoring Service (ARMS)

- Alluvio Aternity's Digital Experience Management (DEM)

- BMC's TrueSight Operations Management

- Chronosphere

- DX Application Performance Management

- eG Enterprise's Epsagon Professional

- Google Cloud's Operations Suite

- Grafana Labs' Grafana Cloud

- Honeycomb

- Intergral's FusionReactor

- ITRS Group's ITRS

- JenniferSoft's Jennifer APM

- Lakeside's SysTrack
- LogicMonitor's LM Envision
- Logit.io
- Logz.io's Open 360 Platform
- Netreo's Retrace
- OpenText's Micro Focus AppPulse Web
- Oracle's Application Performance Monitoring
- Sentry
- ServiceNow's Cloud Observability
- Splunk APM
- VMware's Tanzu Observability

You could read many books on what is necessary for the topic of performance management, proactive observability, and monitoring.

A best practice is to not over-collect and analyze telemetry that is not actionable

Build robustly so that the need to find out what's happening or what happened does not have to be dug out of mounds of collected telemetry. Be wise in how you build your logic and its operational footprint. Build it in layers, and do not skimp on the hardcoding that provides a fundamental framework for your solution. Make it a robust product-izable platform (even if it is never sold as a standalone product), and you will have a layered, rock-hard foundation to build the data pipelines of your data factory. This all makes sense, doesn't it? But with cloud service integrations, your solution will be under constant stress to take design shortcuts, leading to fragile implementations to just get it all working. You still have to implement intelligent management features with "right-sized" intelligent observability.

Hybrid or multi-cloud choices!

What will really complicate your goal to build a robust system will be mandates to use hybrid on-premises hosting together with cloud hosting for the core of your solution. Worse yet, you may be mandated to implement a multi-cloud architecture or a single-cloud architecture with cold standby to another cloud provider. Often, these are due to corporate risk managers standing up and asserting that the cloud introduces a vendor lock-in, may be too costly, or may not be within proper corporate control. Questions arise, such as "What if there is a security breach? Will we be able to recover if cloud or cloud network failure occurs?" You can just say, "Well, there is insurance for that, you know!"

But realistically, you must yield at times to external forces and just add an architected solution that addresses the business needs, even risk-based needs.

Hybrid choice issues

Hybrid approaches will need to be discussed; however, the movement of data contradicts data gravity truths and, as such, will cost more if hybrid or multi-cloud solutions are politically mandated.

If you are faced with a hybrid solution of on-premises and cloud solutions, you have some options with Microsoft's cloud services where they can extend their cloud operating environment into your own. Deployment services via Azure ARM and IaaS management are unified as a result. This will make some services and servers run locally as if they were cloud services. Management, provisioning, and observability remain uniform for this type of deployment. It is an option that you will want to investigate.

Where the aforementioned is not possible, you have to sit back and focus on the people, process, and technology impacts of all your architecture choices and be ready to build three times as much: one for each cloud (or on-premises) approach and another to unify them into a consistent solution.

Multi-cloud choice issues

Multi-cloud approaches will run counter to the cloud provider's customer lock-in goals, and they all have these. They are, of course, fine-tuned and quite encompassing. What you will want to do is to integrate at the edges of your architecture: ingest zones and consumption zones. If you try to integrate in the middle or on a service-by-service basis, I can assure you of high operational costs, configuration complexity, and operational faults. Proper training will be essential for both clouds since that effort is doubled when using multi-clouds. We've found The Cloud Bootcamp, LLC {https://packt-debp.link/uldKaY} to be an excellent source for this type of multi-cloud training. Also, the *2023 Flexera State of The Cloud* report (https://info.flexera.com/CM-REPORT-State-of-the-Cloud) provides much food for thought when implementing cloud solutions.

Once the business and IT leadership have decided on the level of risk that cloud architecture poses in order to gain the benefits of the cloud, you can move forward with logical architecture, but not before.

Solve political governance issues and avoid costs!

Since we will have to absorb complexity and fight against the clear cloud provider lock-in goals to get an architecture produced, you as a data engineer will need to know that the domain owners and senior leadership are on board with the strategy, architecture, and designs. Some organizations have strong governance processes, but where there is flexibility and negotiation room, then use it for the benefit of the organization. This is also where taking career risks and expending personal capital will be beneficial for everyone.

The benefits of a multi-cloud strategy

Organizations are embracing multi-cloud strategies! Embracing multi-cloud designs runs contrary to the cloud providers' wishes but is our reality in many enterprises. Google's GCP and Oracle's Cloud lag behind Amazon's AWS and Microsoft's Azure's rate of adoption considerably. Imagine the fear that could arise when your cloud infrastructure is to be stopped or that cutting-edge third party software does not or will no longer run on the cloud provider of your choice. The impact is personal and impactful to the business. Public cloud adoption still continues to accelerate, and cloud initiatives grow. Challenges should be handled by taking a centralized approach to the cloud (such as by building a **center of excellence (COE)**) and forming core reusable operational blueprints if/when supporting a multi-cloud strategy.

The top challenges are going to be unified security and cost controls since various risk factors, including the risk of budget overruns and expertise spending, will rise. Your organizations will struggle to control growing cloud spend, so give them features to control, alert, and cap all spending.

Cloud costs and accounting must be clear

Make costs and accounting clear to each DevOps and DataOps team. Make their implementation choices traceable to monthly operating bills, and alert FinOps to deviations from agreed-upon baseline budgets.

When deciding to implement cloud PaaS services and cloud integration services, do so with a clear vision of how they will be provisioned, released, configured, change-managed, and monitored for SLA and data contract compliance.

If your organization wants to launch a private cloud (and some are bold enough and big enough to try this approach), you will have lots of best practices, blueprints, and tools to develop based on the lessons learned from the public cloud providers. If you can obtain that information or hire personnel who have built those solutions in the past, it would be even better. A private cloud can play an important role in your organization, and one can be built effectively, but expect an ever-growing chorus of *Why sayers* who want to know the answers to a few questions in the following section.

Why build what can be used so easily from the public cloud?

The answer is that with less vendor lock-in, there is more negotiating power versus the costs. Cloud Service provider (CSP) lock-in will reduce your enterprise procurement leverage in price negotiation. So, you want to work with the business price negotiators and give them options up until a deal is signed. Usually, this takes place at the senior IT leadership level of your organization, but they and you should be providing what is necessary to leverage a better deal than the one seen in the cost estimators of the SP's billing consoles.

Do not ever pay the list price! Everything is up for negotiation; remember this when you develop your architecture. There will be aspects of the multi-cloud approach that increase complexity and your engineering costs.

Multi-cloud approaches increase complexity and engineering costs. Be very transparent with FinOps groups, and make sure they have a seat at the table when making choices going into the logical architecture. If they agree with the choices, risks, and costs, they will not be blindsided later when costs land.

Summary

This chapter brought you on a journey to discover what goes into a logical architecture and an example of a logical reference architecture. You also discovered key aspects of the architecture driving best practices. By forming your own logical reference architecture diagram, you can create an architecture that may be communicated to technical staff and gain feedback prior to gaining funding. Via structured and guided solution architecture processes, data engineers and software engineers may make proper choices for tool and solution design decisions. There will be many choices. The physical architecture will be completed next during many agile sprints. You learned that conceptual capabilities have become real features that are needed in the solution. You provided details for the development of the physical architecture to become a strawman next and then incrementally enhanced through development epochs.

The data-as-a-product principle became real in the logical architecture where data quality and dataflows became more than a concept but real componentry and processes of the solution. Dataflows between the zones were shown to be more standardized, leading to the later development of robust CDC blueprints for the data factory/data pipelines for curated data delivery.

Also, choices were made that define cloud versus on-premises versus hybrid data hosting, with significant points made about data migration costs between on-premises hosted data and the cloud, which are not cost-effective solutions. Business security risks and/or political governance issues were also touched upon. A framework for what is to be agreeable dataset SLAs via contracts will guarantee delivery at a reasonable cost.

Lastly, multi-cloud architecture design cautions were identified that could and often do violate individual cloud specific PaaS best practices and will increase development cost and runtime effort for your chosen cloud providers. Picking tools that do not cut across the cloud provider lines is in your interest, even though the SaaS solution in another cloud looks very appealing.

In the next chapter, you will see how to develop a physical architecture that aligns with the logical architecture and how that develops in collaboration between software and data engineering staff, with oversight from architects, business stakeholders, and IT management.

7

Architecture Framework – Physical Architecture Best Practices

The need for a more detailed architecture framework will be elaborated upon in this chapter as the **physical architecture**. As stated earlier, the framework is documented as a three-level conceptual, logical, and physical architecture and in this chapter, you will be exposed to the need for a formal way to depict the physical components of the system design that will implement the features identified at the logical level of the design. This will enable you to clearly communicate the deployed state of the IT solution to all operational teams, such as the following:

- **Information Technology Infrastructure Library** (**ITIL**) **Ops** (also referred to as traditional IT operations)
- Agile **Site Reliability Engineering** (**SRE**)
- **DataOps** (renewed focus on data as product operations)
- **DevOps** (Agile software's hybrid development/operations teams)
- **TestOps** (Agile test lead design teams)
- **Machine Learning Data Operations** (**MLOps**)
- **FinOps** (cloud financial operations/oversite)
- **SecOps** (security, security incident response, privacy, audit, and compliance teams)

It is important to note that this chapter is not going to flesh out a fully functional physical architecture for your solution. However, it will elaborate on the need for capabilities to become deliverable and operable features, functions, configurations, and deployed services that constitute your vision of the concept of something being **well architected**. Your need to document the solution blueprint and clearly identify the design's foundational framework using modern architecture tooling will be emphasized. The need to keep the architecture up to date will be clear; since it's going to be an important tool for collaboration, communication, and reduced operational risk.

You will also want to focus on the operational aspects of your solution. Without solid operations, your **Service Level Agreements** (**SLAs**) and data contracts will be viewed as unachievable. Your solution could be deemed fragile if attention is not given to the development of necessary non-functional requirements. Getting into a proper engineering mindset for success is going to make your physical architecture best practice a reality, and that can be a difficult shift in your thinking. Operational knowledge to avoid service issues comes with years of experience wasting time fighting battles that could have been avoided with some up-front design thinking. We hope to enable you to leapfrog over these difficulties with the best practices necessary to achieve solid success.

Physical architecture overview

The *physical architecture* provides justification for choices made at the logical level where **Non-Functional Requirements** (**NFRs**) and various decisions have to be made. A significant focus will be on the **data contracts** and service level agreements (SLAs) for all deployed software and scripted PaaS integration blueprints that appear to the casual observer as if they were custom-coded implementations. Architecture processes will clearly drive the operational characteristics of your design. The questions of *what* and *how* will be answered regarding various technology choices. This will be very clearly and consistently defined as integration becomes institutionalized in the physical architecture. The configuration management approach you decide upon will be very important when formulating this physical layer of the architecture. This layer will produce the formal as-is state of your solution and fully align with the implementable blueprint that the logical architecture declares as the solution design. As with the logical architecture, the physical architecture is formulated in collaboration with the engineering staff as well as the operational staff as various **proofs of concept** (**POC**) are conducted. Best practices are to be applied with a focus on obtaining an acceptable state of operational readiness. Obtaining formally satisfactory acceptance testing results is a key goal. In the process of developing the physical architecture, there will be important operational considerations to address regarding the integration and runtime operations of the technologies selected. These will be reflected in a **runbook** that should be concise and able to explain (or have links to enable drill down) what is necessary to do the following:

- Provision
- Start
- Stop
- Manage
- Diagnose
- Remediate errors
- In all ways, operate your solution using commands and controls that you have built

If you do not formalize the process and technology as-is state, the people operating the solution will have to dig in real time for information that is essential to maintain, troubleshoot, run, and manage the system produced - a situation that is both error prone and time consuming.

So, *what's in the physical architecture?* The answer is that the formal definition of the as-is state of the technology, process, and people roles for features implemented, integrated, and specified as blueprints in the logical architecture. In order to support feature-based development with their releases, the operational aspects of the solution require a dashboard of these features. You can even envision each feature that is exposed to an end user or consumer of data services having red, yellow, and green states. The yellow state enables SREs to dive in and execute processes to correct anomalies in system operations for data quality before they become overtly visible to the consumer. This is important so that SLAs and data contracts are maintained proactively in order to buy time for resolution. It is in the implementation of the yellow operational processes that there is much work to do for the designer of a modern data system.

Note

Often, DevOps teams are *not looking to build for this type of feature driven operation* and the architecture, and Ops, and DataOps teams will be responsible for making sure it is acceptably part of the requirements.

Orchestration of the data pipeline (or runnable instances of generic data pipelines), through the physically deployed services and software components are depicted in the physical architecture, to implement the flows of the logical architecture.

Again, a picture is worth a thousand words, but the words are still required! The words in the case of the physical architecture will be the parameters of the configuration-managed design. The picture is what you create as the physical architecture diagram for others to review. As an example, it is not good enough to just state that a PaaS, SaaS, or IaaS service is being used in a data flow or feature. How integrated components failover, in a high availability integration, is just as important as how they recover after that failure time period ends! You have to be able to see the orchestrated steps. All these steps made during state transitions between primary and secondary high availability failure are processes subject to orchestration tracking. Therefore, the physical architecture is not just a static design; it also supports the processing steps to maintain a healthy system and it is also the formalized definition of the *"as-is"* configurations of the solution.

Like with the logical architecture, the physical architecture diagram reads from left to right and denotes a set of processes. You must be able to do a deep drill down of the modes of operations as segments of the diagram undergo primary, secondary, or high availability (parallel) operations. So many real-world tradeoffs will be made when selecting one or more of the approaches for each segment of the physical architecture that it will be hard to say which is best (note that skill and experience help here). You will have to live with the quirkiness of the cloud PaaS services, SaaS integrations, language choices, scripted integrations, and tool provider offerings to find a way to maintain the prescribed SLAs and data contracts of your features. This is what will be elaborated upon in the chapter.

A single page representative diagram provides the best high level physical architecture of the data engineering solution. This is similar to the one page definition of the logical architecture. Configuration management and architecture modeling tools selected for those options, in the logical architecture, will help formalize the physical design as it is developed. This way the artifact remains a living document/ repository of the architecture. Keeping the architecture up to date and in sync with deployed configurations is going to be a mandate, which is often lost after the Agile team is disbanded or as the design matures over time. Keep the **finacial operations** (**FinOps**) teams happy by making sure that the business as usual budget retains a talented and budgeted slot as part of system maintenance overhead, in perpetuity!

The goals of this chapter are to do the following:

- Define the contents of the physical architecture and illustrate them.

- Provide the best practices for that physical illustration and how it can be rendered to enable effective stakeholder communications.

- Provide best practices for future-proofing the physical architecture with an emphasis on the configured and operational features mapped components or services of a data factory.

The three goals for this chapter align with elaborations of the conceptual and logical architectures of your solution in the previous chapter.

Best practice organization

You will observe that we have adopted the following structure for the physical layer:

- Best practice composition of the physical architecture:

  - What is the structure of a physical architecture, including an example?

  - What as-is/current state formalizations are required of the physical architecture?

- Learn how features are operated in the context of the system's NFRs across the data processing zones and how datasets interact with various segments of the physical architecture that is deployed with their configurations so that SLAs and the various data contracts are maintained. Each feature will have operational characteristics and process definitions aligning with the following operational responsible areas:

  - **ITIL Ops**

  - **SRE**

  - **DataOps**

  - **DevOps**

  - **TestOps**

  - **MLOps**

  - **FinOps**

  - **SecOps**

- How to implement a modern, future-proof architecture:

  - Configuration management (with the cloud's structural metadata)

  - Runbook

  - Feature-based seleases

  - Operational orchestration

  - Summary service dashboard

  - Forensic data analysis

- Learn best practices regarding how to tie the logical architecture to the physical architecture.

With this structure, a data engineer will be able to hand off the solution to the operations teams with the operational handles required to satisfy the SLAs and data contracts of the businesses and business data owners.

How does the physical architecture align with the logical and conceptual architecture?

Where the rubber meets the road is where the solution designs become physical technologies that are deployed, the processes that govern how that technology is going to operate, and the skilled people who put the system together (and maintain it over time). All three have to be in alignment and meet the **Objectives and Key Results** (**OKRs**) of the solution. The physical architected designs have to be aligned with the logical *architecture's features* and with the conceptual architecture's capabilities.

With all that complexity regarding alignment, you will want a way to explain what is *"to be,"* from *"what was"* previously defined for each architectural iteration. This way, key stakeholders are able to guide the implementation. You will also want to create a breadcrumb trail for those wanting to know why choices were made and how to help in regard to providing funding, marketing, support, maintenance, and all the fringe requirements that make the functional and non-functional requirements implementable.

> *"Garbage doesn't turn into gold, no matter how much math you throw at it."*
> *Cassie Kozyrkov {https://packt-debp.link/GegWgq}, CEO at Data Scientific, Google's*
> *first chief decision scientist, decision advisor, keynote speaker (http://makecassietalk.com),*
> *LinkedIn top voice*

You want the data product of the solution to not be garbage! A balance between data needs, operation's needs, and cloud software needs has to take form (or gel) into a roadmap plan to steer the Agile implementation, which leaves a proverbial wake in the water as the ship sails toward its destination. The Agile iteration's results, or the wake as the metaphor espouses, are going to be the physical architecture of your design. They can't be brushed off as irrelevant or excused away because they take too much time to generate. This will be a constant pressure to be released that will arise between the architects and engineers of the project.

Define the technologies of the current as-is state and prepare for iterative change to the new to-be state!

You will always want to be able to align the physical architecture diagrams, structural cloud infrastructure metadata, and configuration management tool with the actual deployed as-is infrastructure. To do this, one could and should explore the use of a third-party tool. It is advisable to use such a tool as **LucidChart,** also known as **LucidScale** {https://packt-debp.link/e2mh2Z}.

Alternatives to this are other cloud discovery or plain diagramming tools, such as the following:

- AWS Workload Discovery {https://packt-debp.link/76Ll4t}

- Brainboard {https://packt-debp.link/3h5zDQ}

- Cloudcraft {https://packt-debp.link/ad3xmH}

- Cloudmaker {https://packt-debp.link/ZKABVU}

- Cloudockit {https://packt-debp.link/8iE7cJ}

- Creately {https://packt-debp.link/KMGgGZ}

- Diagrams.net (Draw.io) {https://packt-debp.link/cwOzss}

- EdrawMax {https://packt-debp.link/eJFSw9}

- ExcaliDraw {https://packt-debp.link/kG5dhc}

- Glify {https://packt-debp.link/pAO6R4}

- Holori {https://packt-debp.link/7CrSkd}

- Hava.io {https://packt-debp.link/kxjIL0}

- LucidScale {https://packt-debp.link/jMSCho}

- Mindgrammer {https://packt-debp.link/3YPhqA}

- Miro {https://packt-debp.link/7vgtfH}

- OmniGraffle {https://packt-debp.link/U5zoki}

- Terraform Graph {https://packt-debp.link/PocB9Z}

- Visio {https://packt-debp.link/Tv4NA4}

- Visual Paradigm {https://packt-debp.link/x7kzCY}

The endnotes to each tool listed provide details on each of the products. Some tools' capabilities are better than others and especially if you are operating in one of the primary cloud vendors (AWS, Azure, or GCP). You can find one or more tool that does the job for you as you develop a current state *"as-is"* physical architecture. The architecture is refreshed and kept up to date without the pain of manual recreation each day as the CI/CD pipelines produce streams of change. Two features that are particularly important are: cloud configured item discovery and cost/billing integration. The use of billing tags with a structured tagging design has to be upfront in your mind. Many engineering team leads have failed because they did not know nor manage their costs during and after development. The *FinOps* team can stop project momentum in a minute if they see that the requirements have busted the project's financial assumptions, or that the cloud vendor costs have not been constrained for the use case envisioned. An easy-to-understand, transparent, and effective *"as-is"* cloud design and cost tracking and estimate capability will clear the path for the engineer and must not be omitted from the physical solution's design.

Consider the need to define, discover, and report on each physical environment based on these environment types:

- Development environment

- System integration environment

- UAT/pre-production environment

- Production environment

- Forensic or audit environment

- Cold (or warm) production environment

- High availability environment

Development environment: There may be many of these development environments, sandboxes, or dev/test proof of concept areas. Some Agile teams require one per developer. They are easy to spin up and are often erroneously left up or fail to be accounted for in cost estimates. For a best-in-class data engineering approach, assume one for each team with nightly Gitflow check-ins to the master and CI/CD-enabled regression testing process so the overall build function is never broken. The virtual dunce cap gets to be worn by a team offender who breaks the system after a failed check-in and prevents mergers the next day.

System integration environment: This is where various teams hand off their developed solutions for integration and further stress, performance, and IT acceptance testing, including security, OSS compliance, and operational readiness. This is expected to follow a **blue/green** environment switchover/deployment methodology. It is possible to stream changes to production but beware of the impact on data quality. This can arise because a software engineer can agree that streaming software changes into production is fine, but a data engineer's testing has to take into account larger sets of data to verify results. Remember that there is a need for **Data Quality Management** (**DQM**), and part of that testing involves trend analytics! With an unstructured stream of software changes operating against a quantum of data, the end-state data quality goals will be hindered.

User acceptance testing (**UAT**)/**pre-production environment**: This is very much like the production environment with its cloud configurations that may be able to scale up to full user load if needed. This mirror of production enables the blue/green environment swap via the same configurations as UAT. If the physical environment has this operational capability, much NFR testing can also take place in the UAT environment rather than in production after deploying software.

Production environment: This is the environment where all the magic really happens. Users are exercising the features that they are entitled to perform because they appear on a service dashboard and have been mapped to OKRs that the business understands and can take ownership of. All too often, dashboards are created for technical components since their creation is easy, but feature and data contract dashboards involve threads of activity through many cloud components. It is the job of the engineer to make technology transparent to the business and implement what is hard to implement, thereby making it easy to understand that the objective is *met, always met, and never fails to be met*. Real-time SLA tests are created, run, tested, and then validated as well as the data contracts enforced in the production environment.

Forensic/audit environment: Forensic tasks for problem resolution and drill-down audits are to be expected in a data factory system. A downscaled or degraded green production environment serves this purpose since it is anticipated that a green/blue transition (a reverse transition) will take place to bring such an environment online. This way, users are not affected due to the production load having been shifted off to a blue-labeled environment for analysis.

Cold (or warm) production environment: This is a live production environment that does not service data consumers or end user service loads but can be activated within a defined timeframe. It's kept warm in that it may be kept partially in sync with the running production environment or brought up to readiness within a short but specified period defined by the **Mean Time to Repair** (**MTTR**). A cold environment takes longer to bring up to speed and make ready for service than a warm environment, but either one must be tested to make sure that they are always ready for activation. There have to be clear operational commands and controls. This is because not only the primary functions but all the NFRs such as monitoring, alerts, dashboards, data quality, security, and data entitlements have to be brought online. They must all be functional even in the cold or warm production environment once user load is allowed in for service.

High availability environment: This is the environment type that is a best-in-class way to implement 99.9-to-99.99% availability. It is the best way to realize various cloud provider service's 99.999% stated availability that can only be achieved if those services are implemented in conformance with the cloud provider's well architected blueprint. Notice that without a high availability design, a 99.99% solution SLA target is a more realistic goal for your customer facing SLA because it leverages the cloud provider's service that cannot be exceeded without some necessary planning and cost allocation. This planning involves the implementation of high availability design patterns into your design. To begin, your solution's overall effective SLA must be measured and reported for each customer facing feature. This high availability design approach will use cross region and/or cross availability zone data replication with automated failover with load sharing. Your solution's *effective SLA* can now be designed to exceed the cloud providers' standalone service SLA! High availability designs are best practice recommendations! They will come with increased operational cloud costs, but lower manpower costs because cloud service faults will not become visible to your data consumers. The solution should now be able to account for cloud service outages in its design. The solution's feature based SLAs still need to be measured in real time and reported against baselines as part of your developed failure and stress testing processes. The strength of your cloud environment to withstand faults and preserve your end users' SLAs and data contracts is only as good as the weakest link you add to your solution's infrastructure that is glued together with your code.

Knowing the weakest link in your cloud infrastructure is essential to being able to maintain and manage what you have produced as a result of your Agile efforts. All too often I've heard a dev lead say that if only they had turned something on, there would not have been an outage. Or, at other times, they might say something to the effect of: *"I did not know that was still running! It was only spun up as a quick test environment."*

This is a problem when it happens after a $20K billing line item appears on the IT budget from the cloud provider's monthly billing report! Knowing what is running, where it is running, *who is running it*, and *how it is expected to incur what charges* in the future is essential for the physical architecture to be able to answer. The physical architecture will be leveraged again and again over time to optimize the solution. The as-is quickly changes to the to-be and then that later becomes the new as-is solution. The architecture's design documentation cannot be allowed to become an old paper tiger. So, you want to wrestle with that issue upfront. Plan for its definition to remain a living sub-system for structural metadata along with your integration glue. Expect it to be leveraged for many other purposes, so make sure it is humanly readable and end-to-end consistent as a complete set (with version controls, all clutter cleaned up, and so on). At least it will always be a pictorial as-is artifact depicting what is deployed. If it is tied into a configuration management tool, that is all the better. However, beware of overthinking how a legacy IT operational configuration management repository ties in with the physical architecture, because it can lead to a failed integration effort.

Next, let us dive into the process definition needed for the physical architecture.

Define the processes of the current as-is state and prepare for iterative change to the new to-be state!

So, now that a physical design diagram exists (you can see examples in the endnotes), as well as a way to keep it in sync with your environments, you need to define: *how it is to be kept in service and changed over time?* These process definitions should begin in some logical order. The creation of a *"current state assessment"* for your solution will provide a foundation for collaboration. It is an approach that we as authors have used very successfully in the past. Asking questions and collaboratively working as a team to get responses to those questions builds consensus across all teams. Knowing what is being agreed upon definitively provides a foundation for the methodical approach to refining future *"to-be"* architecture states. Creating grounded and documented changes to problems encountered during solution development will yield buy-in, complete understanding, and then agreement. At least, that is, until the first substantive argument arises! Expect some heated discussions regarding process definitions and roles required to complement the processes being defined. Notice that we have *not* fleshed out roles yet. That will come in the next section.

The process definition stage begins with an understanding of the **ITIL v4** {https://packt-debp.link/2l1w24} processes. This alone will cause cloud providers and Agile DevOps purists to flinch! We will not get into all the details of all the ITIL3 versus ITIL4 nuances, but by grasping the need for the answers that ITIL V4 services provide, we will level-set on what it means to *not* cut corners on IT operational NFRs. When reviewing the services, from the referenced endnotes, you will see that the processes directing your solution's NFRs are going to be a mix of ill-defined, semi-defined, or undefined pressures that you will want to clarify early on. Even if you approach Agile development methodologies and think that your DevOps teams will handle it all, they will not! Five years from now, when those teams are on to bigger and better things, new team members will have to be onboarded! These processes will either have been applied without you or by you, but they will be applied eventually. A few service processes stand out as essential:

- ITIL release management

- ITIL change management

- ITIL incident management

- ITIL problem management

ITIL release management: Release management practices make new or changed services and features available. This can be done with a communications wrapper around your blue/green deployment practice. Solution deployments can support streaming changes from your CI/CD pipelines if those deployment workflows have human and **Quality Assurance (QA)** or DQM gates for acceptance and approval. Some DevOps teams try to trivialize release management processes and do not put the workflow configurations into their DevOps pipeline's CI/CD. They then lose the approval gates needed to preserve quality and reduce error. Part of this process is provisioning a solution through automation with the ability to release backward-compatible changes and then, if necessary, revert to the last known good state if failure to release with quality occurs. Do not fail to define processes that buy your teams time to react to release issues, conduct post-mortems, revert, and finally reapply the release after corrective action is taken.

ITIL change management: You have to weigh the need for speed against the inherent risk evidenced by any change! Change-related incidents can create unwanted service disruptions. Unmanaged changes increase cost. Various types of change must be supported such as scheduled, emergency, routine, and large change types. Various role definitions and formal change process policies need to be created and adhered to, such as **Request for Change (RFC)**, **Change Managed Chain of Approval** (which is a management role for orchestrating the end-to-end string of events required to effect a change until it is applied and completely released to a consumer), and **Post-Mortem Reporting** for efficacy reporting purposes.

ITIL incident management: Incident management is not **Problem Management**. It is the process of escalating and responding to incidents so that a **Mean Time Between Failures (MTBF)** and **Mean time to Recovery (MTTR)** are maintained over time. Restoration of service is conducted as fast and as accurately as possible with an effective incident management process.

ITIL problem management: Problems are identified as anomalies to the standard SLA, data contract, or observed trend as collected statistics start getting close to some defined violation level. It is not good enough to go from a green to a red state for a features service. The problem has to be detected when it is in the yellow state. Setting up monitored thresholds buys time for teams to react to problems before they become critical. It is advised that you create this yellow state and reflect it in your dashboards. There will be a root cause for any problem after it is identified, but at the start of a problem, it will be unknown. Problems will arise from incidents, but not all incidents are problems. They can be explained away before escalation. Once it is acknowledged as a problem, an incident is flagged and managed in an **IT Service Management (ITSM)** tool and related to other incidents, forming the justification for it becoming designated as a *problem*. Problems may have much longer MTTR times than incidents that often require communication SLAs to end users at some interval. It is not uncommon to have a workflow for problems to report to consumers, customers, or users within 20 minutes, and then every 20 minutes thereafter until problems are resolved. Problems will have a root cause and often some new entry to a project's backlog as a new objective and key result (OKR) for future remediation.

There are many other services you will want to consider as an architect and data engineer; however, the preceding ones are essential and really not optional.

In the next section, you will read about the people and roles necessary to affect a proper future-proof physical architecture. You will at least learn how best to adapt to change. This is where the rate of technology change is going to exert a lot of pressure on your current teams, and that will force you to consider keeping them formed for maintenance tasks far into the future, thus raising your project's **Total Cost of Ownership (TCO)**.

Define the people (or rather the roles) of the current as-is state and align to the new to-be state!

So much work goes into the Agile design process that you can easily forget what roles are needed. You can even get fixated on the success of startups and think that success can easily scale for your large organization's various departments without modification. On one end of the Agile process, there are role definitions such as the **Spotify Agile Process** {https://packt-debp.link/dQNYGK} and at the other end, you can find enterprise TOGAF {https://packt-debp.link/jkIFRJ}, with the Open Groups Open Agile Architecture {https://packt-debp.link/6E3Ebd} being a compromise.

In an organization that chooses to leverage an Agile development methodology, employees are accountable to their peers, their manager, and their clients. A key failing is that as a professional, you will want to remain accountable over time even after team reassignment to other projects. The Agile management system is built to allow goals to trickle down to all levels of the enterprise and, while doing so, produce collaborative dialogue leading to self-developed accountability. That accountability is a professional responsibility! As an architect and stakeholder, you want to not just give lip service to the Agile requirement for personal accountability but rather make it possible for that to happen. Provide maintenance and business-as-usual support teams after a cross-functional development team activity ends. Often, this is forgotten and break-fix costs skyrocket. Worse yet, fixes may not be made in a timely manner and customer satisfaction stats will fall through the floor, all because key staff have been reallocated to the next project. You will want to create a reward system to accompany this approach in order to recognize individual performance and overall collaboration. As a result of adopting Agile practices, the organizational structure is flattened. Cross-functional teams (**Feature Teams** or **Squads**) are formed to implement OKRs. Cross-functional roles will organically be created by the self-organizing teams to construct the solution. In the *Spotify Agile Process* model, they are named **chapters** or **guilds**. Resource allocation is driven by the demand or activity level of the teams. However, note that the refinement of high-level guesstimates (on scope and funding) will still be required when the OKR is established. This reality is a glitch in the concept of fully self-organizing egalitarian team management since some planning was done for the team in advance. We need to fill this gap which appears. People need to be empowered and self-organizing, and the FinOps team still requires the solution to be conceived, funded, and handled by three (underappreciated) roles:

- **Enterprise Architect (EA)**
- **Chief Engineer (CE)**
- **Site Reliability Engineer (SRE)**

You can see that these individuals can provide a broad brush outline that enables a solution to be launched within a +/-10% error factor for time and effort estimation. This will cause professional friction in the Agile teams since they have no real voice in the process, but it is a reality of Agile in an enterprise setting. It is also why each enterprise needs its own version of an Agile manifesto that is enforced by policies in order to clearly provide the governing constitution level power to remove the stress before it is experienced.

Enterprise Architecture role definition

The **Enterprise Architect** (**EA**) role includes the discipline needed to manage consultants who design business and operating models. The Open Agile Architecture role definition identifies the following:

- Strategic marketing/marketing research
- **User Experience** (**UX**)
- Design thinking
- **Lean Product and Process Development** (**LPPD**)
- Socio-technical systems
- Organizational sociology
- Operations strategy
- Software architecture

You will also want to add the important, needed data architecture skill requirement to this list, since without a refocused intensity on proper curation of data, a solution will not remain in production for long. Questions answered by the architect include *what, how*, and often *when* the physical components of the system are to be delivered. The architect controls the roadmap to success, and it is 100% aligned with the business strategy and the OKRs that become team objectives.

Chief Architect role definition

The **Chief Engineer** (**CE**) is a role that combines entrepreneurship with system architecture skills. A CE position should be filled by an individual with the following traits:

- Ability to define a clear and compelling product vision.
- Ability to inspire engineers.
- Ability to design the solution architecture for a product and its value stream.
- Skilled and knowledgeable in many, if not all, disciplines needed to remediate cross-disciplinary problems that will arise.

The engineer and the architect are to work closely together through the Agile process to steer proofs of concept, designs, and delivery. Team leads are accountable to the CE for what they deliver and how it conforms to the architecture. The CE can cut across teams, manage people issues when they arise, and confound the egalitarian goal of Agile development.

Site Reliability Engineer role definition

The **Site Reliability Engineering** (**SRE**) role is a job role with a set of practices from production engineering and operations. It was defined and started at Google and has largely been adopted by the industry. This role represents the IT operation's policies, processes, and services needed to *"keep the lights on,"* for a solution that has been developed and currently in production. The role has much to say about the NFRs that must be built. A senior SRE will set the runbook requirements, keep track of *"tech debt"* and report it to the chief engineer and enterprise architect.

In *The DevOps Handbook: How to Create World-Class Agility, Reliability, and Security in Technology Organizations* {https://packt-debp.link/gYWwTZ}, Gene Kim observes the following:

> *"... most DevOps organizations were hobbled by tightly coupled, monolithic architectures that*
> *– while extremely successful at helping them achieve product/market fit – put them at risk of*
> *organizational failure once they had to operate at scale."*

> **Note**
>
> A key success criterion is to evolve your architecture to have sufficient componentization to support your organizational evolution on an ongoing basis.

The **strangler pattern** {https://packt-debp.link/8XKvPM} can be the key to the approach needed to build out a system solution and to keep it refreshed over time. This kind of evolution is made possible by enabling the system components to evolve behind an unchanging API to be leveraged by the data consumer, user, or other system component.

It is the role of the SRE to point out what NFRs have to be seriously considered when a team is delivering an integrated technology, even a new one that has yet to be 100% proven to work at scale in an organization's production environment. The SRE is a member of potentially many teams to represent this operational interest. As an example, operational issues created by Cassandra's eventual data consistency can arise particularly with data handling so that real-time access does not impact data access patterns before that consistent state is attained. Users can see massive delays in summary financial portfolio calculations as a result of not accounting for eventual consistency after making changes to owned balances. This is where customer service calls will go through the roof. After a change, one would expect the summary portfolio totals to adjust in real time. Refer to Vanguard's new UI, where summary totals for external portfolios can take up to five minutes to appear on the real-time dashboard of a user without any pacifier indicating that a recalculation is in progress. An SRE would and should identify this type of design issue well in advance. An SRE will ask development teams to build command/control primitives, dashboards, and assertion logic to support the system's NFRs and accommodate fixes to odd functional aspects of the technologies or cloud provider services used. These will cause operational issues if not mitigated and correctly integrated. SREs will also point out when orchestration features of core technologies such as container management services (such as Kubernetes) are not up to par with expectations. They will insist that Kubernetes clusters be managed with a tool such as one of the following:

- Rafay {https://packt-debp.link/P6tLsW}

- Rancher {https://packt-debp.link/PCaTe5}

- OpenShift {https://packt-debp.link/xyoIpl}

- Tanzu {https://packt-debp.link/tbviZW}

- Mirantis Kubernetes Engine {https://packt-debp.link/hmbRWN}

If one of these has not already been purchased, integrated, or implied in the logical architecture, the SRE will still seek immediate remediation of the design error with the chief engineer or enterprise architect.

How should the physical architecture align with the operational processes/capabilities of the solution?

There are many operational functions required of your solution. There are also many best practices that you can read about before implementing your solution. Not all will align with each other or fit your needs, yet you will be forced to work with people who will not yield to other perspectives. Their way was either the only way they knew how to be successful, or they just would not consider others' needs. Worst of all, they only promote the approaches provided by the cloud providers – which *will* conflict with your needs. This is all going to happen at some point in the project's life cycle – that much is guaranteed! This tension will be felt when rationalizing the motivations for each role and making your case for an operational approach that is good enough to begin. Each ops type sees their facet of a modern cloud-based information processing system from the lens of their unique perspective. As a data engineer, you cannot implement a perfect solution from a single perspective but must assimilate each and produce one that works well enough. Here are those perspectives again from the beginning of this chapter:

- **ITIL ops**: Refer to *ITIL Service Operation: Phases, Functions, Best Practices* {https://packt-debp.link/sZbm0V}.

- **SRE**: Refer to *Google – What is the Role of an SRE?* {https://packt-debp.link/inhstq}.

- **DataOps**: Refer to *What is DataOps? A comprehensive introduction* {https://packt-debp.link/KKN6n3}.

- **DevOps**: Refer to *ITOps vs. DevOps: what's the difference?* {https://packt-debp.link/OUJMtl}.

- **TestOps**: Refer to *Understanding TestOps Best Practices and Working Architecture* {https://packt-debp.link/j0PMDy}.

- **MLOps**: This is a reason that this book even exists. Modern data engineering has to produce analytic datasets for all consumption modes: machine learning, analytics, and transactional processing purposes. Refer to the following resources:

  - *AWS: What is MLOps?* {https://packt-debp.link/FjAKDU}

  - *Databricks: What is MLOps?* {https://packt-debp.link/9MUaM3}

  - *Microsoft: What is MLOps?* {https://packt-debp.link/NknNBK}

- **FinOps**: FinOps is not just about counting costs but also about how you measure the metrics that go into the rollup and rolldown accounting for the enterprise. Refer to *What is FinOps?* {https://packt-debp.link/ZcbnLd}.

- **SecOps**: Security operations are not a trivial need, and much care has to go into this operational design. Refer to the following resources:

  - *Sumo Logic: What is SecOps?* {https://packt-debp.link/F3mve8}

  - *ServiceNow: What is SecOps?* {https://packt-debp.link/0VZ3N6}

  - *Crowdstrike SecOps 101* {https://packt-debp.link/nqTAUY}

  - *Palo Alto Networks: What is SecOps?* {https://packt-debp.link/a7bI77}

There are many blogs, expert opinions, and written guidelines in each of these disciplines. Cloud solutions can be tough, especially since the cloud platform is 80% done and always evolving. This is felt acutely when running solutions that have not applied your operational best practices and cloud changes. The solution you choose will impact your end user SLAs or data contracts – and those must be met to avoid revenue drop-off. Each cloud provider has implemented well-architected architecture and design guidelines:

- AWS Well-Architected Framework {https://packt-debp.link/YffNrv}

- Microsoft Azure Well-Architected Framework {https://packt-debp.link/VvnsIp}

- Google Cloud Architecture Framework {https://packt-debp.link/y9FVEC}

- IBM Well-Architected Framework {https://packt-debp.link/ZcXpYy}

- Salesforce Well-architected {https://packt-debp.link/TNZ6On}

- Best practices framework for Oracle Cloud Infrastructure {https://packt-debp.link/rw5dAb}

Each cloud provider clearly realizes that the problem space is a difficult one to wrestle and they have produced training and documented guidance that is commonly available. Please take these cloud provider approaches with a grain of salt and be ready to justify your deviations and not drink the provided Kool-Aid without thinking things through first. Not all advice will remain as written, and it has changed many times in the past few years. The advanced solution approaches will also not always work as directed. The caveat often noted is that you need to run your own proofs of concept. Only *you* can iterate through the Agile process's epics, sprints, stories, and spikes to build *your* solution effectively.

Examples of physical reference architectures

It is always a good idea to check with the cloud providers concerning their recommendations for your solution since these recommendations will change every time you ask a question with a new spin in order to clarify your objectives. It all makes sense since nobody is a perfect communicator, and the technology available is always changing. You will want to build your solution with the newest best API and not always the latest API in order to get and then keep your design working. You will also not necessarily want to work with the latest released, cloud service offered until all the kinks are worked out. A good source for AWS solutions is found in the *AWS Solutions Library* {https://packt-debp.link/Ly5VW8} referenced architectures. A good example of a well architected solution is *Multi-Region AWS Data Lake with Analytics on AWS* {https://packt-debp.link/Id963Q}, where a backup and restore pattern is leveraged for replicating data lake data and metadata for failover to a second region during disaster recovery. This is a cost-effective and elegant physical architecture. It is not as detailed as required to estimate full costs, but those details can be added as an exercise. What makes it a physical architecture is that it addresses the NFRs that make the solution bulletproof in the event of a regional failure at AWS, and these do happen at times.

Like AWS, Azure has a set of well-architected solutions to be referenced. They are located in the Microsoft Azure Architecture Center {https://packt-debp.link/Kaz0j8} so you can see that a large collection of patterns exists. You can leverage them as you put together niche areas of your overall solution. Browse Azure Architectures {https://packt-debp.link/TllON5} with one that can be useful when considering the need for analytics at the end of the data factory: Enterprise Business Intelligence (Analytics) on Azure {https://packt-debp.link/ICA2LV}.

When it comes to prescriptive recommendations for *your* data architecture, we have to stop and say that knowledge and advanced training, as well as hands-on trial and error, are key. If you ask two consultants, you will get two answers and a rousing debate. If you ask ten, you will get a hundred different answers and options with some commonality between them, as well as a whole lot of heated discussion! If you think that a prime consultant or certified partner can provide you with a 100% guarantee of success, you will also find that they suffer from a plethora of **Subject Matter Expert** (**SME**) confusion. This happens even with the architects of the cloud providers themselves.

The best recommendation from us to you is that you first form your strategy (such as a data-engineered factory) with the leading roles required and then let those leads hire the right people to architect your strategy. Let the leaders hire the Agile teams as self-organizing teams is a best practice. Then secondly, set up the governing and operational processes of your physical architectures, and lastly, integrate the technologies of that architecture, leveraging a mix of blueprints (also known as patterns) from the well-architected framework of your cloud provider(s). Do all of this while keeping the vision, mission, and principles clear in your mind, so you are not swayed by the barrage of necessary choices that are made to mold the desired capabilities into revenue-driving features. There you have it in a nutshell; the design will be right when it withstands the test of time since it will remain future-proof only when you make informed and wise choices.

Summary

This chapter brought you on a journey where you were able to discover what goes into the definition of physical architecture with a few examples of physical reference architectures. You also discovered key aspects of the architecture driving the best practices. By forming your own physical reference architecture diagram, you learned that you can create an architecture that may be communicated to various operational staff to obtain the desired correct operational readiness state. Proper choices for physical architecture were made in a complete and comprehensive way. No corners were cut in the process so that what was released remains workable in the future. A properly engineered solution should be based on what has been proven to always work and never fail, not on what looks the most convincing at the time. The architecture you produce will be complete and, even if it is incrementally built over many Agile sprints, it will *not* deviate from the goal. This can easily happen as pressure to deliver causes shortcuts to be considered.

Logical features that were driven from a list of desired conceptual capabilities become operationalized in the physical architecture. The data-as-a-product principle became real again in the physical architecture, where the product was produced under formal processes driven by the contracted goals for service. Integration choices became formalized configurations with tested and assured SLAs with reasonable cost models produced for financial compliance (all too often, this is a missed aspect of cloud based solutions).

In the next chapter, you will learn about software best practices to be applied when building out your solution's physical architecture.

8

Software Engineering Best Practice Considerations

Nothing is built in a day and never by one person. The world of data engineering is akin to a symphony, where different elements blend seamlessly to create a harmonious outcome. The mission, vision, architecture, and **objectives and key results** (**OKRs**) constitute the song that the software data engineering teams play. Harmony is achieved when all are focused on the clear future state and not a series of meanderings toward the abyss. In this chapter, you will learn the essence of software engineering best practices, which serve as the cornerstone for effective data engineering. Understanding and adhering to these best practices enables data engineering teams to navigate the complexities of organizational objectives and architectural frameworks, ensuring a melody of synchrony rather than a cacophony of discord.

In this chapter, you will learn how software engineering practices should be applied to data engineering. You will explore key areas such as Agile methodology, OKRs, and implementing data as a product. Additionally, you will gain knowledge of critical aspects of testing, error handling, and the importance of a well-structured operational DevOps wrapper. Finally, you will explore thoughtful language selection, configuration management, and the prudent pruning of dead code. This voyage will not only elucidate the essential best practices but will also showcase their pivotal role in achieving the overarching goals of a data engineering endeavor.

By navigating the pages of this chapter, you will arm yourself with a robust framework of best practices that will serve as a compass in the vast sea of data engineering challenges. This expedition will offer a treasure trove of insights empowering you to foster a culture of excellence, drive organizational agility, and build robust, scalable data engineering solutions. The following key topics will serve as our guiding stars:

- Embracing and tailoring Agile methodology
- Cultivating a product-centric approach to data
- Prioritizing test driven design and **shift left testing** (**SLT**)
- Navigating operational DevOps
- Selecting the right tools and languages
- Ensuring scalable and secure configuration management

- Understanding and overcoming **Platform-as-a-Service** (**PaaS**) limitations

- Exploring microservices

- Delving into Cloud ITIL **non-functional requirements** (**NFRs**)

- Unveiling the essence of data journey journaling and pipelining

The following list has been provided to give you a snapshot of the engineering best practices that you will encounter in this chapter:

- Follow the architecture.

- Implement Agile methodology for your organization.

- Generate OKRs.

- Implement data as a product.

- Implement SLT processes.

- Implement the difficult first.

- Avoid premature optimization.

- Automate Cloud code snippet deployments with standard deployment scripted wrappers; code must be in version control under a flow (that is, Gitflow {https://packt-debp.link/zD4fb7}).

- Define and implement cloud ITIL and NFRs first!.

- Implement data journey journaling to facilitate future problem resolution.

- Implement data journey pipelines that are experimental first!.

- Choose languages with solid reasoning.

- Drive scripting and PaaS code with parameterization using a secure configuration management repository tool with a configuration management database to hold the configured artifacts under management.

- Be prepared to prune dead code over time.

- If it doesn't fit, don't force it – implement a custom microservice when you encounter a PaaS limitation.

From the preceding list, you can gather a general understanding of what will be elaborated upon in the following sections. The details will provide firm guidance concerning **software best practices** (**SBPs**) in detail.

SBP 1 – follow the architecture!

In the labyrinthine world of data engineering, one guiding star shines brighter than most: the principle of *following the architecture*. The architecture should guide software development, not the other way around. But why does this axiom command such respect and prominence among software engineering best practices? In this section, we will elaborate on the core benefits of following the architecture.

The core value of architectural integrity

At the heart of data operations lies the essence of reliability. By steadfastly adhering to a predetermined architectural design, every data process and transformation can be expected to produce predictable, desired outcomes. Just as a physical architect meticulously plans a building for optimal utility, data architectures are thoughtfully crafted with efficiency in mind.

When we walk the path set by the architecture, benefits in system performance naturally emerge. This same commitment ensures ease in maintainability. When we operate within the known confines of a well-followed architecture, it streamlines troubleshooting, future upgrades, or even significant modifications. Furthermore, in our modern interconnected digital landscape, no data solution exists in isolation. Architectural integrity facilitates seamless interactions between diverse components and services, fostering a cohesive ecosystem. And in an era where data breaches make headlines, architectural fidelity becomes even more crucial, weaving mechanisms to protect and ensure data privacy.

The downstream impact of deviating

Venturing off the prescribed path of architecture isn't just a minor detour – it's a trek into uncharted, often perilous, territory. The following are possible problems you may encounter:

- **Data inconsistencies**: Diverting from the architecture may produce aberrations, leading to compromised data quality and trust

- **Performance bottlenecks**: Unplanned alterations can introduce hitches, disrupting the smooth data flows conceptualized in the original design

- **Increased costs**: Resource utilization can spiral out of control due to redundant or mismatched processes, hiking operational expenses

- **Troubleshooting challenges**: Diagnostic efforts get stymied when issues arise from areas outside the architectural blueprint, making them elusive and time-consuming to address

With the potential negative impacts mentioned here comes the understanding that they will need to be minimized. To effect proper engineering discipline, you will need to ensure adherence to the architecture and its processes within the engineering team(s).

Ensuring adherence in your data engineering team

One way to ensure fidelity to architectural design is through regular reviews.

Periodic checks, akin to architectural audits, can spot any deviations or misalignments early, ensuring timely course corrections. It is also essential for you to keep a detailed chronicle of architectural decisions. These decisions aren't merely technical edicts but informed choices with underlying rationales. Committing them to documentation ensures they don't get lost over time. Equipping team members with the knowledge of not just the *what* but the *why* of architecture can also make a notable difference. Lastly, leveraging modern tools and automation can add a proactive layer of oversight, with monitoring solutions that detect and alert about potential architectural breaches.

Continuous evolution and architecture

While the merits of architectural integrity are undeniable, it doesn't denote rigidity. As technology and organizational needs evolve, so should architectural approaches. The challenge lies in striking a balance between unwavering adherence and the necessary flexibility to innovate, ensuring that the architecture remains relevant and effective.

Conclusion

In data engineering, architectural designs represent more than just plans – they encapsulate a broader vision. Committing to this vision, by rigorously following the architectural design, brings forth not just immediate efficiencies but promises long-term excellence in performance, reliability, and innovation. As we carve the future path for data solutions, a renewed pledge to architectural integrity will be our guiding light.

SBP 2 – implement Agile methodology for your organization!

In today's rapidly evolving technological landscape, organizations must be Agile to respond swiftly to changes. One approach that has gained substantial traction in addressing this need is the Agile methodology. By embedding Agile principles into your organization's DNA, especially in data engineering, you can drive improved outcomes, foster innovation, and ensure that your teams remain adaptable in the face of changing requirements. In this section, you will learn about Agile development's origin, its core principles, and its significance in the world of data engineering.

Introduction to Agile methodology

Agile, a term now synonymous with modern software development, has its roots deeply entrenched in the quest for improved software delivery processes. Originating from a gathering of 17 software developers in 2001, **Agile methodology** arose as an antidote to the cumbersome, linear approaches of traditional waterfall methodologies. At its heart, Agile propounds four core tenets:

- Iterative development
- Fostering collaboration
- The nimbleness to adapt
- An unwavering focus on delivering tangible value

Agile principles and their significance in data engineering

In data engineering, these principles find a renewed resonance. Iterative development advocates for shorter, frequent releases, allowing data solutions to be refined progressively. This approach stands in stark contrast to major releases that might be riddled with unforeseen issues.

A collaborative environment is another cornerstone, underscoring the importance of regular interactions. In a data-driven world, it's imperative to be in sync with stakeholders, data consumers, and fellow team members. With the volatile nature of data sources and user requirements, the responding to change principle is crucial. Gone are the days of static requirements; agility in adapting to changes is a must.

Lastly, the focus on value ensures that with each iteration, the most critical features or fixes are delivered, maximizing the benefit to the end user.

Benefits of implementing Agile in data engineering

Adopting Agile within data engineering projects brings a host of advantages. One of the most tangible is the faster time to insight. With Agile's quick cycles, data processing and analytics are expedited, equipping businesses to make informed decisions promptly. The methodology also offers remarkable flexibility, enabling teams to swiftly adjust to emerging data challenges or new opportunities. This dynamic approach leads to higher quality data solutions as continuous refinement spots and rectifies errors and inefficiencies. Lastly, the iterative nature of Agile fosters consistent stakeholder engagement, ensuring solutions are always aligned with broader business objectives.

Challenges and considerations in Agile data engineering

However, Agile's journey in data engineering isn't without hurdles. Data, by its very nature, has certain immutable characteristics. Unlike traditional software code, some decisions around data, once made, have lasting consequences. Then there's the challenge of managing dependencies. Data workflows are often intertwined and parsing them out in iterative cycles can be complex. Also, to ensure Agile's success, there needs to be keen attention to team dynamics. A balanced mix of roles and a rhythm in communication are paramount.

Steps to implement Agile in data engineering

For teams considering this shift, starting small is wise. Begin with a pilot, selecting a specific project or team to experiment with Agile practices. To lay a strong foundation, invest in training and workshops, orienting teams to Agile's principles, particularly focusing on nuances unique to data projects. The choice of an Agile framework, be it scrum, kanban, or another, should align with the project's specific needs. As the journey progresses, regular reflection sessions, such as retrospectives, can be invaluable in assessing and fine-tuning the approach.

Tools and Agile practices tailored for data engineering

In today's tech landscape, a slew of tools facilitates Agile in data engineering. While a detailed discussion on version control might be deferred, it's noteworthy to mention its critical role in tracking changes. The principles of CI/CD, traditionally seen in software development, find their place in data engineering, enabling automation in testing and deploying data solutions. Furthermore, various collaborative platforms are available, aiding everything from backlog management to sprint planning {https://packt-debp.link/HdLg6E}.

Conclusion

Agile, in the context of data engineering, is not just about a set of processes; it's a mindset. It empowers teams to be more responsive, adaptive, and ultimately more effective. As the data landscape continues to evolve, embracing Agile offers teams both the structure of proven processes and the flexibility to navigate future challenges.

SBP 3 – generate objectives and key results (OKRs)!

Bring the business along and generate OKRs. You can let them drive so long as *Agile methodology and architecture best practices are communicated and agreeable.*

OKRs are transformative and facilitate clarity of purpose, measurable outcomes, and organizational alignment. Their utility is not limited to general business operations. In the area of data engineering, OKRs provide the blueprint for obtaining results and measurable impacts. Let's elaborate on the value of setting and complying with OKRs.

Introduction and deep dive into OKRs

Originating in the 1970s at Intel under the guidance of Andy Grove and John Doerr and later popularized by companies such as Google, OKRs have since become a staple in modern management methodology {https://packt-debp.link/BsaVRO}. At their core, OKRs consist of two main components:

- The *objectives*, which are qualitative and capture overarching goals.

- The *key results*, which are quantitative and specify how you will measure the progress toward these objectives.

These are set with a specific frequency, typically quarterly or annually, promoting a cyclical and iterative approach to goal setting and evaluation.

OKRs bridge the gap between high-level organizational vision and ground-level execution. By transparently setting and sharing objectives, departments, and teams understand their role in the larger company mission. This transparency fosters collaboration and helps prevent siloed operations. A unique aspect of OKRs is their cascading nature-starting from overarching company OKRs – they trickle down, allowing even individual contributors to align their tasks with the company's broader vision.

OKRs play a pivotal role in ensuring that data initiatives are not just technical endeavors but strategically align with business objectives. By quantifying outcomes through key results, the impact of data operations becomes tangible. This clarity aids in setting priorities – helping teams discern which projects to tackle first. Furthermore, given the dynamic nature of data, OKRs provide a framework for teams to swiftly adjust priorities, ensuring adaptability in the face of new challenges or requirements.

They instill a sense of clarity and focus, enabling teams to dissect expansive goals into concrete, actionable tasks. By establishing quantifiable outcomes, they promote accountability and ownership, pushing individuals to take the helm of their respective roles and results. Furthermore, OKRs act as a catalyst for collaboration, fostering a bridge between data engineers and other stakeholders, ensuring that projects align with business needs.

Crafting data-centric OKRs

Formulating effective OKRs for data engineering involves a blend of ambition and realism. An example of a compelling objective might be to *improve data reliability across all pipelines*. This could be complemented by a key result such as *reducing data pipeline failures by 20% this quarter*. The essence of OKRs lies in challenging the team while ensuring the goals remain attainable. This ensures motivation remains high without leading to burnout or disillusionment.

Potential challenges with OKRs in data engineering

Like any tool or methodology, there are nuances when applying OKRs to data engineering. Data's unpredictable nature can throw a wrench into well-laid plans. Addressing data anomalies or unexpected shifts in data quality can sometimes clash with established OKRs. Additionally, it is essential to avoid superficial metrics. Key results should genuinely mirror meaningful outcomes and not just look good on paper. Striking a balance between technical deliverables and business-driven results is also pivotal, ensuring that OKRs resonate both with the data engineering team and the larger organizational stakeholders.

Reviewing and iterating on OKRs in a data context

The cyclic nature of OKRs emphasizes reflection and iteration. Regular retrospectives post each OKR cycle offer invaluable insights into what worked, what didn't, and where adjustments are needed. These learnings, coupled with technological advancements or evolving business objectives, provide the blueprint for the subsequent OKR cycles, ensuring the methodology remains Agile and effective.

The OKR framework offers a potent mix of vision and pragmatism. It ensures that data engineering teams are not working in isolation but are intrinsically tied to business outcomes. Adopting OKRs translates to a clearer organizational vision, enhanced alignment, and a continuous drive toward improvement, all crucial ingredients for success in the evolving world of data engineering. Embracing OKRs not only clarifies objectives but fortifies the bond between data operations and overarching business goals, driving organizations toward sustained success.

SBP 4 – implement data as a product!

Implementing your data as a product enables flows through your data engineered platform. SLAs, cataloging, versioning, corrections, lineage, time series, and quality metadata are all capabilities of the logical architecture! In this best practice, you will want to refocus the software engineers on the data product and not just the data input and output of a system. This means you (the data engineer) are responsible for helping the software engineering team see data as a product and then build software solutions that accommodate that need. The effort will be to encode semantics with data and create various knowledge-aware data flows. This results in more software development effort because many open source and third-party tools do not yet support that goal today. However, they will over time as support grows for this need. The requirement to curate data ready for ML/AI use cases will help drive this effort.

Organizations are constantly exploring strategies to unlock the full potential of their data assets. While traditionally, data has been viewed merely as a byproduct or a supporting element, there's an emerging philosophy that challenges this notion: viewing data as a product. This shift in perspective isn't merely semantic – it is transformative, changing how businesses approach, manage, and derive value from their data. In this section, we'll explore this transformative mindset, outlining its significance, benefits, and best practices to operationalize it.

he transformative benefits of viewing data as a product

Positioning data as a product brings forth an array of benefits. For starters, it ensures end user alignment. By crafting data solutions tailored to specific user needs, organizations can markedly enhance satisfaction and utility. This viewpoint also champions streamlined life cycle management, ensuring that data retains its relevance, freshness, and value as it traverses its life cycle. What's more, when data is seen as a product, consistent **quality assurance** (**QA**) becomes paramount, naturally boosting the trustworthiness and reliability of data outputs. And, not to be overlooked, this mindset significantly improves collaboration, acting as a bridge between data producers and consumers. It facilitates seamless communication and establishes essential feedback loops.

At the heart of the data as a product philosophy is the end user. User-centric development remains paramount, ensuring that user needs and preferences are prioritized at all development stages. Furthermore, foresight is exercised, especially when combined with *SBP 13 – drive scripting and PaaS code with parameterization,* enabling future growth and adaptability. Last, but certainly not least, transparency and documentation are emphasized, assuring that end users maintain a clear understanding of data sources, transformations, and overall reliability.

To transform data into a truly valuable product, several components stand out. Adaptive data ingestion mechanisms evolve to accommodate changing data sources and formats. Advanced processing leverages the latest tools and techniques, transforming and enriching data effectively. The selection of efficient storage solutions seeks a sweet spot, balancing cost implications with speed and reliability. And finally, data is made available via user-friendly access and delivery systems, designed with ease and versatility in mind.

Yet, like any paradigm shift, the task of productizing data isn't without its challenges. The continuous evolution of both data and user needs demands agility. Meeting varied expectations necessitates carefully juggling diverse stakeholder requirements, ensuring no compromise on quality or efficiency. A central challenge lies in harmonizing access and governance, balancing the need for data accessibility against the ever-important backdrop of regulatory compliance.

Interestingly, the *data as a product* philosophy doesn't exist in isolation. Its effectiveness is amplified when seamlessly integrated with other best practices. Combining it with the Agile methodology offers agility, integrating with OKRs ensures goal alignment, and coupling with architectural adherence guarantees robustness.

In conclusion, viewing *data as a product* isn't just a perspective shift; it's a competitive edge. Organizations that truly internalize this philosophy stand to gain in alignment, collaboration, and overall success. For data teams aiming to climb new pinnacles, embracing this mindset is not just recommended; it's imperative.

SBP 5 – implement shift left testing (SLT) processes!

It's all about the data, but what data if it's not cataloged and tested to a degree of correctness that downstream derived data, insights, and metrics are correct? Test-driven design, shift left testing (SLT), and quality SLAs are essential to be able to track errors in data, trends, and any derived use cases so that *trust* can be established.

Making sure our data solutions are top-notch and reliable is paramount. SLT, a game-changing approach in software engineering, is just as critical when it comes to data engineering. By integrating testing earlier into the development process, organizations can anticipate and mitigate potential issues, enhancing both the quality and efficiency of their data solutions. In this section, you will learn how SLT processes will benefit your engineering goals.

Understanding SLT

Shift Left Testing (SLT) is more than just a technique – it is a philosophy. The core tenet is to initiate testing procedures earlier in the software development life cycle, as opposed to traditional methods where testing is predominantly a later-stage activity. In essence, the testing activities shift toward the *left* on the project timeline, emphasizing proactive issue identification and resolution.

The foundational logic of this methodology is straightforward: the sooner you identify a problem, the easier and cheaper it is to rectify. *Left* in this context points toward moving the testing phase earlier in the development sequence, ensuring a more seamless and efficient production process.

Benefits of SLT in data engineering

Applying the shift-left approach in data engineering has several advantages:

- **Early detection**: By focusing on early-stage testing, data engineers can identify discrepancies, anomalies, and potential issues long before they make it to production. This proactive approach minimizes disruptions and ensures smoother deployments.

- **Cost-efficiency**: Addressing issues in the initial stages is often more cost-effective. It reduces the extensive rework and potential downtimes that come with late-stage or post-deployment fixes.

- **Improved product quality**: With a shift left approach, data products and pipelines undergo rigorous testing from inception, ensuring their quality, reliability, and efficiency from the outset.

- **Faster time-to-market**: Fewer late-stage errors means reduced back-and-forth between the development and testing phases. This efficiency translates to quicker releases and faster value delivery.

Along with the benefits of SLT, you will want to know how to implement the approach. This will be defined in the next section.

Implementing shift left testing

The SLT paradigm demands a synergistic collaboration between data engineers, quality assurance personnel, and other relevant stakeholders right from a project's inception. By espousing continuous integration – where code changes are frequently merged and subjected to tests – issues are promptly spotted and rectified. Employing automated testing tools specifically tailored for data transformations, quality checks, and other data-centric concerns becomes indispensable. An efficient feedback mechanism, which swiftly channels testing insights back to developers, further enhances the process.

Specific shift left testing strategies for data engineering

Within data engineering, certain tailored strategies accentuate the effectiveness of SLT. Rigorous data quality checks at early stages validate data consistency, authenticity, and overall quality. Schema validation tests become pivotal in confirming adherence to predefined data structures. It's also crucial to verify the accuracy of data transformations that are integral to ETL processes. Integration testing takes precedence to ensure that various data sources and sinks interoperate without hitches.

Challenges in shift left testing for data engineering

While the approach is replete with benefits, it's not devoid of challenges. Data engineering often grapples with diverse data sources, intricate formats, and voluminous data, which makes testing a daunting task. The *shift left* approach might necessitate the deployment of additional resources, both in terms of tools and personnel, earlier in the project. Additionally, a cultural transition might be required to disband traditional development silos and foster early-stage collaboration.

Tools and technologies to facilitate shift left in data engineering

Numerous software solutions have emerged that cater specifically to early testing, automation, and continuous integration within data projects. These tools, when judiciously chosen and implemented, can substantially augment the *shift left* approach in data engineering. Some of these tools are as follows:

- Open source tools:

  - Cucumber (MIT License)

  - FitNesse (Common Public License)

  - Jasmine (MIT License)

  - Mocha (MIT License)

  - RSpec (MIT License)

  - Robot Framework (Apache License)

  - Selenium (Apache License)

  - Watir (BSD licenses)

- Third-party tools:

  - Appium

  - Eggplant Functional

  - Katalon Studio

  - Micro Focus Unified Functional Testing

  - Ranorex

- TestComplete

- Silk Test

- SOAtest

- Test Studio

- Tricentis Tosca

- Rational Functional Tester

- Rational Performance Tester

Synergy with other data best practices

SLT doesn't operate in isolation. It synergistically interacts with other best practices, such as data as a product, Agile methodology, and OKRs. When cohesively implemented, they jointly amplify the efficacy of the data engineering process.

The emphasis on adopting SLT in data engineering cannot be overstated. As data solutions become increasingly integral to business success, proactive quality assurance methodologies such as shift left stand as cornerstones for fruitful and efficient data projects {https://packt-debp.link/8UMl3u}.

SBP 6 – implement the difficult first!

In software engineering, there's a prevailing adage:

"Eat the frog!"

"Eat That Frog!: 21 Great Ways to Stop Procrastinating and Get More Done in Less Time" (https://www.amazon.com/Eat-That-Frog-Great-Procrastinating/dp/162656941X).

This metaphorical advice suggests tackling the most challenging task of your day first, freeing you from the looming dread of facing it later. Similarly, in data engineering, we adapt this notion to our processes, emphasizing the importance of confronting the most complex problems at the outset. It's not just about getting the hard tasks out of the way; it's a strategic maneuver with profound implications. Why you should tackle difficult issues first will be explained in the following sections.

The philosophy of tackling the hard tasks first

Tackling the hardest problems first may appear counterintuitive, especially in a discipline as nuanced as data engineering. However, it is deeply rooted in prioritization and risk management.

By dealing with the most challenging parts early, teams can evaluate the feasibility of their methods and proactively handle potential setbacks. This acts as a safeguard, ensuring that projects don't suffer late-stage disruptions, which tend to be costlier and more time-consuming to rectify.

For data engineers, this methodology is invaluable. Laying down a solid foundational architecture from the get-go – through intricate data integrations or optimized big data processing – not only showcases the team's expertise but also ensures smoother execution of subsequent tasks. This approach aids in refining data architecture, obtaining early feedback, bolstering system scalability, and assuring stakeholders of the team's prowess.

How data engineers can prioritize difficult tasks

Discernment is paramount. It starts with requirement analysis, where tasks are sifted based on their complexity. Dependency mapping then pinpoints tasks with numerous dependencies, ensuring future bottlenecks are averted. Risk assessment, although often overlooked, can serve as an early warning system, highlighting tasks that might have hidden challenges.

Implementing difficult data tasks

Certain situations in data engineering vividly underscore this approach's value. Whether it's the intricacies of establishing effective real-time data pipelines or ensuring rigorous checks for inconsistent data sources, prioritizing these challenges facilitates smoother navigation of real-time data processing and data quality landscapes.

However, the path is not devoid of obstacles. A concentrated effort on formidable tasks might quickly deplete initial resources and even risk team burnout. Maintaining equilibrium is crucial, ensuring that while tackling giants, smaller yet pivotal tasks aren't sidelined.

Synergy with other data best practices

This principle aligns seamlessly with other best practices in data engineering. From aligning early problem detection with SLT to drawing synergies with Agile's feedback loops and the data as a product paradigm, this strategy reinforces other methodologies.

Conclusion

In essence, *implement the difficult first* encourages teams to be proactive, not reactive. For those willing to face their most formidable challenges head-on, the ensuing journey is often more streamlined and foreseeable.

SBP 7 – avoid premature optimization

Use profiling to find bottlenecks. Avoid unnecessarily adding complexity to components that aren't bottlenecks.

To avoid premature optimization, only optimize those parts of the systems that are actually on the critical path.

Optimization can be a double-edged sword. While it's essential to ensure software operates efficiently, there's a nuanced art in discerning when and where to optimize. In this section, you will be guided by some common wisdom and learn how to avoid premature optimization of your solution.

"Premature optimization is the root of all evil."

Donald Knuth, one of the most influential figures in computer science, famously stated this cautionary axiom (https://dl.acm.org/doi/10.1145/356635.356640). It not only underscores the hazards of misplaced priorities but also propels us to revisit our strategies with a critical lens. Why did Knuth place such emphasis on this matter, and how does it resonate with the modern challenges in data engineering?

The true cost of premature optimization

Every developer knows the joy – the thrill – of producing clever, streamlined code. While the desire to optimize for its own stake is especially common among inexperienced engineers, this exhilaration often tempts even seasoned professionals. However, intricate optimization, especially when done prematurely, can introduce numerous pitfalls:

- **The thrill of clever code**: There's an undeniable appeal in crafting sleek, efficient code. It's a demonstration of a developer's prowess. However, this can be a siren call leading to convoluted solutions where simplicity would suffice.

- **Hidden costs**: While optimizing a piece of code or a data pipeline might seem advantageous, it often results in a complex system. This complexity can waste development time and make the entire system harder to maintain, debug, or extend.

- **The irony**: In a twist of fate, premature optimization, instead of enhancing performance, can often degrade it. The added layers of complexity might prevent future, more impactful optimizations.

Before plunging into optimization, you must first diagnose where the genuine bottlenecks lie:

- **Using profiling to find bottlenecks**: Profiling tools shine a spotlight on performance hotspots, guiding developers to areas that genuinely benefit from optimization.

- **Targeted optimization**: Once identified, developers can then focus their energies on refining these specific components, ensuring that optimization efforts yield tangible results.

Missteps in optimization can have cascading effects:

- **Misallocation of resources**: Optimizing non-critical parts siphons away valuable time and resources from areas that might demand attention.

- **Reduced flexibility**: A system entangled in premature optimizations becomes rigid, making it resistant to changes or future improvements.

Recognizing and avoiding the trap in data engineering

Data engineers, working with vast and varied datasets, must be particularly wary of over-optimizing the following:

- **Performance versus complexity**: While performance is paramount in data processing tasks, it's pivotal to balance it with the overhead of added complexity.

- **Iterative development**: Begin with a functional, straightforward solution. As the system matures and actual bottlenecks emerge, refine and optimize based on concrete evidence and feedback.

Balancing performance needs and over-optimization in data engineering

The landscape of data engineering poses unique challenges and opportunities:

- **Performance benchmarks**: Establish measurable criteria that guide optimization endeavors, ensuring that they align with actual business needs.

- **Cost-benefit analysis**: Any optimization should be weighed against the intricacy it introduces. Does the performance gain justify the potential maintenance and flexibility costs?

Synergy with other data best practices

The principle of avoiding premature optimization harmoniously aligns with other best practices. For instance, when paired with implementing the difficult first, engineers can craft robust foundational systems while sidestepping unnecessary complexities. Similarly, with SLT, ensuring that code is functional and bug-free takes precedence over early-stage intricate optimizations.

Knuth's advice remains evergreen. As we forge ahead, building and refining data systems, let's heed his wisdom. Prioritize clarity, maintainability, and genuine performance imperatives over the ephemeral allure of unnecessary optimizations. In doing so, we ensure that our engineering efforts stand the test of time.

SBP 8 – automate cloud code snippet deployments with standard deployment scripted wrappers

Code is code and must be in version control under a flow (such as Gitflow).

PaaS service provisioning can be niche and custom. Avoid manual deployment by scripting major operations and configurations such as deploy, redeploy, QA, pause, run, stop, and blue/green releases, all of which need to be automated operations.

The significance of deployment automation is undeniable. It's not just about delivering code anymore; it's about ensuring its reproducibility, traceability, and, most importantly, consistency. This is where the philosophy that code is just that – *code* – and it all must be retained in a version control system under a development/approval flow. This is a given for a modern software system. It makes common sense since any work product should be considered a business/enterprise asset and that asset must be put under some process control. In this section, you will learn that operational wrappers and configurations are complex, and they also need to be retained in version control systems after being scripted.

The importance of deployment automation

Automated deployments in cloud environments are no longer a luxury but a necessity. The diverse services and tools that the cloud offers, when paired with the fluidity and dynamic nature of data engineering, demand streamlined processes. Any code, be it the primary application or the deployment scripts, should be treated with the rigor of version control, ensuring traceability and collaboration.

Platform as a Service (PaaS) has become the go-to for many developers, eliminating the need to manage underlying infrastructures. However, with customization comes the challenge of deploying applications seamlessly. Manual deployments, in such intricate environments, open doors to inconsistency and errors.

Every Platform as a Service (PaaS) environment comes with its own set of operations. Deploying, managing, scaling, and toggling between production and testing environments in PaaS can often be monotonous and error-prone when done manually.

The remedy? Scripting – an automated, standardized process that eliminates guesswork and human-induced errors. The technique of blue/green deployments benefits significantly from this approach. In a blue/green deployment, there are two identical production environments – one active and one idle – scripting to facilitate the swift and safe transition from one state to the other, thus ensuring continuous delivery with minimal downtime. What you will want to take away from this discussion is that you want to automate your deployment. You also want to clearly communicate the deployment model that's been selected for the solution and any deviations for critical subsystems.

The deployment model choices

There are options that you will want to choose from when selecting the deployment model. These are driven by the following factors:

- The complexity of the deployed solution (or part of the solution) that needs to be deployed will drive deployment model decisions. Simple, small, compartmentalized, easily integrated, or isolated subsystems that do not have a lot of dependencies best fit the blue/green deployment model. Try not to force your primary deployment model on all your solutions subsystems, and do not have too many of these deployment models, even if each were to be automated. That proliferation of models produces operational management and maintenance issues. Larger complex subsystems will need an advanced strategy, such as a *canary* deployment approach.

- Your business's tolerance for functional availability at scale should be considered when selecting a deployment model! Some critical businesses need to keep functioning during solution deployments and during any potential rollbacks. In the eventuality that a system and its data state have to be reverted to a last known good state due to an error, you will have to design each solution deployment as if it were a mini-disaster to be recovered from within a fixed **mean time to restore** (**MTTR**).

- The project or overall solution budget versus the TCO of your desired deployment model's costs have to be factored in and justified and then communicated to all. You will want to *right-size* your deployment model and be able to justify the process when it is put into operation.

- Any deployment model choice you make has to take into account that some or all users could be impacted during a deployment model's cycle of operations. Defining the acceptable level of impact is a planning exercise that then requires testing to make sure that level is achieved. Choosing the acceptable impact level and creating tests aligned to measured stats over time validates your deployment model's outcome. These sustained results and solid reporting build trust in the business. A *Big Bang* deployment model causes long downtimes but is cheap to implement; however, your business may not appreciate the user impact unless the business knows that users will *not* be seriously affected during long downtimes during off-peak hours. As an example, insurance company IT solutions often still enforce many hours of maintenance window downtime on weekends.

To summarize, the top choices for your deployment model are as follows:

- **A/B deployment** is a continuous deployment model with sets of users routed to an A or B version of a feature supported on an A versus B version of the servers, or services, supporting the feature's version.

- **Big Bang deployment** is the opposite of continuous deployment. It is a legacy approach that requires a lot of static testing.

- **Blue/green deployment** is a semi-continuous deployment model that requires two versions of the system to exist. Changes may stream into the green environment, after which the feature is fully rolled out after testing in the green environment is completed. When looking at systems with huge data volumes where the correctness of the data matters most, this model is often a safe approach to use. However, there will be resistance from the cloud providers' published best practices that often propose the canary deployment model.

- **Canary deployment** is a continuous deployment model that requires you to develop an advanced automated test discipline. Since features are rolled out with initially zero user feature traffic allowed, progressing to 100% user traffic is allowed at the end of testing. This ramp-up approach is often used with feature-based development in the cloud.

- **Continuous deployment** is the root of modern deployment models for a system deployment where changes stream into an environment in an automated manner. Testing must be automated, but the state of data quality *cannot* be assured until all software changes and their impacts settle after a deployment activity is applied in an environment. Reversion to a last known good data state is difficult at best. Additional compensating application transactions are required if data errors are encountered after code deployment, rather than a reversion to a last known good.

- **Ramped deployment** (the same as **rolling deployment**) is a continuous deployment model that targets an increasing number of servers/services until all are ungraded.

- **Recreate deployment** is a flavor of the non-continuous Big Bang deployment model where there is a lot of user impact during downtime. This deployment model is used when a system is to be radically upgraded from a prior version. In this case, the user experience is so radically changed that the clean break in access is just a prelude to the clean break in UX experience to be handled immediately after access is restored.

- **Shadow deployment** (also called **dark launching deployment**) is a continuous deployment model often aligned with functional A/B testing, but users have no access to the new feature. The user traffic is routed to both A and B versions in parallel to see how the system behaves. Users are unaware that this routing is happening in the background.

Benefits of using scripted deployment wrappers

With scripted deployments, we introduce a layer of standardization, making each deployment predictable. This not only minimizes errors but also accelerates deployment cycles, a crucial aspect for time-sensitive data workloads.

Version control – ensuring consistency and traceability

It's essential to ensure that every piece of code, configuration, or script is under version control. Gitflow, for instance, provides a systematic approach to versioning, making it easier to manage, collaborate, and deploy without hitches.

Relevance to data engineering in cloud environments

In data engineering, where data sources and workloads might change frequently, being Agile is paramount. Automated deployments ensure that changes can be rolled out swiftly, ensuring timely data processing and analytics.

Practical implementation steps

The first step toward automation is scripting your deployments. Every operation and every trigger in the PaaS should be scriptable. Once scripted, integrating with a version control system such as Git ensures traceability. Adopting methodologies such as Gitflow {https://packt-debp.link/AyD6oD}, can further structure and streamline these processes.

Challenges and precautions

While automation is desirable, it's not without its challenges. Ensuring that scripts remain updated with evolving infrastructure changes is crucial. Security, in particular, should never be compromised during automation, requiring vigilant practices.

Synergy with other software and data best practices

Automated deployments aren't standalone. They gel well with practices such as avoiding premature optimization and ensuring that performance enhancements are done in a controlled, versioned manner. Similarly, SLT, when combined with automation, ensures that any code, be it data processing or a deployment script, is always of the highest quality.

Embracing deployment automation paired with structured version control is no longer optional. It's a strategic move, one that ensures consistency, efficiency, and traceability, paving the way for successful data engineering projects.

SBP 9 – define and implement NFRs first

NFRs delineate how a system operates, rather than the specific functionalities it performs.

In essence, while **functional requirements** (**FRs**) might dictate what a system does, NFRs specify how well it does those tasks. Common examples include performance benchmarks, scalability thresholds, system reliability, security standards, and maintainability guidelines. You will learn more about why NFRs need to be built first and then the functional requirements (FRs) as part of the effort to implement your OKRs.

Distinguishing functional (FRs) from non-functional requirements (NFRs)

FRs are explicit about the functions a system must perform. For instance, a data analytics application might need to generate specific reports from input data (an FR). On the other hand, NFRs might dictate that these reports be generated within a specific time frame, that the system can handle a certain volume of data inputs without degradation in performance, or that data is encrypted to safeguard confidentiality.

The inherent dynamism and scalability of cloud environments make NFRs particularly significant. In the cloud, where resources are often abstracted and can be scaled on demand, ensuring that a system not only functions but does so optimally, becomes pivotal. For instance, while a cloud solution might scale to accommodate increased data loads, it's the NFRs that will dictate how swiftly and efficiently this scaling occurs.

Relevance to data engineering

For data engineers, the cloud has transformed many aspects of data storage, processing, and analytics. But the importance of NFRs becomes accentuated here. Imagine a data pipeline that processes massive volumes of data. Its functional requirement is to process this data. However, NFRs will determine how quickly this data is processed, the pipeline's responsiveness as data volume grows, and the measures in place to ensure data accuracy and integrity during processing.

Key NFRs in cloud data engineering

Among the myriad of NFRs, some stand out in data engineering. Performance benchmarks determine desired processing speeds and acceptable latencies. Scalability ensures the system can handle surges in data, both in terms of volume and complexity. Reliability touches on consistent operations and data availability, ensuring that any processed data is accurate and retrievable. Security is about more than just breaches; it concerns data integrity, privacy, and regulatory compliance. Lastly, maintainability ensures the system can adapt to evolving data landscapes, accommodating new data sources or analytics tools with minimal disruption.

Defining and implementing NFRs

Stakeholder engagement is a critical first step, ensuring that NFRs are aligned with business goals and user expectations. Clear and comprehensive documentation is a must, serving as a roadmap for developers and a standard against which the system is tested. It's crucial to verify these NFRs early and often, using tests and monitoring tools to ensure they're being met as development progresses.

Risks of neglecting early implementation of NFRs

Being reactive with NFRs often leads to steeper costs down the line. Retrofitting a system to meet overlooked NFRs can be significantly more challenging and costly than building with these in mind from the outset. Plus, there's the risk of system inefficiencies, outages, or even failures if NFRs aren't adequately addressed.

Data engineering in cloud environments is a complex endeavor, with many moving parts. Central to ensuring that these systems not only work but work optimally is the early definition and implementation of NFRs. By placing NFRs at the forefront of cloud-based data engineering projects, developers can ensure robust, efficient, and resilient systems that meet the demands of modern data operations.

SBP 10 – implement data journey journaling to facilitate future problem resolution

Data journey journaling is an emerging concept that resonates deeply within the data engineering world and has garnered attention in the broader scope of SBPs. This methodology focuses on meticulously documenting the path data traverses, from its point of origin to its final consumption. In this section, you will gain an appreciation for developing data as a product by implementing the data journey so that data issues (or problems in the curated datasets) may be remediated quickly and efficiently.

A data journey can be defined as the comprehensive pathway data takes, encompassing stages such as ingestion, processing, transformation, storage, and, ultimately, presentation. Each stage offers unique challenges and insights, necessitating a clear understanding of the journey for efficient system management. By enabling detailed tracing of data journeys, you can glean in-depth analytical insights. Such rigorous observability streamlines the identification of issues, leading to swifter resolutions.

In the web domain, **clickstream analysis** is akin to keeping a meticulous log of user journeys, gathering valuable insights from each click to better understand user behavior and interaction patterns. Similarly, in data engineering, tracing the flow and interactions of the data's journey with *journaling* provides crucial visibility into how data moves and is transformed, revealing bottlenecks or inefficiencies. Both practices are fundamental in diagnosing system issues and enhancing the user or data journey, respectively.

Relevance to data engineering

Data engineering today is a complex web, often involving intricate pipelines where data undergoes multiple transformations. This complexity accentuates the importance of journaling. Understanding the lineage of data – its journey from source to presentation – enhances trust among stakeholders, ensuring that data is reliable, verifiable, and hasn't been tampered with. With the aid of data journey journaling, troubleshooting becomes markedly more efficient. When data anomalies arise, having a clear journal outlining the data's journey can swiftly pinpoint the issue's origin. Additionally, archived journal entries serve as a historical record, enabling data engineers to understand recurrent problems or to contrast data states across various timelines.

Challenges and considerations

While the advantages of this methodology are evident, there are inherent challenges. Chief among them is ensuring data privacy and adhering to regulatory standards. Journaling massive datasets also introduces storage implications that need careful consideration. As data engineering landscapes are continually evolving, the journaling processes must be routinely revisited and updated to stay relevant.

Data journey journaling is an indispensable tool in contemporary data engineering. Its parallels with practices such as clickstream analysis for the web reiterate its importance. As data becomes increasingly central to decision-making in various industries, ensuring its reliability, transparency, and traceability through journaling becomes paramount.

SBP 11 – implement data journey pipelines that are experimental first!

There's an emerging emphasis on adopting an experimental first mindset within the data engineering community. This innovative approach promotes the idea of treating data releases with the same precision and caution as software code releases, underscoring the importance of iterative refinement and rigorous testing. Agile development thinking can be applied to how you create your datasets, and in this section, you will be exposed to that process as your data pipeline solution takes shape.

A data journey delineates the progression of data from its inception at the source to its final destination of consumption. With today's multifaceted frameworks and workflows that define modern data pipelines, the impetus for an experimental approach becomes glaringly evident.

An experimental-first stance brings with it several benefits. By emphasizing rapid iterations at the outset, teams can harness the advantages of prototyping and early-stage testing. This proactive approach not only minimizes overarching issues by validating solutions in contained settings but also ensures that the pipeline remains Agile, primed to adapt to the mutable nature of data structures and shifting business needs.

Enabling data pipeline experimentation as datasets are readied

As datasets are marshaled for release, it's crucial to accentuate experimentation. This involves crafting prototype pipelines that are receptive to feedback, evolving insights, and the unforeseen quirks of data. Integrating robust feedback loops is pivotal – these mechanisms offer a wealth of knowledge from automated tests, invested stakeholders, and end users of the data, ensuring a gold standard of quality upon release.

Releasing data like code

The proposition is simple yet revolutionary; release data as you would release code. By weaving in version control, teams can preserve a granular history of data iterations. Datasets, much like software, should be subjected to exhaustive testing before release. And if post-release anomalies emerge, having the infrastructure for swift rollbacks ensures the sanctity and integrity of the data.

Challenges and considerations

Yet, this journey is not devoid of challenges. It's imperative to transparently communicate to stakeholders, setting clear expectations about the experimental nature of data releases. Balancing resources is another tightrope act, ensuring that the focus on experimentation doesn't eclipse other pressing needs. And amid these shifts, the edicts of data privacy and security remain sacrosanct.

To conclude, championing an experimental first approach is more than a mere best practice – it's a paradigm shift in data engineering. By mirroring the meticulousness of software code releases, we elevate our data pipelines to unparalleled heights of reliability and excellence.

SBP 12 – choose languages with solid reasoning

Diversity in language choice brings forth a richness of capabilities but also comes with the inherent challenge of managing complexity. It's a delicate balance between leveraging the strengths of various languages and not overwhelming the engineering process.

Key languages in data engineering and their roles

Within data engineering, specific languages have established their significance:

- **Python (3.10.x specifically)**: Dominates modern data processing due to its vast ecosystem. Its simplicity and the powerful libraries available for data analysis, machine learning, and automation are unparalleled.

- **SQL**: Remains indispensable for data querying and management, transcending the test of time.

- **Java**: Known for its robustness and portability, Java plays a significant role in large-scale data processing systems and big data ecosystems. It's particularly valued for its performance in distributed computing environments.

- **Shell scripting (Bash, KSH)**: Shell scripting remains a potent tool for orchestrating data pipelines, automating mundane tasks, and gluing together various components of a data system.

- **Scala**: Often found in big data processing due to its functional programming features and seamless integration with Java ecosystems. Scala is a powerful choice when working with Apache Spark and other JVM-based data processing frameworks.

- **Others**: Numerous other languages cater to specialized use cases, and each deserves consideration based on the task at hand.

The pressures and limitations imposed by PaaS offerings

Often, the selection of a platform can dictate language decisions. There are inevitable compromises to ensure compatibility and seamless integration when working with specific PaaS offerings.

While diverse language choices can introduce versatility, it's paramount that all code, irrespective of its language, adheres to the designated development framework. You must consistently meet the NFRs of the pipeline, ensuring the harmony and efficiency of the entire data process.

The journey to a judicious language selection involves several steps:

- **Define clear objectives**: It's essential to match the language choice with the project's specific needs.

- **Evaluate integration needs**: A thorough assessment ensures the chosen language synergizes with other tools, platforms, and existing code bases.

- **Balance team expertise**: The team's existing skills should be weighed against the potential advantages or challenges a new language might introduce.

- **Be ready to compromise**: External factors, such as compatibility with PaaS offerings, might influence what might otherwise be the ideal language choice.

Now that we've considered these choices, let's look at some pitfalls to avoid.

Pitfalls to avoid

Several traps lie in wait during language selection:

- **Bandwagon effect**: Merely opting for a language due to its trending status can be a perilous choice.

- **Overlooking maintenance**: Future challenges can manifest if you choose a language that lacks substantial community support or sees a decline in popularity.

- **Ignoring integration needs**: Do not neglect how a chosen language integrates with essential tools or platforms as this can lead to cumbersome workarounds later.

The path of data engineering demands that deliberate and informed choices be made concerning language selection. While the richness of available languages adds to the toolkit, a structured approach and adherence to a common framework ensure the smooth functioning and scalability of data processes.

SBP 13 – drive scripting and PaaS code with parameterization using a secure configuration management repository tool

… and preferably a configuration management database.

Harnessing the power of parameterization in scripting and PaaS code is a crucial practice for managing the complexity that comes with scaling systems. As configurations become increasingly intricate, the use of secure configuration management repositories and databases is no longer a luxury but a necessity. This approach not only introduces much-needed flexibility but also safeguards against security vulnerabilities, ensuring our systems remain both adaptable and secure. As you develop a cloud solution, you will want to read about the best practices for scripting PaaS service configurations and parameterizing those scripts and configurations.

The power of parameterization and configuration management

The increasing complexity of software systems demands a shift toward more dynamic and adaptable configuration. Static or hardcoded settings are the progenitors of future rigidity and potential security pitfalls.

Parameterization allows configurations to adjust gracefully to varying environments and requirements. The configuration management repository and database serve as the cornerstones of this adaptable framework, mitigating the impact of complexity and enhancing security by externalizing configuration details.

Minimize the impact of configuration complexity with a configuration management approach so that all scripting is flexible and does not contain data security breaches.

The growth of configuration complexity

As systems unfurl their capabilities over time, they amass a bewildering tangle of configurations. This complexity is a harbinger of the chaos that can ensue without meticulous management. A dynamic and flexible configuration management system is imperative to ensure that as the system evolves, its configurations remain manageable and coherent.

Why parameterize?

PaaS solutions, with their diverse and specialized requirements, underscore the necessity of dynamic configuration. Parameterization confers the ability to separate code from its operating environment, enabling a single script to be performed across multiple platforms with simple parameter adjustments. This separation is essential not only for flexibility but also for maintaining security as it removes the need to store sensitive data within the script itself.

Configuration management repositories and configuration management databases (CMDBs)

The centralization and versioning of configurations in a repository simplify the management of complex systems. Such repositories also play a pivotal role in safeguarding sensitive parameters, allowing for secure storage and controlled access. **Configuration management databases** (**CMDBs**) are a more advanced solution. A CMDB brings structure and clarity to configuration management. By mapping relationships between configurations and integrating them with deployment tools, CMDBs provide a sophisticated solution for managing the complex configurations of large systems.

Best practices for secure configuration management

The stewardship of secure configuration management is a disciplined practice, rooted in the principles of encryption, rigorous access control, and ceaseless vigilance through audits. To encrypt is to shield, to control access is to guard, and to audit is to consistently reaffirm the security of our configurations. These practices are not mere protocols; they are the very essence of trustworthy software engineering.

The implementation of parameterization and the adoption of secure configuration management practices are fundamental to managing the ever-increasing complexity of modern software systems. To parameterize is to prepare for the future, to manage configurations is to tame the wilds of complexity, and to secure them is to honor the trust placed in our systems. By embracing these principles, we equip our systems with the resilience and flexibility to adapt to new challenges while maintaining the highest security standards.

SBP 14 – be prepared to prune dead code over time

PaaS service configurations can create mounds of dead code. Manage the complexity and do not allow vestigial code to build up in deployed systems. Purge it if it is not used.

Maintaining code quality is an ongoing battle. Pruning dead code becomes not just a routine cleanup operation but a necessary discipline to avoid the perils of code base stagnation and to maintain the agility of systems. This section explores the importance of identifying and removing dead code, especially within PaaS service configurations, where the agility of the platform can often lead to a quick buildup of unused code.

The accumulation of dead code in software and PaaS systems

The digital world's life cycle is riddled with the remnants of past coding endeavors. Dead code, or code that is no longer executed in the live environment, has a notorious propensity for creeping into software and PaaS systems, lurking unseen and often ignored. This redundant code can arise from various sources: deprecated features, unused variables, or obsolete functions. In the context of PaaS, where configurations can be as transient as the services they orchestrate, dead code can pose significant risks, such as security vulnerabilities, resource inefficiencies, and increased maintenance overhead.

The unique challenge of PaaS service configurations

PaaS environments are particularly susceptible to dead code accumulation due to their inherent dynamism. Services within PaaS can evolve quickly, leaving behind configurations that are no longer relevant. These outdated configurations contribute to technical debt, complicating the system and obscuring the clarity of its operational structure.

Dead code is an artifact of change. As PaaS services respond to the ever-shifting demands of the business and technological landscapes, code can become obsolete almost overnight. Refactoring efforts might leave behind old code paths that are no longer reached. Temporary fixes or features may become permanent fixtures, contributing to the clutter without serving a functional purpose.

The presence of dead code is not a benign oversight; it carries a real cost. Maintenance efforts must extend to the lifeless parts of the code base, squandering valuable time and resources. For PaaS configurations, this is even more pronounced as the platform's efficiency depends on streamlined and active configurations. Furthermore, the cognitive load on developers increases as they sift through the detritus to find and understand active code paths.

Combating the specter of dead code requires vigilance and the right set of tools. Code coverage analysis can reveal unused paths, while static analysis tools can flag potential redundancies. For PaaS systems, monitoring configurations for activity can help identify candidates for pruning.

Pruning dead code

The act of pruning should be systematic and careful. Using version control allows for safe removals, making it possible to revert changes if necessary. Testing is crucial, especially after removing code that might affect PaaS configurations. Documentation can aid in understanding the rationale behind the pruning decisions.

Cultivating a culture that values regular maintenance and cleanup of code is essential. Routine reviews of PaaS configurations, open communication about the necessity and impact of dead code, and recognizing the efforts of those who keep the code base lean can all contribute to this mindset.

Not all unused code is dead code. Especially in PaaS configurations, some code may be intended for future use or may serve a purpose not immediately evident. Overzealous pruning, without a thorough understanding of the system, can lead to functionality loss and service disruption.

Dead code pruning is not just an act of cleaning up; it's a proactive measure against the entropic nature of software development. It ensures that PaaS systems, which are particularly vulnerable to rapid obsolescence, remain efficient, secure, and easy to understand. This practice is integral to maintaining the health of the code base and the efficacy of the software services it supports.

SBP 15 – if it doesn't fit, don't force it; use a microservice

If an application or process outgrows the operational parameters of its current environment, it should not be constrained by force. Instead, microservice architecture offers a scalable solution. In this section, you'll learn how to discern PaaS limitations and the strategic deployment of microservices to extend, enhance, and evolve the capabilities of your software solutions. We'll explore the indicators for when to make this shift, how to implement it effectively, and the potential pitfalls to avoid, ultimately guiding you toward making informed decisions in the complex interplay between PaaS services and microservices.

PaaS and its boundaries

PaaS provides a wealth of resources to streamline application deployment, yet it has its operational boundaries. Understanding these limitations is crucial as they often delineate the sandbox within which developers can play. This introduction sets the stage for recognizing when PaaS may restrict an application's growth and how adopting microservices can provide the necessary room to maneuver, ensuring that the system architecture can evolve with its requirements.

The boundaries of PaaS often manifest as computational ceilings, stringent storage capabilities, and bandwidth thresholds. Each of these could be a bottleneck for scaling and performance. Understanding these limitations is pivotal, especially when your application demands grow unpredictably. The consequences of hitting these ceilings can be dire in production, leading to service degradation or complete outages, which, in turn, can erode user trust and incur financial costs.

Microservices as a contingency strategy

Enter **microservices** – an architectural approach designed to overcome monolithic constraints by decomposing an application into smaller, independent services. Each microservice is a self-contained unit, managing a specific operational function. This strategy not only enables teams to bypass the operational ceilings of PaaS but also encourages a more resilient design by distributing responsibilities across discrete services that are easier to manage, scale, and update.

Invoking microservices is a strategic decision. Proactive system analysis can highlight when operations approach PaaS limits, allowing for a timely pivot to microservices. This shift is particularly critical when you observe performance bottlenecks or when scalability becomes a pressing concern.

The incorporation of microservices into a PaaS-driven architecture requires meticulous planning. It is essential to ensure that PaaS components and microservices coexist without friction. Data integrity across these services must be preserved, and a robust failover strategy should be in place to address any potential downtime in the PaaS environment. This ensures a seamless experience for end users, regardless of the underlying architectural complexities.

Challenges and considerations of this dual approach

Melding PaaS with microservices is not without its challenges. It introduces a new layer of complexity in monitoring and potentially increases costs. Additionally, it's crucial to ensure that microservices do not introduce significant latency, which could negate the performance benefits they are meant to provide. Balancing these factors requires careful consideration and ongoing optimization.

Pitfalls to avoid

Over-relying on external microservices can become difficult to manage and secure. It's important to keep abreast of PaaS enhancements that may render certain microservices redundant. Integrating microservices necessitates a thorough review of security practices and compliance measures to ensure data remains protected during transit between services.

Leveraging the strengths of PaaS while embracing the flexibility of microservices requires a nuanced understanding of both paradigms. As a developer, the onus is on you to not force a solution within the confines of PaaS when a microservice could elegantly solve the issue at hand. It's a delicate balance of understanding the operational limits, recognizing when to expand beyond them, and ensuring that the solution fits the problem without introducing undue complexity or cost. Being informed and adaptable is key to harnessing the true power of both PaaS and microservices.

Summary

Throughout this chapter, we have underscored the significance of adhering to software engineering best practices, an arsenal of guidelines that serve as the bedrock for creating resilient, efficient, and secure systems. These practices are not merely recommendations; they are the distillation of years of experience and insight from industry veterans, offering a roadmap to navigate the intricate landscape of software development. By integrating these practices into your workflows, you're equipped to confront and mitigate common pitfalls, enabling you to deliver solutions that stand the test of time and scale.

Embrace these best practices as your allies; they empower you to craft software that not only meets the current demands but is also poised to adapt to future challenges. From meticulously managing configurations to strategically pruning dead code and intelligently leveraging microservices, these practices collectively uplift your engineering acumen. In adopting them, you become part of a lineage of developers who not only aim to meet specifications but also aspire to exceed expectations, delivering not just functional but exemplary software products.

In the next chapter, *Key Considerations Regarding Agile SDCL Best Practices*, you will dive into very important best practices introduced in this chapter.

9

Key Considerations for Agile SDLC Best Practices

The **software development life cycle** (**SDLC**) is the backbone of any successful software project. In this chapter, we'll address the **Agile SDLC** – a dynamic and flexible approach that fosters rapid iteration, collaboration, and adaptation. Adhering to foundational SDLC practices is not just recommended, it is compulsory – particularly when managing complex, distributed systems. As we explore the intricacies of Agile methodologies tailored for data engineering, we will begin by laying a foundational understanding of what Agile SDLC entails and its significance in contemporary software projects that are often challenged by rapid changes and high complexity. By laying out the fundamentals of Agile SDLC, we provide a primer into an environment where change is the only constant, and rigidity in the process is the antithesis of progress.

In this chapter, we will firmly establish what constitutes an effective Agile methodology for your organization. A key consideration is that start-ups and small organizations need different processes than large- and medium sized companies: Agile works best for start-ups and small organizations while large medium sized organizations do better with **Lean**.

Prevent Agile from being fragile

Agile can be fragile! You must first develop an organizational Agile Manifesto that defines a workable Agile methodology for the organization. This is necessary to discourage teams from adopting any flavor of the day Agile approach that some genius decides is going to be good for everyone! Mitigating the chaos that misaligned Agile frameworks produce, such as **scrum** or **Kanban**, is essential. Disagreements in methodology can transform a healthy team's workflow into confusion and result in ultra poor performance. There are many flavors of Agile! See *7 Types Of Agile Methodologies* {https://packt-debp. link/7fixFt} and *12 Agile methodologies: Pros, cons, and when to use them* {https://packt-debp.link/3mXhBA}. Scrum and Kanban are two of the most prominent Agile methodologies, each with its own principles and methodologies. Scrum, known for its structured approach and time-boxed sprints, is often favored in projects requiring clear milestones and deliverables. Kanban, with its focus on **continuous delivery** (**CD**) and flexibility, is well suited for projects with evolving requirements. Each methodology has its strengths and weaknesses, particularly in the context of data engineering projects.

Regardless of the core Agile methodology you chose to model your development processes on, it is through peer reviews, Agile ceremonies, and a strong adherence to the chosen SDLC that you need to align team efforts with sprint and epic goals. This maintains high team performance and system quality. In order to promote clear processes for these high performing teams, we'll also advocate for an **Objectives and Key Results (OKR)** model over the traditional **Key Performance Indicators (KPI)** {https://packt-debp.link/i49qFk} model. OKRs are more aligned with business understanding and demonstrability.

Emphasizing the significance of documentation, this chapter underscores that the only standard that matters is the one that is written and enforced by IT management, ensuring clarity and consistency. Your reading of this chapter will produce an awareness that you have an ambitious target to achieve: an effective Agile framework and methodology that ensures your data engineering projects are not only efficient but also robust and aligned with strategic business objectives. You'll learn to develop a customized Agile process that aligns with your organization's size and delivery needs; understand the criticality of business strategies within the Agile framework; and appreciate the operational and security strategies that ensure the delivery of high-quality, secure data products. You will not only grasp the theoretical aspects of Agile SDLC but also be equipped with practical insights and the knowledge needed to evaluate trade-offs for immediate application in your environment.

By getting to grips with the following key topics presented in this chapter, you will understand is needed when defining an effective Agile SDLC in the context of a software data engineering effort:

- Agile methodology
- SDLC processes
- OKRs
- Documented Agile methodology
- Business development strategy
- Test/quality strategy
- Operational strategy
- Data security strategy
- Customized Agile methodology

Agile methodology: Why should you define a methodology? Can't you just pick one and use it? We have never found even one organization that has done this. All methodology is tunable and even if one picks a well-known, off-the-shelf methodology, it must be customized with workflows supported by tooling. Exploring how to define and select Agile methodologies is essential. This way, the final methodology is aligned with your organization's governance and policy needs. Understanding the impact of these choices on the data engineering team is crucial. The methodology selection process should go beyond mere frameworks and focus on identifying Agile best practices that resonate with your organization's core principles and address specific concerns such as process scalability and power structures within the organization. Some will see that they lose influence, yet they are exactly the ones you will want to buy in on your development approach.

SDLC processes: It is essential that you adapt the Agile SDLC for building cloud based distributed systems. The customization should reflect your organizational Agile manifesto. Specifically, you will want to address the scale and complexity specific to your enterprise. This could involve, for large organizations, making sure that there is an OKR setting steering committee set up and making sure that the integration of 10, 20, or 30 teams' work products can work together effectively to form a **minimum viable product** (**MVP**). Or, it can be as simple as a data product owner making sure OKRs are set up to support a single team. The complexities vary. This is especially true for businesses that deal with sensitive PII data in a financial or health setting, but this can also arise when building device software and data systems that are regulated by government standards (ISO and so on).

OKRs: Favoring OKRs for their clarity and business alignment over the KPI model is a clear objective in your future. This will be a difficult topic for human resource groups that have embraced the KPI model. Additionally, since they report to the CEO and CFO, the flexibility to switch to an OKR model from a KPI model will be difficult. There has to be a balance between the need to hold an individual responsible for a goal expressed as a KPI and the data product owner's need to have the OKRs of the team achieved. The OKR has to be composed of line items that support the HR KPI model. Weave your future methodology carefully through the gauntlet of discussions required so as not to go splat against that windshield of that HR freight train.

Documented Agile methodology: We emphasize that there needs to be a focus on creating a documented Agile methodology that resonates with your organization's concerns and priorities. This documentation is crucial for maintaining standards and ensuring that Agile practices are implemented consistently and effectively across the organization. What goes into this documentation are workflows and well-thought-out, codifiable process flows that only an experienced Agile coach can produce once that individual understands your organization's modus operandi. Your Agile methodology documentation must not be an Agile sales brochure!

Business development strategy: **Proofs of Concept** (**POCs**) and **Minimum Viable Products** (**MVPs**) will have to be stressed as essential deliverables with your Agile strategy. You will want to ensure that these tools support risk-adverse smart development, that is consistent with the organization's principles.

Test/quality strategy: **Quality assurance** (**QA**) baselines and **service level management** (**SLM**) across your data factory's data zones will be supported in the Agile framework, underscoring the importance of conforming to audit requirements.

Operational strategy: Enacting data delivery strategies through transformational coding practices and lineage tracking, tailored to the operational scale and structure of your organization will be essential.

Data security strategy: Establishing a resilient, zero-trust security mindset to protect your investment, without paralyzing system operations complements your Agile practices. Meeting your organization's specific security requirements is a side benefit of adopting Agile practices because the rate of change that Agile release practices support nicely fits into the needed rate of change required to keep a system secure.

Customized Agile methodology: Documenting and enforcing the Agile methodology and its processes, specifically tuned to the needs of your organization, will ensure that you can effectively address key goals of your organization even if they are not all initially specified when your projects are funded.

By the chapter's conclusion, you should feel confident in developing an organizational Agile manifesto and applying it as your company's Agile methodology to ensure your data engineering projects are responsive and efficient. With a solid framework for continuous improvement, this will be possible. Let's begin our journey to elevate the standards of software engineering and show how Agile supports a well-engineered future-proof data engineering effort.

Agile methodology

Agile, in its essence, is about iterative development and responsiveness to change. This is a paradigm shift from traditional, sequential approaches in software development. Its genesis in the early 2000s, marked by the **Agile Manifesto** {https://packt-debp.link/7BSJWU}, represented a collective yearning within the software community for methods that embrace the fluid nature of software projects, particularly resonant in the area of data engineering.

Agile's roots can be traced back to a dissatisfaction with the rigid, linear (waterfall) approaches of the past. The Agile Manifesto emerged as a crystallization of this thinking, underscoring values such as individual interactions, working solutions, customer collaboration, and responsiveness to change {https://packt-debp.link/RYoejS}.

In this section, you will learn the core principles of Agile and how to best apply them within your organization. You will learn how to select the Agile methodology for your organization while avoiding inconsistent *flavor of the day* approaches.

Core principles of Agile from the Agile Manifesto

As mentioned in *Chapter 8*, these principles pivot around flexibility and efficiency:

- **Iterative development**: *Early and CD of valuable software* and delivering *working software frequently, from a couple of weeks to a couple of months* are two principles you will want to keep in front of your thinking as you implement your SDLC.

- **Unwavering focus on delivering tangible value**: *Continuous attention to technical excellence and good design* is a battle fought each day and a result of real-time events taking place during development.

- **Fostering collaboration**: *Business people and developers must work together!* This is important as the complexity of the cloud, machine learning, artificial intelligence, and data volumes are becoming more complex.

- **Nimbleness to adapt**: *Welcome changing requirements, even late in development* and *regular reflection on improving team effectiveness* will support the need to develop just enough and just in time!

The core principles of the original Agile Manifesto must be encapsulated into the manifesto used by your organization and your teams. However, you can't simply pick a process off the shelf. It must be integrated and developed to fit your purpose within your organization.

Agile and data engineering

Agile's iterative nature fits perfectly in the ever-changing world of data engineering, where being adaptable is imperative. Data projects often face evolving requirements and handle complex, large-scale data, making rigid plans impractical. The variability of data sources, the sheer volume of data, and the intricacies of distributed systems architectures in data engineering demand an approach that's not just flexible but strategically adaptive, which the correctly chosen Agile methodology can provide.

Why Agile?

Before delving into the benefits of Agile, let's consider a real-world scenario from a known start-up in our past that followed a waterfall approach. In this environment, large, monolithic releases were the norm, often culminating in chaotic pushes to meet deadlines. One core developer was tasked with two years' worth of work in just six months. While he miraculously delivered shippable software, the unsustainable workload led to his burnout and departure, leaving behind an undocumented, bug-ridden code base.

This approach's shortcomings were evident. The lack of iterative development and inflexibility in the face of delays led to significant stress on the team and compromised code quality. The organization's unwavering adherence to their initial product vision, despite market and development realities, ultimately led to their downfall, selling at a fire sale price while their Agile-embracing competitor achieved a significant and profitable start-up exit (in other words, a company sold with a significant market capitalization).

Reflecting on this start-up's journey, we're starkly reminded of why Agile methodologies have captured the interest of so many in our software data engineering field. Agile, with its iterative and flexible nature, could have been the life raft that this start-up needed amidst turbulent development seas. Yet, it's imperative to understand that Agile is not a universal panacea. The efficacy of Agile is contingent upon its proper management and customization. Agile methodologies, if not judiciously applied and managed, can become susceptible to breakdowns, rendering them fragile in the face of organizational challenges. In the next section, we will discuss selecting the right Agile methodology first, and then we'll get into the needed customizations.

Selecting the right Agile methodology

Agile is not a monolithic single methodology! It is a family of methodologies with differences between members. Agile encompasses several frameworks, including **Scrum** {https://packt-debp.link/RiyRUF}, **Kanban** {https://packt-debp.link/hNOozd}, **Lean** {https://packt-debp.link/pZxVsf}, and **Extreme Programming** (**XP**) {https://packt-debp.link/m9h9Oc}. Each brings its own nuances. Scrum with its sprints and roles, Kanban with its continuous flow and visual management, and XP with its focus on technical excellence and frequent releases. We've put together a few resources for you to look at when deciding on Agile methodologies:

- *Parabol. (n.d.). Agile Frameworks: A Complete Overview.* (https://www.parabol.co/blog/agile-frameworks)

- *Agile Alliance. (n.d.). What is Extreme Programming (XP)?* (https://www.Agilealliance.org/glossary/xp)

- *Sealights. (n.d.). The Agile Process: Scrum, Kanban and XP.* (https://www.sealights.io/quality-metrics/the-agile-process-scrum-kanban-and-xp)

- *Rose, S., Borchert, O., Mitchell, S., & Connelly, S. (2020). Zero Trust Architecture (NIST Special Publication 800-207). National Institute of Standards and Technology.* (https://nvlpubs.nist.gov/nistpubs/SpecialPublications/NIST.SP.800-207.pdf)

The selection of the right Agile methodology for your organization is critical for success. The choice of an Agile methodology for data engineering efforts should be informed by factors such as team size, project scope, and the intricacies of data handling. For example, Scrum might suit projects with clear milestones and deliverables, while Kanban could be more suitable for projects requiring continuous delivery (CD). Scrum, Kanban, and XP each have their strengths. Scrum's structure aids in managing complex projects, Kanban's flexibility suits continuous delivery (CD) needs, and XP's focus on technical excellence benefits technically demanding projects.

Agile in start-ups and small companies

In start-ups and small companies, Agile methodologies are not just a choice; they're an essential strategy, enabling these smaller teams to pivot quickly and respond effectively to the evolving market demands. It's this context-driven application of Agile that fosters a culture of innovation that's closely aligned with business realities.

Agile and OKRs – a symbiotic relationship

Within small-scale operations, the integration of OKRs within Agile frameworks becomes a critical element. OKRs in this context are not mere metrics; they are the navigational tools that guide the Agile process, aligning it with tangible business outcomes. This convergence of Agile practices with OKRs ensures that the flexibility and speed of Agile methodologies are not lost in chaos. A chain of OKRs directs development toward the strategic business goal. They drive growth and deliver value, and they can be changed within a reasonable scope via acceptance of the engineering team.

Challenges and tailoring Agile for small entities

However, adapting an Agile development methodology to the unique ecosystem of small companies comes with its own set of challenges. It's a delicate balance of maintaining Agile's core principles while modifying its practices to suit the specific constraints and needs of the small business, which is often a growth-focused culture. The key lies in embracing a flexible version of Agile by emphasizing rapid iteration and practicality over rigid adherence to methodologies. There will be hyper stress as growth and innovation are rewarded since that is a key driver for small businesses and start-ups.

Transition from Agile to Lean in medium and large companies

Agile suits start-ups and small companies, offering the agility they need. In medium and large entities, the ability to implement an *effective* MVP *on time* and *with a given set of capabilities* will be highly valued. For medium and large companies, **Lean** is a better match. Lean's systematic approach to processes and efficiency caters to the complexities of larger organizational structures. This shift from Agile to Lean isn't just about changing methodology; it's a strategic move that integrates Agile's collaborative strengths into Lean's efficiency-driven framework.

In larger companies, Agile's role in breaking down silos, fostering teamwork, and a culture of open communication sets the stage for Lean. By retaining Agile's spirit of collaboration while embracing Lean's focus on minimizing waste and maximizing value, organizations can effectively manage their growth. This approach ensures that the shift from Agile to Lean bolsters decision-making and maintains organizational efficiency, aligning with the evolving needs of growing companies.

Lean development processes that you develop will need to focus on Lean's five principles:

- Defining value

- Creation of a value stream

- Creation of development flow

- Adhering to a pull system

- Remaining in a continuous improvement mode of development, deployment, and operations

As a company grows, there will be pressure to transition the development process from Agile to Lean development. This will pose transition challenges.

Challenges and best practices in transition

In the transition from Agile to Lean within medium and large companies, the key challenge is balancing Agile's customer-centric and adaptable ethos with Lean's efficiency and systematization. A pragmatic, phased implementation of Lean is essential, underpinned by clear communication and continuous team training. This ensures that the agility and innovative spirit of Agile are retained even as Lean principles are integrated. For a transition to be effective, you will want to first train employees to be adaptive and remain highly skilled at many levels. You want to say that the organization's members are *fungible* and can take on different roles when necessary. You also want to be able to create cross-functional teams (composed of members from across the enterprise organization) to develop solutions. Beware that a named or assigned cross-functional team member may not be skilled or empowered by their hierarchical manager; so, test for cross-functional team efficacy early! Key members can be used to pollinate cross functional development teams with best practices across the larger effort since the cross functional model will enable new talented stars to shine.

Then comes the need to manage projects with a bit of top-down design structure, so that the overall project appears as a planned breakdown of well-thought-out production steps. Setting OKRs at this stage makes sense! Since you are going to be stressing the teams (a lot), you have to be able to balance team members' workloads. To achieve this balance, you need to obtain honest and frank feedback from all team members, and you need to listen and then plan carefully.

You then have to have processes to execute a *pull system* that relies on customer demand to initiate the creation and modification of production deliveries as part of your Lean framework. To empower cross-functional teams, provide for continuous improvement methods, and develop mentoring managers and teachers, your Lean framework has to be documented. The Lean framework must become a process all senior management buys into and HR supports with diligence! The documented process makes it possible for you to enforce compliance with the governance codified in the workflows of a product design tool chosen. This customized process is codified in the tooling's workflows to allow product managers to guide and align their teams to success.

Managing the accompanying cultural shift is just as crucial. Companies need to foster a culture that equally values innovation and process-driven efficiency, ensuring that their project management methodologies evolve in sync with their strategic and operational growth. This transition, if strategically planned, deeply understood, and communicated effectively, can maintain a dynamic and efficient project management approach that is crucial for the company's ongoing success.

Agile and Lean should not be seen as conflicting but as complementary. Each is suited to different company growth stages. The skill is in identifying which methodology aligns with the company's current size and objectives. As a company grows, its project management approach should mature, keeping methodologies relevant and aligned with its evolving goals.

Avoiding inconsistency and flavor of the day Agile practices

You will want to plan an approach to push back on those suggesting each team develop their own Agile approach; since, after all, they are *self-organizing*, right? The answer has to be *No*! There are SDLC rules to be created and followed by all teams and team members. You have to stress the fact that all are in the game to succeed together! There are some serious risks to be addressed:

- **Risks of Agile inconsistency**: Inconsistency in applying Agile methodologies can lead to confusion and inefficiency. This is particularly true in data engineering, where the complexity and scale of projects demand a coherent and stable methodological approach.

- **Dangers of methodology hopping**: Frequent shifts in methodologies can severely disrupt project timelines and erode the intrinsic benefits of Agile development. It leads to a lack of clarity in roles, processes, and objectives. It disrupts project timelines, compromises the quality of deliverables, and diminishes the inherent benefits of a consistent Agile approach. In data engineering, this inconsistency can lead to critical delays in data pipeline implementations and the production of flawed datasets, ultimately impacting the reliability of data models and broader business goals. The key to avoiding these pitfalls resides in establishing and maintaining a consistent Agile approach. This involves setting clear guidelines and nurturing a shared understanding of the chosen methodology among all team members.

- **Different Agile methodologies cater to varied project needs**: For instance, Scrum is often suited for projects requiring firm delivery timelines and high-quality outputs, while Kanban may be more appropriate for research-oriented teams where flexibility and continuous delivery (CD) are key. However, it's essential to align these methodologies with the organization's OKRs. This means ensuring that even teams operating in a more flexible Agile environment adhere to critical delivery deadlines and quality standards, aligning their efforts with the organization's overarching goals and timelines.

Consistency is critical but achieving it requires the support of leadership as well as Agile coaches, which you will learn about in the next section.

Role of Agile coaching and leadership in fostering consistency

Effective Agile implementation hinges on skilled coaching and strong leadership. These roles are key to guiding the team and ensuring understanding and adherence to Agile principles and practices. You will find that when you form a team, each member may have developed his secret sauce to success using an Agile approach you can't adopt. They are married to those alternative approaches. You have to begin by breaking down that implicit resistance to a new way to move forward with an agreeable Agile SDLC. Codify the agreement in your methodology documentation, workflows, tooling, and QA (data and software) processes. This prevents backsliding to alternative Agile approaches when the pressure mounts to deliver.

Strategies for maintaining discipline in Agile practices

Discipline in Agile is not about rigidity; it's about a shared commitment to the set of practices and principles in your organization's Agile manifesto. Regular retrospectives, ongoing training, and a culture of continuous improvement are essential in maintaining this discipline. Even with these, there will be team member failures. Know how to handle these and do not blame the methodology or allow the process to become the scapegoat; some will surely try! This is one big reason why your Agile manifesto has to be well thought through.

Standardizing workflows and encoding them in tools such as **Jira** or **Azure DevOps Pipelines** brings effective coherence to Agile processes as they are put into action within the teams. A unified toolset streamlines processes, aids in effective communication, and brings a level of consistency that is crucial for the smooth operation of Agile data engineering teams. It reduces the cognitive load of team members, allowing them to focus on the core aspects of data engineering challenges, and helps them not miss import development steps.

A unified toolset is important, as you will learn in the next section, because a unified Agile approach is necessary for team performance and system quality.

The impact of a unified Agile approach on team performance and system quality

A **unified Agile approach** is a consistent approach to implementing the Agile methodology. It is grounded in the organization's Agile manifesto. A unified Agile approach significantly boosts team performance. It fosters a collaborative environment within and between teams and reduces the complex dynamics to optimize development. This leads to enhanced productivity and efficiency. Consistency in Agile practices leads to a more cohesive team environment and makes team member *fungibility* possible.

Imagine if each team set their own Agile processes up and senior management had to make a member change. It would be like dropping in a person from the planet Mars! The culture shock will hurt the overall effort!

At Elsevier, before Agile took root, we observed this happen between teams where one executed Kanban and the other a modified version of Scrum. When the sprint closed and the member who used to develop in Kanban was asked for his code to integrate, the response was *"It's not ready!"* Just a prototype was produced. The team set the expectations clearly, but the Kanban developer was not looking at the right info on the Scrum ticket (in JIRA) and the "deliver what you can" approach allowed too much latitude with the feature's capability. The developer dropped the scope down from what was expected to make any delivery. This can be a blessing or fault with Kanban. The resulting code could not be integrated! The HR mess that followed was profound after this seriously misaligned delivery.

With a unified Agile approach, team members develop a deep understanding of their roles and the Agile process, leading to improved collaboration and job satisfaction. Team members are fungible and not easily confused. A data engineering team adopting a consistent Scrum approach will see more predictable sprint outcomes, quicker adaptations to change, and a more streamlined development process.

Maintaining high-quality systems

A unified Agile approach is instrumental in ensuring system quality. It nurtures an environment where **continuous integration** (**CI**), testing, and delivery are part of the norm, leading to robust and reliable data systems.

The relationship between Agile practices and system reliability requires clarity. Agile practices such as CI and CD are critical in data engineering. They ensure that systems are not only up to date but also resilient to changes and scalable. These capabilities accommodate the business's needs for rapid development and responsiveness to evolving business requirements subordinate to the overall OKR. It is in the creation and alignment of the OKRs across teams that complexity in Agile development arises and huge MVP delays can arise when the number of teams exceeds three and there are no integration Agile teams envisioned for the product.

We have seen upwards of 20 teams in a large project within an even larger organization work on big projects. 20% of the teams are usually specialized integration teams with great authority to adjust the mandatory, high priority, development backlog of any other team when bugs or changes are needed to the Agile team's completed work product.

This is how a Kanban team can be forced into a **Scrumban** team or a Scrum team overnight; roll in the cots, please! Nobody's going home! Why? Because the process was not thought out completely! The data factory with its pipelines that you are solutioning will have many dependencies that require timely delivery, high quality, and full functionality as planned! This is why one often sees **Agile-fall** (a mix of Agile and waterfall development processes) in large organizations and the reality is clear. The organization modified the Agile manifesto to their needs as all enterprises must do. Do this wrong and you get Agile-fail; pardon the pun, but it is not fun to discover this reality at the 11th hour of your project.

Using Agile to ensure CI and CD for data engineering

In data engineering, CI and CD ensure that data systems are consistently aligned with evolving business requirements and technical advancements, a cornerstone of a successful Agile implementation.

In summary, adopting an Agile methodology in data engineering is not just about choosing a framework; it's about embedding a philosophy of adaptability and continuous improvement within the team. It requires a strategic approach, considering the unique challenges of data projects and the nuances of different Agile frameworks. The ultimate goal is to enhance team performance and ensure the delivery of high-quality, robust data systems that align with strategic business objectives.

If you're excited about adopting methodologies in data engineering, you will want to read on to the next section, where you will learn how to discuss and apply them to the SDLC process.

Software Development Lifecycle (SDLC) processes

In this section, we'll discuss the nuances of SDLC processes, tailored to meet the unique demands of distributed data systems. Here, we explore the critical need for a well-defined customized SDLC framework that accommodates the complexities and peculiarities of procuring product datasets in a modern environment. We will examine the role of Agile methodologies, dissecting their core principles, and understanding how they can be effectively integrated and adapted to enhance SDLC processes.

This exploration includes a thorough analysis of how Agile principles contrast with traditional methodologies in handling data complexity, the significance of selecting the right Agile methodology based on team size, project scope, and data complexities, and the importance of maintaining consistency in Agile practices. This section is a deep dive into the synergy between Agile methodologies and effective SDLC processes, highlighting the impact of a unified Agile approach on team performance and system quality, and underscoring the necessity of CI and CD in the field of data engineering.

Establishing a defined SDLC tailored to distributed data systems

The SDLC is a critical part of software engineering. It encompasses stages such as planning, designing, building, testing, and deployment. When applied to distributed data systems, which often involve complex, auto-scalable architectures, the development of the SDLC needs special consideration. These systems demand an approach that accounts for their distributed nature, focusing on consistent reliability, availability, scalability, and performance.

Creating a bespoke SDLC for distributed environments

Tailoring the SDLC for distributed data system development involves integrating data engineering specifics into each phase of the life cycle. This customization requires a deep understanding of the unique challenges and requirements of distributed systems. These include considerations for data volume, data velocity, and data variety. The goal is to develop an SDLC framework that not only supports the technical demands of these systems but also aligns with the overall business strategy's objectives. Early on, we mentioned the need to develop with the mindset of *"IT has to work, always work, and never fail!"* This is still the case. Supporting that goal in the SDLC, so that all teams are guided into that mindset, is essential. There is a lot of fragility a cloud architecture can burden the developer with. Relieving some of that burden is one of the data engineering manager's tasks.

The role of peer reviews and Agile ceremonies in maintaining alignment with project goals

The key benefit of using Agile processes in your development is being able to quickly respond to changing conditions to maintain alignment with the project goals (OKRs). This section describes two features of Agile that enable this alignment to be achieved:

* Peer reviews
* Agile ceremonies

The importance of peer reviews

In software and data engineering projects, constructive peer reviews, encompassing code reviews and collaborative design sessions, are imperative for maintaining high standards of code quality and system design. They offer opportunities for knowledge sharing, identifying potential issues early, and ensuring that the development aligns with the project's objectives. Best practices for conducting effective peer reviews include clear guidelines, regular scheduling, and fostering a culture of constructive feedback. A team that has a caring mentorship and coaching attitude toward its members is a team you will want to be associated with.

Agile ceremonies tailored for the SDLC

Integrating **Agile ceremonies** such as sprint planning, daily stand-ups, and retrospectives into the SDLC enhances project alignment and team communication. Customizing these Agile ceremonies to fit the specific needs of data engineering projects ensures that they effectively support the various stages of SDLC, from planning through to deployment. Here's a straightforward approach:

- **Sprint planning**: In data engineering, sprint planning should zero in on concrete data goals. Break down complex tasks and set achievable targets per sprint, focusing on clear, manageable data objectives.

- **Daily stand-ups**: Stand-ups should be used for prompt issue resolution and progress checks on development tasks, including data delivery tasks. By encouraging open discussion about data processing challenges and collaborative problem-solving, much progress can be made toward the team's OKRs.

- **Retrospectives**: Reflecting on work done and how it was done puts a focus on data quality and output integrity. These sessions are opportunities to identify what's working and what needs improvement in handling data, ensuring each sprint brings enhancements.

- **Customization for data engineering**: Being able to adjust ceremonies to fit the unique demands of the data engineering workflow is valuable. They support deliverables such as data quality checking and integrating new data sources. Ceremony customization ensures that the effort remains relevant and focused for the team.

- **Continuous improvement**: By leveraging insights from these ceremonies, the manager can refine processes, aiming for better collaboration and data quality in subsequent sprints.

This adaptive approach to handling ceremonies ensures that Agile ceremonies don't become routine meetings but remain effective tools for project team success.

Maintaining project alignment

Ensuring that all team members are aligned with the sprint and epic goals is crucial for the success of the project. It is essential to track workflows established in a tool such as Jira. Additionally, techniques to maintain this alignment include clear communication of goals, regular check-ins, and adaptive planning to accommodate changes and feedback.

Addressing the issues with Scrum versus Kanban conflicts and promoting team unity

As you learned earlier in this chapter, having a unified Agile methodology is crucial. But unfortunately, methodology conflicts will arise. In this section, you will learn how to navigate and resolve these conflicts.

Resolving methodology conflicts

Addressing conflicts between Agile methodologies within teams (see earlier discussion regarding the Scrum versus Kanban debate) involves understanding the unique demands of the project and the team's working style. Strategies include adopting a hybrid approach or choosing the methodology that best aligns with the project's needs.

> **Note**
>
> We focus on Scrum and Kanban because of their popularity but the same advice applies when resolving conflicts involving other less commonly used methodologies.

The key is to harmonize these methodologies in a way that benefits the team and project, fostering a flexible and adaptive approach to project management.

Fostering team unity

Building a culture of collaboration and flexibility is essential in resolving methodology conflicts and promoting team unity. This involves encouraging open communication, respecting different perspectives, and fostering a team environment where collaboration is valued over rigid adherence to a specific methodology.

Integrating SDLC processes in data engineering projects, particularly in distributed systems, requires a tailored approach that accounts for the unique challenges of these environments. It is critical to incorporate peer reviews and Agile ceremonies into the SDLC to maintain alignment with project goals and maintain high standards of quality and collaboration. In particular, the retrospective is needed to point out what's not working well. Additionally, story points must be policed so that they accurately reflect the proper scope and do not turn into junk from teams treating them as fuzzy numbers. Finally, burndown velocity should be actively managed and then fed back to the developers.

The goal is to develop an operational framework that is not only technically sound but also adaptable and aligned with the strategic objectives of the organization.

Objectives and Key Results (OKRs)

Performance measurement in Agile and data engineering is crucial, and it's here that OKRs and KPIs come into play. These methodologies, rooted in management science, differ significantly in their application and focus. KPIs, with their quantifiable targets, emerged from traditional business operations, whereas OKRs, born in the tech industry, focus on setting ambitious, outcome-based goals to drive strategic initiatives.

A detailed comparison between OKR and KPI handling within the organization's Agile framework is needed. They are different! When we compare OKRs and KPIs, the key differences lie in their scope and application. KPIs are typically narrow, measuring specific operational outcomes. OKRs, in contrast, are broader and encompass strategic goals, encouraging innovation.

In the Agile development process and data engineering context, while KPIs offer clear metrics for specific tasks, they lack the strategic context of OKRs. OKRs align well with Agile's adaptable nature, offering a more holistic approach to performance measurement.

HR often wants teams to have KPIs. The idea is that when issues arise, the organization has an HR/FinOps requirement to find *one throat to choke!*

> **Note**
>
> The origin of the phrase "one throat to choke" comes from Roman history where Gaius Julius Caesar Augustus Germanicus (https://www.quora.com/What-is-the-origin-of-the-phrase-one-throat-to-choke) used the phrase when threatening the people. It has been adapted by some to illustrate a philosophy in cutthroat IT development where accountability and responsibility have to be centralized to control correct outcomes.

The enterprise can block the rollout of any OKR setting processes that can't produce policy driven compliance paperwork. We've seen this too often. Since the KPI is a measurement and the OKR a result, one can trace successes and failures in the KPI to the OKR to show the tie-in. Be creative if overarching OKR setting and management are not yet fully accepted in your enterprise.

Integrating these measurement methodologies within the Agile cycle shows that OKRs can align with Agile's sprints and milestones. This offers a flexible framework that adapts to changing project needs.

The benefits of using OKRs for data engineering projects

OKRs are particularly advantageous in data engineering settings. They foster strategic thinking and innovation, aligning technical efforts with larger business objectives – this is crucial in data engineering, where technical complexity can overshadow strategic importance. OKRs ensure business objectives are at the heart of technical goals, translating complex data tasks into clear, measurable business outcomes.

Implementing and measuring OKRs in the development cycle

In implementing OKRs within a data engineering environment, specificity and pragmatism are key. A clear definition of objectives and key results (OKRs), aligned not just with project goals but also with overarching business strategies, is foundational. This process isn't a one-off event; it demands continuous dialogue across the team, facilitated by leadership. Leaders should not only articulate these OKRs but also ensure their relevance and comprehension at every level of contribution.

Regarding the cadence of review, in a data engineering context, particularly in large organizations, *periodic* should translate to a more structured timeline. For instance, adopting a bi-weekly review for smaller teams and a monthly rhythm for larger groups can be effective. This structure fosters a rhythm of reflection and adaptation, critical for maintaining the agility of OKRs.

For larger organizations with multiple Agile teams, the governance of OKRs could entail establishing a central OKR oversight committee. This body, possibly comprising a mix of senior management and Agile practitioners, would be responsible for ensuring the alignment of OKRs across different teams, thus harmonizing departmental objectives with the company's strategic vision.

In conclusion, OKRs offer a strategic, dynamic approach to performance measurement in Agile data engineering projects. Their business alignment and adaptability within the Agile framework make them invaluable for driving innovation and ensuring data engineering efforts align with broader business goals. Effective implementation requires clear definitions, communication, and regular adjustment, guided by leadership, and embraced by all team members.

Agile methodology (tuned to the organization)

OKRs are a pivotal tool for aligning technical prowess with strategic business objectives. In this section, we'll explore OKRs, focusing on their significance and application within the Agile framework. As we navigate through the nuanced differences between OKRs and traditional KPIs, we uncover how these performance measurement methodologies, though rooted in different origins, play crucial roles in shaping the trajectory of data engineering projects. The journey here is not just about understanding these tools in isolation but appreciating how their integration within Agile cycles can profoundly impact project outcomes, fostering a culture of strategic thinking, continuous improvement, and alignment with overarching business goals.

We invite you to reimagine performance measurement – not as a mere tracking tool, but as a strategic compass guiding teams toward innovation, efficiency, and alignment with the ever-changing demands of the business world.

The importance of creating and documenting customized Agile methodology

In the dynamic world of Agile methodologies, a tailored approach is essential. It's not just about adopting common frameworks such as Scrum or Kanban; it's about dissecting and reconstructing them to fit the unique contours of an organization.

Customization over adoption

This customization extends beyond mere surface adjustments. It cuts to the heart of the foundational aspects of Agile, reshaping its principles to align with the organization's specific needs, culture, and business objectives. This process of customization ensures that Agile practices are not just implemented but are deeply embedded within the organizational fabric, enhancing both efficiency and effectiveness.

Documenting the methodology

Creating an organizational Agile manifesto is akin to charting a map for a unique journey. This documentation process involves outlining each tailored component – from sprint lengths and specific roles to customized ceremonies. The manifesto serves as a guiding document, providing clarity and direction for the Agile methodology's application within the organization. It's a living document, one that evolves as the organization and its Agile practices mature.

Stakeholder involvement

The involvement of stakeholders in shaping the Agile methodology is critical. Their insights ensure that the methodology aligns not just with the technical teams but also resonates with the broader business objectives. This engagement is a delicate balancing act, harmonizing the agility and adaptability of Agile practices with the stability and predictability required in business operations. It's about creating a methodology that bridges the gap between rapid technical innovation and steady business growth.

The role of IT management in enforcing the documented Agile processes

Having well-designed, documented Agile processes isn't enough – IT management must be willing and able to enforce them. This section discusses how IT management can make or break Agile processes. We focus on three key areas:

- Leadership and governance
- Training and support
- Monitoring and compliance

Leadership and governance

The role of IT leadership in embedding a customized Agile methodology within an organization is multifaceted. Leaders must not only champion Agile practices but also actively govern their application, ensuring that the organizational Agile methodology is faithfully translated into everyday operations. This leadership is about setting a vision for Agile implementation, fostering a culture that is receptive to Agile methodologies, and ensuring that these methodologies are consistently applied across projects and teams.

Training and support

Maintaining Agile effectiveness demands more than initial training; it requires a continuous learning approach led by IT management. Training initiatives should be constantly updated, ensuring teams are adept in the organization's specific Agile methodology and ready for evolving project challenges.

For Agile coaching, the key is being selective. Choose coaches with substantial Agile experience, thorough knowledge of the organization's Agile approach, and strong mentoring abilities. These coaches should interact closely with various teams, engaging in regular sessions and Agile ceremonies, and offering personalized mentorship. Their role is not just to teach but to actively foster skill development and Agile practice improvement within teams. By carefully selecting qualified Agile coaches and embedding them into the organizational fabric, the sustained effectiveness of Agile practices is significantly enhanced.

Monitoring and compliance

Ensuring adherence to an organization's Agile methodology involves ongoing monitoring and regular compliance checks. IT management should establish metrics and conduct audits to assess the effectiveness of, and adherence to, the Agile methodology. This monitoring is not just about enforcement; it's about understanding how Agile practices are being applied and identifying areas for improvement. The approach to monitoring and compliance should be adaptive, aligning with the organization's culture, and responsive to feedback.

In conclusion, crafting a customized Agile methodology that resonates with an organization's unique environment is a journey that requires careful planning, stakeholder involvement, and strong leadership. It's about creating a framework that not only supports technical agility but also aligns with and advances business goals. The role of IT management in this process is pivotal, encompassing the governance of Agile practices, the provision of training and support, and the monitoring of compliance and effectiveness.

Business development strategy

In this section, you'll explore the integral roles of Proof of Concept (POC) and Minimum Viable Product (MVP) in data engineering projects. You'll learn how POCs serve as crucial preliminary tests for new ideas, offering a safe space to validate hypotheses before significant resource allocation. In contrast, MVPs represent the initial, functional iteration of a product, essential for garnering early user feedback and guiding iterative development.

This section also highlights the significance of fostering collaboration between business development and engineering teams, ensuring that Agile practices are not only technically proficient but also strategically aligned with business objectives.

By the end of this section, you'll gain insights into effectively integrating business strategies within the Agile SDLC, balancing technical agility with business acumen.

Elaborating on the role of POCs for risk mitigation and MVPs for business delivery

In the Agile framework, especially within data engineering, two crucial concepts are POC and MVP. These approaches, while distinct, play critical roles in the life cycle of a project.

Understanding POC

A **POC** is akin to a preliminary sketch in an artist's portfolio. It's a small-scale endeavor, often experimental in nature, designed to test a specific hypothesis or idea within the broader canvas of a potential project.

In the context of data engineering, envision a POC as a trial balloon – testing a new data processing algorithm or piloting an innovative data storage method, for example. It's not about building the final product; rather, it's a litmus test to gauge the viability of an idea, a way to ascertain whether a concept has the wings to fly before investing substantial resources into its development. The POC effort also has to vet all the unknown unknowns in the technology being used. This happens a lot in the cloud where the services do not always work as you would expect or at best, they have quirky behavior around the operational edges of their capability (scale, volume, or time).

Note that team members have quoted the time needed for their sprint work efforts based on their as-is understanding, which is often subject to limited experience/understanding.

You can't have a spike before each ticket; so, there has to be some slack time allocated for cloud service/ technology research. Without understanding this, the team members will feel that the team lead is playing whack-a-mole with them, and they are the mole! One step away from being hurt by the process rather than being helped by it. POCs serve as a critical checkpoint in the journey of project development, particularly in data engineering efforts where data handling approaches are complex and evolving. POCs are a prudent path to validate any hypotheses and assess the feasibility of integrating new technological advancements into larger-scale projects.

Understanding MVPs

The desire to bring a product to market leads developers to first produce an MVP. After all, businesses don't want all the frills built first, they want the core built first. They want to test it in the market! Then, they evolve the product to be just right, so the **general availability** (**GA**) version is profitable. The MVP takes us a step beyond the first exploratory steps of a POC toward the development of a real solution. It's the first tangible draft of your vision, akin to the first of many draft revisions of a technology book before its publication.

This MVP is not just a rudimentary version of the product but a carefully crafted iteration that, while basic, is sufficient to engage early adopters. In the field of data engineering, think of an MVP as the initial release of a data analytics tool, stripped down to its most essential features. With that tool comes the data as a product with it the semantic context metadata structure to enable it to be self-serviced. It may not have all the context instance data specified but the structure is 100% present. These core functionalities are not mere placeholders; they are the backbone of the analytics tool, and data classes designed to fulfill the primary user's needs.

An MVP is more than just a product; it's a learning tool to gauge market acceptance. It can also be used to gauge internal integration acceptance as part of a larger effort. It is put into a user's hands early in the development cycle. This enables you to gather valuable insights and feedback as early as possible, and then sculpt the product's future direction. This approach aligns perfectly with the Agile philosophy of iterative development, where the product is not just built but also grown, and nurtured by real-world use and user interaction.

Understanding the nuances between POCs and MVPs is a key necessity. While a POC is a smaller experiment to test a hypothesis and result risk, an MVP is a minimal form of the product intended to attract early adopters and validate a product idea in the market. Essentially POCs serve to mitigate risk while MVPs enable early business delivery:

- **Role of POCs**: POCs are instrumental in mitigating risk in data engineering projects. They allow teams to explore the viability of new ideas or technologies without the full investment required for complete development. This early testing phase is crucial in identifying potential issues and evaluating the practicality of integrating new technologies into larger projects.

- **Using MVPs**: MVPs play a pivotal role in early business delivery. They allow for the launch of a product that, while not fully featured, is sufficient to meet the basic requirements of users. This early release gathers valuable feedback for iterative development, which is central to Agile methodologies. MVPs help validate the product idea in a real-world setting, ensuring that development efforts align with market needs and user expectations.

Strategies for collaboration between business development and engineering teams

Effective collaboration between business development and engineering teams is essential for the success of Agile projects. Techniques for facilitating this collaboration include regular, transparent communication and the establishment of cross-functional teams. These teams act as bridges, translating business needs into technical solutions and vice versa.

Collaborative planning and goal setting are crucial. Joint planning sessions should aim to align business objectives with technical capabilities, ensuring that both sides understand and agree on the project's direction. Shared goals and metrics for success not only align efforts but also foster a sense of shared purpose and accountability.

How to integrate business strategies within the Agile SDLC

Aligning Agile with business strategies involves adapting Agile methodologies to support not just technical execution but also broader business objectives. This means integrating business milestones into the Agile planning and review processes, ensuring that technical progress is always aligned with business goals.

Measuring the business impact of Agile projects is another key aspect. Tools and techniques for assessing this impact include Agile metrics such as velocity and **burndown rate** as well as other metrics that directly measure business outcomes, such as the following:

- Customer satisfaction

- Net promoter score

- Market penetration

- **Return on investment (ROI)**

These metrics should be part of the regular review and adaptation cycles within Agile methodologies, ensuring that the project remains aligned with its intended business outcomes.

The integration of business development strategies within the Agile SDLC in data engineering projects is not just about aligning technical and business goals but about creating a symbiotic relationship between the two. This involves understanding and leveraging the roles of POCs and MVPs, fostering effective collaboration between business and engineering teams, and ensuring that organizational Agile methodologies are adapted to support and measure business strategies and outcomes.

In the next section, you will learn how to develop a testing and quality strategy to ensure data integrity and system reliability.

Test/quality strategy

In this test/quality strategy section, you'll gain insights into the essential testing approaches and **quality assurance (QA)** measures that are pivotal in data engineering. This includes delving into **unit testing**, **system performance testing**, and **user acceptance testing (UAT)** within various data zones and understanding how they collectively ensure data integrity and functionality. You'll also learn about establishing QA baselines, managing **service level agreements (SLAs)**, and the critical role of QA tooling and audit conformance in data pipelines.

This section is designed to equip you with a comprehensive framework for ensuring data integrity and system reliability, all while handling the evolving challenges and complexities of the field of data engineering.

Approaches to testing in data zones

Testing takes on a multifaceted character, with data zones demanding specific testing approaches. These include unit testing, system performance testing, and UAT, each addressing different aspects of data integrity and functionality. Unit testing in data pipelines is about verifying the smallest testable parts of a data processing application, ensuring each segment functions correctly before integrating into the larger system. It requires tools and frameworks that can handle the unique challenges of data workflows, such as managing large data volumes and complex data transformations.

System performance testing goes beyond mere functionality to assess aspects such as scalability, reliability, and efficiency. It involves evaluating the system under various conditions to ensure it meets the expected performance benchmarks, focusing on metrics such as data throughput rates and resource utilization. User acceptance testing (UAT), conversely, brings the end-user perspective into the equation, creating real-world scenarios and test cases that reflect the user's environment and requirements. This step is critical to ensure the system not only works technically but also aligns with user expectations and needs.

Establishing QA baselines and SLA operations management

Quality Assurance (QA) in data engineering is not just a process but a cornerstone of project integrity. Establishing QA baselines involves setting thresholds for acceptable data quality, which is pivotal for maintaining consistency in data processing. These baselines are instrumental for ongoing monitoring, encompassing techniques such as automated checks to ensure data accuracy, completeness, and consistency. SLA management complements this by defining expectations for data delivery and processing. It's about ensuring that data services meet agreed-upon standards and timelines through clear criteria and compliance tracking.

Balancing the quality requirements of raw and derived data adds another layer of complexity. QA practices must be tailored to address the nuances of both. While raw data requires integrity and accuracy checks, derived data, often the product of complex transformations, demands additional validation to ensure the transformations accurately reflect in the final output. This balancing act is crucial for the integrity of the overall data life cycle.

Necessity of QA tooling and audit conformance in data pipelines

QA tooling is indispensable in a data factory's data pipeline design. The right selection of tools for data validation, verification, and anomaly detection ensures the continual maintenance of data quality. These tools need to be woven seamlessly into the data engineering workflow, providing automated checks that span the data life cycle. Alongside tooling, ensuring compliance with industry standards and regulations is paramount. This involves implementing comprehensive audit trails and logging mechanisms, which are essential for tracking data lineage, maintaining transparency, and ensuring regulatory compliance.

Adopting best practices and lessons learned from industry leaders is a strategy that should underpin all QA efforts. This approach fosters a culture of continuous improvement, ensuring that data quality and pipeline reliability are not just maintained but progressively enhanced. In an environment where technology and data are continually evolving, such a strategy ensures that data engineering practices remain at the cutting edge, delivering reliable and high quality data solutions.

A comprehensive test and quality strategy in data engineering is a multifaceted endeavor, encompassing a range of testing approaches and QA measures. From the detailed focus of unit testing to the broad lens of SLA management, each aspect plays a crucial role in ensuring data integrity, system reliability, and compliance. This strategy is underpinned by a continuous improvement ethos, drawing on industry best practices to navigate the ever-evolving challenges of the data engineering field.

Please continue with us to the next section, where you will learn to develop an effective operational strategy ensuring not just high quality but also an efficient and strategically aligned data engineering operation.

Operational strategy

In the *Operational strategy* section, you will unravel the essential elements that drive the efficiency and reliability of data engineering processes. This includes understanding how transformational code and data lineage tracking enhances data delivery and integrity, and the significance of data zoning in aligning technical operations with business goals. You'll also learn about establishing and managing SLAs and the important role of monitoring and alert systems in maintaining data quality.

By the end of this section, you'll have insights into creating not only efficient but also strategically aligned data engineering operations.

Operationalizing data delivery through transformational code and lineage tracking

In an Agile data engineering environment, operationalizing data delivery is a critical aspect. It involves automating and streamlining data processes to ensure an efficient and reliable data flow. Transformational code plays a key role in this process by automating data transformation tasks, thereby enhancing the agility and efficiency of data operations. Similarly, data lineage tracking is fundamental for maintaining data integrity and traceability. It refers to the process of understanding, recording, and visualizing the data's journey from its source to its final form.

Effective lineage tracking helps in pinpointing the origins of data, understanding its transformation history, and ensuring the accuracy and reliability of data. Tools and techniques for lineage tracking range from simple logging mechanisms to sophisticated software that can map and visualize complex data flows.

Dataset availability in labeled zones

Dividing data factory processing into zones (discussed extensively in *Chapter 5*) is a strategy employed in data management to categorize and label data based on its type, sensitivity, and usage. This organization of data into zones, such as raw, curated, and consumption zones, helps in managing access, optimizing storage, curating value-added data with quality, and ensuring data security. Implementing this data zoning architectural pattern facilitates efficient data handling and plays a critical role in aligning data management with business objectives.

SLAs for data

Establishing SLAs is essential in data engineering. SLAs define the expected level of service, particularly in terms of data availability, quality, and timeliness. Creating and managing effective SLAs involves aligning them with business goals and ensuring that they reflect the needs and expectations of both the data engineering team and its stakeholders.

Alert mechanisms for quality deviations and their impact on operations

Simply establishing data quality standards isn't enough. You must ensure that you are always meeting them. This means that they have to be part of the toolsets supporting the SDLC and the deviations handled as part of the development process as the team iterates toward its GA product launch.

In this section, you will learn about two important tools to consistently maintain high-quality data monitoring, data quality, and alert mechanisms for data quality deviations.

Monitoring data quality

Continuous monitoring of data quality is essential to maintain high standards and comply with defined SLAs. This involves implementing techniques and automated alert systems that can detect and notify teams of any deviations from the established data quality parameters.

Responding to quality deviations

Upon detection of quality deviations, swift and effective response strategies are crucial to mitigate their impact on business operations and decision-making. This response might include immediate data correction measures, analysis of the root cause, and adjustments to prevent future occurrences.

Best practices for handling data deviations

Setting up robust alert mechanisms and learning from industry best practices are crucial for managing and mitigating quality deviations. This includes guidelines for creating effective monitoring systems and strategies to promptly address any identified issues, thereby ensuring that the data engineering operations remain reliable and trustworthy.

Operational strategies in data engineering encompass a range of practices from operationalizing data delivery to ensuring dataset availability and managing SLAs. The implementation of lineage tracking, effective data zoning, and robust alert mechanisms for quality control are fundamental to maintaining the efficiency and reliability of data operations. These strategies, when effectively executed, ensure that data engineering processes are not only Agile and responsive but also aligned with the strategic objectives of the organization.

In the next section, you will learn to apply an effective data security strategy, ensuring that the valuable data and data pipelines that you have developed are well protected.

Data security strategy

Your organization's data and data pipelines are valuable assets (especially if you have been following the advice in this book). Because of this, organizations are frequently targeted by attackers attempting to steal their data or disrupt their operations. Security is best addressed early in the SDLC. In this section, we'll arm you with techniques to protect these valuable assets.

Implementing zero-trust architecture in data engineering environments

Zero-trust architecture has produced a paradigm shift in cybersecurity. The approach operates by default on the principle that nothing inside or outside the network is trusted. It's supported by a developed framework built on the premise of *never trust, always verify*.

In the context of data engineering, where data flows are extensive and often cross organizational boundaries, the relevance of zero-trust security becomes paramount. It's about constructing a security model that adapts to the complexities of modern data environments, where traditional perimeters will no longer be sufficient. Please be aware that fully integrating zero-trust architecture into data systems is fraught with complexities. It's not merely about adopting a set of security protocols but about rethinking the entire security framework from the ground up.

One of the significant challenges lies in the comprehensive assessment and restructuring of existing data pipelines and infrastructure. Zero-trust principles often demand a granular level of control and monitoring, which can be challenging to implement in systems not originally designed with this model in mind. Additionally, the integration of zero-trust principles must be strategic, ensuring security enhancements strengthen rather than disrupt data operations.

In essence, while the rationale for moving toward a zero-trust architecture in data engineering is compelling, its full implementation is a complex endeavor that requires a thoughtful, nuanced approach. It's about balancing the ideal of a zero-trust environment with the realities of existing system architectures, operational workflows, and the inherent challenges of retrofitting a new security model onto established data infrastructures.

Strategies for preventing system penetration and maintaining continuous operations

Zero-trust architectures provide a strong data security foundation. In this section, you will learn the importance of building on that foundation to achieve even more secure and robust systems by applying comprehensive security measures.

Comprehensive security measures

In data engineering, securing systems against intrusions and breaches demands advanced methodologies. This includes techniques such as robust encryption, stringent access control measures, and effective network segmentation. Each of these plays a pivotal role in fortifying data systems against unauthorized access and potential security threats.

Ensuring operational continuity

Balancing security measures with the need for uninterrupted data operations is a delicate act. It involves implementing security protocols that do not hinder data flow but ensure its integrity. Key to this balance is a solid disaster recovery and business continuity plan, specifically tailored to the data systems in question. This planning ensures that, even in the event of a security incident, data operations can continue with minimal disruption.

SecOps considerations in the design and operation of data systems

SecOps, the integration of **security** and **operations**, is a concept gaining traction in the field of data engineering. This approach involves embedding security considerations into every stage of the data life cycle, aligning it closely with Agile methodologies and SDLC processes. The benefit of this integration is a more resilient, secure data environment where security is not an afterthought but a foundational aspect of data processing.

Best practices in SecOps for data engineering involve a proactive stance toward security. This means incorporating security measures at every stage of data processing, from initial data collection to final analysis and storage. Tools and techniques for continuous monitoring and incident response are essential, allowing for real-time detection of security anomalies and rapid response to potential threats. This continuous vigilance ensures that security is an integral part of the data engineering process, woven into the fabric of daily operations.

A comprehensive data security strategy in data engineering encompasses zero-trust architecture, robust security measures to prevent system penetration, and the seamless integration of SecOps practices. These strategies ensure that data systems are not only secure from external and internal threats but also resilient and capable of maintaining continuous operations even in adverse scenarios. The integration of security into every aspect of data engineering, guided by best practices and real-world applications, is key to building asnd maintaining trustworthy, secure data environments.

Summary

As we draw this chapter to a close, let's revisit the pivotal role that an Agile SDLC plays in data engineering. Agile methodologies, renowned for their adaptability and iterative nature, are indispensable for effectively managing the complexities of modern data systems. Tailoring these practices to fit the nuances of data engineering operations brings forth the level of flexibility and responsiveness needed in today's fast-evolving tech world.

In this context, the role of leadership and the cultivation of an Agile culture within organizations cannot be overstated. Leaders play a pivotal role in advocating for and embedding Agile principles, ensuring that these methodologies are not merely adopted but are ingrained within the team's ethos and practices.

Reflecting on the core concepts discussed, it becomes clear that a customized approach to Agile methodologies and SDLC processes is crucial. These tailored practices, woven into the fabric of data engineering operations, bring a level of flexibility and responsiveness that traditional methods simply cannot match.

Agile's synergy with data management, emphasizing cross-functional collaboration, leads to improved project outcomes and alignment with business needs. Here, the importance of cross-functional collaboration comes to the fore. Organizations embracing Agile thrive when data engineers, business analysts, project managers, and stakeholders collaborate seamlessly. Breaking down silos and encouraging cohesive teamwork is essential for realizing the full potential of Agile methodologies in data engineering.

Encouraging a mindset of continuous improvement and adaptation in software engineering practices is indispensable. This mindset, supported by Agile methodologies and a collaborative, culture-centric approach, contributes significantly to the ongoing success and relevance of data engineering practices.

As we wrap up this chapter, the message is clear: take the concepts and strategies we've discussed to heart, focusing on leadership, cultural change, and cross-functional collaboration. Your journey of continuous learning and adaptation isn't just about staying up to date with technology; it's about leading the way in data engineering. Agile SDLC, with its focus on flexibility, resilience, and teamwork, is your guiding star on this path.

In short, Agile SDLC in data engineering is all about continuous evolution, fueled by teamwork, strong leadership, and a culture that welcomes change. It's a journey toward creating data solutions that are not only technically robust but also strategically aligned and adaptable to the ever-changing demands of business and technology.

In the next chapter you will be provided with best practices for **Software Quality Testing**.

10

Key Considerations for Quality Testing Best Practices

In Fred Brooks's classic work *The Mythical Man-Month: Essays on Software Engineering*, which some have referred to as the bible of software engineering, you can read the following:

> *You can't measure productivity by hours; there is only one metric for productivity: the quality and usefulness of the software produced. (Fred Brooks in The Mythical Man-Month: Essays on Software Engineering, (*https://packt-debp.link/MRc9dc*))*

In this chapter, you will be exposed to the many aspects of solution testing with a focus on preserving data quality. We begin with an overview.

Quality testing overview

According to the late Fred Brooks, the only quality metrics that should be considered, among the myriad of possible metrics, are the measure of productivity based on a system's operational quality followed by how much the product is effectively used. The cost/benefit ratio and the solution's effectiveness should justify the amount of time and effort spent on the development of your solution. The focus on quality is a key takeaway. When we read Brooks's work years ago, the point that most stood out to us was that 40% of a project's effort will go into maintenance and that includes the iterative testing effort. This effort could increase drastically if a late project were to have many new resources added at the end stage when development delays could no longer be glossed over, resulting in a poor upward facing management report. You will want to make it an important goal to apply methods to reduce defects and the cost of testing your solution. The IBM Systems Sciences Institute reported the following:

> *… it costs 6x more to fix a bug found during implementation than to fix one identified during design. The cost to fix bugs found during the testing phase could be 15x more than the cost of fixing bugs found during design. (Integrating Software Assurance into the Software Development Life Cycle (SDLC) from the Journal of Information Systems Technology and Planning; January 1, 2020)*

First, we begin with an elaboration on what *not* to test, followed by a primer on how to evolve a test discipline within your organization. Lastly, we will dive into the key terminology that level sets everyone on how testing should be discussed and debated as testing processes mature. Once the foundation of a proper testing mindset is created, we'll dive into a good example of how a test framework may be applied in data wrangling and **machine learning** (**ML**) areas of a data pipeline solution.

How not to test!

All too often, data engineers will focus on *what* to test rather than *how* to test! Hiding testing costs or trying to ignore them by thinking that they can be reduced by sheer force of will is lunacy. Hiring experienced subject-matter experts may be great, but they still must work as a team, and even if they have a huge capacity to produce quality code under great pressure, trivializing the test effort is lunacy! Yet, it is very tempting to omit the necessary time to perform testing from a project schedule or to not formalize the testing discipline. Failing to account for the errors that need to be mitigated due to improper or inefficient project management is foolish! The end users will tell you that even if the organization allows a product to be designated as minimally viable, that in order to get to the state of being a **generally available** (**GA**) product, more work is still needed! Testing in production is still an approach that is often applied in financial software development shops. Why is this, you may ask? Because, by ignoring the test cost issue and pushing responsibility downward to the lowest level developer, *bugs become personal*. With a healthy dose of fear of a defect impacting customers as a world problem, developers make sure it *all* works, even if the system does not provide a minimally viable safety net to efficiently produce and maximally innovate for the good of the team and the product. Achieving maximum personal programmer efficiency, overall team efficiency, and product usefulness (with sustained quality) will require a balanced, formal test methodology approach. Without this, there will be a great toll taken on the morale, job satisfaction, and innovation of the team.

Another way we have seen test costs trivialized and removed from a solution's schedule is when the philosophy of failure is applied. This is when a senior leader observes that in the organization, projects have been failing at a rate of ~30% {https://packt-debp.link/dhu5AY} and since the organization is flush with cash, you can mitigate risk due to IT failure by launching three competing projects with the same goal, where each is blind to the competing others. The winner takes all, and although IT delivers, the cutthroat competition will have two failing teams dismissed. Yes, this really happens, and even within organizations.

The key here is that *people* are leveraged against *people* to squeeze out quality with extremely aggressive motivation. We don't mention this to get you thinking that this approach is a good idea! It is shown to you so that you know how some managers will go to extreme ends to cut the cost of testing and push quality down to the individual. Why do the developers stay in such a Machiavellian development shop? Greed. The reward for winning the competition and producing an acceptable MVP is often astounding. However, like for most gamblers, probability does catch up and eventually, the individual loses. The winner is the manager, so long as an inexhaustible supply of raw materials (developers) is available. It is much better to establish or fix your test methodology processes so that they dovetail into your **software development life cycle** (**SDLC**).

Evolving the test discipline

You will want to evolve the test discipline within your organization. It can take a long time, even years to get things just right, but most companies do *not* want to spend 40% of the effort on testing and quality within each project. Is there a better way? The answer is yes! A way forward can be found if you are willing to think in reverse.

Apply the wisdom

"In all things consider the end" {https://packt-debp.link/7X2itk}.

If you consider the state of the end product and build out a framework that the product must operate within, you are creating **service level agreements** (**SLAs**), data contracts, software quality assertions, performance goals, **high availability** (**HA**), volumetric, and scale-up (or scale-down) criteria that have to be met, forming the non-functional acceptance criteria of the solution. These requirements form a picture of the proverbial square peg getting inserted into a square hole. The round peg is just too big; the square peg, if it has rough edges, may go in, but with a lot of friction, as does the triangle and any other shape you may envision trying to be pushed into a frame of your acceptance test framework. Developing a great quality methodology will enable you as a data engineer or engineering manager to *know* that the system developed will satisfy its **objectives and key results** (**OKRs**).

Test terminology

Modern software testing processes consist of a number of stages. Estimating each stage is a key to success. The **International Software Testing Qualifications Board** (**ISTQB**) {https://packt-debp. link/9FIxTI} specifies five testing stages as follows:

- Test planning and control
- Test analysis and test design
- Test implementation and execution
- Evaluating exit criteria and reporting
- Test closure activities

You will want to understand what each of these means to you as you design your future-proof data solution. You will also want to expand the application of these test stages with the same discipline for the data your solution curates. So, testing is not just for the software operating on the data it curates. The testing methodology should be applied to the data as a product itself. After all, you are curating data as a product, just as the factory is the product when developing the software solution.

What you will find is that the discipline of software testing has a lot of complexity and with it, over 570 terms (from ISTQB) {https://packt-debp.link/mRZjfO} that need a degree of understanding as you make planning decisions for your solution's implementation. These terms include the following:

- Black box versus white box
- A/B testing
- Capacity testing
- Causal-loop
- Cause-effect
- Closed-loop
- Combinatorial testing
- Compliance testing
- Concurrency testing
- Confidence interval
- Acceptance testing
- Regression testing
- Coverage
- Crowd testing
- Cyclomatic complexity
- Data obfuscation
- Data driven testing
- Defect density
- Functional testing
- Fuzz testing
- Heuristic evaluation
- Hypothesis testing
- Load testing
- Metamorphic relation
- Model based testing
- Multiple condition coverage
- Negative testing
- Pairwise testing

- Path testing

- Penetration testing

- Performance testing

- Perspective-based reading

- Phase containment

- Pre-condition versus post-condition

- Prisma

- Ramp-down versus ramp-up

- Random testing

- Reliability growth model

- Requirements-based testing

- Root cause analysis

- Salting

- Session based testing

- Shift left testing

- Smoke testing

- Spike testing

- State transition testing

- Statement versus structural coverage

- Stress testing

- Summative evaluation

- Requirements traceability matrix

- Transcendent quality

- Unit testing

- Usability testing

- Visual testing

All of these have to be understood. Data engineering management will need assistance in this test effort since testing methodology is a niche area in IT. It is vital and often underrated in importance. As you can see from the short preceding list, there are many things to learn when trying to implement a cost effective test implementation for your solution. Not having to incur that 40% project maintenance impact that Brooks referred to as being a recurring theme over the years is essential.

ISTQB has great training programs and partners (such as Global App Testing {https://packt-debp.link/Koh5Bk}) to help with this effort, but you will want to form a center of excellence to address this problem with subject matter experts who are trained and certified. These testers are expected to be aligned with *your* test methodology. Effective consistency, proper coverage, and minimized costs are *your* goals. Address the need to always show progress on objective software and data quality metrics in an auditable dashboard to be evaluated before holding team level retrospectives and setting future OKRs. Never let the tech debt that is identified through testing and maintenance exceed your ability to manage that debt!

Key test definitions

There are a few areas of test methodology that need explanation before you take a deep dive into the creation of a pragmatic test framework for *your* solution. These are the key concepts that drive *your* direction:

- **Test driven development (TDD)**: Drive development from tests developed before coding solutions:

    - **Acceptance test driven development (ATDD)**: Drive TDD tests from the perspective of the end user's acceptance criteria

    - **Developer test driven development (DTDD)**: Drive TDD tests from the perspective of the developer's desired outcome, which supports technology related criteria not necessarily tied to the end user's acceptance criteria

- **Behavioral driven development (BDD)**: Drive development from the perspective of the desired solutions and behaviors.

- **Shift left testing (SLT)**: This is a testing approach generalization of the need to move testing to the beginning of the development life cycle. Testing is then re-applied via orchestrated waves of regression testing when changes or new code is checked in and CI/CD DevOps pipelines are triggered. With SLT, your testing is not left as an afterthought once a solution nears its MVP complete stage, nor is it applied when the product is made generally available after development completion. SLT testing will not fit into traditional development waterfall SDLC practices; therefore, it fits into your Agile processes.

The following sections will dive deeper into the categories identified; however, when formulating a quality program for *your* data solution, you will want to consider something to live by, as quoted earlier: "In all things consider the end."

It is true that you are creating a solution for something! Never lose track of that fact. Test for correctness when first developing the code for your solution, and then incrementally, so it maintains its correct functionality in the future. Proper testing provides assurance for data contract preservation, code efficacy, and end user acceptance. You want to create software/data contracts and assert on defined criteria when component/object execution starts. This guards against poor data inputs or IT operations that badly reflect on your work output and the overall system that is being created. The chaos monkey {https://packt-debp.link/9NsGGY} is still alive and well! All too often, a system melts down because data tests are not implemented to defend against garbage processing.

Likewise, a system fails because a critical component was just never started. If it were started, all that fancy monitoring and observability you built would have worked just fine! When errors do get encountered, the easy path is to just abend (execute an abnormal termination)! You may think, *"Let the monitoring system just restart my failed component,"* but the reality is you get restart after restart because you did not think the scenario through completely and did not implement corrective action between restarts with tests to validate that the next restart would lead to correct processing. You need to restart from a *last known good data checkpoint*.

Manually patching datasets together is disruptive and time consuming. IT operational testing should be applied to complex distributed cloud solutions so that they gracefully degrade. This buys time for issue research and, if necessary, manual corrective action before catastrophic failure becomes visible to even *one single* end user. Live by this motto:

"If you can't make it 100% perfect, then make it look like it is 100% perfect!"

You will make mistakes, introduce bugs, and cause issues – it's natural. You will want to implement offline static tests, run-time executed dynamic tests and be able to use *your* testing approach to find root causes to buy yourself enough time to correct errors before a solution meltdown arises. You will want to never let *your* real-time data get corrupted. For that to be possible, runtime production running tests are going to be essential, hence the need for TDD.

Test driven development (TDD)

Test driven development (**TDD**) and behavioral driven development (BDD) are similar in the way they approach systems development. Both TDD and BDD place testing and collaborative development at the top of the list of requirements when building a solution. With the TDD approach, teams first test and then use any results to direct downstream development plans. TDD is a software technology driven approach with the correctness of the developed software as its central focus.

Acceptance test driven development (ATDD) versus developer test driven development (DTDD)

Programming language specific frameworks have arisen that use the TDD approach. Two subtypes are: **acceptance test driven development** (**ATDD**) and **developer test driven development** (**DTDD**).

ATDD's focus is on the *accuracy of requirements*. It is somewhat like the BDD approach with its focus on the *user's behavior*; however, they both emphasize user requirements encapsulating any technology requirements. In DTDD, the creation of test development precedes systems development. Often, unit test definitions are completed before code writing starts for any production software.

The benefits of TDD are clear. The solution's degree of test coverage is very high, with user functionality covered from the time the design starts. Also, lower defect rates are attainable since in TDD, additional tests are created after defects are detected to prevent future recurrence. Lastly, you can expect development productivity to rise with TDD, since testing scenarios are additive and executed in the CI/CD DevOps pipelines after the code is checked in. Systems integration, smoke, stress, performance, scale, and many other downstream testing steps are all flexible and most assuredly executed as part of your regression testing approach.

There are some issues related to implementing a TDD approach. A significant issue is dealing with test scenario maintenance. There is the potential to *over-test* your solution, or worse yet, lose track of why a given test even exists in the first place. If tests are not well documented, categorized and tracked, then efficacy metrics will fall and defects will rise. The baggage of poorly orchestrated TDD testing also bloats costs and creates a cloud of uncertainty as the end to end time to complete all tests exceeds normal operating parameters. After all, you need the system ready for the next day's wave of check-ins and CI/CD launched testing!

In the past, we have created testing categories that have to be completed in 20 minutes, 2 hours, and 4 hours, therefore giving developers the necessary time to fix issues before the next test cycle begins. Having at least three cycles possible before the next day begins is in *your* interest.

Behavioral driven development (BDD)

In the **behavioral driven development** (**BDD**) approach, a system's desired behavior is formalized and then used to develop software.

Gherkin Syntax {https://packt-debp.link/u7Vl5X} is often leveraged for expressing BDD test scenarios. BDD is a user-centric approach that focuses on the system's behavior. Collaboration between stakeholders is fostered early on in the development process.

Smartbear's Cucumber {https://packt-debp.link/wKhGwd} promotes the creation of living documentation {https://packt-debp.link/S66Pyy} which is a way to share feature definitions that support effective communications between business and development teams.

In CucumberStudio, you are able to co-design acceptance tests with the business, which creates a shared understanding of exactly what the product should be doing. You are able to reuse test steps with action words so test maintenance is minimized, and important test functionality is kept in a single place. Cucumber natively integrates with Atlassian Jira {https://packt-debp.link/IM0vJo}.

Whether you select TDD or BDD as your testing approach depends on many factors. Choosing BDD will get you into test syntax and toolsets that allow you to effectively communicate the value of your testing to others. The TDD approach provides a lot of technical flexibility, which is also very important. Both approaches support the concept of shifting testing to the left of the development life cycle. This brings testing efforts closer to the daily output of the software and data engineer.

Shift left testing

Shift left testing (**SLT**) is the term used to define testing as being moved earlier in the software development life cycle process flow; testing is therefore *shifted left*. The mental picture here is that you should read a development flow from left to right on a printed page. Moving testing as far upstream or to the left in the development process as possible makes it clear that testing is to take place early in the development pipeline. The shift-left approach makes it possible to identify and then resolve software defined data quality issues earlier than in traditional waterfall development processes. The end-to-end solution's quality measures are improved, and there should be an observable reduction in testing cost (time and money) and, ultimately, in development progress (development velocity).

So, you may ask, why hasn't everyone adopted this agile approach? The answer is that not all development performed is agile development! Software release cycles have to be created as iterative releases into a running system supported by CI/CD-automated DevOps pipelines for building, testing, and deploying in your framework. Legacy development shops have had a hard time with agile, mainly due to scaling it out effectively across multiple teams. Budgeting issues confound the agile approach since the business sets an OKR and has to fund it; it can't be late without cost or revenue impact, and each quarter, those numbers have to be reported to stakeholders, the board of directors, and stockholders if there is a miss in expectation. Communication and multi-team progress tracking can't be wishy-washy, and corporate financial controls confound agile approaches and push large organizations back toward waterfall or hybrid agile-fall development life cycles.

Companies that embrace agile but fail to address problems that a new process integration presents will fail in any process re-engineering effort. This has been seen time and time again. For further reading, look at *Why Agile Fails in Large Enterprises* {https://packt-debp.link/xHDieq}.

Implementing SLT helps cement the benefits of agile development and is a lever that gets attention on the key issues to be tackled:

- Developing correctly

- Developing what the consumer wants

- Developing it fast

- Developing it cost effectively

- Developing it to last, yet remain flexible enough to change as the customer changes

- Developing it to withstand the next technology disruption cycle that could balloon costs for yet another re-engineering effort to rebuild what was already working

If SLT has struck a resounding chord, then you will want to know how best to build out an SLT framework and what some of the pitfalls to be expected are. Please read on to the next section, where we will discuss some key capabilities and best practices.

Test framework example

Now that you have become aware of some of the approaches to testing, it's time to look at a practical example and for that, we have to begin with key drivers. Please look at *Figure 10.1*, where you can see a test framework that you can take into account as you develop your solution. It should be formalized and made available to all those on the solution's various project teams to utilize. In *Figure 10.1* you can see that your test framework, integration, and best practices work together to form the scope of your test effort:

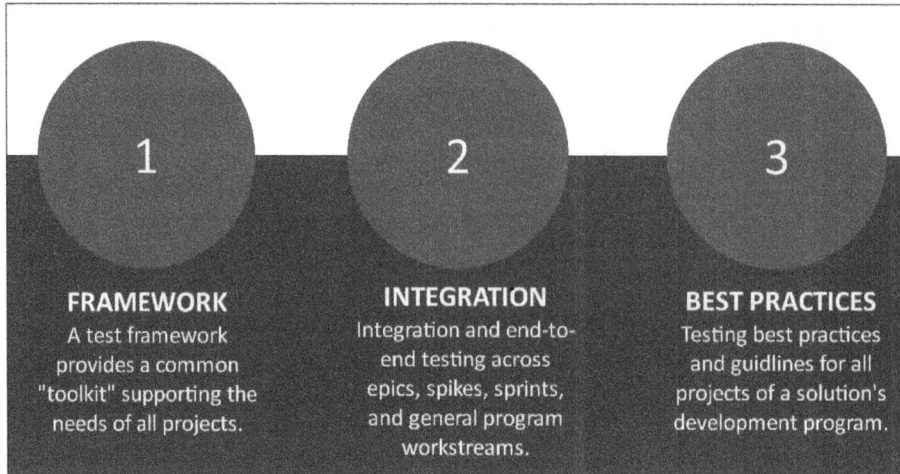

Figure 10.1 – Azure DevOps scope

Remember that the goal is to reduce the overhead of testing and make it as easy as possible to run tests. The developers and testers need to be able to catch deviations from the status quo as development sprints release code into various developed environments. You will know that the system is minimally up and running based on integration tests that are running in the DevOps pipeline and you will be implementing testing best practices from the prior sections of this chapter.

Not all tests can be run after completing early development sprints, but for later sprints, all testing is expected to be cumulatively applied and run in parallel as a regression test suite builds up over time. It is essential that testing is created in such a way as to provide for parallelized execution of all tests in the suite. This will coerce the normally single threaded cloud DevOps CI/CD systems to support the performance needs of your test framework. Just running all of your tests in a reasonable time will be a challenge. Expect thousands of tests!

You will sometimes have to stop parallel execution, assess, and then only when a stable point is reached, continue. You will create metrics and dashboard entries, and then in an automated way, resume test scenario execution when corrections are made. This will also involve AWS cloud formation or Azure ARM template launches to supply test infrastructure automation supporting your parallel testing goals. For this, you will need cloud infrastructure to run on as well as a test budget with costs attributable to frequent defect offenders (high velocity, poor quality development/developers). This process flow has to be part of the test framework's definition and be evident and transparently presented to the test manager through the framework's configuration capability.

Often, the simplistic single threaded Cloud DevOps tools allow for too much flexibility when developing stepwise test processing without central configuration or dashboards. These should exist as the number of tests increases. Testing can fail to progress across the configured process flow for many reasons. An effective testing framework makes it easy for developers to just plug in new/adjusted tests of various types with the assurance of their timely orchestrated execution, often in parallel with other tests. These functional software tests should also extend to data quality tests as part of any advanced **user acceptance testing** (**UAT**) of the data factory pipeline.

As was noted earlier, early stage agile software development sprints will not be capable of running the full regression test suite. However, these tests still have to be developed as part of your SLT goals. This is evident in the flow depicted in *Figure 10.2*, where **systems integration testing** (**SIT**) activations only start being run after a system is physically available for integration. Then, they are run later when data can be run through the system as minimally viable user acceptance testing (UAT). Later yet, only after full data sets can be processed will full UAT be possible with all regression test steps applied. These regression tests will also launch stress, scale, and performance test scenarios:

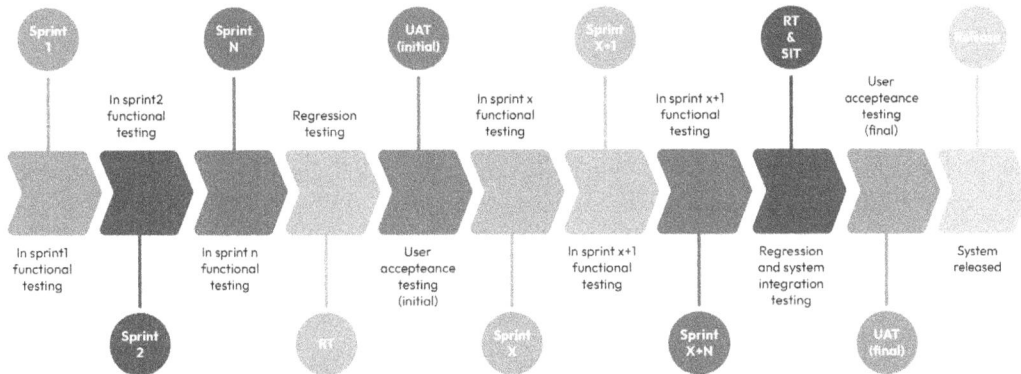

Figure 10.2 – Azure DevOps Agile test integration

Testing begins with unit testing and progresses through functional integration testing, SIT, and ultimately, UAT. Data quality testing can and should be performed fully at UAT test time, but that testing has to include metrics, derived data, and measures so those datasets have to be complete and accurate for this testing to be complete. Refer to *Figure 10.3* for details:

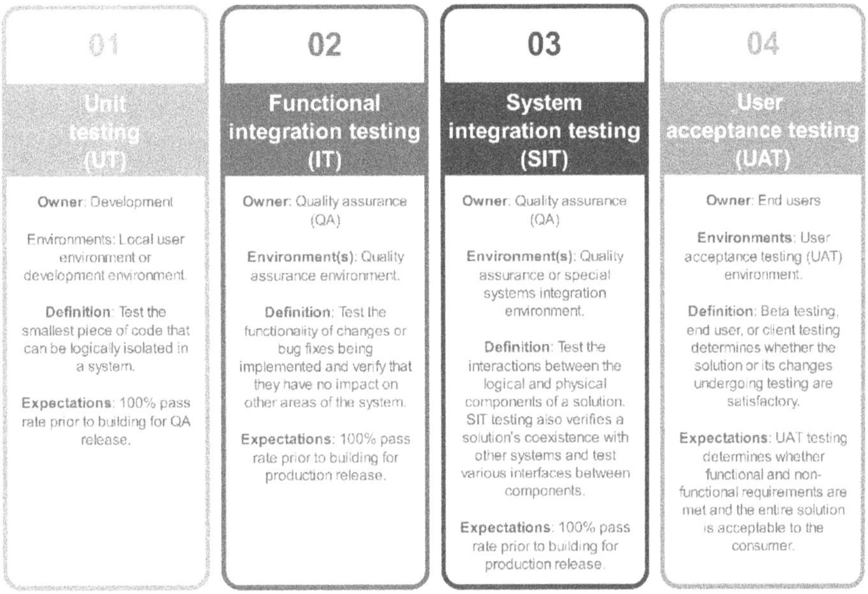

Figure 10.3 – Azure DevOps test levels, environments, and owners

Stabilizing and then freezing test datasets for testing purposes can be problematic when they are not managed for cost impact and quality. Using prior recorded data as a seed for the development of comprehensive data driven testing via canned test datasets can be helpful, but this test data should be obfuscated for privacy and other security reasons. Captured data is then tweaked for testing purposes.

Just make sure the test data generator is programmatically driven and not a manual process – you will have to regenerate canned test datasets periodically, and manually curating test data does not scale!

> **Note**
>
> We are not trivializing the effort needed to create good test data. We are just pointing out that your testing approach has to include created or production-modified data that reflects your real datasets. We'll also get into the topic of synthetic data creation supporting MLOps for training set data in *Chapter 16*. Also, using machine learning (ML) data set generation approaches can help reduce the burden of manual test dataset creation.

Captured data should even be expanded upon to support valid features of the solution that path testing requires, since these paths are commonly driven by conditions found in the data to account for all processing edge cases. Handcrafting new test data from a green field for each new software or data processing feature developed can be daunting if the cost-benefit is not accessed. That being said, know when testing is *good enough* and move on to other testing threads, or push the details back down to the developer's unit testing level when appropriate.

When you freeze data in a dataset for a given test, it should still be registered for use in your test orchestration framework so that it can be repeatably applied and reused in regression testing. The solutions quality process then evolves over time. We've seen test data get tossed all too often. As a result, too many downstream test scenarios are invalidated. Don't let this happen to you! Be aware that even test data drifts over time and its efficacy degrades. Test data may only be valuable in a finite timeframe, and using previously crafted, now frozen test datasets can adversely affect your current test quality. Frozen test data occasionally has to be thawed out and reconstituted so that future regression testing remains effective. With time series datasets (such as financial datasets), this issue is a really important one to consider.

In the processing of tests related to preserving the integrity of the solution, **data quality management (DQM)** goals will require the creation of synthetic transactions for testing purposes. These include trend preserving fake data necessary to enable performance, metrics, measures, and trend tests to complete successfully in the absence of authentic data. Be ready to catalog synthetic data and synthetic transactions and be able to track the impact via test data lineage tracking in the final curated data visible to users as consumable data. This will enable you to create superior data quality tests.

The core engine of a test framework consists of the orchestration of object handlers, common utilities, common test solutions, output components, and a repository tool such as NuGET {https://packt-debp. link/Nua0PN}. Test data management is featured prominently in the following framework, as is test reporting (via dashboards). Azure DevOps pipelines are shown in *Figure 10.4* for the Azure cloud provider:

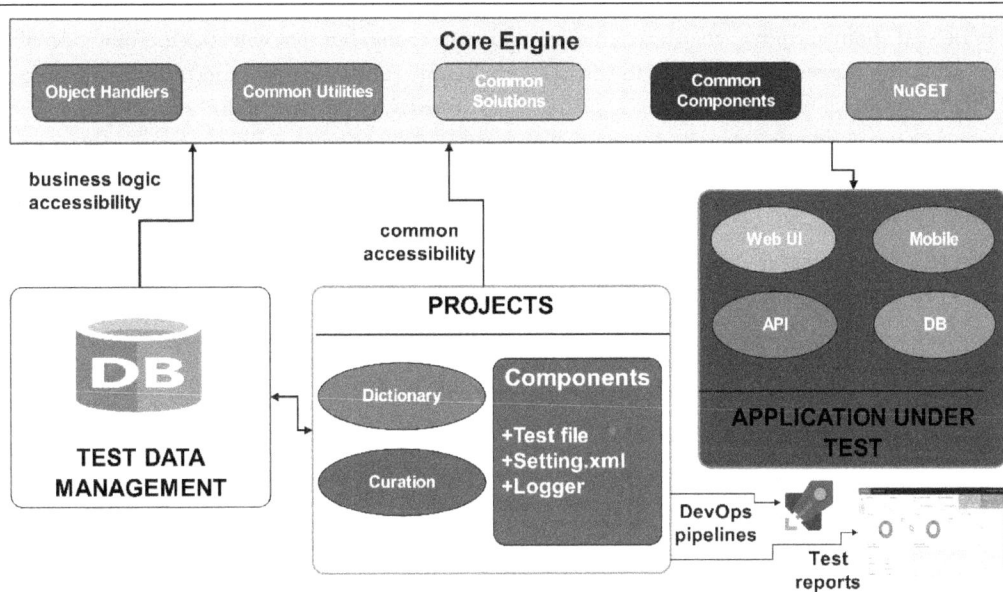

Figure 10.4 – Azure DevOps core framework

It is used for test orchestration, but note that AWS has its equivalent in the AWS DevOps {https://packt-debp.link/YKUOOY} toolkit. Lastly, the **PROJECTS** box lists two core projects:

- **Dictionary**, which is the *at rest* metadata for your data factory solution

- **Curation**, which outputs factory data between zones

Curation represents the solution's developed code (which is a gross simplification of the effort but for the purposes of this test framework illustration, it is sufficient). Test reporting features are not illustrated in depth; however, they are going to involve the creation of numerous outputs. These reports and related anomaly alerts to affected developers and stakeholders form the test system's data error or anomaly detection system. Reports are to be sent to affected developers and stakeholders. These errors are detected as they are encountered after code, test data, or configuration check-in. Error checking occurs after repository check-in (such as GitHub), which launches the configured cloud CI/CD DevOps pipelines.

These are not our data factory's processing pipelines, but rather the DevOps development pipelines that run after each new code check-in event is triggered. Code is deployed, proven to pass all tests, and validated across many steps before starting final UAT testing. Test failures should be grouped into categories such as *fatal*, *recoverable*, *warnings*, and so on. It is when multiple failures occur that these failure reports will be used as a tool to enable engineers (software or data) to triangulate root causes easier than with test scenarios that halt immediately after a single output is produced. It is best to run as many tests as possible so long as the test pipeline sequence can continue to run viably.

Additionally, security testing will be necessary and for that, you should leverage specialized tools. As an example, look at Veracode Dynamic Analysis {https://packt-debp.link/aqyfTc} which will require code to be scanned and test data generated by data test engineers using homegrown test data generators. A lot can be said about security, data privacy, and compliance testing, but that will take too much detail for this book. We suggest that you look to consult with security subject-matter experts in this area and also look at some real-world test use cases. DevSecOps {https://packt-debp.link/EvhnUH} is a key approach to forming an effective agile security testing approach. An example is Bright Security {https://packt-debp.link/GrwnVt} which won the 2023 CyberSecurity Breakthrough Award {https://packt-debp.link/fSOGUN}.

In *Figure 10.5*, you can see the application of *pre-conditions* and *post-conditions* that form the test steps in the overall scenario:

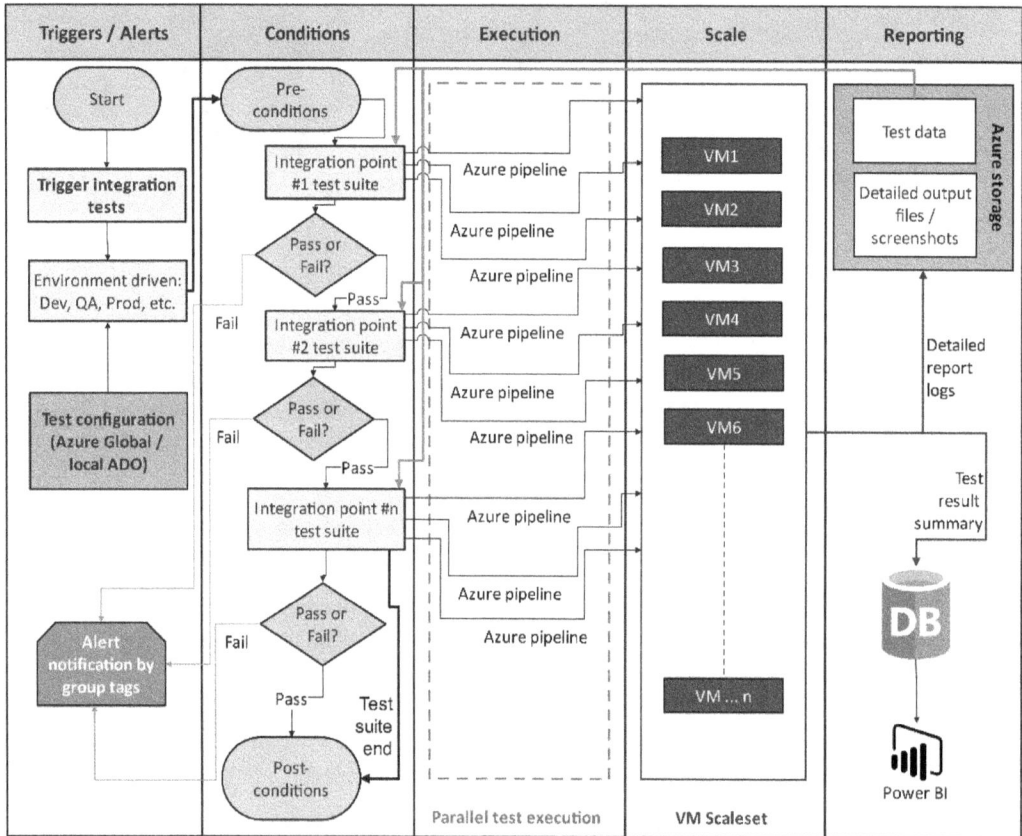

Figure 10.5 – Azure DevOps test flow example

There will be hundreds of tests in your fully developed data factory solution, but each test is an instance of the generic test model in regard to the format of its definition. This way, they fit into the test framework cleanly, and their outputs are reported as contracts that are either *met* (*test passed*) or *not met* (*test failed*). The **parallel test execution** environment has the capability to execute tests in parallel and, at times, wait for some parallel steps to complete before moving on to the next testing stage. Test scenarios should be configurable and/or scripted as part of being configured into the TestOps {https://packt-debp.link/3mZr2F} scenario orchestrator for management within the test framework.

Unfortunately, out of the box, the cloud provider's DevOps pipeline CI/CD services will not offer the full scope of parallel execution, process stepwise execution in branches, or wait for the convergence event processing necessary for your solution. You have to build these into your test framework. Cloud instances, pools of virtual machines, and the launches of **infrastructure as code** (**IaC**) environments (refer to earlier references in *Chapter 5* and defined fully later in *Chapter 11*) are to be automated in order to keep operational cloud costs down and still finish testing before developers begin their next check-in cycle.

Developers will want to know whether they broke the build before going forward each day. This type of practical testing speeds up agile development and minimizes the waste often seen in projects that have less structured automated testing, or worse yet, manual testing. Note that test results are stored and not just reported on from test run to test run. This history is important since the data results should drive a report and not just flat information from logged outputs where the test engineer is left to manually see results. Reports sum all post-condition outputs and have assertion checks to validate outputs that are stored and reported to the developers.

In *Figure 10.6*, you are able to see how these pass-fail condition checks should be applied as part of the test scenario driver's operation:

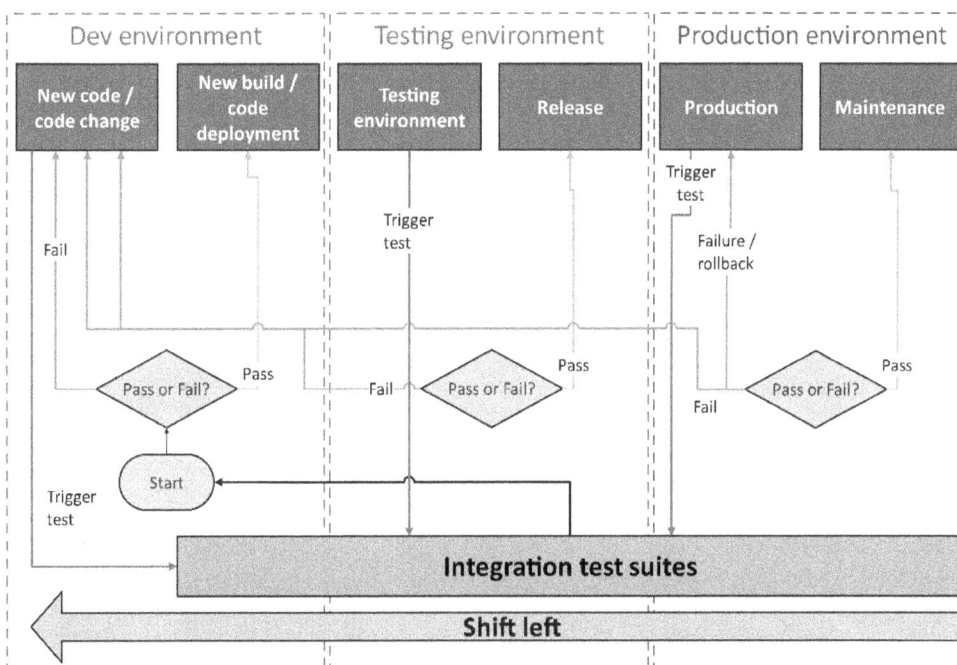

Figure 10.6 – Azure DevOps shift-left flow example

Testing should progress from the check-in event that triggers the test scenarios to begin, but you should be able to start at a time that is not going to cause excess work when an item is checked-in multiple times in a row. Test scenario instance runs can be aborted and restarted if and when repeated test triggers take place. The check-in itself should never be held up, but the testing kickoff can be queued up as a result of these frequent event triggers.

The first check of the test scenario is for the completeness of the package and the pre-conditions for a successful build, followed then by the solution build itself. The impact of the check-in will determine the scope of overall testing to be triggered. Then, the testing environment is spun up. The release packages are generated and then pushed to a staged production environment (usually a blue/green deployed environment). Lastly, the maintenance environment can be created if a failure arises that causes the production environment to have to be reverted. The very intent of shift left testing (SLT) is attained in the flow example in that testing is moved very close to the code check-in release event trigger and the **integration test suites** kick off after that checking trigger event.

In conclusion to this section, you can see that testing is not going to be easy, but you want to make sure testing is maximally affected and works to accelerate the development and quality levels needed to deliver on the solution's OKRs.

Next, we'll discuss the data wrangling effort, which requires a complete understanding of the data profile so that the data's conceptual shape can be understood and then processed by your solution.

Data wrangling and profiling

Often, testing efforts focus on the quality of the software developed; however, in a data factory solution, the data is the product and needs to be tested for contract compliance. These tests require the data to be profiled and the shape of that data assessed for being fit for purpose. This is called **data profiling**, when ingested data is wrangled as part of its ingestion into the data factory. The wrangling of data is a mix of deterministic and stochastic based processes:

- Transformations (to coerce data into a normalized processable form)

- Redactions (to remove detected errors)

- Reductions (to remove restatements)

- Data masking/obfuscation (for security and privacy)

- Summarizations (facilitating downstream analytics)

- Alt-data expansions (to use tables to deterministically expand data using known reference sources)

- Metadata enrichments (to classify and provide semantic context)

- Semantic alignments (to enable data shape fitment and inferencing leading to downstream embeddings)

- Other value-added enrichments (to provide hooks needed to facilitate the keys needed for generating measures, calculations, and metrics in later processing steps before end user or application consumption)

Many of the wrangling steps listed require data to be profiled before or during the wrangling and they are often rule based data profiling processes.

Deterministic data profiling

Deterministic data {https://packt-debp.link/Fvzzjw} methods (rule-based methodology) are best suited to many of the profiling tasks required in the initial wrangling effort. Most traditional data processing involves the normalization of data and assessment of that data to the entity relationship model for correctness and fit. Where not fit, data can be coerced into a normalized form after enrichment and mapping. However, that being said, there are other times when patterns across large datasets are best captured using stochastic- (or ML-) driven methodology. **Probabilistic data** is assessed differently from **deterministic data**.

Machine learning driven data QA

Data patterns appear in large datasets as clusters of similar data features. These remain unknown until they are mapped to known patterns that have previously been assessed as being a correct fit for processing. The cluster's data shape will be assessed for its match to known clusters, making classification possible within a threshold of certainty. **Clustering** is an unsupervised machine learning approach best suited to use cases where training the machine learning model is not possible or where a trained model would not be effective when classifying the source dataset. Algorithms such as **K-Means** clustering {https://packt-debp. link/uIVkjm} or **Latent Dirichlet Allocation** (**LDA**) clustering {https://packt-debp.link/2oHnuq} are often applied for clustering but they operate differently. Only with machine learning ensemble techniques are they effective together, but they can stand alone for niche data processing use cases.

Pattern detection and matching is an important semi-supervised machine learning algorithm approach used in data profiling, where **support vector machine** (**SVM**) {https://packt-debp.link/pSgdjc}, **linear regression** {https://packt-debp.link/F8zaVw}, and **logistic regression** {https://packt-debp.link/RPciql} algorithms come to mind.

The data pattern that the SVM represents can be matched to the incoming data and all data not matched is therefore new to the system. The issue is that there is often a lot of unmatched data on day one, when the data profiling starts running. That will require a lot of training datasets to be crafted and then SVMs trained for use so overall data coverage is acceptable to the business. Along with this comes retraining the SVM model to get it right for ongoing business operations. After all, the data product has to be 80% ready for downstream processing steps to be effective, and even then, expect reprocessing of existing data to reflect newly refreshed models over time.

Data clusters may be discovered via unsupervised algorithms leading to entirely new categories to be identified in the ingested data. That can be an advantage! Also, with supervised models, known patterns can be applied to filter out matching clusters (a.k.a. topics) from incoming data; therefore, leaving only what is *new* to the system (to be classified) or the *anomalies* (to be discarded).

There are many other pragmatic use cases where machine learning is used in data profiling and wrangling. It can be very effective in handling localized data mapping to your system's neutral form. If English is the standard and your NLP resources and dictionary (or semantics) are written for English, then mapping item descriptions (for example, in a retail scenario) to English makes sense. Retaining the original local language is often essential, but having the neutral form available makes matching work well.

Other use cases arise in trying to detect incoming data that could break existing trends before being applied to the system as a truth. Another example is where a graph defining a dataset needs to be traversed and fit (or found to fit) within another graph (subgraph isomorphism problem {https://packt-debp.link/t8fXR9}) of a similar but not exact match. This is similar to the **Traveling Salesman** problem {https://packt-debp.link/FQoW2J}; both are NP-complete problems. As such, they are great fits for enhanced ML algorithms. Note that the term **NP-complete** is a shortened term meaning **non-deterministic polynomial-time complete**, indicating that any algorithm applied to this class is a non-deterministic **turing machine algorithm**. A Turing machine algorithm is a mathematical formalization of a brute-force algorithm. A problem being nondeterministic means that there is no solution possible via brute force algorithm and as such stochastic algorithms are required.

> **Side note**
>
> We'd like to see how GPT-4 can handle the development of needed solutions to many NP-complete problems. Solving these really tough processing issues with new innovations will resolve key problems that have plagued computer scientists and mathematicians for decades.

The data profiling, wrangling, and preparation of data to be designated fit for purpose have an underlying theme. The answers to the following questions are needed:

- Is it the *truth*?

- Is it a *true fact*?

- Do the semantics hold true for that data in a given domain?

- Can new insights be harvested now that the ingested data is semantically enriched and understood within the system?

If all the answers to the preceding questions are *yes*, then within a subject domain, data may be used, and its inferred relationships can be used to glean insights. The observed data patterns can even be used as a template in other domains, leading to great leaps of insight that cross domain boundaries. The implementation of this vision, where context sensitive data patterns carry intrinsic meaning, makes a solution future-proof because it enables the data to come alive. All this is made possible by testing data for syntactic and semantic correctness, then organizing it for direct use, inferred use, and recurring pattern-transitive reuse across domains.

Summary

In summary, you have learned that testing is difficult and costly. Leaving testing until the end of a project incurs high costs and is the least effective approach possible for agile development efforts. Building a TestOps orchestrated test framework that shifts testing responsibility to the left (SLT) is optimal for most modern development teams. When implementing testing, a TDD or BDD approach implants a test mindset into the stakeholder's thinking, since they will want to know how their OKRs are going to be validated. Lastly, you have been exposed to the data profiling and wrangling needs of a data factory that uses deterministic and probabilistic machine learning techniques to assess and then prepare data for being staged as fit for purpose in your data factory.

In the next chapter, you will be provided with best practices to build out your IT Operational Services.

11

Key Considerations for IT Operational Service Best Practices

What does it mean to be operationally effective? The answer to this question is: It depends! In reality, it is based on many factors, but it all comes down to running and managing a solution during and after it is done. So, in this context, does *"it"* really mean *"information technology"* (*IT*) or does it mean just *"it"* – your system? We'll let you decide as you read this chapter and see that without correct operational best practices IT devolves into a thing (*"it"*) that is not reliable, available, scalable, or performant.

With Agile and Lean development methodologies, that state of *being done* may never really occur. Alternatively, a product's stage of being done may be attained only after it is ascribed a versioned release number and made generally available within one pass through the cycle. How *IT* gets done is iteratively defined by these cycles through the DevOps **Software Development Life Cycle** (**SDLC**).

In this chapter, we will explore the following topics:

- Service level agreements and service level management of (various) contracts
- Data contracts leading to quality data factory outputs
- Solution monitoring with observability capabilities
- System and data contract anomaly detection
- Blue/green deployment vs other release/deploy processes on operations (what are the tradeoffs?)

With this information, you will be able to build an operational framework with processes that are sustainable when your solution enters its run/manage phase.

IT operational best practices overview

Just looking at the technology stack reported by *SD Times* in 2017, from Atlassian {https://packt-debp. link/sV36qr} in *Figure 11.1* {https://packt-debp.link/TdzvqG}, you can see that the number of products associated with the DevOps cycle is daunting:

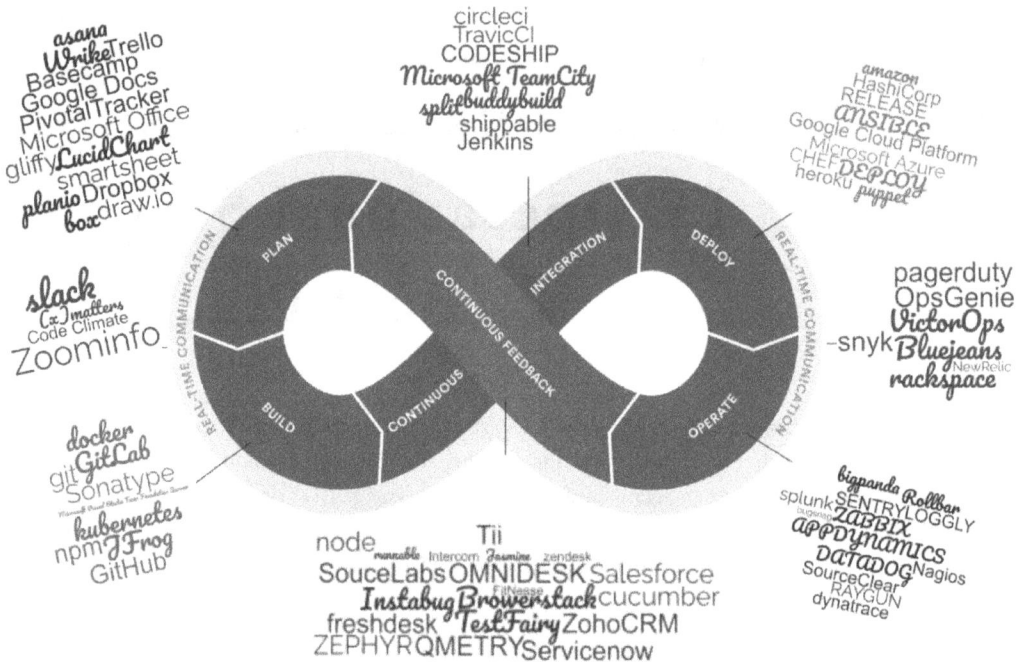

Figure 11.1 – Atlassian Stack

What you will need is a documented operational manifesto driven by best practices that will help make tool selection and process definition for your ecosystem possible. For this manifesto, you will not want to begin from a green field. So, let's begin with the **Information Technology Infrastructure Library** (**ITIL**).

From other chapters, you will recall that we mentioned the need for ITIL when building out your **Information Technology Service Management** (**ITSM**) process framework. The following quote from Gene Kim will underline this need:

> *"The DevOps Movement fits perfectly with ITSM , ITIL and ITSM still are the best codifications of the business processes that underpin IT operations, and actually describe many of the capabilities needed in order for IT operations to support a DevOps-style work stream."*
>
> *Gene Kim* {https://packt-debp.link/iG7VKC}, *author of "The Unicorn Project"* {https://packt-debp.link/EIS7F3} *and coauthor of "The Phoenix Project"* {https:// packt-debp.link/DF2Vjf}

You may want to pause at this point in your reading and research what was just said, or maybe even read Gene's books to get a better appreciation for this area. ITIL/ITSM is essential for your solution. After grokking the service definitions, many non-functional requirements will be conceived in your mind, which is exactly what the study should produce. These will become OKRs for the development teams. We mentioned earlier how important the capabilities identified from this ITIL legacy service approach are still relevant with today's cloud technology. Refer to these sections of the book:

- *Chapter 7, Architecture Framework – Physical Architecture Best Practices*, under the *Physical Architecture Overview, Best Practice Organization, Define the Processes of the Current as-is State, and Prepare for Iterative Change to the New to-be State!* section

- *Chapter 8, Software Engineering Best Practice Considerations*, under the *Software Engineering Best Practices Overview* section

- *Chapter 9, Key Considerations for Agile SDLC Best Practices*, under the *Software Development Life Cycle (SDLC Processes), Objectives and Key Results (OKRs)* section

We are not trying to force you to take a different operational approach than you want to take. However, learning from the mistakes and successes of others is our recommendation. You must be ready for a reset on your current understanding of IT operational processes needed to maintain a robust reliable solution.

IT operational best practices – introduction

Our goal is to integrate DevOps Agile practices into your business solutions development life cycle with an adaptation of the disciplined best practices set forth in the ITIL framework.

> **Note**
>
> DevOps is not a framework. It's a process that is vague but clarified if you flesh it out and develop your IT operational definition. The process should have teeth and be enforced with **policy statements**. It will need to be communicated to all teams since it forms a binding set of non-functional requirements that can't be truncated from a release's feature set (which often is attempted). The long-since validated ITIL service definitions provide the structure needed to accelerate delivery while limiting risk and encouraging continuous improvement with reduced effort. However, once again, note that the framework should pose questions that you need to answer for your organization.

Refer to *Figure 11.2* for the service objectives needed in your IT operations framework:

Continuous Improvement

Service Strategy

- Planning
- Change
- Configuration
- Release
- Validation
- Evaluation

Service Transition

- Catalogue
- Service Level
- Availability
- Capacity
- Continuity
- Security
- Supplier

Service Design

- Generation
- Portfolio
- Financial
- Demand
- Validation
- Evaluation

Service Operation

- Event
- Incident
- Request
- Problem
- Access

Reporting

Measurement

Demand

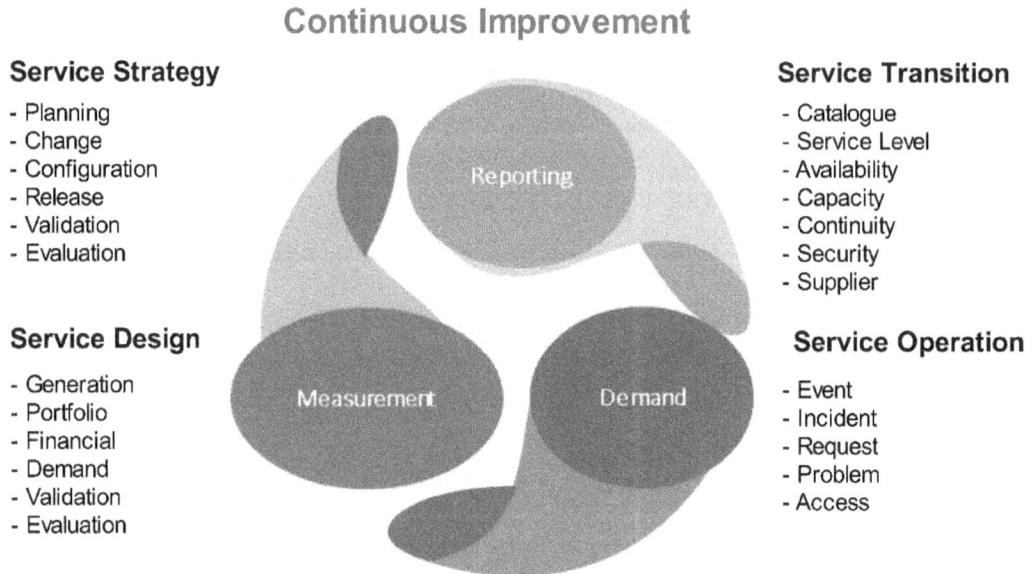

Figure 11.2 – ITIL framework objectives

You will naturally want to minimize the complexities of the various flavors of Agile, DevOps, TestOps, DevSecOps, FinOps, MLOps, AIOps, and DataOps processes found in use today across many businesses, products, and tools. Not all vendor tool approaches to Agile development will lead to easy integration. You will want to select a foundation set of capabilities to build on. Then, use the ITIL framework as a checklist to make sure that the capabilities you desire are not being omitted or trivialized. Where that begins to occur, you want to augment the service (or tool) with value-added integration.

Let's look at an example. In configuration management, where cloud services require configuration, or where Terraform is used with a secure KeyStore, there needs to be a way of providing for cloud service parameters/configurations *without* having these hard-coded into Azure ARM, AWS CloudFormation, or Terraform Templates.

Many vendor toolsets do not all play nicely in the sandbox of your IT operational environment. This will complement the build-out of your cloud ITIL processes. The various cloud providers offer quick paths to *their* processes, driven by *their* best practices, not yours.

The first IT operational best practice!

Know what you want from an IT operational environment and build to that expectation from a core set of integration services selected from a vendor. Perform comprehensive due diligence backing up that decision. Also, provide process solutions to answer questions your particular organization will have in regard to effective IT operations. Note that if the best practice recommendation to first define a logical architecture was followed, then the logical components with their interfaces will remain effective over time. You will need to select or reintegrate new vendors, situations, or integrations, as they are going to be subject to a lot of change.

Examples of this include Microsoft Azure DevOps, AWS DevOps, GCP SRE, or many of the third-party tool providers. Atlassian will try to minimize the impact of integrating tools with a generic open source Ops tool ecosystem. An integration product enables you to implement a flexible *you integrate what you like* approach (to a lesser or greater degree) in order to tackle toolchain integration complexity. Cameron Deatsch recognized this complexity issue and Atlassian has responded to the need for DevOps tool integration.

> *". . . many of their enterprise customers use Atlassian products to break down barriers of development and operations, but until today, they didn't have a way to centralize and consolidate DevOps tools."*
>
> *- Cameron Deatsch, 2017 (formerly Atlassian's head of server and enterprise marketing and chief revenue officer, now an industry independent advisor)*

The various approaches available will obviously be researched completely by you, the responsible data engineer or manager. Cloud provider service groups or independent consultants will attempt to baseline your knowledge by first assessing your DevOps maturity. For example, consider the following:

- Atlassian has a DevOps Maturity Model {https://packt-debp.link/itFmAI}

- The SourcedGroup {https://packt-debp.link/izqDov} conducts a four-week assessment with their Azure DevOps Maturity Assessment {https://packt-debp.link/aJcOm2}

- Microsoft Azure produces the Azure DevOps Capability Assessment {https://packt-debp.link/fAMSth}

- Google has the *"2023 DORA State of DevOps Report"* {https://packt-debp.link/QjsP6T}, as well as **Site Reliability Engineering** (**SRE**) {https://packt-debp.link/FWhVgU}

- Oracle has **Oracle Cloud Infrastructure** (**OCI**) DevOps {https://packt-debp.link/my73vI}

There are many features in a smooth-running and managed systems solution. Development groups attempt to think through operational requirements even if they were not well documented before release. Later, IT operations and others will add additional non-functional requirements, which form a huge set of gotchas that were missed by developers. This is a common occurrence. You, as an engineer (data or software), will want to focus on trying to fit the *"big rocks"* into the limited amount of space within your metaphorical jar (your bailiwick within the budgets of funding, time, and effort). Start by defining **service level agreements** (**SLAs**), as well as your approach to maintaining them through observability and automated situational response handling.

Data contracts are similar to SLAs. Data contracts have a scope that pertains to the curated data of the solution. Data contracts can be validated with assertion checks. We like to call these **data probes**, which are designed to be run after important stages of the data factory are completed or before a given stage is attained. These act as gatekeeper functions or circuit breakers, thus preventing data errors from being compounded by subsequent downstream processing. The data contract is therefore enforced and validated by the **data quality management** (**DQM**) system which, when put into real time production, can create data contract aligned telemetry metadata. Telemetry data helps when diagnosing reported data quality issues that are able to get through automated testing. Monte Carlo Data {https://packt-debp.link/Olvca2} defines the data contract as follows:

> *"A data contract defines and enforces the schema and meaning of the data being produced by a service so that it can be reliably leveraged and understood by data consumers."*
>
> *– Monte Carlo Data {https://packt-debp.link/HUH8VS}*

Note that the **Pareto Principle** (see the following note) is very relevant to the topic of data contracts.

Pareto Principle

80% of *consequences* are the result of 20% of *actions*.

The principle is very relevant today. You will be able to apply it as justification for your future schema development. Over time, the number of tables and their derivations that you designed will have morphed into a complex data mesh with the only way to unravel the mess (not mesh this time) being through the use of data contract setting. These contracts align with the data mesh architecture that you design, which leverages data lineage metadata tracking. The *actions* you take in the design have a positive multiplying *consequence* as the system grows over time. This provides great value as you journey toward the goal of futureproofing your data engineering solution.

We have found that when building a system, all building steps should be automated. There should not be any manual steps. From building the Linux kernel in an on-premises server or cloud instance all the way to service configuration with its security settings, all steps need to be automated. The benefits of automating building and deployment in the cloud are many. Firstly, the approach enhances business operations. Automation also increases team efficiency and productivity by reducing repetitive tasks that may take a lot of time. Freeing up resources for strategic efforts is always prudent. Achieving faster **Time to Market** (**TTM**) is now supported by better IT delivery service. Agile methods are not just for developers but also for the operational folks in your organization. These benefits provide greater cost savings as a key side effect. The costs of finding and fixing human errors are reduced. The business impact of automation is that there is greater solution reliability. Operational consistency and repeatability is a great goal that can be achieved with automation. Lastly, your system's security is enhanced since your solution inherits the fixes applied for other systems as templates. These gaps can't be rolled out once detected and fixed because they use the same automation. You can read about a deployment horror story here:

> *"During the deployment of the new code, however, one of Knight's technicians did not copy the new code to one of the eight SMARS computer servers. Knight did not have a second technician review this deployment and no one at Knight realized that the Power Peg code had not been removed from the eighth server, nor the new RLP code added. Knight had no written procedures that required such a review."*
>
> *– SEC Filing (Release No. 70694, October 16, 2013) (https://dougseven. com/2014/04/17/knightmare-a-devops-cautionary-tale)*

The choices you make in tool selection, Agile DevOps integration, and cloud configuration are dependent on your performant CI/CD pipeline. This SDLC-required DevOps CI/CD pipeline affects the velocity and quality of your software solution along with its data-engineered capabilities and features. SDLC Agile processes can result in data being released in an inconsistent state. This comes as a result of packaging and deployment processes that do not consider the data's set state of correctness. When software components are released, they affect data processing, and that can lead to unknown state errors after release. Data correctness has to be tested in the CI/CD pipeline as well as software correctness, but often it is not!

In a complex data processing system, **logging** is expected, as is **trace** output, and the need to assess all for correctness as determined by operations. Implementing observability goals for logging, tracing, and metrics generation *together* has helped the industry move beyond the approach of collecting stuff (in a raw format, meaning *everything*) and then trying to find the needle in the haystack. It is true that you may have logged an event when it occurred, but by not recording its context, you have produced a result that is not observable. It was garbage when logged the moment the message hit the storage array if the message's context was not also retained. So, why log what is not usable? The answer is: don't log anything without thinking about capturing its context and whether it could lead to an observable conclusion. Your observable data has to be just that – interpretable information.

Lastly, as a preface to the deep dive in later sections, you will want to consider the process and methods for deployment into any given environment. Your solution has to be smoothly handed off and deployed, but that smoothness is often interrupted by real-world events, bugs, business changes in direction, and so on. All of these will cause you to have to roll back at some point.

A solution's packaging, deployment, and change processes are to be automated. They need to be scoped and, if necessary, broken up into feature-based releases to lessen their impact on overall stability (so no big bang deployments, please!). The identification of what is, and what is not, part of a deployment has to be designed upfront using the type of deployment model selected. The solution's features form logical divisions for releases. Also, technology developed reusable components and framework code can form logical divisions of a solution. The subsets can be grouped into releases; just not all of them, and certainly not all at once, please!

We'll begin a deep dive into IT operations best practices with a discussion on SLAs.

Service Level Agreements (SLAs)

Setting an SLA is essential to maintain non-functional data product requirements. Begin with defining *what* you are trying to achieve, and then break that into subgoals that can be managed by applying observability. By writing the goals down as agreements, you are going to be forced to comply with those agreements and build measurements, metrics, and monitoring to verify that you are meeting those goals. Then run the system and observe the system's metrics, outputs, and trace logs so that agreements can be managed.

A service agreement can be just about any processing assertion that stakeholders and end users view as being important. You will also be a stakeholder since you probably know intuitively what may go wrong. You designed the system, after all! You will also want others to weigh in on a system's operational readiness criteria. Constructive feedback will add to your already large list of SLAs that have to be met in order for you to say: *IT's* working (i.e. your IT developed solution is working)!

You will want to maintain service levels and data contracts in a formal manner. You will also want to know when your solution is starting to deviate from the expected behavior, which means diverging from your contracted SLA goals. Just like when you are driving and you encounter an amber light, you are being given a warning to slow down. Your car will have to be stopped in just a short time. Be prepared! You want to know when an observed deviation from the norm is happening, not after the deviation has already occurred. See the later *Observability with proactive alerting* section, where we elaborate on observability, for more information.

The formula for success is in *planning*; so, you will want to carefully think about the pre-canned response scenarios associated with any degraded state of operations that you envision. This systematic approach to error handling is tied into the **fast fail/auto-restart** pattern often seen in object-oriented designs' **structured error handling**. Low-level handling is taught in university (or through life experience), and from our experience, we can tell you to implement **systems error handling** as part of **Service Level Management** (**SLM**). Link any SLM processes to programmatic **structured error handling** as part of your architecture. This unifies the error handling and response design needed for robust operations. It is often not implemented well and causes complex distributed systems to melt down during operations.

It is so very important that you think proactively! This way, the number and severity of user impacts to observed anomalies are kept to a minimum. It should become normal system behavior to self-heal or, if necessary, activate problem management or support level escalations processes early in order to get subject matter experts (often the developers) involved. These automated, event-driven system state transitions need to be activated well before problems go critical.

In production, when problems become visible, SLAs are broken. With those breakages come major corrective actions, which are all conducted when there is an ongoing outage. Build time to respond to detected issues before they become critical and do so in an automated way as part of your IT operational processes.

Data contract service level agreements/data contract management

Data contracts are similar to SLAs but for data. They can be validated with assertion checks. We like to call these data probes, which are designed to be run after key stages of the data factory are complete or before a given stage is attained. These act as gatekeeper functions, preventing data errors from being compounded by subsequent downstream processing. The data contract is therefore enforced and validated by the data quality management (DQM) system which when put into real time production, can create data contract aligned telemetry. The telemetry metadata helps when diagnosing reported data quality issues that may get through automated testing.

Based on our writing about the Monte Carlo Data vendor earlier, the vendor has a good grasp on the data contract issues, which you will need to address in your data engineering solution. Data contract management requires the following capabilities:

- **Measurement** (testing, time to resolution, number of data incidents) to reduce data delivery latency.
- **Preventing schema changes** that are needed in downstream processing.
- **Preventing quality issues** with circuit breakers.
- **Preventing recurring data problems** via detecting *data health* issues through metadata and lineage.

- Recognizing the need for real-time **data quality telemetry** and a *data observability* mindset.

- Adopting **data conventions** that service data contract implementations, such as the following:

  - **Clear SQL** is necessary; this includes standard formatting, commentary, style, and conventions.

  - **Data incident processes** are escalation processes and workflows to resolve data issues, as well as structured postmortems in evolve processes and code quality in order to prevent the same issue from recurring.

  - **Data literacy** enables to provide an explanation of data values in context from acquisition through curation in plain written language.

  - **Data ownership** will drive you to consider the data mesh principles regarding data domain ownership in earlier chapters; refer to *Chapter 4*, *Architecture Principles* under *Architecture Foundation*, *Data Mesh Principles*.

  - **Data SLAs** will identify *when* data is complete/ready, *who* curated it, *where* it is to be acquired, and *how* to get help with issues.

  - **Documentation**, when written in plain language, will provide an overview, as well as clear information on the curation, transformation, and end goals.

  - **Monitor drift** will show, in the lineage trace, where acceptable data value ranges have drifted and account for schema/model drift as well.

  - **Naming conventions** for columns, tables, views, materializations, and so on promote consistency and understanding.

  - **Semantic layer** is needed to understand *data at rest*, *data context*, and *concept connectivity* to provide meaning.

With these key capabilities, data contracts may be created, maintained, and verifiably enforced. You will be able to track all compliance and remediate issues associated with deviations from the agreed-upon contracts or SLAs quickly and effectively. You will also want to build out an analytics management subsystem to maintain your data contracts, SLAs, and metrics. This way, compliance with the contracts is visualized and auditable. The system validates the openness and assurance of data contract compliance.

The concept of a data contract is further advanced in the BITOL {https://packt-debp.link/TNGaNB} Open Linux Foundation Sandbox Project. In the project GitHub, there are templates to begin with so that your data contract can leverage the advancements others have been making in this area. It's an open source project, so expect some differences from what Monte Carlo ships as a generally available product. A data contract is defined by BITOL as the agreement between a data producer and consumer, and it contains a number of data contract categories:

- **Fundamentals**: Name, version, and description

- **Schema**: Physical and logical implementation details

- **Data quality**: Governance, policies, and rules

- **SLA**: Latency, retention, and frequency (arrival/delivery)

- **Security and stakeholders**: Roles, attributes, groups, users

- **Custom**: Properties for extending the data contract

In *Data Contract 101* {https://packt-debp.link/a3p5IN}, JG Perrin refers to the data contact as follows:

> *"A data contract acts as an agreement between multiple parties; specifically, a data producer and its consumer(s)."*

> *– JG Perrin (*https://medium.com/@jgperrin*)*

For deeper dives into data contracts, we invite you to explore *Implementing Data Mesh* {https://packt-debp.link/Xj4kGV} by JG Perrin and Eric Boda, as well as *"Driving Data Quality with Data Contracts: A comprehensive guide to building reliable, trusted, and effective data platforms"* {https://packt-debp.link/4U29NN} by Andrew Jones.

In the next section, we'll take a deep dive into the DevOps CI/CD pipelines needed to implement software and data contracts, which will constantly be under threat of violation due to change.

Continuous integration/continuous deployment (CI/CD)

In modern system development, the complexities and rate of change can pose a wilting set of problems to the engineering manager. It is essential that the end-to-end DevOps pipeline (or, the CI/CD system processes supporting the SDLC) implement waves of software regression testing for new check-ins. Additionally, data contract assertions are needed to ensure that new software does not negatively impact the existing state of system data. The assertion outputs are observable outputs of the system that themselves map to success or failure states. Some types of data contracts can only be enforced after a period of time, a data arrival initiated trigger event, a calendar schedule, a data sensitivity threshold, or a measure/metric trigger has been reached.

A key point here is that the assertions and the checking of those assertion outcomes form gates to be unlocked in the CI/CD pipeline. Passing these gates allows forward progress in the process flow to continue. This ultimately leads to a package release with placement into the release set as part of the code being deemed operationally ready for integration and deployment.

Observability with proactive alerting

Observability at the highest level is the degree to which a system's status can be obtained from externally available data. Observable systems provide the capability to understand various issues as they arise. **Observability pillars** {https://packt-debp.link/1VY5gk} are as follows:

- Metrics

- Logs

- Traces

Additionally, note that observability's scope usually extends to the user experience, cloud, internet, and internal networking operations, as well as the state of any data contracts. From the perspective of an end user, a data consumer should never see that your system has failed to deliver contracted (or agreeable) set expectations. If you know that the system you built gracefully degrades, handles errors, and is built to not impact end users when recoverable failures do occur, then you are designing properly to handle any event.

If handling unexpected errors is as important to you as delivering proper results, then you have bought yourself enough time to respond to issues. Catching issues when they first begin arising is a goal of your observability capabilities. This is a key part of an effective operational system that will withstand the test of time and circumstances as they arise.

In some solutions that we have seen, graceful degradation is implemented by tossing users from sessions in order to keep the high-value sessions running. A similar flavor exists whereby the system allows sessions to first complete a complex transaction before getting abended (which means it is abnormally terminated). This prevents self-inflicted retry overloads, thereby inflicting downstream loads upon retry. This is especially painful in microservice ecosystems.

You know that things are going to go off the rails sometime in the future. Why not build in the capability to manage errors, with the collection of clear telemetry data? Consider that your system is a big **finite state machine**, with smaller subsystems that are also finite state machines. You will be able to affect correct state transitions within your system as responses to errors aligned to predefined scenarios within a context that the automata defines. So, error handling is no longer just print, error, then fail fast processing. See the following code block:

```
try {}
catch {
 print error log message;
 <fail-fast>
};
```

Rather, it is a transition to error-recovery state logic, as follows:

```
try {}
catch {
 <transition current to error-recovery state>;
 <gracefully degrade>
};
```

These state transitions are driven by conditions that are events, which drive the finite state machine event transitions. If you adopt this approach, you are on the path to designing **systemic structured error handling**. This enables the system to automatically respond to detected events, handle them, and even correct data anomalies in output data, then finally transition back to a normal operational state. Structure is given to system error processing pathways as if they were normal event handling processes; they are not left to disconnected processing brought about abrupt component fast fail logic.

Many modern systems are built with **Inversion of Control (IoC)** {https://packt-debp.link/fUKxaw} software patterns (including Java Spring Framework {https://packt-debp.link/UuM74f}, Servlets (deprecated), and Microsoft Windows). Events can be used to drive processing aligned to your finite state machine and not through top-down code execution. Scalable, distributed code execution and orchestration can get really complex, but the effort is worth it in the end. When weaving a network of PaaS cloud services, microservices, and scripts sprinkled around your cloud account, this design pattern has even more value.

Processing this disconnected mess in a highly distributed system does not lead to robust operations unless operational characteristics are aligned to allowable states. Deviations to those states will be handled, corrections will be made, and process flow will lead back to the correct processing state to continue. All of this should be automatic! You will observe that it is not natural for cloud services to be integrated this way. However, you will want to design your solution to avoid sleepless nights spent patching your data back after a system meltdown just because it unnaturally entered an operational state from which there is no exit state transition.

If you do not handle most, if not all, possible operational conditions, then errors will become globally visible. All correct as well as anomalous operational states can be handled by design. This may be visualized in a state transition diagram.

This visually appealing operational management approach has been used successfully in the past. Run time-collected observable telemetry data can be put into a tracking dashboard. Responding without manual intervention is also made possible since event triggers are fired off when the system enters degraded states of operation. Since the operational states have been designed and then codified as state transitions, the system can still adhere to set SLAs. This is possible because the state transitions are the only known pathways to correct operation.

You have coded all states to the operational plan and the system follows that plan rigidly in order to drive operations. You can be assured that the software is correct and that all erroneous pathways are being closed off from occurring. If the system crashes now, there will be bugs, and they should have been removed in the quality control process before production. If you have tested and proven the software to be correct and aligned with the functional OKRs, as well as the non-functional OKRs, you have achieved success. You have removed bad code before it is promoted to production and tested it for operational readiness. Also, you have asserted contracts for correctness before code execution after the system passed all functional testing.

In 1986, Bertrand Meyer {https://packt-debp.link/2sNKng}, while designing the **Eiffel programming language**, elaborated on the need for software contracts. In his book, *"Object-Oriented Software Construction"* {https://packt-debp.link/HrPpa2}, Meyer defined the concept of contracts as being necessary for the development of correct software. Software should be **Designed by Contract (DbC)**. In DbC, assertions are created and performed before any logic is executed. A software contract consists of preconditions, postconditions, and invariants. These specifications are referred to as contracts. A Hoare triple {https://packt-debp.link/UQoc6H} formalizes software contract obligations. These obligations can be summarized in assertions that validate the three parts of the contract:

- Expectation (the data contact)

- Guarantee (the SLA for correctness)

- Maintenance (the SLA for availability, timeliness, and so on)

What is great about the DbC approach is that the task of ensuring software correctness is moved back to the design stage and not left as an afterthought. Software and operational quality are the basis of the formalized contract. This will complement your operational goals for running and managing a robust system. The developed code is guaranteed to work as expected before being rolled out at the end of your CI/CD pipeline. Note that the following is true for contracts:

- They may be written as code comments, or formatted in such a way as to be leveraged in your test framework that reads those code comments as test parameters that execute DbC assertions

- They are evident in a number of languages today such as C, C++, and C#, as well as Java and Python (use the -**O** option)

- They are seen in modern tooling such as in Microsoft's Code Contracts {https://packt-debp.link/6bG6Jg}, Microsoft SPEC# {https://packt-debp.link/ZrNxt8}, Java Modeling Language {https://packt-debp.link/UlVSu2}, and UML's Object Constraint Language {https://packt-debp.link/BiA70f}

The DbC approach cannot replace software testing processes or data quality testing. However, the DbC approach *can* ensure that the software executes correctly when run because it was designed to preserve all necessary assumptions and then check results for operational correctness. The result is that there are fewer release iterations needed to get a solution working 100% correctly; the benefits are obvious to IT operations professionals.

Automated data and system anomaly detection and remediation

With the foundation that has been presented so far in this chapter, you are equipped with knowledge of the following:

- The need for ITIL cloud processes

- Inversion of Control (IoC)

- Design by Contract (DbC

- Data contracts

- Service Level Agreements (SLAs)

- Formal, static, and dynamic testing

- The need for observability over generic systems monitoring

- Awareness of the benefits of design time operational state management

- A clear pattern for the creation of a DevOps pipeline

- Implementing your check-in triggered CI/CD processes

Implementing each of these into your designs makes it possible for you to engineer an operational system that is robust. Together, they enable you to build vast amounts of business logic quickly and efficiently once the operational and development framework for the code's system is available. Building the tough stuff first, with minimal cloning of code and with maximum robustness, will make it cost-effective. Building a framework once and reusing it causes you to centralize the operational features of the solution.

The capabilities to scale and perform at scale, without having those capabilities implemented differently across the system, are necessary deliverables for your efforts. Seeing operational code sprinkled throughout the code base and often built differently by every new member of the team is disheartening. Placing non-functional requirements into a core framework removes it from being entangled with your business logic.

There are other capabilities that you will want to add to the framework as you flesh it out. However, we invite you to read through the various ITIL service definitions and apply the same questions that the ITIL framework asks so that you can add the answers to your operational approach. Only you can determine what is most important to your unique enterprise and then decide what is the most important priority to add first.

Data system anomaly detection

One particularly important feature of robust IT operations is data anomaly detection.

Being able to detect data anomalies is a key framework capability. Applying anomaly detection to the telemetry data collected is also wise. It is the logic that you create that decides whether what you observe will fit known patterns (or not). Additionally, if an unknown pattern is detected, then you have a potential anomaly. We say *potentially* because what is observed may be a new permissible characteristic of the data or system being observed. Some new patterns should become part of a set of filters so that over time, the monitoring for observed patterns improves. Separating out acceptable behavior (the known and known unknown patterns) leaves only the unacceptable behaviors (the unknown unknowns) to deal with. This applies to all data: observable operational telemetric data, core business data, consumable data (the output of the data factory you are building), and all the data's metadata.

In *Chapter 10*, *Key Considerations for Quality Testing Best Practices*, we elaborated on the need for deterministic and stochastic methods for finding anomalies in data. Quality testing is to be applied to all data comprised of core business data, gathered telemetry data outputs from the running system, or system-derived data (trends, metrics, or measures). You may want to move beyond building this capability yourself for telemetry data and instead leverage tools developed by others.

The **OpenTelemetry** (**OTel**) {https://packt-debp.link/JkRgTO} project integrates with many third-party observability products and offers some unique capabilities. OTel is the result of two merged projects: **OpenTracing** and **OpenCensus**. The lack of clear standards for code instrumentation and gathering telemetry has given rise to OTel. The standard's goal is to provide for the generation, collection, management, and exportation of telemetry data for storage and visualization in other tools. The **OTel primer** {https://packt-debp.link/imZELH} is a great place for you to dive into the standard and what needs to be instrumented. The standard has also begun to incorporate common cloud **Function-as-a-Service** (**FaaS**) offerings such as AWS Lambda Auto-Instrumentation, AWS Lambda Collector Configuration, and AWS Lambda Manual Instrumentation.

The **OTel Collector** is a generic non-vendor-specific data collector implementation to receive, process, and export telemetry data. It has scalability features, and it supports popular open source observability data formats, such as the following:

- Jaeger {https://packt-debp.link/MuVMV9}

- Prometheus {https://packt-debp.link/vnEmdY}

- Fluent Bit {https://packt-debp.link/xtRDil}

Refer to *Figure 11.3* for a diagram of the collection mechanism:

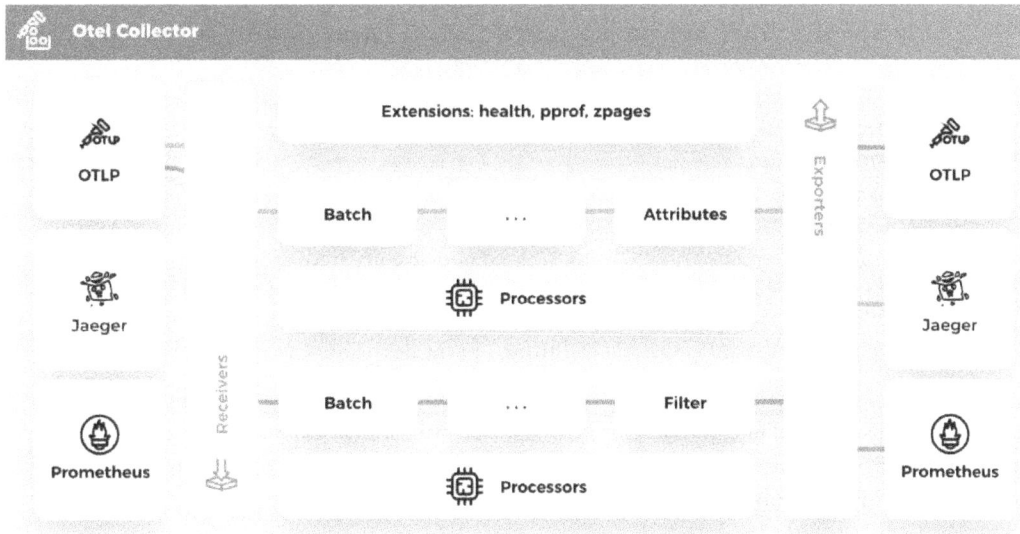

Figure 11.3 – OpenTelemetry (OTel) Collector

Here are a few statements from cloud providers regarding OTel.

Microsoft Azure's position on OTel in the Azure Cloud is as follows:

> *"While we see OpenTelemetry as our future direction, we have no plans to stop collecting data from older SDKs. We still have a way to go before our Azure OpenTelemetry Distros reach feature parity with our Application Insights SDKs. In many cases, customers continue to choose to use Application Insights SDKs for quite some time."*

> *– September 12, 2023, in "Data Collection Basics of Azure Monitor Application Insights"*
> {https://packt-debp.link/gr2yWP}

Amazon supports OTel as identified here in AWS Distro for OTel:

> *"… is a secure, production-ready, AWS-supported distribution of the OpenTelemetry project. With AWS Distro for OpenTelemetry, you can instrument your applications just once to send correlated metrics and traces to multiple AWS and Partner monitoring solutions. Use auto-instrumentation agents to collect traces without changing your code. AWS Distro for OpenTelemetry also collects metadata from your AWS resources and managed services, so you can correlate application performance data with underlying infrastructure data, reducing the mean time to problem resolution. Use AWS Distro for OpenTelemetry to instrument your applications running on* **Amazon Elastic Compute Cloud *(EC2)*, Amazon Elastic Container Service *(ECS)*,** *and* **Amazon Elastic Kubernetes Service *(EKS)*** *on EC2, AWS Fargate, and AWS Lambda, as well as on-premises."*

> *– Amazon Web Services, 2023* {https://packt-debp.link/KArclh}

Google also supports OTel as indicated here:

> *"As a developer, IT operator, DevOps engineer, or SRE (site reliability engineer), you are responsible for the performance and health of applications that you build or operate. The information that you will use to determine whether an application is healthy and performing as designed is called telemetry data. While technology providers have created agents to collect telemetry data, using these agents can tie you to those providers. OpenTelemetry creates both a single open standard for telemetry data and the technology to collect and export data from cloud-native applications so it can be monitored and analyzed, Learn about how 'OpenTelemetry works with Google Cloud's operations suite' (https://cloud.google.com/trace/docs/setup) for monitoring and analyzing cloud-native applications and infrastructure running on Google Cloud."*

> *– Google Cloud Platform, 2023* {https://packt-debp.link/Cy4SjA}

Collecting observable data rather than monitoring data is essential. OTel has advanced the field through collaboration and successful integration. A system that runs well at first will not always run well in the future. A complement to the creation of observable telemetry is the addition of **Application Performance Monitoring** (**APM**) for performance management.

Application Performance Monitoring

Application Performance Monitoring (**APM**) is an area of great concern for an IT operational group manager and for the **Site Reliability Engineers** (**SREs**) reporting to that manager. APM should also be an area of great concern for data engineers. All too often, it is taken for granted that a query or schema scales in order to support the volume levels needed in the future. They don't do this because tests were not provided that enable a window of correct operations to be set at the time the queries were built. Even when database statistics are regenerated at periodic intervals during the day or week, the non-functional scale requirements can be violated, or the overall performance SLA may be violated at apparently random times. These observable random query timeouts, delays, or slowdowns will affect the chain of query executions supporting application use cases. The total time of that use case is subject to an SLA for acceptable behavior.

APM tooling helps identify baseline performance and deviations from that baseline over time. APM tooling helps the engineer know when delays are being encountered and when optimizations are required. It's a best practice to detect scale and performance issues early and then automatically execute probes to see whether performance assumptions are being met in production.

With performance regression testing and incremental small optimizations, you can expect to see positive effects. Data and software engineers are still computer scientists at heart. The flow of hypothesis, test, observe, validate, and repeat is ingrained in our thinking. Our suggestion is: do not forget your roots!

In the process of taking baby steps, negative effects will appear in the solution that you thought should work as expected. If you create an ability to collect the effects of your trial and error, you will also appreciate the need to collect these as observable data in real time when in production. A small change in a query, the creation of an SQL query hint, the use of a new index, the removal of compound column indices, or data cache reorganization have drastic effects on the total observed performance of a use case. The individual who makes these changes may not be aware of potential negative side effects. You will want to look at the total use case scale and performance needs of your use cases and then be defensive in your designs and programming. Collect the complex system's end-to-end chain of dependencies as needed by the use case, and then render a dashboard summary performance across all use cases. Put red, yellow, and green dashboard lights on each use case and create tests that run within the assertions that preserve the contract required for correct operations.

Even with proper recording and scientific analysis, reassessment of change impact needs to result in a positive result after each data reorganization or query optimization. Often, the cost of maintaining a fully loaded production twin test environment (as in, a blue/green deployment ecosystem) is expensive but necessary. A twin with 100% or even 50% production data copy (obfuscated, of course, for PII and privacy concerns) is optimal. However, if this is not possible due to cost, then you will need a lot more people and strict processes to not affect end users when enacting data reorganization.

Data optimization testing in production will affect end users. Being able to revert quickly is the only way to recover from what could be world-facing problems when testing in production. Knowing the tradeoffs of testing in production and getting everyone in agreement with the highly controlled process is essential.

Revalidating all DataOps scale and performance agreements when senior management changes are made is doubly essential. IT operational management teams, as well as data engineering teams, will be held accountable for violations in data contracts or SLAs. Both are responsible and accountable for making sure data is serviced, even when the following happens:

- Data volumes increase (or decrease)

- User load increases (or decreases)

- Cloud IaC auto scaling (or scale back) events are triggered

- State transitions such as site failover, availability zone failover, or application (with schema) deployment (or rollback) are ongoing

You will want to place APM tooling into your design to help preserve scale and performance SLAs. The tool could gain a life of its own if its features begin to drive your operational processes rather than your IT operational designs drive its integration.

Application Performance Management (APM) tools

Some of the best APM tools are identified here:

- *Amazon CloudWatch*

- *Cisco's AppDynamics*

- *Datadog*

- *Dynatrace*

- *IBMs Instana*

- *Loupe*

- *Microsoft Azure Application Insights*
- *New Relic*
- *Pinpoint*
- *Scouter*
- *Stackify Retrace*
- *StageMonitor*
- *Sumologic APM*

Details on each of the product's capabilities follow.

Amazon CloudWatch {https://packt-debp.link/QOKypS} monitors AWS resources and, in real time, the applications run on AWS. It is used to collect metrics that enable you to measure AWS resource consumption and application usage. Amazon CloudWatch's monitoring page depicts a summary of metrics collected and used in the account. Dashboards are highly customizable and there are many standard metrics that you can choose from when building custom dashboards. What is advantageous is that alarms can be created and used to send notifications or activate logic to affect resource changes (such as scaling up or down via **Infrastructure as Code (IaC)**) as thresholds are met. Amazon CloudWatch provides visibility into application-specific cloud utilization and application performance, as well as a system's health.

Cisco's AppDynamics {https://packt-debp.link/0RCCUl} provides capabilities to manage and monitor application solutions. Mobile and browser traffic is mapped to backend databases/server processing for end-to-end monitoring coverage. AppDynamics is distributed, and agents collect and then forward metrics to a core AppDynamics Controller that aggregates data to form a baseline for your desired metrics. Performance deviations to an established baseline are alertable. The alert captures a snapshot of the context that contains details down to the code line causing the detected anomaly.

Datadog {https://packt-debp.link/UDyZaj} allows you to collect and then visualize both frontend use cases and backend service utilization in a single pane. Many services and application-specific functionalities are united in the APM metrics and on the dashboard. The tool provides deployment tracking and detects performance regression after code deploys. This tool understands that a solution changes and that interpreting observability output between deployed versions is necessary to determine the operating correctness of a solution.

Dynatrace {https://packt-debp.link/tlIaEv} is a large APM vendor with over 650 supported technologies. It is considered to be open and extensible. It integrates with major cloud platforms and solutions, such as the following:

- AWS
- Cloud Foundry
- Google Cloud Platform
- Heroku

- Kubernetes

- Microsoft Azure

- Oracle Cloud

- Red Hat OpenShift

- Red Hat OpenStack

- SAP Business Technology Platform

- VMware Cloud on AWS

- VMware Tanzu

IBM Instana {https://packt-debp.link/1dxdA0} provides microservice monitoring to implement a solution's microservices observability requirements. This is accomplished via automation, which includes application discovery, agent deployment, and monitoring configuration. Instana is able to minimize the effort needed to identify the root cause of performance problems.

Loupe {https://packt-debp.link/jBW7fv} is a combined logging and error reporting tool with centralized log management. Via automated error analysis, you are able to perform root cause analysis of runtime errors quickly. With this capability, performance issues can be identified quickly. The tool is optimized for use in a production setting since it records data into memory and then flushes output to a local disk.

Microsoft Azure Application Insights {https://packt-debp.link/hpDCat} enables you to collect, analyze, and respond to cloud telemetry data from various environments. Azure Monitor can scale well. It can help maximize a system's performance and availability by proactively identifying problems before they arise.

New Relic {https://packt-debp.link/GQVFs7} offers an all-in-one observability solution. New Relic is a well-known tool for APM and observability. Data for engineers is provided to monitor, debug, and improve a deployed stack. New Relic may be configured to leverage OTel. It also comes with more than 700 integrations. New Relic APM 360 is the New Relic APM tool. Data can be ingested in many ways, including via New Relic agents or via OTel.

Pinpoint {https://packt-debp.link/kOTvaw} is an open source APM tool for large-scale distributed systems. It is written in Java, PHP, and Python. Pinpoint was inspired by Dapper {https://packt-debp.link/tPd1vh}. Pinpoint provides the capability to analyze the overall structure of a system and how components are interconnected. It enables you to trace transactions across distributed applications.

Scouter {https://packt-debp.link/DdBVGG} is also an open source APM tool. The monitoring targets supported by the Scouter Agent are as follows:

- A web application (running in Tomcat, JBoss, Resin, and so on)

- Standalone Java applications

- Host Agents (running on Linux, Windows, and Unix)

Stackify Retrace {https://packt-debp.link/uSY9YA} is an APM and observability platform with the capability to obtain code-level tracing. It is a full life cycle APM. Stackify Retrace will do the following:

- Aggregate application and server logs and provide enhanced search across them with drill-down

- Consolidate web requests, SQL queries, and HTTP calls

- Associate logs with transaction trace outputs

- Provide views for searching and log drill-down

- Analyze log tags and structured logs for analysis

- Configure/monitor customized log queries for proactive alerting

StageMonitor {https://packt-debp.link/fNvz2l} is another open source solution for APM of Java server applications. It deploys as a Java monitoring agent that integrates tightly with time series databases such as Elasticsearch, Graphite, and InfluxDB. It is used to analyze metrics and Kibana data for request-to-call-stack dependency. Preconfigured Grafana and Kibana dashboards are provided. Since all code is available, the tool may be customized by the user.

Sumologic APM {https://packt-debp.link/xtQNYx} is an application performance monitoring solution to monitor user activity, span analytics, service maps, and transaction traces between microservices.

In addition to APM tools, there are observability and monitoring tools that are pressing into the APM space.

Observability and monitoring tools

Other observability and monitoring tools are identified here:

- Better Stack

- Dotcom-monitor

- Graphite

- Splunk

- Sematext

- SigNoz

- Site 24x7

- Zabbix

Some more details on each of the product's capabilities follow.

Better Stack {https://packt-debp.link/QqgiQO} has its features rooted in technologies supplied by ClickHouse. As a result, it is very fast. The integration with Better Stack Uptime allows you to set up synthetic monitors. Collected information is sent to Grafana for UI visualization signal management.

Dotcom-monitor {https://packt-debp.link/V3U2uf} is a tool that provides actual web browser-gathered application monitoring.

Graphite {https://packt-debp.link/7PBXg2} is an open source monitoring solution with three components: Graphite, Carbon, and Whisper. The capability to render graphs is provided via the Graphite Web. Whisper stores data collected by Carbon, which collects data in real time as a time series.

Splunk {https://packt-debp.link/zLgbDC} offers a solution to collect *all* trace data instead of sampled subsets. Splunk APM correlates cloud infrastructure and microservices log data. An AI-driven approach sifts through trace data to highlight any microservices responsible for an observed error. Splunk provides you with the capability to segment transaction data based on many factors, including the cardinality of running resources. OTel integration is also available.

Sematext {https://packt-debp.link/youF5B} has capabilities for uptime and API monitoring across various locations or private networks. Monitoring capabilities include DNS, TCP, SSL, and HTTP protocols, as well as customizable monitoring for website performance. Sematext provides a single pane of glass UI that makes data correlation easier.

SigNoz {https://packt-debp.link/3LnoSz} is an open source observability platform built with OTel and Clickhouse, in a single pane of glass.

Site 24x7 {https://packt-debp.link/OpIy9C} is powered by AI and provides an easy all-in-one solution for website monitoring. It has basic functionality compared to its competitors, but it can get you started! It's reliable and quick but it does not deliver the same depth of capabilities or features as the competition.

Zabbix {https://packt-debp.link/VN1kcU} allows you to collect metrics from many sources. It allows you to automatically detect problem states within the incoming collected metrics data flow. Intelligent threshold definition options enable a tunable alerting capability. Zabbix also has a large partner network to make your integration easier.

There are many options to select from when choosing an APM tool. Choices will be directed by your budget as well as the type of observability you need to process given the architecture of your solution and the platform selected for IT operations. Many APM problems in a data factory can be designed out if more thought is given to the solution's non-functional requirements – but not all of them! There will always be unknown unknowns that will cause issues; many of these come from bugs (features) that arise when using cloud, third-party, or home-grown dependent code. A good APM tool will be worth its cost *when*, not *if*, problems arise in your products' IT operations.

You will want to address IT operational issues when deploying systems in the next section. There are various schools of thought on this topic and many books, postings, and blogs exist. The advice needs to be sorted through but we have some suggestions to consider in this area.

Blue/green versus continuous deployment trade-offs

Regarding the deployment of a solution, you will find that the IT operations folks sometimes insist on a big switch process. This is where the initial release and all subsequent release iterations are fully tested, able to process data in a 100% backward compatible manner, and can be reverted back to a *last known good* state within a given burn-in period. This enables IT operations to literally throw a big switch forward and then backward with ease if required. IT operations management will want DevOps folks to develop a solution in their own environment (the *green* environment) and then flip out the *blue* environment as part of a *blue/green* deployment. Note that both environments are considered to be production-ready with the hope that after deployment, the vestigial environment will be shut down to save costs. This approach checks off many operational readiness criteria that you need to develop, such as the following:

- Developed code

- Scripting

- Cloud function scripting

- Regression testing

- Acceptance testing

- Infrastructure as software (via IaC)

- Runbooks run/manage analytics scenarios

- Rollout/digression scripts

- Data schema change/digression scripts

- Transaction capture during change log handling

- Backward compatible observability (log formats, trace, metrics (dashboards), and so on)

What is nice about the *blue/green* deployment model is that it is a clear demarcation between developer (DataOps and DevOps Software Engineers) responsibilities and IT operation (FinOps, ITOps, DevSecOps, and SRE) responsibilities. The picture for a clean, tested handoff is very attractive, although it is an expensive way to implement production handoffs. There are benefits to this type of deployment – you can carve off a third environment for disaster recovery and deploy to any cloud-provided region with ease, and you can also assess a valid **Mean Time to Repair** (**MTTR**) OKR with this approach.

When performing forensic data analysis, a new environment can be cloned from the production environment (running a labeled release; call it *orange*) and launched for DevOps QA, as well as DQM analysis. IT operations professionals rarely want DevOps folks to access their production environments for security and privacy concerns. Large enterprises will forbid access or tightly constrain access to view -only access when working with production systems.

There are ways to avoid blue/green deployment for solutions that are not data volume intensive. When the pain of data corruption goes up, compounded by the presence of a huge volume of data, you will see that creating solution segments that can be deployed using different models is optimal. This drifts from having a single deployment model for the entire solution (blue/green deployment). This allows you to embrace the model best fit for some components or sub-solutions (such as analytics GUIs versus backend data factory zone processing versus data profiling at ingest, and so on).

Some parts of a solution are best serviced with a *continuous* deployment because small deltas and a continuous stream of these changes are not disruptive to the process flow. Streaming deployments can enhance the user experience through the application of a rapid and non-destabilizing series of changes that positively enhance the user experience. This is not the case where large volumes of data have to be changed and validated in bulk.

The choice between *continuous* deployment and *blue/green* deployment is a natural one for releases that affect vast volumes of data. Data needs to be released as consistent sets, so the blue/green deployment approach works better. Testing is required as small, frequent changes still require large data access and long running validation logic that is best accommodated by the blue/green deployment model. After all, who can guarantee that a streaming set of features affecting a huge data set will not adversely affect overall data quality? Data engineering requires data to be correct when consumed; software changes can take some time to affect changes in a data state as algorithms wrestle the data down to a final, steady state.

When sub-components have guaranteed backward-compatible interfaces, then choices may be made in the deployment model. So long as changes are not going to affect the rest of the solution, old and new versions can be exchanged without adverse effects. Your data factory is built with this in mind since data propagates through zones and the processing in each zone is unique to that zone. This way, a zone could be updated without affecting another.

We have not gotten into other deployment models such as **canary release** {https://packt-debp.link/uhj85M} because that is a twist on continuous deployment. Also, it poses problems for data factory solutions or just about any backend solution where data has to be segmented to support parallel versions of the code that affect that data. With this segmentation, you also have to offload old versions of code while still trying to meet your overall performance OKRs.

So, what are some continuous deployment tools available? This is a list of some of the options:

- Atlassian Bamboo
- AWS Code Deploy
- Azure DevOps (Pipelines)
- ClickUp
- CircleCI
- Codeship
- GitLab
- Jenkins
- Octopus Deploy
- Puppet
- Travis CI

Reading through the capabilities that the vendors offer will provide you with a good idea of what is possible with a proper tool and process definition.

Atlassian Bamboo {https://packt-debp.link/xCMCxC} is a **Continuous Delivery** (**CD**) tool with capabilities such as workflow automation, native disaster recovery, high availability, and elastic scalability to maintain performance. Atlassian's Bamboo, Bitbucket, and, Jira Software are fully integrated so there is traceability from a features request to deployment.

AWS Code Deploy {https://packt-debp.link/s1NwHm} is part of Amazon Web Services. AWS services can be deployed with your code quickly. CodeDeploy also has the capability to work with on-premises servers.

Azure DevOps (Pipelines) {https://packt-debp.link/aZvqbn} is part of Microsoft Azure Cloud Services and it allows you to build, test, and then deploy Node.js, Python, Java, PHP, Ruby, C/C++, .NET, Android, and iOS apps. Integration exists for container registries such as Docker Hub and Azure Container Registry, or deployment for containers on individual hosts or to Kubernetes. Azure Pipelines support is now available for other cloud providers such as AWS and GCP.

ClickUp {https://packt-debp.link/4oh6BR} allows you to manage work from many different angles (aka 15+ views). It is primarily a project management tool and through its integration with other cloud DevOps and CI/CD tools, it can be turned into a powerful CI/CD tool.

CircleCI {https://packt-debp.link/F1szbs} is a performance-optimized solution for deployment with integration to version control systems. It has Docker support to enable micro-service developers to keep development and production environments consistent. Workflows can be enhanced to provide flexibility, which is required for most CI/CD process implementations.

Codeship {https://packt-debp.link/giA989} is a hosted CD service. It can support complex workflows, but it is a relatively simple tool. **Continuous Integration** (**CI**) can be set up in just a few steps. Auto deployment takes place after tests have been completed successfully. GitHub and various test framework integrations exist for Codeship.

GitLab {https://packt-debp.link/5cIgQB} (versus Microsoft's GitHub) has been enhanced into a DevOps tool by the open source community for various phases of the SDLC. A CI/CD pipeline now exists so that when developers commit changes to GitLab, a native CI/CD process is called. Therefore, you don't need to rely on external CI/CD tool providers. New features are always coming from the GitLab open source teams.

Jenkins {https://packt-debp.link/hhlT2R} is an extensible automation server used as a simple CI or as a CI/CD hub for your projects. Jenkins can be extended via its plugin architecture. There are more than 1,800 community-contributed plugins developed to date. Jenkins on Google Cloud {https://packt-debp.link/7J9Mfe} and Jenkins on AWS {https://packt-debp.link/Njlk5o} provide information about cloud service integration.

Octopus Deploy {https://packt-debp.link/gOT4KX} is a release management tool for CD solutions. It simplifies the deployment processes. Automation and release management capabilities are some of the product's focus areas. Deployments across testing, staging, and production environments are managed so that advanced deployment processes, such as canary release and blue/green or CDs, are possible. Octopus implements DORA {https://packt-debp.link/NyOVxA} metrics and supports automated runbooks.

Puppet {https://packt-debp.link/OCgG7n} is a continuous deployment tool that allows DevOps engineers to configure and deploy IaC. Its capabilities focus on automated provisioning. Reporting features provide insight into the CI/CD process for issue resolution. Integration with common cloud platforms is available. **Role-Based Access Control** (**RBAC**) is a key feature.

Travis CI {https://packt-debp.link/cI0YNv} automates application deployments with advanced features such as matrix builds (multiple environment testing), as well as automatic code deployment to target environments with highly customizable workflows and integrations. It provides Docker and multi-language support.

The choices for CI/CD, with emphasis on the *D* for *deployment*, are prolific. Each tool is very customizable. A deployment goal is that the final built package is to be handed off to the IT operations manager at the end of a pipeline. That handoff must include the pipeline configuration itself because deployed code can only be understood and recreated in the context of the configured pipeline. We have seen systems built and deployed without this clean packaging and deployment definition. As a result, deployed code could never be recreated the same way twice. You do *not* want that to happen to your solution. The IT operations manager has to agree that what is deployed meets the enterprise/company's IT policy standards and that the production IT support functions do not cross the mandated organizational separation of responsibility standards.

Cloud services and Agile process fanatics sometimes do not make the choices easy. Engineers *want* and *need* full control of the DevOps pipelines running in some CI/CD platform offerings. With that, they need the ability to adjust and reconfigure workflows as a solution matures. At some point, the solution and its CI/CD processes should be frozen (by taking a snapshot) and not be so flexible as to not provide the handle needed for IT operations. This handle is required when the built and packaged solution needs to be transitioned to others.

This is part of the operational readiness assessment usually required by the IT managers' SREs. These SREs are responsible for provisioning and then the first (**L1**) and maybe even the second (**L2**) level of support aligned to the ITIL problem management service guidelines. To accomplish this task, SREs will need to know what is in the package and how it was built in order to take control of the solution in the designated production environment.

Key takeaways

As a takeaway from this chapter, you will want to keep IT operations in the front rather than the back of your mind as you build out your future-proof data-engineered data factory solution that can stand the test of time. What your vision is for a *good* IT operational model will be stressed when you have to balance various stakeholder demands and divergent technical cultures even within your own company.

Everyone will want to leverage *their* own experience. Part of your job will be to get them to see the big picture first, and then knit together a new plan of action based on the shared vision. Some preexisting organizational policies and governance models will constrain your mission, and for that, we can only wish you much luck, as will you have to make compromises and sell those compromises to all your constituents. You will win some battles and lose others. For sure, change will remain constant, so be Agile, but just do not build something that is fragile.

You will have to select products that fit your IT operational processes and balance the need for development velocity against SLAs and technical contract (software and data) demands in order to preserve quality through your orchestrated operational processes. You will have to fight against detractors who just want to get IT done and forget about operations, observability, security, and audits. They will have *loud* voices and the business may even back this cowboy development rhetoric, so beware. You may want to do things *right* but you will have to make adjustments. Then you will have to stand behind suboptimal processes brought about through compromise. This will happen with toolset selections, as well as OKR meeting implementations and core designs. Just come back the next day, pick up the banner again, and fight the battles that need to be fought in order to keep delivering. Keep up with the deluge of releases that will form the update stream of your evolving solution.

You have learned that not all tools fit your IT operational process vision and that you will have to select a set of tools that does! Then you will have to define how to use and how not to use that set of tools in order to meet the architecture's goals. You will even want to use terms from the past, such as ITIL, when the cloud vendors state other approaches in their well architected frameworks.

Our advice is to *stick to your guns!* You're the new sheriff in town, and as sheriff, you need to keep to the agreed-upon IT operational governance standards and processes. Never depart from them. Otherwise, if you blink, they will take that moment to undermine the very processes that lead to successful IT operations. You will need to be part of the effort to distill the essence of the questions posed by the legacy ITIL services. These still need to be answered for present cloud technologies. Do not compromise until the answers map to your needs. You can have many dialogues with consulting experts, but if you ask two or more people, you will get two or more answers, even from the same cloud vendor! Beware of answers that do not fit your need – there will be many of them.

Focus on your needs. The first and most important one is to *keep the lights on* after the DevOps, DataOps, and SecOps folks have declared their victory (and achieved a degree of doneness) and launched a set of packages over the IT operational wall. Note that there *will* be a wall. Most enterprises will not let the cowboys into the corral until they can show that they are conducting their business in a well-behaved and consistent manner. This involves building all those non-functional requirements that the ITIL services demand to be built, such as audit, security, compliance, and a PII development firewall. This issue has arisen in *every organization* we have encountered over the years of our combined careers. The DevOps and DataOps processes need to account for this reality. They should not be built to support startups, small businesses, or even medium-sized businesses, but rather, they have to be able to mature to support large (and huge) multinational organizations.

SLAs, data contracts, and design-by-contract mandates will be delivered to DevOps teams from some central data, software, finance, and security architecture groups. These center of excellence groups set up a solid wall of non-functional requirements that should be mandated by policy-driven governance and enforced once they are adopted. Testing is essential to preserve data quality, but this testing needs structure and has to be *shifted left* as well as enforced operationally (at runtime) by contracts. The CI/CD pipeline has to be automated to drive these tests and eventual packaging and deployment to an environment for regression testing. The real time/run time execution of these tests results in *assertions of correctness*, which enforce the contracts that define the agreement for correct operations. IT managers and SREs should not just accept IT releases but rather, they need to be part of the development SDLC process and co-develop IT releases. This way, they know what they are getting at the time of operational readiness and have steered the solutions releases to meet their non-functional operational expectations. One key takeaway is that the solution must be *observable* in order to effectively implement problem management, change, and release management services of the IT organization. These non-functional service requirements will be enforced in a system's primary mode of operation and then in a disaster or failover degraded mode. Therefore, there is time to react to real IT issues before they become customer-facing issues. Build solutions that directly address the real-world problems that you know will occur. Do not leave these requirements as afterthoughts.

Treat data as a product, and with that emphasis, it will become reliable! The effect of a bad package deployment is that the data can become corrupt. Never allow manual data tweaks to take place. Also, make sure that package deployment *progression* and *digression* scripts are created and executed. If and when a data fix is needed, then that fix can only be made with a new software release that, itself, has been tested and proven to be able to set the data back to a known good state. Be ready to detect data anomalies in data products by executing the proper level of DQM testing. This way, data contracts are enforced. Note that these tests will form a safety net to catch released software packages that could break existing data contracts. Having them run in a production environment saves you a lot of pain.

When you finish with the development of testing, and in particular DQM testing, you will want to capture the baseline scale and performance characteristics of your solution. You will want to be alerted to deviations from this baseline norm. APM tooling will help you detect and diagnose these deviations, and they will occur. Data volumes increase, data drift is constant, and software errors sometimes appear much later than the burn-in period after a release has passed full-scale rollout. It is your IT operational responsibility to build the operational characteristics of the solution to address this kind of issue early. If you do, then you will not have to worry when issues arise, and they will. You will not have customers walking away from your enterprise solution when you build out your IT operational processes to be sustainable.

After your solution is in production and the cloud ITIL processes are working and effective, the tools set selected and working as expected, and all IT operational non-functional requirements are complete, you will think that you have been successful. This is true until the DevOps/DataOps folks release something new, yet again! IT operations SREs have to validate each release for operational readiness, for each and every release, that is! Change is good; too much is bad because it raises the probability of failure, and just enough change at just the right time is the optimal solution. Automation, tooling, and processes reduce the SREs' pain. It is common knowledge that as humans, when we walk, *each step causes us to teeter on the edge of disaster*. Our body is a fine-tuned organism with feedback loops and command and control that we take for granted. If you are in IT operations, be prepared to be taken for granted as well. What keeps us upright in our body is a brain that is constantly sending micro-corrections to our feet in order to forestall the disaster of a fall. This thought may make you uncomfortable, but it works for our body, and you are building, evolving, and constantly changing an IT system to work in a similar manner. IT even has (or will have) AI capabilities in the future. IT will be subject to a multiplicity of releases (steps) and corrective changes (adjustments) in order to remain operational (upright). One final thought for the data engineer: *help them help you!* Work as a team! IT operations is not the enemy!

Summary

In this chapter, you explored many operational topics and were exposed to the best practices for SLA setting and management for important contract terms. Data contracts were defined, and references were provided for further reading. All of this led to the development of quality data factory outputs. Solution monitoring with observability was stressed and the necessary capabilities were elaborated upon. You were also exposed to the essential need for data anomaly detection so that you can detect data drift and gross data errors before they get into the factory's curated data product. Lastly, you saw how blue/green deployment versus other release practices affect the data in the factory and why operational tradeoffs will have to be negotiated with the software developers as part of your data engineering solution.

In the next chapter, you will be provided with best practices to build out the framework for your data services.

12

Key Considerations for Data Service Best Practices

When thinking about the complexities of data engineering, you may want to grab hold of a compelling metaphor in your mind as you dive into the details. Your goal is to press the processed information and derived knowledge out of the raw data so that you give the consumer the ability to glean new insights. It's necessary that you create effective data services with today's technologies and prepare to handle future technology innovation as it arises. Data services insulate data and information from the consumer. The consumer could be a subsequent processing step in a data factory's data flow, an algorithm, or an external data analyst. Collecting curated data can be thought of as a multifaceted diamond.

The processed diamond speaks to both its inherent value and the complex craftsmanship required to unlock its potential. This craftsmanship, essential to the data service design, demands more than just an appreciation of the diamond's facets. It requires a deep engagement with the material, enabling practitioners to both navigate its complexities and utilize its strengths. True artistry lies in your capacity to walk around the diamond, understand its structure, and use it by leveraging its properties. This duality encapsulates the essence of mastery in data services.

As we embark on this exploration, we'll delve into the intricate dance of shaping, polishing, and ultimately employing data to serve a myriad of purposes. From securing sensitive data to navigating the complexities of global compliance, the challenges are as diverse as they are imperative.

Data service best practices overview

This chapter begins with an appreciation for the facets that bring out the beauty of a diamond. In our case, the collected knowledge of the curated data release sets and the associated metadata form the raw material. This has great worth, but only if it is accessed through refined perspectives is the insight teased out of the material. Think of it as a lump of coal if it's pressed, solidified into a lattice, and formed by the chemical properties of carbon molecules (the hexahedron). Your data, information, and eventually knowledge are formed under the pressure of the data factories processing it into its fit-for-purpose form, which is either explicitly built as a **semantic graph** or implicitly aligned with the business domain's semantics, which could be modeled as a semantic graph. The result is the creation of the data's **knowledge graph** (**KG**). This knowledge graph is analogous to the hexahedron lattice of a lump of coal but aligned with your definition of the domain. Repeating instances of objects formed by the graph are called the knowledge base. This knowledge is a formal population of objects according to the rules defined in the domain's semantics. For example, red is a color and an attribute, and it can be given to a fruit, such as an apple, as can the color green. The semantic rules dictate that purple cannot be ascribed to an apple. Likewise, the color black may be ascribed if the variety attribute type is "black diamond," with synonyms such as "Huaniu apple" and "Chinese Red Delicious apple." Dark purple can be ascribed the color so long as the modifier "dark" is ascribed and the other conditions are met. You can see how much logic goes into defining the color of an apple! Imagine the coded logic needed to make sure data values are set correctly for your simple attributes. Data should be smart and contain these rules if we are to reduce semantic errors and ensure information is understood and used properly. The data is valuable so long as it is defined correctly, and semantics is preserved. But the metaphor of the diamond is even more powerful as we consider the need to create slices through knowledge. These are called **facets**. If two-dimensional directed walk criteria are defined, and a walk of the graph is conducted for one of these criteria, the output will be a **taxonomy**. In our example, that taxonomy could look like a list of fruits with subsections under each fruit being potential colors for that fruit. Various other attributes are omitted from this first taxonomy but can be included in other walk definitions and result in other taxonomies, such as fruit/synonyms, fruits/sweetness, or synonyms/origins. Any number of simple walks can be conducted for a knowledge base, so long as there are no repeating cycles. These facets can be intersected and used to reduce result set options and tease out insights. This faceted intersection is used in many webshopping sites to reduce thousands of product offerings to just the few that are pertinent to the buyer. Who hasn't selected a product on the Amazon shopping site by starting with a broad search and refining the result set by the exposed facets, such as size, color, cost, customer review, and so on? We use this approach daily! Why not commoditize it for all data and make the data smart? This is at the core of what we will be discussing in this chapter regarding building effective data services. It does not dictate any technology direction for building the knowledge base; however, in later chapters, you will see that there is an advantage to having data organized as a semantic graph, and in parallel a scalable data store to accommodate cost and performance concerns until the cloud providers catch up on this need.

There are a few foundational premises that we need to adopt to understand the content in this chapter. In the *"Zero trust versus defense in depth"* section, data national localization, privacy protection, and the management of **personally identifiable information** (**PII**) are not mere afterthoughts – they are integral to the very architecture of your data services. These need to be incorporated at the genesis of your system design effort rather than being tacked on as solutions to emergent problems. This proactive approach is critical in today's digital expanse, where the ubiquity of data not only mandates stringent protections but also strategic utilization.

As we unfold the various layers needed to design your data services, from architectural blueprints to operational strategies, we'll emphasize the need for foresight in design. Whether you choose to implement a data mesh over a data fabric, craft data pipelines needed to curate data for analytics, or navigate the intricacies of machine learning, each decision reflects a commitment to embed security, privacy, and compliance into the core of your system.

This journey through data services is not just a survey of best practices; it is a deep dive into the ethos of data craftsmanship. Every tool selection, every DataOps process, and every governance policy is a deliberate stroke in crafting a polished gem that not only meets our immediate needs but also anticipates future challenges. Thus, this chapter is an invitation for you to approach the design of your data services as a craft. This craft is part art and part engineering. The approach we'll take will challenge you to see beyond the rough exterior of raw data, envisioning the potential within. By designing your system with the metaphor of your knowledge as a diamond, we will design valuable service methods to expose beauty in ways that only craftsmen can envision. The power of best practices for data services will also become evident. So, let's proceed with both vision and meticulous care, guided by the understanding that the most effective use of our data diamond comes not from external adornments but from the quality and foresight of its initial cut.

As we progress, we will pivot away from architectural and operational scaffolds that underpin effective data services. Instead, we will focus on the data to be serviced, create understanding from it, and build domain logic into the service if it doesn't already exist. Since most data we process today is rather raw and not defined with semantics, it is just not smart. The decision to implement a data warehouse, data lake, data mesh, or data fabric, and the intricacies of building data pipelines tailored for analytics versus machine learning, are strategic considerations that determine whether your data services merely exist or excel at their intended purpose. It's here, in the melding of theory and practice, that the craftsmanship of data services comes to life.

Furthermore, we'll extend our exploration of tactical tool selections, DataOps standards, data access, and quality management into the mechanism required to service data effectively. Each decision and standard that we adopt is a stroke of the chisel, shaping the raw data diamond into the shape we require data to be in so that it can become a polished gem that reflects the light of our strategic objectives.

Software and data engineering drivers for best practices

Implementing data services requires you to think differently about data. Instead of knowing what the data is and just acquiring it (as with SQL in an RDBMS), you have to discover how it is stored and what it means and then set up `getter` criteria for access, followed by defining how you want it delivered. Then, you need to ask for it to be supplied and evenly paced so that you aren't drowning in terms of volume. It's a bit like orthoscopic surgery. You can look at what's there – you can even probe it. But when you get it, you can do so only after all the preconditions are set and you better not acquire what you are not permitted to see (or even know exists). You better get it right and know how to interpret the results. Data services can be complex and not just in the development of the API syntax and rudimentary semantics. A lot of your work will go into the definition of how you set up the diamond cuts, followed by polishing (data rendering) before the light of the query reflects what your data consumers want. Data needs to flow outward from the master source in a manner of your choosing and at a rate that you establish for the consumer.

Did you know that darkness reflects no light? It is the absence of light. If you are not permitted to even know that data exists, you better not be able to reflect on the data structure; otherwise, you can infer that it does exist, and you may not be permitted to know that either! You can inadvertently expose some truths via your data service designs if you aren't careful. This can be a violation of privacy, data protection, or corporate security. This happens in some domains, especially finance and health data. Imagine if you knew insurance companies were correlating all health data to assess your health score. Imagine if you knew which enterprise held that type of data. That company is an immediate target for extreme hacking attempts, just because the structure to hold that information was exposed. So, you'll want to implement **zero trust**, **defense in depth**, and **data privacy** very carefully.

Zero trust versus defense in depth

Before we begin, we need to define zero trust and defense in depth (also known as **castle-and-moat**). Your security design pattern and tool choices could lead to insecurity if you let the tool vendors start defining your solution rather than your **chief information security officer** (**CISO**). If you do not have an individual in this role, your enterprise is strongly advised to consider establishing one. With that role should come the power to govern and enforce auditable, standards-compliant security safeguards. A zero trust architecture should be capable of supporting **zero trust principles** {https://packt-debp.link/RNEWFs}. These include the following:

- **Verify explicitly**: This means that you always want to authenticate and authorize access based on available security attributes, such as the user's identity, security profile, location, accessing devices, health and service version, type of data, and prior anomalies.

- **Use least-privilege access**: This will limit a user's access to be **just-in-time** (**JIT**) and **just-enough access** (**JEA**). It also includes risk-based policies and data protection.

- **Assume breaches**: This will make it possible for you to respond quickly to minimize any effects of a security incident. The incident alert triggers an automated network segmentation response. You also will need to verify end-to-end encryption pathways and use collected metrics to ensure your security posture, coordinate all threat detection actions, and improve security responses.

Zero trust is meant to provide the most secure architecture because it does not assume any access path is trusted. Every resource request requires authentication so that access management, device and user authentication, and strong segmentation approaches are assured. Zero trust is different from defense in depth (or castle-and-moat) architecture, which provides an implicit trust for all activities in the protected area once secure access is granted. **Zero trust network access** (**ZTNA**) is a special zero trust use case that secures data access when consuming users, analytic applications, or datasets outside your security zone, or when you are providing access to others external to your service infrastructure. You want to build data services to meet the **ZTNA** requirements. Incorporating zero trust security designs into your data services architecture will fortify security. It demands meticulous design considerations to maintain cloud portability, balancing coding simplicity with the complexity of operational and development landscapes. Microsoft's *"Zero Trust Essentials eBook"* {https://packt-debp.link/vmuwdq} is a great place to start your journey toward implementing a cloud-based zero trust data services architecture. The following security layers are identified by Microsoft:

- **Identity** with authorization and access

- **Endpoints** that provide device authority for smart devices

- **Applications** service access authorization, including **SaaS**

- **Network** segment and connectivity authorization

- **Infrastructure**, including **PaaS**, **IaaS**, and on-premises hosted environments

- **Data entitlements**, rights, fair use, and so on

The five pillars of zero trust {https://packt-debp.link/ZUfVpB} (identity, devices, network, data, and applications and workloads) lead to the creation of two **objectives and key results** (**OKRs**) that are considered to be essential in your designs. These are as follows:

- **Do not disrupt end users**: This means don't let your security design be so onerous to use as to frustrate the user or the established user's flow of information.

- **Provide end-to-end visibility of your secure landscape**: You can do this to ensure you have threat intelligence, risk detection, and command/control over all your conditional access policies (this is no easy task in the cloud).

Implementing a comprehensive modern security design can be frustrating because it's a changing technology landscape. Companies such as ZScaler {https://packt-debp.link/00xGqt} have produced platforms that help resolve the problem, but whether you use a party tool or build zero trust support, study various tools' capabilities first. The integrated solution you implement will be driven by your tolerance for risk, your budget (costs), and the expected time-to-market for your solution. Zscaler's top tier alternatives are listed here:

- Akamai {https://packt-debp.link/BTRRGz}

- BetterCloud {https://packt-debp.link/HnQByH}

- Cisco Cloud Security (Duo and CloudSOC **CASB**) {https://packt-debp.link/xxvd11}

- Citrix Gateway {https://packt-debp.link/csoL7x}

- Check Point Software Technologies {https://packt-debp.link/1JhX2A}

- Cloudflare {https://packt-debp.link/TG6wED}

- Forcepoint Secure Web Gateway {https://packt-debp.link/Z8bWFi}

- Fortinet **CASB** {https://packt-debp.link/0OUAWo}

- Lookout Cloud Security {https://packt-debp.link/wdf8No}

- Netskope {https://packt-debp.link/Ee8Z9a}

- Palo Alto Networks Prisma Access {https://packt-debp.link/MBf1zY}

- Proofpoint {https://packt-debp.link/XWUaNI}

- Skyhigh Security {https://packt-debp.link/85ys9j}

- Symantecs (Broadcoms) Zero Trust {https://packt-debp.link/E7RuE1}

- Trellix {https://packt-debp.link/9zqXvD}

- TrendMicro {https://packt-debp.link/YzMrHQ}

- Zscaler

Additional alternatives to **CASB** product vendors include the following:

- Cato SASE Cloud {https://packt-debp.link/UYteBr}

- Cohesity {https://packt-debp.link/qKXSsK}

- Delinea Secret Server {https://packt-debp.link/4FtauD}

- Forescout Platform {https://packt-debp.link/EYsPKV}

- F5 security solutions {https://packt-debp.link/p7c3pV}

- Kaspersky Security for Internet Gateway {https://packt-debp.link/StbYbQ}

- Rubrik {https://packt-debp.link/GCdAgW}

- Seraphic {https://packt-debp.link/o80nF6}

- Sprinto {https://packt-debp.link/O3DkHg}

- ThreatLocker {https://packt-debp.link/Zaw7aF}

A side effect of adopting and communicating your security architecture as part of your data services architecture is that you will simplify the engineer's coding. This will prevent later refactoring, which is necessary to plug design gaps that will arise in the absence of a solid security design. You will also be preventing cloud portability issues that arise with a unified zero trust design blueprint.

National data localization

Businesses often grow beyond their original national boundaries to become global service providers. When this occurs, they need to be ready to embrace the legal expectations for operating outside their original local constraints. Language, privacy, regulation, auditing, and compliance assumptions are going to have to be flexibly handled in a global business. Even the revenue and corporate tax models are going to be subject to variant tax structures with rules that require data to be collected, processed, and secured in different ways compared with the way they were stored when servicing a single location. You will want to build for today, with an eye on where your business is growing to tomorrow. This affects your data architecture. The granularity of your consumable data has to be such that it can be easily leveraged after globalization requirements drive software data service refactoring. Your data services must be agile and adaptable so that it complies with diverse national localization regulations. These will impact your costs and privacy management strategies as you navigate global data management complexities.

You will be faced with deciding on how to store your data in a language (locale) neutral form or link all localized data versions to a core set that can be manipulated as a single fact representing the underlying concept (otherwise known as semantics). This way, data can be re-interpreted into the data consumer's locale at the data service's edge. This is workable until you have quality issues re-rendering the neutral locale data form into more complex locales for which auto-translation capabilities are insufficient. Keeping parallel locale data versions in your consumable datasets is a practical requirement when the number of locales needed negatively affects overall quality.

Language is not the only neutral form of locale required for transformation. You will be faced with deterministic rules that have to be applied to datasets so that they comply with local government regulations. Various sales and corporate tax rules, accounting, and auditability (record retention) require data to be tagged with locale-dependent metadata if and when locale transformations are applied. This metadata is then used by the locale rules recursively so that rules that are trend or threshold-dependent can be applied correctly. This affects your data structures, metadata collected, and rules base. The best practice is to work with information service providers. For example, business services such as **Avalara** {https://packt-debp.link/Vv9ut1} should be considered for their global sales tax calculation capabilities. The following are similar examples:

- Anrok

- Avalara

- Bloomberg Tax

- Complyt

- countX

- Dutycast

- Fonoa

- KYG Trade

- TaxAct Business Returns

- TaxCloud

- Taxdoo

- TaxJar

- Thompson Reuters ONESOURCE

- Intuits ProConnect Tax

- Sovos Sales And Use Tax Software

- Vertex O Series

> **Note**
>
> We haven't included additional endnotes for these vendor services because they are business-focused rather than engineering-related. As such, they vary in depth and breadth of coverage. You will have to do the research yourself!

There is third-party vendor support for other areas related to global business, such as financial trading, global supply chain, expedited shipping consolidation and planning, and so on. These are beyond the scope of this book, but all require specific data architecture patterns and service integrations that will stress your architecture. You must engineer data service designs to meet the challenges posed.

Privacy protection

Data services require a robust data quality management framework to comply with privacy laws. A key objective will be to ensure that privacy protection is embedded in the data service design and your overall data operations process. These laws need to be interpreted into **data quality management** (**DQM**) tests to ensure compliance. Then, the auditable trace should be retained in production to assure governing agencies of ongoing operational compliance have in-place, but always shifting, privacy laws.

Proper data service designs insulate the consuming user from direct access to released data based on the consumer's profile. That profile should include the user's locale data and privacy regulations. Data privacy violations have risen in punitive severity to the point where sovereign legal protections have been created to govern technology offerings. In the EU's **General Data Protection Regulation** (**GDPR**) and the USA's **California Consumer Privacy Act** (**CCPA**) laws, you can see that consumers are gaining the right to know and decide how their data is collected, stored, processed, and sold. There are so many new laws popping up around the world that you need to know what they are well in advance before violating one inadvertently by allowing a consumer from a controlled region to access your data service.

Knowing all the laws, the rules needed to comply with them, and the auditable user entitlements necessary to comply with these controls is essential. You also have to know when a new law arises that doesn't sound like a privacy law but is! Consider the following laws, all of which constrain the privacy-preserving capabilities of your system:

- **Children's Online Privacy Protection Act** (**COPPA**) is a law that restricts children's (<13 years) personal information from being collected

- **Fair Credit Reporting Act** (**FCRA**) directs how consumer information is collected and used

- **Family Educational Rights and Privacy Act** (**FERPA**) ensures that student records from schools receiving US Department of Education funding are protected

- The **Gramm-Leach-Bliley Act** (**GLBA**) (also known as the Financial Modernization Act) states that financial corporations need to protect customers' data and be able to explain what is done with sensitive data

- **Health Insurance Portability and Accountability Act** (**HIPAA**) requires providers to tell you how protected health information is used

- United States federally scooped or US state-specific laws:

  - **The New York SHIELD Act** protects the personal, private information of New York state residents

  - **Privacy Act of 1974** protects federal agency staff's personal information

To stay current, you must have a **chief data officer** (**CDO**) in your organization who can stay on top of the impact of current and new data privacy legislation in the areas the company operates *and* where citizens are serviced, even if they are insulated from direct data access via a common data service wrapper. At least 35 states (and Puerto Rico) have put data regulations in place. Many of these laws specifically address digital data:

- **California Privacy Rights Act** (**CPRA**): Proposition 24 (2023-01-01)

- **Colorado Privacy Act** (**CPA**): SB 21-190 (2023-07-01)

- **Nevada Internet Privacy Bill** (**SB260**): BDR-52-253 (2021-10-01)

- **Massachusetts Data Privacy Law**: 201 CMR 17.00 (2010-03-01)

- **Minnesota Data Privacy Act**: Minnesota State Chapter 13 (1979)

- **Virginia Consumer Data Protection Act** (**CDPA**): SB-1392 (2023-01-01)

- **Other proposed US state laws**: Ohio, North Carolina, Rhode Island, Pennsylvania, New Jersey, Massachusetts (**MIPA**), Hawaii, and New York (**NYPA**).

Companies such as **NetWrix** {https://packt-debp.link/62Mv47} and **AOSphere** {https://packt-debp.link/tZhqWs} can help align development with current and upcoming legislation. Remember, you are a data engineer, not a lawyer! But you need to know what is expected from your designs if you are supporting a business that is operating in controlled locales. The last thing you want to have happen is that you are told just before going live that your data services solution does not meet legal requirements.

Personally Identifiable Information (PII)

Handling personally identifiable information (PII) data in data services will require special consideration in your designs. There are ways to obfuscate all PII data and there are ways to mask it. A balance in managing PII incorporates both obfuscation and masking techniques. You will also need to develop and maintain an audit trail for compliance purposes. These capabilities ensure an individual's privacy without compromising information accessibility and integrity. Some approaches do remove direct access to raw data and that can't be avoided. Additionally, when a person is to be forgotten, as required by the EU's GDPR, you may be obliged to remove any obfuscated data or data that was derived from the protected entity. You will need to be careful when using information that was derived from protected data.

When obfuscating PII data for privacy compliance, the mechanism that's applied must be non-reversible. However, you may use reversible obfuscating techniques when they're not subject to privacy regulation, such as when you need to create test datasets and you don't want developers to see real data but can under the supervision of the CDO, who can view real PII for compliance or debugging purposes.

In the next section, we'll depart from the discussion on the thorny legal and privacy issues and dive into the best practices needed to build a world-class data service ecosystem.

Data service engineering best practices

As we dive into the best practices, please keep in mind the thoughts presented in the prior sections of this chapter. The best data service is a service that enables data to be identified (queried), manipulated, value added, and massaged into becoming just what is desired (made fit-for-purpose). Then, this data may be emitted as a **change data capture** (**CDC**) flow of information, and in the rare cases where the data service is smart, it can be exposed as knowledge output. In the very rare case where analytics processing takes place behind the API, a single optimized answer serving as an insight may be produced. The best practices in these sections will lead you to a better understanding of how to accomplish what today is rarely made possible with API-enabled data services. We'll be using a few example implementations that are worth studying as we emphasize key results that the best practices need to deliver. These have been taken from a few outstanding examples:

- **Xignite** data services are exposed as APIs that provide coverage to acquire and engage investors. Xignites Market Data as a Service was one of the first market data platforms built in the AWS cloud.

- **Dun and Bradstreet** (**D&B**) provides data services in over 120 countries, with more than 300 million companies, and more than 430 million contacts mapped to business work locations. D&B is another example of a data provider offering superior data service offerings.

- **Precisely** (originally **Urban Mapping**, acquired by **Maponics**, then **Pitney Bowes**, and now **Precisely**) is a provider of embedded geographic technology data services that provides mapping functionality and on-demand data services for online mapping applications.

- **Datarade** {https://packt-debp.link/08VG5f} is a hub for collecting **Data-as-a-Service** (**DaaS**) and **Data-as-a-Product** (**DaaP**) data suppliers.

We'll elaborate on data service best practices in the sections to follow.

Data engineering best practice 1 – implement a data mesh, not just a data fabric

Cloud providers are implementing data fabric architectures that are advertised as being a superset of the data mesh architecture pattern. While we agree that we need the cloud providers to adopt the data mesh principles, we think it wise to consider the need to treat data as a product very seriously. For that view, you will want to assess whether the following statement lives up to your expectations:

> *"Microsoft Fabric's data mesh architecture supports organizing data into domains and enabling data consumers to be able to filter and discover content by domain."*

> *Naama Tsafrir, Microsoft, 2024* {https://packt-debp.link/Heql6T}

We envision the ability to define the domain semantically as a necessity. This is needed to future-proof your solution. How this is done appears to be an issue with some cloud provider offerings. You still need semantics to be made easy and part of your solid design pattern. That pattern sets up a lot of guardrails on your data services when they're exposed to end users. The adoption of a data mesh architecture for your data services design enables more flexibility than legacy approaches found with RDBMSs or data warehouses/data marts. This is because the mesh exposes data as a product with its metadata at rest with its metadata lineage. Data lakes and lakehouse architectures currently miss semantic linkage requirements and push metadata into external tools. That being said, **Dremio** {https://packt-debp.link/WZIKbp} has a formula for building out a data mesh on an Open Lakehouse design, which may be useful for your effort. Dremio Sonar is used for universal semantics in their approach.

The data mesh approach can support your time series, real-time, and static curated data factory architecture. The future requirements of a modern data ecosystem need data to be viewed in context, be self-describing, discoverable, and defined by unambiguous semantics.

D&B has a **D&B Direct+ Documentation** {https://packt-debp.link/2JBB57} site that contains valuable documentation on their data services API. There are several **Postman** {https://packt-debp.link/t9RvCr} examples located for review in D&B Postman {https://packt-debp.link/Io7wOi}. Whatis significant is that there is a monitoring feature in the JSON request to support real-time updates for requested information. A best practice is to follow this D&B example and enable collaborative development with Postman and deliver requested information in a monitored Pub/Sub manner. A picture is worth a thousand words. The D&B data service is a model that you will want to replicate in your thinking. Incorporating their vision is a great complement to D&B. Just as an artist uses an example to paint a picture, the model that D&B created provides you just enough context to get the picture solidly focused in your mind's eye.

D&B's Data Cloud has a vast amount of high quality data, and it uses a master data architecture pattern to curate data via customer processing data flows. These logical data pipelines transform raw content into business information. D&B Analytics Studio {https://packt-debp.link/P3vzwH} uses the **Databricks Runtime** {https://packt-debp.link/xYpy0S} libraries for Python, R, Java, and Scala, and they are pre-installed via the service offering. The D&B system transforms mastered, raw data into insights for risk assessment and credit analysis. These pipelines are crucial for aggregating, cleaning, and structuring data for analytics purposes. Integrating your real-time data analytics pipelines further enhances their service by providing immediate insights into business trends and risks. With **Data Blocks** {https://packt-debp.link/nD0yoj}, much of the manual work of connecting business relationships is eliminated in the **D&B data service**. As an example, **Data Blocks** provides a connected view of business relationships. Also, with the **D&B Data Exchange** {https://packt-debp.link/B854HX}, you can become a domain owner for enhanced data on the D&B system.

Xignite implements **Xignite CloudAlerts** {https://packt-debp.link/06Xh2V} to produce real-time outputs for financial data used in many analytics use cases. This enables customers to perform market analysis and forecasting. **SoFi** made use of **Xignite CloudAlerts** to drive users to the **SoFi** platform at the right time, such as when a watchlist stock performs well or poorly above an alert threshold. Also, Xignite's auto-maintaining collections are configured and then automatically adjusted by rules that are via the API for the **Xignite Market Data Cloud Platform**. This is a compelling capability of a modern data service. Xignite can tune up, configure, and then launch requests such as a BOT into the data service platform and then receive just the outputs needed, as defined by the request. This includes auto-adjusting the scope of a collection, so as not to have to come back for manual refinement via a customer's non-existent handlers. This is an excellent understanding of what we call orthoscopic manipulation of data criteria before executing a streaming Pub/Sub request for the results.

Optimizing your data pipelines for low-latency processing and incorporating advanced analytical models such as D&B and advanced BOT-like requests/responses (for example, Xignite) improves the usefulness of the information provided. The results are more nuanced and timely market predictions.

Data engineering best practice 2 – implement data pipelines (for analytics)

A lot of our focus in this book has been on creating an architecture and the engineering designs required. This leads to a data factory approach to creating consumable datasets. There are three main types of consumers:

- **Big data processing consumers**: These must be serviced with non-traditional data processing approaches to support the required data scale. This is due to the huge size of today's datasets (with a high number of rows) or complex datasets (with a high number of columns).

- **Analyst users**: These users come with a myriad of analytics algorithms that require coordination (more on that topic in *Chapter 15* when we elaborate on the analysis workbench):

  - **Predictive analytics** will let you know *"What will happen?"*

  - **Prescriptive analytics** provides you with information related to *"What should you do?"*

  - **Descriptive analytics** answers questions such as, *"What happened?"*

  - **Diagnostic analytics** addresses the need to determine the cause: *"Why did something happen?"*

- **Machine learning/generative AI processing algorithms**: These need to use data as part of a machine learning operational flow (see *Chapter 16* for more detailed best practices), including the following:

  - **Predictive machine learning**: Supervised, unsupervised, semi-supervised, and reinforcement algorithms

  - **Representative deep learning**: Neural networks, deep belief networks, recurrent neural networks, **convolutional neural networks** (**CNNs**), and transformers

  - **Generative AI**: Types of algorithms based on **large language models** (**LLMs**)

Data services that have been built to provide services for these types of data consumers need to consider that huge datasets can't be moved much after being curated (unless you want your cloud or network costs to rise considerably). That means a new approach is needed to access that information in place.

> **Key takeaway**
>
> In your future-proof solution algorithms are often going to be brought to the data, rather than the data being brought to the processing location of the algorithm.

Any implementation of this inverted pattern where algorithms are brought to the data is complicated by your data security and entitlement requirements. Also, business concerns over **intellectual property** (**IP**) will be a headwind to be overcome as trust in your data service needs to be established. Nevertheless, the trend to move curated data around less often continues.

Key trend

The trend to move curated data around less often continues!

To future-proof your data services, you will have to support this new pattern as well as the legacy pattern for data extraction via your data services. Properly leveraging the cloud provider's **PaaS** services will keep costs down and lay the foundation for efficient, scalable, and adaptable data processing architectures.

When you offer consumers the capability to operate on data remotely in your sandbox (build a walled garden), you need to give them data service features to manipulate logical views that facilitate teasing out insights. Since consumers don't have direct access to underlying data, they must have access to curated test datasets made available to test algorithms. Then, when ready, they can inject a final version of their algorithm into your production environment, with results rendered back for final analysis and reporting. This approach makes some business consumers and their developers very uncomfortable because data is not in hand and as a data engineer, you want to have hands-on access to all your data. Managing any data leading to the insights that you are responsible for delivering is now subject to a trust agreement with another. This is just one of many obstacles that you will have to overcome when designing your data services. You can see examples of how this is done by both Xignite, with their Xignite Market Data APIs {https://packt-debp.link/HwOGJo}, and D&B, with their D&B Analytic Functions and Blocks {https://packt-debp.link/hnWHij}.

Data engineering best practice 3 – implement data pipelines (for machine learning)

In the preceding best practice, we identified a need to curate data in a data factory pipeline manner for machine learning/generative AI processing algorithm consumption. We can build on this need and state that data and metadata need to be available specifically for this class of algorithms. You can see that cloud providers and tool vendors also recognize this need and have put great effort into providing features for their data catalog, metadata capture, and data lineage analysis tooling. This is evident in Microsoft's Fabric (Azure Purview), Databricks (Unity Catalog), and others.

What you will need to do is build your data service to best service *data with its metadata* as part of your solution. Keep this purpose fixed in your mind! Data services have to be servicing both, and both must be leveraged together when presenting knowledge to the RAG and prompt engineering tuning processes (see *Chapter 16*) to make LLMs work for your business domain. This way, data in its context is used in generative AI tuning. Special attention has to be given to maintaining semantic data integrity over time. Machine learning pipelines must be specifically designed to handle data drift where meaning can change but the coded analytics processing and machine learning models haven't! Your data flows have to reflect underlying semantics and solve the unique challenges required by **MLOps** life cycles.

The machine learning life cycle necessitates that you offer data services that support your consumers' use of those services in their pipelines. Complex insights exposed in your data services outputs will support predictive business risk and creditworthiness assessments (as with Xignite). Users will want to leverage your data services for their predictive financial modeling. Ensuring that your services support scalable pipelines with diverse data types is going to be crucial to your success. Incorporating automated data factory curation processing with feature engineering will significantly enhance the consumer's machine learning model performance.

Data engineering best practice 4 – use equivalent production and staging environments

Maintaining equivalence between production and staging environments when developing data services is paramount for minimizing deployment risk and guaranteeing reliable data handling and service delivery. A digital twin of the curated datasets is needed; however, issues arise when interpreting this best practice's term equivalent. Cloning all production data to a staging environment is just not realistic. Also, to maintain security and privacy, some classes of production data will need to be masked and/or obfuscated. However, some sort of periodic synchronization production into staging environments is required for pre-production systems evaluation for stress, scale, performance, acceptance, regression, and final smoke testing. This all takes place in a staging area. Data flows have to be created so that the effect of new curated arriving datasets can be evaluated for impact before the next system version goes live. The cost of the staging environment should be estimated, and the tool costs also be deeply discounted when discussing the cost with cloud providers and tool vendors (all part of vendor management processes). Cloud costs could also be negotiated and reassessed for cost/benefit value over time since the PaaS, SaaS, and IaaS services in staging look similar to production but do *not* carry any production loads. Without vendor negotiation, you are going to be paying production costs for what is not production value.

Lastly, concurrency and distributed computing are vital for handling the data volumes and real-time demands of your system. This is clear from the Xignite and D&B business intelligence service examples. Implementing efficient, scalable, and trustworthy distributed systems improves the data consumer's experience and data timeliness.

Data engineering best practice 5 – a pipeline's concurrent threads should run and scale in a distributed manner

When designing data services, think about **reliability, availability, scalability, and performance (RASP)**. To obtain these four measurable qualities, you will have to build, specify, and meet the expectations set in these areas. You might be wondering, *"Why are just these four important?"* Well, to be frank, there are many more capabilities your solution will exhibit, but these stand out as being the most important for your success. As an example, **resiliency** is a mix of availability with reliability that's experienced when the system you develop is put under stress. You can focus on these four capabilities or more. Make sure all key capabilities are measured and reported consistently. Remember, if it's not measurable, it's an aspiration and not a capability of your solution. Let's consider each individually:

- **Reliability** measures whether your system has delivered correct outcomes within specified times. For this metric to be as high as possible, your software engineering testing approaches and DQM processes have to be auditable. Reliability standards by industry exist. They should be drivers that you use when you build your system. For example, international, reliability, and quality standards can be found in the following standards:

- **American Society of Mechanical Engineers (ASME)**

- **British Standards (BS)**

- **Deutsches Institut für Normung (DIN)**

- **Gosudarstvennye Standarty (GOST** – Russian state standards)

- **Japanese Standards Association (JSA)**

- **Korean Standards Association (KSA)**

- **National Bureau of Standards (NBS** – United States standards)

- Others, such as the **International Commission for Standardization (ISO)**

- **Availability** measures your system's uptime. When building a reliable data service, it cannot be done from the perspective of the data consumer. This includes times that you said you can and should be down when building your system, such as during backups, or even cloud service provider outages. What you need to do is implement **high availability (HA)** features into your design so that the effective SLA of a consumer experience is greater than 99.999% available (also known as **5 nines**), which means your system is allowed to be down no more than 5 minutes, 13 seconds per year. Realistically, most cloud provider services are at 99.99% availability SLA (or 52 minutes, 9.8 seconds per year downtime maximum allowed), or less when a 99.9% SLA is specified (8 hours, 41 minutes, 38 seconds per year downtime maximum allowed). There is not enough room in this book to explain how to obtain these SLAs in each of the cloud provider systems! However, you will have to implement your solution's HA design in one or more of these clouds and then implement several failure tests to measure and report on your maximum SLA.

- **Scalability** is a characteristic that has to be measured holistically across your entire data services solution, including services, instances, networks, algorithms, networking protocols, PaaS services, and SaaS applications. Load scalability is particularly important to your data service designs. Obtaining a high level of scalability requires cloud service auto-scale features to be automatically managed for upward growth based on increasing load *and* effective down-scaling when that load fads. This capability to expand and contract to handle various loads is essential. You also get no rest from meeting your availability and reliability SLAs when adjusting infrastructure during in-progress auto-scale activity. Nor do you get a reprieve in your SLA targets between the time you detect the need to start scaling (up or down) and when the system finishes adding or taking out resources.

- **Performance** is the measure of how well your system is operating. You can add many other axes with this measure, such as the rate of ingest transactions, service delay, service response over time, end-to-end throughput, and so on. If you can create a **key performance indicator (KPI)** – and you will – then you can measure performance. For data services, you want to keep your eye on throughput and latency. Basically, how much can the system output and how long does it take to start arriving and then finish?

With metrics aligned with these RASP measurements, you can graph the overall compliance to your targeted goals on a Kiviat chart to show how well (or not) your solution is providing data services to consumers. Presenting a clear dashboard for IT operational status is going to help your solution when you go live and then stay alive in a production environment.

Data engineering best practice 6 – data pipelines should run as streams making use of PaaS services where possible

When you design a data service, you will want to look at how datasets are presented to a consumer. Sometimes, the quantum of response data can fit nicely into a record, suitable for streaming records to a consumer in a Pub/Sub manner. At other times, the data response unit is so big that it needs to be placed in a file (bulk) with only the event placed in a queue for the consumer to pick up the data from the file placed somewhere else on the network. Then, there are the times the response needs to go into a bucket (a collection of files)! You want to know when any big data deliveries are ready since they can take a long time to set up for access, and a result can be fetched with a callback much later. This last pattern is where analytics spaces should be set up for the analyst's consumption based on descriptive criteria that the data service used to create the analyst's sandbox (sometimes called a walled garden).

Even batch processes can be made to look like streams by implementing micro-batches, where the granularity can be tuned either statically or dynamically (on-demand based on system loading or volume of data in a query response). So, streaming is great, and the cloud providers have a lot of support for streaming data services; however, you have to decide on the granularity of the stream record and that has to be flexible. You also must be aware of what happens when a cloud provider's account's configured capacity is nearing exhaustion when unpredictable queue behavior takes place.

For example, think about what can go into an outbound JSON packet. There can be a wrapper containing any number of objects. Design your streams for flexibility and use cloud provider PaaS services with your design in mind and put time into reusable framework patterns so that others can easily reuse them. When building an activity stream-based collector system for a major publisher's health product, streaming data collection was a big feature. If we hadn't accounted for an adaptive payload, the system would have suffered unbounded queue growth upon ingesting raw data being sent from many web services.

Data engineering best practice 7 – create DataOps standards for data pipeline development

The term **DataOps** was coined in 2014 by Lenny Liebmann {https://packt-debp.link/P2yVf4} and is a moniker for *Data Operations*. Many have created solutions aligned to the hype that the term evoked. See *"Gartner Hype Cycle for Data Management Positions Three Technologies in the Innovation Trigger Phase in 2018"* {https://packt-debp.link/h1lJPB}, where it was pointed out that there were few standards or frameworks available in 2018 for the valuable principles presented by the concept. Today, cloud providers and companies such as **DataKitchen** {https://packt-debp.link/bXQMWG} have created DataOps frameworks based on the *"DataOps Manifesto"* {https://packt-debp.link/k9AuQW}, including the following:

- *"The DataOps Cookbook"* {https://packt-debp.link/34i9W8} by Christopher Bergh, Gil Benghiat, and Eran Strod

- *"Recipes for DataOps Success"* {https://packt-debp.link/9lx0a6} by Christopher Bergh, Eran Strod, and James Royster

With that framework comes several tools:

- Apache Airflow {https://packt-debp.link/m7Rlud}

- Dagster {https://packt-debp.link/6ogara}

- DataKitchen DataOps Observability and Automation Software {https://packt-debp.link/LEurky}

- High Byte Intelligence Hub {https://packt-debp.link/QukKbI}

- Papermill {https://packt-debp.link/F39Rng}

If you implement DataOps processes in your data factory solution, the pipelines will conform to unified observability and not some hodgepodge {https://packt-debp.link/9WtLtz} of divergent monitoring and workflow orchestrator designs. Look at the endnotes for additional reading on this topic and you will become acquainted with the processes needed to implement this best practice.

Your data output will be compliant with standard processes that you design and enforce across the teams so that your solution's data services operate well. Your DataOps framework process ensures consistent SLA management and dataset readiness, streamlining pipeline development and service operations. DataOps practices are essential for improving the quality and speed of data pipeline development, deployment, and runtime efficiency. It may take a bit of effort to implement DataOps initially, but it is a core enabler for subsequent high-quality and high-velocity development.

Data engineering best practice 8 – implement tool selection criteria with weighted selection toward built-in integration with core components of the architecture

In *Chapter 5* {https://packt-debp.link/KSRllI}, we elaborated on the need for due diligence when defining your solutions architecture. This is even more important when selecting tools and integrating them into your data services capabilities. Strategic tool selection within data services requires you to prioritize integration with other core architectural components to minimize development complexity and foster system cohesion. Selecting the right tools for data pipeline development will become contentious. Toolsets are what engineers get accustomed to using. If they are experienced, they carry a lot of biases toward or against the tools you may wish to consider for use. Use due diligence processes to take subjective bias out of the evaluation and take all engineering inputs and factors into consideration. Let the results be open and honest and you will get past the friction. Be ready to deliver good and bad news to all after the selection process is complete. Represent the perspectives that matter most when performing due diligence. Such perspectives include the following:

- Data

- Operations (infrastructure)

- Applications

- Integration

Additional perspectives for management, business, and information (semantics) can be added if you desire. The choices are yours based on your organization and those who need to buy into the final decision.

Figure 12.1 provides an example of perspectives as they could appear in the output of a formal due diligence effort:

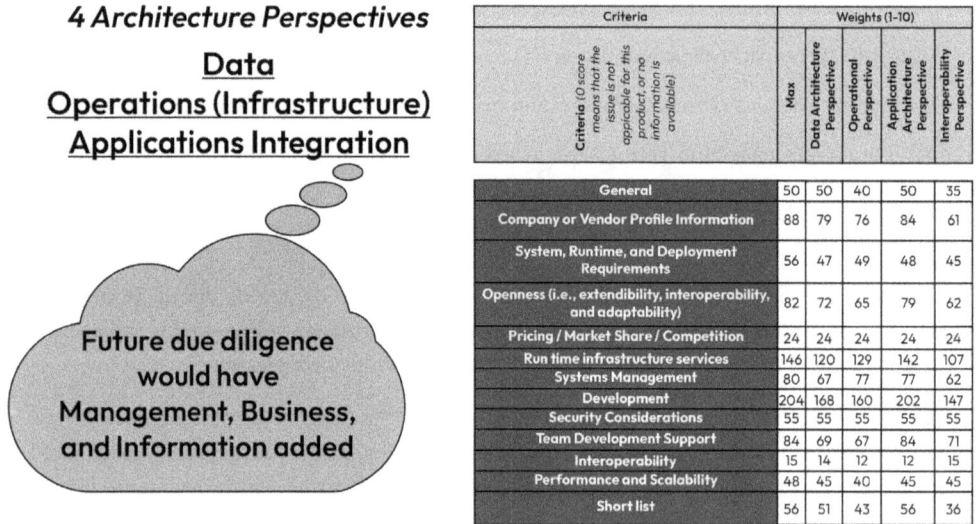

4 Architecture Perspectives

Data
Operations (Infrastructure)
Applications Integration

Future due diligence would have Management, Business, and Information added

Criteria (0 score means that the issue is not applicable for this product, or no information is available)	Max	Data Architecture Perspective	Operational Perspective	Application Architecture Perspective	Interoperability Perspective
General	50	50	40	50	35
Company or Vendor Profile Information	88	79	76	84	61
System, Runtime, and Deployment Requirements	56	47	49	48	45
Openness (i.e., extendibility, interoperability, and adaptability)	82	72	65	79	62
Pricing / Market Share / Competition	24	24	24	24	24
Run time infrastructure services	146	120	129	142	107
Systems Management	80	67	77	77	62
Development	204	168	160	202	147
Security Considerations	55	55	55	55	55
Team Development Support	84	69	67	84	71
Interoperability	15	14	12	12	15
Performance and Scalability	48	45	40	45	45
Short list	56	51	43	56	36

Figure 12.1 – Four architecture perspectives for due diligence

Taken individually, the weighting and scoring criteria represent the importance of a service feature from your subjective experience and current corporate agendas. Viewed from the overall enterprise perspective, the optimal solution is the average of each of the architecture perspectives. You will also want to summarize the collected data to drive objectively scored and weighted results, as shown in *Figure 12.2*:

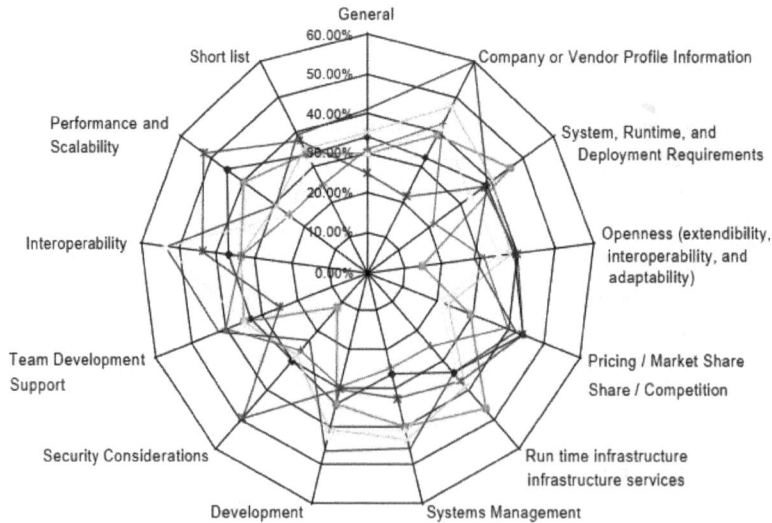

Figure 12.2 – Due diligence results in a Kiviat chart

We've omitted the actual tools that were selected from the preceding diagram since that isn't important for what we seek to convey – that is, the need for due diligence and the process to obtain consensus. Building a modern data services echo system with up-to-date technology still needs capabilities similar to the older monolithic application server products available 25 years ago. The due diligence processes for product selection are not all that different today. The tool selection outcomes after analysis are very different given current data volumes and cloud provider capabilities. You need to address huge data storage, distributed/HA, and pipeline processing needs to handle scaling, and then there are more integration considerations. Prioritizing tools that seamlessly integrate with your architecture will make it possible to maintain consensus on the tools selected to bring your solution to market quickly.

Data engineering best practice 9 – implement Pub/Sub models for economy of scale when supporting customers with the same dataset subscription

In *Chapter 14* {https://packt-debp.link/Jc435z}, we will elaborate on the significance of Pub/Sub streams; however, for this chapter's best practice, we will focus on the need to service repeating workloads from customers. In the past, when two requests came in for the same data, we serviced the second request from the cache. It makes sense to do this since we don't want to overload the data origin (source) with our duplicate request load. So long as the cache was trusted and kept up to date with write-through transactions, the cache approach worked well. We even implemented **content delivery networks** (**CDNs**) such as Akamai to cache web objects, pages, and script (such as JavaScript) artifacts at the edge of the internet to speed up browser and data services.

You'll want to implement a pattern for data services request fanout when request data overlaps and then integrate that approach with your caching designs as a practical implication of your data service's Pub/Sub architecture. It sounds simple, but you will have to know when to bust the caches and warm them after redeploying your data services code. Our advice is to always design and build a **cache manager** and not leave this as an afterthought. The smarter the cache manager, the better your service performance with optimized cost impact.

One last thought: if you spend time on the cache manager design, you'll see that the data service is driven by logical data changes. Hook the cache manager command's logical data changes, not low-level physical data changes. This type of logical cache control will make caching understandable to businesses. Xignite and other financial data service providers spend a lot of time optimizing their data services to allow users to receive tailored data updates with the least backend duplication of effort.

Also, as an example, in the early days at Bloomberg, I (Rich Schiller) was able to show that data could be distributed to 100 institutional desktops faster than data could be supplied to a single Bloomberg Terminal (1994 timeframe) using a custom gateway, written by myself, using intelligent caching, on SunOS (UNIX), and the new yet to be released Bloomberg API. As a side note, this was built with a laptop sitting on a cardboard box behind Michael Bloomberg's desk next to Tom Secunda. It was a challenge! What was really being codeveloped was the yet-to-be-released Bloomberg API, which Bloomberg LP's Art Finkelstein was developing while sitting next to me. I was testing and developing at the same time. The outcome was the very real Micrognosis/Bloomberg Gateway. The gateway was impressive, to say the least! As an architect at FD Consulting, before being acquired by Micrognosis, the flagship trading information system was very fast. It used UDP/IP data transfer rather than TCP/IP or multicast IP data transfer protocols. The Micrognosis/Bloomberg Gateway tied slower (but valuable) Bloomberg data to a fast financial data distribution system. The gateway's proof of concept showed that if you build a smart caching gateway, you can also build a way to scale out the Bloomberg system to handle the performance issues posed by exponential user growth, this being caused by a limit of around 30,000 concurrent terminal users without a fan-out gateway to fix the scale issues. Needless to say, Bloomberg's internal terminal gateways later provide this fan-out service today for their terminal systems users. Smart controller-based caching services reduce origin load and provide for a global multi-region scale.

Data engineering best practice 10 – data wrangling tool selection to create a clean gold zone copy of datasets

In *Chapter 13* {https://packt-debp.link/nUPfEd}, we elaborate on data profiling (data wrangling) in a lot of detail. Regarding data services design in this chapter, your best practice can be summarized as follows:

> **Key takeaway**
>
> Implement **master data management** (**MDM**) in nontraditional ways!

This means you need to address many of the capabilities of a traditional MDM solution without purchasing one. Like with **enterprise resource planning** (**ERP**) systems, MDM solution vendors have been given a bad reputation for producing excessively pricey products, a high degree of lock-in, and budget risk that most enterprises can't afford. We'll leave the list of failed, acquired, or floundering MDM vendors to your research. There are several corporate success stories, but they all come with legacy mindsets and big wallets.

Selecting appropriate data-wrangling tools is critical when designing your reliable data services. The approach that we suggest you build toward should focus on internal data consumers. Then, after testing, you should scale the services with measured quality before exposing services to external data consumers. Use the data factory pipelines as your test bed.

As an example, **Pricing Excellence** {https://packt-debp.link/0ABdze} developed their **Data Ladder** {https://packt-debp.link/c8Oqxr} and **Data Match** {https://packt-debp.link/IyAUGc} products for their web scraped retail data business products. This was an internal product that was good enough to be made an independent product. Pricing Excellence's goal was to provide clean retail data for analytics and, as many in the field know, retail data is anything but clean!

Use your wrangling capabilities to curate your architecture's gold zone datasets and then the consumable zone datasets to ensure mastered data remains of the highest level of quality and usability.

Data engineering best practice 11 – a data catalog is part of an essential metadata implementation

Chapter 13 {https://packt-debp.link/OVNBOu} and *Chapter 16* {https://packt-debp.link/aTJDRq} address the need to create catalogs for data and analytics self-service. In this chapter, we're emphasizing the need to catalog data lineage and data versions, which are internally retained traces that are left over as data progresses through your factory pipelines. These are only needed until a final curated form is produced that is accepted into production. After this event, the metadata lineage versions will need to be cleaned up, purged, or placed into an archive for eventual roll-off. Only what's needed should be retained. Here, you must retain just enough for what is needed if a data rollback is ever to take place. You do not want to purge the older lineage metadata versions immediately but archive them just in case!

Many metadata, dictionary, and catalog tools can't handle the versioning described previously and their data is sticky (once it hits the catalog store, as with **Azure Purview**, it cannot be cleaned up). Unneeded cataloged metadata brackets, after a data product is marked final, must be taken out of service to effect good metadata management processes.

Only the metadata for a final release set of datasets should be retained in the dictionary/catalog. You must label all metadata in your design with descriptive life cycle tags. This metadata about metadata is how you can implement these implied life cycle requirements of the data mesh. Even if a third-party tool or cloud provider is not supporting this capability yet out of the box, you still need to build support for dictionary and metadata catalog management.

Data engineering best practice 12 – define data owners, security, rights, and access for consumers upfront

Earlier in this chapter, in the *"Software and data engineering drivers for best practices"* section, under *"Zero trust versus defense in depth"*, we pointed out that you want to think about security early in your solutions design. This provides more than just access control capabilities since privacy, entitlements, and rights management features are also part of a future-proof design. You want a clear upfront definition of data ownership and security for the data services you implement. Then, you want the data service itself secured. This is essential for maintaining a trusted data management practice. There is no single prescriptive best practice today, other than to say that even if you use a cloud provider's best practices and anticipate everything possible, you will still have to learn to dance when a nefarious actor finds a way to circumvent your control. Anticipate this! Build **ZTNA** into your solution. But also build a command and control capability to observe the current state security of your production solution and execute probes that alert you to an actor's plan as it is detected and before it starts to unfold. Implement gates that shut access to gold zone data and put all this into a security runbook to be called as a last resort to halt damage to enterprise data.

Use your data catalog to determine what assets have to be protected and how. This assumes qualitative metadata exists for the datasets under management! The data's value to the enterprise should be captured in the dictionary. After all, you have a data product to protect, not just a collection of enterprise data. Also, any value-added direct or derived data given to the domain owner to steward for a user's consumption needs protection. Drive your scripted and software-coded security assessment processes so that they support your data security contracted obligations. This is a crucial role in effectively managing your data assets.

Example 1 – how D&B benefits from a data catalog

When it comes to D&B (https://www.dnb.com), there are several capabilities worth elaborating on:

- **Enhanced data discoverability and accessibility**: D&B, specializing in business information and analytics, handles extensive datasets for companies worldwide. A data catalog facilitates easier access to this data for analysts and clients, enabling them to find the specific information they need quickly. Metadata descriptions, tagging, and search capabilities within the catalog reduce the time spent searching for data, increasing efficiency and user satisfaction.

- **Improved data governance**: Given the sensitivity and regulatory implications of D&B's data, particularly around credit risk and business intelligence, a data catalog aids in implementing robust data governance policies. It helps define data ownership, access controls, and usage policies, ensuring compliance with regulations such as GDPR and CCPA (see the earlier elaboration). The catalog tracks data lineage, providing transparency into data origins, transformations, and usage, which is crucial for auditability and trust.

- **Quality and consistency**: A data catalog contributes to maintaining high data quality and consistency across D&B's offerings by cataloging information on data sources, quality metrics, and update frequencies. This ensures that clients receive accurate and reliable information, fostering trust and reliance on D&B's data services.

- **Facilitates collaboration**: By making data easily discoverable and understandable, a data catalog encourages collaboration both within the organization and with external partners. It enables teams to leverage existing data assets more effectively, avoid duplication of efforts in data collection and preparation, and innovate more rapidly.

With D&B, the data services focus is on quality, coverage, and discoverability. Also, with D&B Connect Discover {https://packt-debp.link/VHmJY6}, your data can be mastered alongside D&Bs using their services. An example of the value that can be obtained by this D&B service can be found in this quote from Ricoh:

> *"All customer information is provided to our sales force using D&B data to ensure we are properly optimizing territories, etc. Sales have a more direct line of sight to customers that we are calling on."*

> *Christopher Chando, VP of Financial Solutions, Ricoh Americas Corporation* {https://packt-debp.link/XDdqCy}.

Clean, up-to-date company and contact data is invaluable when managing sales processes across businesses.

Another data services example can be found with Xignite's data catalog capabilities.

Example 2 – how Xignite benefits from a data catalog

Xignite (https://www.xignite.com/products) also has several data service capabilities worth elaborating on:

- **Real-time data management**: Xignite provides financial market data in real time, necessitating immediate access to the latest information. A data catalog helps manage the metadata of various data streams, including stocks, currencies, and commodities, ensuring that users can quickly access and integrate the latest data into their applications.

- **Scalability and efficiency**: For Xignite, scaling to meet the demands of thousands of clients worldwide while maintaining efficiency is crucial. The data catalog supports this by optimizing data retrieval and integration processes, reducing overheads, and improving the response time of data services.

- **Client customization and personalization**: A data catalog enables Xignite to offer customized data solutions to its clients by providing metadata on available data products and services. This allows clients to precisely select the data they need, tailored to their specific requirements, enhancing client satisfaction and engagement.

- **Compliance and security**: Like D&B, Xignite operates in a highly regulated financial industry where data security and compliance are paramount. The data catalog plays a vital role in enforcing security policies and regulatory compliance by cataloging data sensitivity levels, access controls, and compliance-related metadata, ensuring that financial data is handled securely and as per global financial regulations.

With Xignite's data services, you can observe the focus on real-time performance, scalability, and comprehensive data coverage for your financial services data needs.

With both the D&B and Xignite examples we have just seen, the implementation of a data catalog is not merely beneficial but essential for managing the complexity, scale, and regulatory requirements of their data ecosystems. The data catalog is a foundational component of their data management strategy, enabling them to leverage their vast data assets effectively, maintain high standards of data quality and governance, and deliver value to their clients through accessible, reliable, and compliant data services.

Data engineering best practice 13 – train your experts and let them train and retrain others

When building out your data services designs, you will want to balance the people, processes, and technology capabilities. Regarding people, you will want to communicate the technology and process changes that a modern data services ecosystem requires. You will find a new group of legacy dinosaurs in your midst who think they grok the data services design as just a new API, but the design involves a tectonic shift in thinking. That requires the visionary designer (hopefully yourself) to train several technology evangelists to get the message out to various teams. These are technology as well as business teams. You are inverting the data services model and making it possible for data to be moved less. Also, the data will be operated on in your backend with consumer-enriched algorithms, and then the outputs will be emitted for final rendering in the consumer's system. This requires continuous training and knowledge transfer, which are key for this type of data service architecture. This also fosters a culture of innovation and adherence to the best practices you wish to see as outcomes in your data engineering effort.

You need to develop a **center of excellence** (**COE**) for the development of this training. Training staff should involve a *train-the-trainer* approach for your COE staff. It is advised that you create a bootcamp program to get everyone hands-on with your data service API implementations. Build a data services portal such as Xignite so that you can try out your sample applications and get your users experienced. Provide everyone's feedback before making data services generally available to all your data consumers.

Data engineering best practice 14 – handle errors gracefully

When building out data services, you need to address SLAs seriously, and from the data consumers' perspective, always run them according to those specifications. There will be times when this is not going to be possible and the line between strict adherence to the SLA versus some degraded mode is required. You are within SLA tolerance but maybe not in 100% data delivery compliance. You can see this happens when summary analytics results are available but not the drill down into the minutia when operating over the weekend when a detailed fact store is undergoing maintenance (or product upgrade). The services are running and supplying data at the aggregate level but not the detailed level. This kind of functional degradation is common and necessary. Just make sure your data contracts have this written in as permissible. Your SLA will still be 99.xxx% but you will have this caveat in the data contracts.

When real errors are encountered, you need to be open and honest and degrade operations to protect the most functionality possible. As a result, you ratchet down the SLA but in a very structured way so that the SLA can be reported at some degraded level and not via a random sliding scale. You have to structure your error-handling service levels like you do the primary level of service and stay within the service swim lane of acceptable metrics. Then, you must report on the switch as a clear violation but one that is managed. All the associated *mea culpa* {https://packt-debp.link/9QjY9o} reporting will still be required. This builds trust that your data services can run well and run well enough within the parameters of a clearly defined degraded mode of operation.

Implementing structured error-handling mechanisms, including checkpoints and replay capabilities, is crucial in data services to maintain the integrity and reliability of data pipelines.

Checkpoints and replay mechanisms in data pipelines make it possible to sustain an effective SLA for your data service. Even if a data service reset is required, it can be made effective from the last known good state maintained by checkpoints. This means you need a service manager to coordinate the reset and replay logic for the application transaction across the end-to-end data service dependency map. What this all means is that you do not let the developer decide on `try/catch` logic but provide them with a set of recovery patterns that the developer has to adhere to when encountering errors. It is not good enough for them to fail *fast*! It is required that they fail *gracefully* and recover *effectively*.

Graceful error handling is a critical component of robust API design, ensuring that users are provided with clear, informative feedback when something goes wrong. This best practice not only improves the user experience but also aids in debugging and resolving issues more efficiently. Let's examine how this principle is reflected in the design of the APIs from D&B and Xignite, focusing on aspects that are generally observable or inferable from public documentation and standard API design practices.

D&B API – graceful error handling

We'll elaborate the D&B's graceful error handling in this section. Although not perfect, this example works well to illustrate some key features that you will want to see in your designs.

- **Informative error messages**: While specific implementation details may vary, APIs such as D&B's typically return HTTP status codes that indicate the nature of an error (for example, `400 for Bad Request`, `404 for Not Found`, and `500 for Internal Server Error`). Alongside these codes, it's common practice to include a human-readable message or an error code with a place to go to find more detailed documentation on it. This helps the end user or developer to understand what went wrong and how to resolve the issue encountered.

- **Consistent error object structure**: APIs designed with best practices in mind, such as those by D&B, often use a consistent structure for error messages. This structure might include fields such as `error_code`, `message`, and `error details`, thus providing a standardized way for developers to handle errors programmatically.

- **Rate limiting and throttling responses**: Public documentation for many APIs, including those in the financial and business data domains, often mention rate limiting mechanisms to prevent abuse and enforce fair use compliance to established data contracts. When a client exceeds the allotted number of requests or aggregate data limits, the API typically starts to return a specified error code (often `429, Too Many Requests`) along with headers informing the user of their current limits. Also, indications as to when requests can be made again are supplied. This is an example of data contract enforcement. The mechanism informs the user about the limits by preventing server overload or abuse.

- **Retry-after headers**: In situations where requests are rate-limited or when there's an impact on a contracted SLA, APIs should use a `Retry-After` header in their responses. This tells the client how long to wait before retrying the request, guiding them on proper usage and reducing the likelihood of repeated errors.

Xignite API – graceful error handling

With Xignite's graceful error handling, you will see a focus on robustness and what we consider API friendliness. The data is complex enough that the API makes acquisition easy. Data is normalized, and even sparsely populated field data is explainable:

- **Detailed financial data error responses**: Given the critical nature of financial data, an effective API includes detailed error responses for scenarios where data cannot be retrieved or processed as requested. This includes specific error codes for data not being available for a given ticker symbol or date range, allowing users to understand whether the issue is with their request parameters or the availability of data itself. What's important to know is that you want to supply valid data fields for what data exists, and not the universe of nulls for data that does not. Financial data can be sparse, with more than 30,000 possible fields per equity instrument. Return fields that are known to exist and fields that the semantics says must be supplied for a given class of data regarding a financial instrument have to be self-explanatory. This definition of what fields *must* be supplied versus what the technical says *could* be supplied has to be simplified and unambiguous in the API.

- **Documentation and support integration**: Making the data service self-explanatory from a technical and business perspective is an elegant way to make a system maintainable. While not directly part of the API's code, how an API's documentation is structured and how easily developers and data consumers can access support documents is very helpful. Xignite's approach includes extensive documentation on error codes and scenarios, integrated support channels for developers encountering persistent issues, and a community forum where developers can share solutions. The portal enables a data consumer to try out data services APIs with any number of language interfaces, *before* attempting full-scale development.

- **Fallback data and partial successes**: In cases where not all requested data can be delivered (for example, a batch request where some items fail), APIs in the financial sector, such as Xignite's, design responses that include both the successfully retrieved data and detailed errors for the subsystem failures. This allows clients to partially proceed with the available data while understanding the scope of any issues. This is a type of graceful degradation that should be accommodated with collected metrics for long-term analysis and reporting to data consumers. Your data consumers will be tracking it as well, so you better be able to tell them before they remind you of SLA and data contract violations.

- **Transparent communication on service health**: For APIs delivering critical, time-sensitive data, maintaining a service health dashboard or similar communication channel helps users plan around potential downtime or degraded performance. This aspect of error handling ensures that data consumers are guessing if there are or were incidents in your system. Sustaining trust via verifiable dashboards goes a long way toward the creation of a reliability standard.

Both D&B and Xignite, given their roles in providing essential business and financial data, respectively, have a vested interest in implementing graceful error handling in their API designs. While the specifics of their implementations can vary and may not be fully detailed in public documentation, the principles of providing clear, actionable error messages, using consistent error structures, implementing rate limiting with informative responses, and supporting developers through documentation and direct support are universally recognized best practices in API design. These practices not only improve the developer experience but also enhance the reliability and usability of the APIs.

Data engineering best practice 15 – run multiple data pipeline tasks as a directed acyclic graph (DAG)

Knowing what the dependencies are for each logical thread of activity leading to the correct execution of a data request is essential to effective data services orchestration. Processing step dependencies, degraded operations, and system state all affect the dependency graph. When implementing a big finite state machine across a flock of cloud services provider PaaS services, your own data services code, and pipeline processes, your execution flow can get all knotted up! You don't want to get into systematic deadlocks, never-ending service loops, or cascading failures that are self-inflicted. To unravel this potential hairball of a design problem, you want to build a service registration feature for your data services. This service registrar would utilize a DAG {https://packt-debp.link/XJv0YG} for either direct task orchestration in your data pipelines or as a sidekick process that's used to log pipeline execution steps as they take place. This DAG will provide you with a clear way to drive all service actions optimally, thus leading to system recovery after graceful degradation operations are completed. It optimizes the efficiency and effectiveness of your data operations, ensuring a streamlined and coherent data flow. The DAG is a source of truth to be leveraged in your dashboard reporting and any recovery logic. This leads to fast localized recovery action without the overkill that's often required to remediate catastrophic system failures via a full restart. Without the DAG's mapping, the only way to get back to smooth operations is through intrusive brute-force service restarts.

Summary

As we reach the end of our exploration into the intricate world of data services, it is instructive to reflect on our journey with the understanding that an elegant solution is essential for these services. This comes from the need to preserve trust in the data service's promised capabilities and to service the data's semantics via data services that do not frustrate those semantics. The result is that the technology implementation will *not* be easy, but it will be a large effort to deliver an elegant, simple solution for complex data. If you make your data service elegant, your data consumers will know it. They will tell then you that you have been successful in your efforts.

The right solution emerges under conditions of significant complexity and challenge, much like the formation of diamonds under the immense pressures and high temperatures deep within the Earth. This metaphor extends beyond a mere comparison of the results; it encompasses the process, the environmental conditions, and the transformative journey from raw carbon to a gem of unmatched strength and beauty. Data services, from their initial design through to their deployment, experience a transformation akin to that of forming diamonds, molded and honed by the rapid pace of technological innovations and the complexity of regulatory environments.

In this narrative, the foundational principles have been navigated, including trust security frameworks, data localization, and privacy protections. These serve not merely as initial facets cut into a rough diamond but as critical pressures that define the clarity, color, and carat of our final product. This metaphorical image, when applied from the inception of data service designs, ensures that our creation is not just compliant and secure but resilient, adaptable, and capable of capturing the myriad nuances of data's inherent value.

As we navigate best practices in data engineering, an emphasis remains on constructing data services similar to the way a jeweler shapes a diamond. Each facet is carefully planned and crafted to enhance the end products' overall beauty and functionality.

Making the right architecture (data mesh or data fabric) choices, including creating efficient data pipelines and ensuring staging and production environments are aligned, contributes to the creation of data services that are robust, scalable, and capable of meeting the demands of an ever-evolving digital landscape.

Moreover, strategically selecting tools, adhering to DataOps standards, and committing to data quality are not merely operational tasks; they are the polish that brings out the radiance of your data services. These elements elevate data services from functional constructs to strategic assets. Your data services will enable the enterprise to leverage data with unparalleled precision and insight.

In concluding this chapter, it's clear that the craftsmanship involved when building data services is both an art and a science. Like a jeweler who transforms a rough diamond into a masterpiece, we too must apply our skills, knowledge, and vision to mold our data services into structures of enduring value and utility. The journey of mastering data services is indeed continuous, with each challenge and opportunity serving to refine our approach and enhance our capabilities. As you move forward, embrace the complexities of data service design with the creativity and discipline of the master jeweler, always mindful of the ultimate goal – to unlock the full potential of your digital diamond! In doing so, you not only illuminate a path to innovation and success, but you also affirm the indispensable role data services play in sculpting the future.

In the next chapter, we will be elaborating on some key considerations that you will want to address when implementing the data profiling capabilities of your solution. We will also address ways you can properly handle data gaps, restatements, and corrections. We will be stressing the need for proper metadata management throughout the chapter. Lastly, some best practices for handling data trends and calendars will be provided.

13

Key Considerations for Management Best Practices

This chapter will address some niche focus areas in data engineering:

- Data profiling

- Gaps

- Restatements

- Correction

- Metadata

- Trends

- Data calendars

These topics are often hotly debated by systems developers and as a data engineer, you will want to have a voice in the decisions being made. The goal is data that remains relevant and clearly correct within its well-defined context.

The niche areas naturally fall into two types, **data profiling** and **metadata**, because creating a modern data factory will involve processing related to making data fit for purpose. Having data properly positioned for use and reuse in downstream processing is an essential need of the end user/consumer application. Today and going forward, automated intelligence, **machine learning** (**ML**), and knowledge engineering will be how data is leveraged and then later re-leveraged without the need to create new custom software. Data will flow through the data factory and be transformed from its raw ingested form to become a finished data product. This end-state data product will be labeled as an **insight** or at least it will be **smart data** that can be leveraged for new purposes, leading to new insights. Data today will be repurposed in ways never anticipated when that data was first ingested and curated for its original use.

Niche focus areas – best practices overview

So, you will want to create a few mental bookmarks in your plan for building a future-proof data solution. The following niche data topic elaborations will provide you with new thoughts to consider as you make your design decisions.

Data profiling involves wrangling data and recording observations as metadata annotations that are associated with the data being profiled. An important attribute of that data is its context, which is defined by its semantics. Data context also has to be set in data calendars, the derived trends observed and the data's implied source definition. Also, expectations (data contracts) add color to the data, as do the **service level agreements** (**SLAs**). There are other contexts, but these are just a few worth mentioning. The range of acceptable values that can be visualized as **swimlanes** will constrain what are acceptable data values for a given data context. These swimlane demarcations exist above and below the data's **trendline**. Building a net to catch bugs works in an entomological sense and it also works as an analogy for data profiling. Build the acceptance criteria correctly and you will catch data bugs early. This enables data to be retained in a state, leading to its correct understanding. With data profiling comes the need to create a formal definition of contracted interfaces along with the versioned dataset. This way, the data's profile gets locked in place with the dataset. This includes the syntactic and semantic wrappers needed for your system, preventing data from being recast to meanings that pervert its truth. As an example, we recently heard a politician misquote the U.S.'s *real GDP* growth using the *nominal GDP* figure! The meaning sounds similar, but the concepts are grossly different. *Real GDP* factors into inflation. Reporting just GDP requires a qualifying context term such as *real* or *nominal* or it becomes ambiguous and grossly misleading. Even adding a free-form qualifier such as "actual" to GDP is also misleading, since it is imprecise and can be confused with either *real* or *nominal* GDP. You will observe these issues many times in your endeavor to retain a dataset's *true* context. Precision is required, along with an interface to the dataset that locks in the precise contextual meaning in order to make it easy to figuratively get a handle on the data (e.g., its interface in its correct context). All too often with legacy data processing systems, the data's context is omitted or buried in the software processing code, which leads to downstream errors. A future-proof data-profiling capability should address these issues.

Gaps in data arise from legitimate as well as illegitimate detected business reasons. The difference is gaps that are part of the business process, such as when a storefront is closed and no retail sales take place, versus when the IT system fails to produce outputs for a week because of an error. Gaps have to be noted as metadata, and then that data should be properly and completely handled in a reversible manner. That handling logic is subject to change over the course of time. Our minds fill in the gaps when we look at a picture, let's say, in a newspaper. The pixels do not form a solid color, yet we observe the solid black on white since the pattern we see *mostly* fills the void of white space. Datasets should be rendered in a way that ignores any gaps or smooths them over during processing as if they were to form a big picture. The capability to do this is essential when rendering data as information for analytics. Being able to process the data at a zoomed-in level versus a zoomed-out level is essential; otherwise, data minutia will distort the big picture. Real world data is messy, and it needs to be processed on the macro level and not just the micro level. Implementing a metaphorical data telescope to smooth the data gaps is needed. Call this **data lensing**, for lack of any better terminology available today. Being able to run a query that returns fuzzy results is better than having that query return nothing just because the data is not perfect! This is a key enabler when assigning embeddings needed for machine learning feature extraction processes, which can be regenerated over and over until data comes into focus, leading to an insight.

Restatements of data with potentially new *data gaps*, or corrected *data overlaps*, will arise over the course of time. The cost of full data factory reprocessing for a dataset can be very high in terms of business impact and IT cost. The fact that data is messy and can be resubmitted after first being processed is a *tail wagging the dog* scenario taking place in your data factory solution. A data restatement causes a ripple effect of change on derived information, knowledge, insight, and curated data product outputs. In our experience, we have seen data restatements cause huge contract and data SLA misses for an end state curated dataset. Data restatement ripple effects were not first assessed for their processing impacts. Workflow steps were not organized first so that just the changed data is processed rather than the entirety of the corrected dataset. Data factory logic should be organized to handle all processing as if it were processing these data ripples (changes) rather than being driven from full processing of the incoming corrected dataset. Pipeline event processing blueprints should be applied to implement these data changes, which will first involve planning the processing work, chunking it up, implementing checkpoints, and then (and only then!), after the plan is created, executing the plan as a series of events. The alternative legacy model of serially, or blindly, processing data as an input file will not work in a modern data factory. Note further that a special type of restatement is a data *correction* that assumes that a data item exists, whereas a *restatement* can also entirely remove an item from existence.

Corrections, referring to the receipt of the same logical dataset with different values, are to be expected in any modern system. With a correction, you can envision some effect on other data created or adjusted in downstream processing. Handling downstream, correction driven impacts should be performed with an assessment of impact first, then followed by impact categorization, and finally, an event driven plan for execution. Corrections processing is often rule-based and driven by the event plan when applying correction impacts. Correction processing has the potential to cause huge downstream ripple effects for the consumer and data consuming systems. You should provide the capability to gauge the impact of this ripple effect and be able to plan, stop/reverse the correction application if its impact is above a threshold, and then convert the entire correction activity into a full reprocessing action if necessary. Just allowing a correction to be made without these considerations will break SLAs and data contracts. Imagine an analyst who issued a C-level report based on old data and now sees the key insight as not being an insight at all, but totally incorrect. What would that analyst think of your solution? We've seen financial traders put their feet through LCD display terminals when systems displayed data slowly. Imagine if they saw a positive trade signal go negative when they'd just acted on it being positive! This historical correction processing issue is a major one when designing **financial ticker plants** {https://packt-debp.link/OUgn3l} where a change to a price is a great asset, so long as you have not traded on the prior price before a correction is applied!

Metadata is data about data. In prior chapters, you learned a lot about metadata. What is often missed is the need for metadata classes to address **data drift**. You will need to account for semantic changes in data so that you can preserve correct meaning over a long period of time. Metadata is therefore **time series** sensitive! When is the degree of **semantic drift** so great as to cause that data to have to become classified as an entirely new type? The changes caused by business and societal shifts cause data to undergo evolution. This has to be accounted for as the enterprise grows and the underlying data's meaning changes. Think about it for a second; the ability to capture context and reflect context shifts for data is essential.

This way, as data values that are no longer considered equal to their legacy values change, they have to be classified as new types (i.e., stored in a different column, class, or object). Here are a few examples:

- What could be classified as a child's toy today could be an AI-enabled android in the future! Today's smartwatch is a manifestation of the Tribune Media Services' fictional device from *Dick Tracy* in the 1950s!

- Today's smartphone will be a wearable gateway computing device of tomorrow, but we doubt it will be called a wearable gateway; rather, it will be called by a brand name and marketed aggressively!

Trends are subject to breakage, synthetic gap filling, or smoothing. By establishing formal data trends, underlying datasets may be used to train ML models and be useful for any downstream analytics in the data factory. After implementing **data munging** operations, datasets will need to be assessed for trend correctness. If necessary, the underlying data for a trend will need to be augmented or reduced in value (of course, with the proper metadata annotations added) as **synthetic data** layers in order to force **fit data** to meet the expected trend profile. The profile is based on a developed statistical model representing the known truth about the class of information being processed. This is required to clean up messy data and make it processable until it becomes less messy over the course of time. As a data engineer, you will have to handle data that requires layers of adjustment to make it *fit for purpose*. As an example, think about a demographic profile gathered by some survey results that underrepresents a class of voters when compared to the most recent census. If a U.S. East Coast city survey were not written in both Spanish and English, there is a good chance that a vast segment of the population would not respond. How do you handle survey data results that now need to statistically fit into a prior census model in order to get a reasonable representation of the truth in regard to this dataset's trend? You first need to create a series of data point adjustments as layers over the raw data. Being able to access the original raw data at the microscopic level while making it usable at the macroscopic level is needed. This preserves the data's trustworthiness and is required to calculate trends and remediate anomalies in the raw dataset. Overall data trend truth is often more important than the raw data, so it has to be correct at the macro level. Note that the macro level data trend can be true, even when the rules defining correctness at the micro level are violated. If you retain the re-processable raw data, and the synthetic data adjustments as layers, you also preserve the macro trend. You also make it possible for any model-produced synthetic adjustments to be regenerable as the statistical models are changed over time. Note that all layered adjustments must remain auditable and be reversible. If one knows that a trend is correct by common knowledge or provable hypothesis, but the raw dataset does not *yet* support that truth, then preserving that trend requires data adjustment. Incoming data values also should remain in the swimlane of acceptable values or be adjusted. These adjustments over time are applied as correction layers over the original raw dataset values. There may be many correction layers placed over a raw dataset. These **data overlays** cannot change the original raw data but when aggregated over the dataset's values, they make the dataset conceptually correct for the desired trend. The data is now fit for purpose. Being able to audit, remove, or augment data overlays as a result of software enhancement, without negatively affecting other overlays, also provides greater transparency and builds trust in the results. A data scientist or analyst should be able to drill down through the data overlay strata applied over the original raw dataset and get down to the original data if necessary. Being able to gain access to the original dataset with all its overlays provides effective trust through transparency.

Data calendars are a driving factor needed to preserve a dataset's class context. Data refreshes, in part or whole, that arrive according to a calendar are essential when preserving proper meaning. Data subject to calendaring has value for analytics so long as apples are compared to apples. For data to be valuable, you need to expose its context. You conceptually know that retail sales volumes on a holiday differ from sales volumes on a weekday or weekend. You also may know that a sales promotion or a free giveaway was launched on a specific day to run for two weeks. These contexts have to be associated with the data. When analyzing data, these calendars have to result in data being normalized and rationalized alongside the data, potentially forming corrective overlays (data lenses) that enable calendaring differences to not distort the macro truth desired. As an example, if an analyst wants to look at marketing campaign effectiveness, they would filter out all normal weekday and weekend sales as well as any sales *under* the campaign's excess volume trendline. If an analyst wants to look at overall sales trends, the sales volume over the campaign trendline needs to be filtered out so only normal sales before rather than after the campaign time period are taken into account. See how complex calendar handling can get! It gets even worse when having to report numbers by day within a month when the normal days in a calendar month vary from month to month. See also when, in a leap year, an extra day's worth of sales volume gets thrown into the mix every four years, thus offsetting sales volumes for the February reportable month, the first quarter, and then the yearly figures. Implementing a standard data/metadata calendar approach to record correct data context enables apples-to-apples analytics.

The goal we wish to achieve in this chapter is to make you aware that, when building out a future-proof, data engineered solution, you have to address tough issues first! Then, when data issues start to arise and they will you do not see your developers scrambling about to produce a cacophony of separate costly niche solutions. Rather, you contribute to a comprehensive, fully-thought-out solution upfront. This solution will be based on the architecture's design and reflect the experience and vision of the data engineering manager. This design should anticipate the need for solutions in these niche areas that *will* cause stress later.

In the overview of this chapter, we mentioned that you will need to address two problem niche types, defined as data profiling and metadata. You know that you will have to coerce ingested raw data into a form that allows it to be syntactically and semantically processable, correct, and complete. Some in the field refer to this type of activity as **data wrangling**. Afterward, you will have to decorate/annotate this cleansed data with various classes of metadata to preserve the original data's context with semantic meaning. The truth in the data enables it to be used and often repurposed in the future. The effectiveness of the **Data-Information-Knowledge-Insight** (**DIKW**) tree of knowledge will become evident based on how well you decorated it with meaning and context that is unambiguously discernable.

Data profiling

We will begin with what others have said about data wrangling with a formal definition:

> *"Data wrangling also called data cleaning, data remediation, or data munging refers to a variety of processes designed to transform raw data into more readily used formats. The exact methods differ from project to project depending on the data you're leveraging and the goal you're trying to achieve." (Harvard Business School {https://packt-debp.link/dZ7j9f})*

We want to underline that in order to future-proof your data engineered solution, you have to allow data to be useful to the consumer, and often the data scientist, so it is easier to consume. Data should be easily available and not require reworking. Data should be classed and then have interfaces created for it. These are to be retained in a data catalog as part of them being made fit for purpose. The effect of not wrangling data correctly for its use comes out in the following quotation:

> "... If you ask every data scientist they also say, I spend 80%, 60%, whatever, to do data wrangling, to do data cleansing, to change the data. That wrangling and cleansing needs to be delegated and decentralized back to the sources of the data. The data should be really usable when they get it." (Zhamak Dehghani, {https://packt-debp.link/SHc961}).

Putting yourself in the shoes of the data scientist, data consumer, and analyst is necessary when building solutions for them. You want to build interfaces to data that make that difficult task easier. You want to make sure the data is ready for use by them now and in the future, and you want *them* to also include all the future (not yet built) AI or machine readable consumers to be serviced. This way, data is ready for consumption in perpetuity. Feature extraction and the creation of semantically aligned embeddings within the extracted core dataset is part of machine learning operations (**MLOps**). If this is supported with the dataset and available before ML model creation, onerous MLOps tasks are reduced in scope. By decorating the dataset and reducing contextual errors, one gets better ML results. The effort is well worth the work since the outcomes have measurable positive effects on the ML algorithm's outputs. Data errors will also have been worked out in advance since the data factory's pipeline design implemented the best practices set forth in this book.

What you will need to do is create service interfaces for data for the sum of all curated data in your factory. All the capabilities and features invested during data profiling (also called data wrangling) output is consumable in a standard way with these interfaces. Today, a data scientist has to write many data extractors and that is just not very efficient. Examples include JSON, JSON-LD, SQL *du jour*, raw files, RAW JSON, Parquet, Thrift, Avro, custom object serializers (and we've built many), and so on. After legacy extraction, raw data is still not normalized, full of gaps, not smoothed, not normalized for its calendar, not fit for its expected trend, and so on. All the negative comments about how poor non-wrangled data is will come out as a stream of expletives hurled at the data engineer. Often, data scientists have to settle on subpar quality data that could have been transformed by your data factory to be what they expect. Making feature extraction and semantic alignment easy is a primary goal of a data engineer, and it is *not* an easy goal to achieve.

So, we suggest that you put a lot of effort into the structure of the data and its metadata context. This will allow data extractor interfaces (the class object's *getter* methods) to do any final rendering of the data into a form that the data scientist requires. You can't control the extracted data's use, but you can create the equivalent of a semantic result set that allows your pipelined, curated data to be rendered in a unified data and metadata form for whatever use is needed. Like an SQL result set, this result should be in an iterate-able list that can be subscribed to as data is presented for entry in the ML model's training set. It can be streamed into an ML-consumable location for processing (such as a Databricks Delta storage-accessible area, often in cloud storage). If you implement this stream as a pub/sub subscribe-able endpoint (such as **Kafka**, **Kubernetes**, and the like), you will make the data scientist more than just happy. This way, as data change arrives due to the normal life cycle of the system, the ML models and the core datasets the models work on are automagically adjusted by event triggered processing.

Data profiling is a key objective in your data factor solution, supporting the data scientists' activities today, and later, the automated artificial intelligence driven data scientist models of the future. Models delivering models is to be expected, similar to the way AI is being used to generate code via ChatGPT, or code-assist toolsets. Data is therefore expected to be *smart!* Data has to be understandable and fungible as well as carry the intrinsic meaning to be understandable and not misused. The fact that 80% of a data scientist's job is consumed with data wrangling tasks, as identified by Zhamak, has to be a problem that goes away. Some just shrug their shoulders and say *it is what it is!* They choose to ignore this massive problem. There are solutions but few have been applied, nor have successful ones been commoditized for general use by non-programmers and semi-skilled programmers. Data scientists have been forced to become data engineers in the absence of easy to use datasets that carry semantic meaning. Imagine if data could be snapped together as easily as LEGO pieces can snap together. One does not have to be a rocket scientist to build a rudimentary LEGO model. It is exactly this need that data scientists have for their data. The data engineering challenge is to build this snap together equivalent to LEGO. How does one align current and future state knowledge engineering and data engineering processing, code, methods, and tools to the task? The answer is to focus on the interfaces within the data platform and make them smart as well as conformant to a standard pattern. Just like LEGO snaps together, each data facet you create for a semantic class should snap together in a clear way and only in that way. This is why free-range methods of facet alignment provided by topic model **labeled property graphs** (**LPGs**) are just too flexible. They do not conform to semantic standards such as **RDF-OWL**. With semantic alignment not technically possible, each topic graph has to be glued together with custom code. You need to command a dominance in the business arena to set and enforce topic graph standard structures and interfaces. With RDF-OWL-modeled semantics, the knowledge engineer builds a model that enables clear knowledge representation. The reason this *"a-ha"* moment has not taken hold in the industry is that it is not embraced by some of the leading cloud providers or the big data tool vendors whose software engineers do not fully grok the data science and data engineering needs. Some have listened to the data scientists' data wrangling challenges and produced big data platforms, such as these lakehouse platforms:

- Databricks {https://packt-debp.link/qBSMA6}

- Teradata {https://packt-debp.link/0TvCOE}

- Cloudera {https://packt-debp.link/OoN2s3}

- Oracle {https://packt-debp.link/seyIN4}

- MySQL HeatWave {https://packt-debp.link/sWMfn0}

- Snowflake {https://packt-debp.link/wFFwaM}

- Google {https://packt-debp.link/isW81S}

What is missing in all of these solutions is the vision that this book presents – this need for a data LEGO equivalent standard.

Consider this quote:

> *"Organizations that want to build their data lakehouse using open source technologies only can easily do so by using low cost object storage provided by Google Cloud Storage, storing data in open formats like Parquet, with processing engines like Spark and use frameworks like Delta, Iceberg or Hudi through Dataproc to enable transactions. This open source based solution is still evolving and requires a lot of effort in configuration, tuning and scaling."*
>
> *(Google {https://packt-debp.link/r7TVfq})*

We are not trying to take away from any of these fine cloud products' capabilities – not at all! They have all been and will continue to be successful. It is you, the potential next disrupter in the field, who will champion the vision where data stands alone, floats between systems, and is not hindered by the big data platform or vendor lock-ins that each of the previously mentioned systems tries to implement, even if they are marketed as being *open*. One cannot create a common syntactic and semantic interface to support data facets with this divergent soup of technology! It makes the data wrangling efforts *worse*!

So, if you fixate on one of the possible choices that have to be made, but still fear that a year out from now, or maybe three, that choice is going to be the wrong one. Think about those who picked the Hadoop stack in 2010 and what had to be done five years ago just to re-platform what was a new and is now a legacy system. We've been there and done that, and we share your pain. What you will want to do is to lean in on the task of defining the data backplane of your system and not get misled by well-intentioned software engineering mindsets that are being brought in when hosting on the cloud. Those capabilities are being sold to the data engineer as the best thing since *peanut butter*! They are for the unenlightened to consume. Keep asking questions related to making data self-aware, self-serviceable, true, discoverable, and able to snap together with other data. You may get a blank stare back from your consultant, followed by a relaunched marketing speech on high speed, auto-scale, Spark, JSON, blah blah blah. Hold on to your data strategy and demand help in building out the vision of data as a product and not as a collection of JSON stuff the data scientists have to guess at. Help the data scientists not have to figure out your data and munge it together thousands of times in thousands of different ways, only to later learn that the source data was not meeting its data contract to start with. All that expensive, time consuming work could lead to insights that are proven to be shortsighted and potentially wrong. Your customers do not want to do IT again and again and again. This is *not* what we mean by building a future-proof solution.

A predominate reason you need a data architecture to address future-proofing needs is to make data easy. Consider the scope issue as defined in Statistica's research, as follows!

Worldwide data growth rate (2025)!

Between 2021 and 2025, Statistica predicts worldwide data generation will jump from 79 to 181 zettabytes per year, increasing at a 23% compound annual growth rate (Statistica {https://packt-debp.link/JZv09V}).

A data architecture that can scale will be one that is able to coerce data into a fit-for-purpose form without the years of custom coding required to just access that vast amount of data in all its divergent forms. A data architecture that presents normalized semantics for its datasets as information and knowledge in a readily consumable form for ML/AI or analytics use cases will present a compelling TCO and ROI.

This book does not allow us to get into all the details of the technology learning curve needed to transform a data engineer into a knowledge engineer who the data scientist wants as their go-to data person. The body of work for the RDF-OWL technology was produced at Stanford University and the W3C and with it, later, came **SPARQL** query abstraction and **SHACL/SHEX** validation. There are conferences and working groups with many high performing tools such as Stardog, Ontotext (GraphDB), PoolParty, and Cambridge Semantics (Anzo) to name a few. Then there are the labeled property graph (roll-your-own) databases that do not enforce semantic modeling constraints but do provide high performing graph operations. Before any technological breakthrough comes, there is always great frustration, cost wastage, and hyper-competition that drives innovation. What we do know is that the data platform that you build will have to shift to conform to an eventual semantic normal data form that reduces the data scientists' wrangling tasks to almost zero.

Years ago, the movie *Minority Report* {https://packt-debp.link/jC1mU6} portrayed a scene where potentially relevant (but disconnected) data was displayed on a transparent glass wall where all data was summarized and rendered in graphical form. Roughly speaking, each dataset looked like a circular chart with notched edges that could be rotated until it aligned with another and became a new intersected dataset. The protagonist of the movie, played by Tom Cruise, was trying to perform a real time visually rendered analytics reduction task (lots of words here, but this is the best way to describe what he was trying to do). All this data manipulation led to an insight based on the intersection of these data facets. What was interesting was that data was visually brought in, turned with the actor's glove, and cast off to the side or locked together to form this insight. Data was *fit* with other data along *facet intersections*, and as a result, the hero was able to get the answer to his implied question of *who did it?* A solution came out quickly and efficiently. No coding was required and as with all fiction, the end result was just stellar! The answer was found, and it was found quickly. The insight did not require Apache Spark processing, Python or distributed/multithreaded Java coding, a master's degree in machine learning or certified cloud *and* Databricks qualifications! The data facets that were leveraged were conceptually clear and visualizable. They could be intersected based on edges that fit together, and as a result, the intersections produced a result in near real time. The intelligence of a human was still needed and that was applied to the scenario; however, in the future, even that intelligence could be automated by AI. Imagine the possibilities if you could see a way forward with this type of capability for your data factory solution's releasable datasets. Producing these faceted data LEGO pieces and supporting the data scientist and data analyst's needs in a clear and visualizable way will be of astounding value.

Heuristic data analysis

Heuristic is a word coming from the modern Greek language, *heuriskein*, meaning *to discover*. The heuristic approach to problem solving leverages prior knowledge and experience when providing solutions. The codification of prior experience with data when assessing new raw data for analysis includes ML, deterministic processing, prior trends, data context, semantics, and prior judgments that provide feedback on decisions. These factors enable future processing to be accurate, timely, and of high quality. Heuristic approaches also include making educated guesses about data that form hypotheses that are resolved to be *true* or *false* later when additional supporting or refuting data arrives.

Your data-profiling tasks should include heuristic approaches {https://packt-debp.link/5F2597}. You will want to learn about some of these approaches by researching what others have done in various fields {https://packt-debp.link/ca3osh}. It is best to leverage third party tools and integrate these tools as part of your architecture. With traditional through to deterministic data preparation, the software engineer will not be able to supply all the necessary data contexts without first processing the entire dataset as a whole. Running many queries and aligning context can be time consuming, involves hundreds of rules, and result in anomalies that require human remediation. That just does not scale to tomorrow's data volume needs! Granted, there will be deterministic, rule based data profiling steps, but when setting trends and assessing gaps and overall context a heuristic approach is required. Some examples of this are as follows:

- Checking for compliance to the fact that any arriving data value must be within one standard deviation of the mean requires the mean to first be calculated after all the data is ingested.

- Seeing that a retail item count is off by an order of magnitude for a region, caused by a missing attribute on some of those items, causing some items to be confused with others in a different class and, as a result, summary counts were inflated. The error detected for the missing differentiation was caused by that single missing attribute. Detecting this and fixing the issue was only possible after the trend's data contract was broken! That check could take place only after all sales volume data was totaled for the item.

Data wrangling and repair also require a good rule based, deterministic, tool driven approach. Heuristic approaches provide direction and scope to the software engineering tasks required by the data engineer to correct anomalies. If not clarified early, the data wrangling tasks become trivialized, or worse yet, left to the data scientist, who will not have the skills, time, or desire to fix data that should already be clean, normalized, and validated.

Preserving the context of arriving data is an essential requirement for driving wrangling processes. All data is time sensitive, even if not recognized or recorded as such. As an example, the price for a retail item is very important today (t-=0), but 10 years from now, (t=+10) it is *not*, due to inflation and other factors such as the specific inflation factor for that item's class. Capturing inflation rates, and with that, the correct kind of inflation rate for an item class as part of the data's context, is essential. Historical price analysis is worthless unless an old analytic lens is replaced with a new lens and they are designed to be exchangeable in advance. The statistical profile captured with the item's price (as one example of a context type) is important for the future. Dimensions such as the percentage of that item's market share or the price deviation from the item's class mean should be retained. Imagine not capturing enough context when data is first ingested; it becomes useless, or worse, it may be used incorrectly! All data and data context should coalesce into patterns that drive future heuristic processing. These formed patterns can be detected in other data classes with similar properties, so insights just pop out of the data as pattern recurrences.

Capturing data context (time, trend, etc.) is essential as you shift your attention from deterministic processing to processing data heuristically. One final suggestion: treat data as if it always has a time context, even if it is not explicitly handled as time-series data. This way, you always have that feature to leverage as you correct data and account for drift that you know will occur in time.

Trend deviations

Trend deviations are deviations from an expected set of values defining a baseline for a known set of values within a class. A dataset's values should exist in a swimlane of acceptable values for a named trend over a given timeframe. A *standard deviation* (σ) is a measure depicting the degree of data dispersal in relation to a dataset's mean value. Data that is clustered around the mean will be within one standard deviation and data further out will be in the second and third deviation and so on, indicating data is more spread out. Data values of the same class can be charted, with an x-axis being time, and real values on the y-axis above or below the trend mean at any point in time. Visually, this looks like a ribbon, with a center line being the mean and the width of the ribbon being the acceptable deviation for that trend. An example can be found in *Figure 13.1*, where a confidence line is defined to predict what the trendline could be for a similar dataset within a given probability. In the tool, you can select a probability range for the confidence line from 70% to 95%. The following chart is drawn with a 90% confidence range setting:

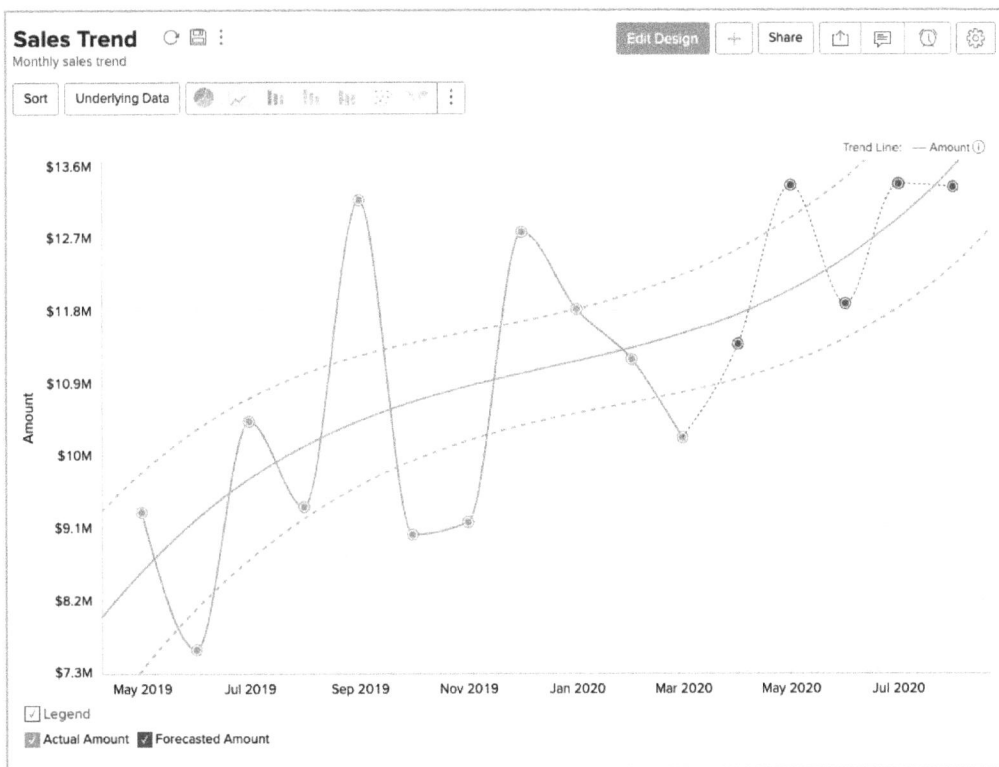

Figure 13.1 – Zoho swimlane example chart for sales trend {https://packt-debp.li1nk/OtxSUz}

When profiling data, you will want to extract the degree to which a data point value adheres to an existing data trend, then classify the degree of adherence into some buckets such as std1, std2, or std3 (distance from the mean). The trend's definition will indicate which deviation is acceptable. There are going to be many other ways to classify deviations, but we have discussed only one method (standard deviation).

There will also be new trends that are discovered through your heuristic processing. Integrating and leveraging third party trend sources and technical tools for this type of analysis is prudent. Some of these vendors and tools are as follows:

- Societal trend sources

- Other non-technology-focused tools

- Technology-focused tools

Some societal trend sources enable you to align your data with external norms. This is prudent to do in order to add credibility to your efforts and pop out new insights that are only possible when these norms are aligned and then augmented with your data. Here are some useful trend sources:

- Exploding Topics {https://packt-debp.link/WoybzT} could be used to discover new trends in data. A trends database enables existing trends to be cataloged for discovery and reuse.

- Axios Pro {https://packt-debp.link/jkmbnM} can discover news trends by topic. It is used by finance, healthcare, media, retail, climate, and government users.

- WGSN {https://packt-debp.link/nxA1f5} is a data analysis toolset for consumer brands, used by marketing teams to create messaging aligned with observed trends.

- Qmarkets {https://packt-debp.link/n2Napt} has trend management features and many other capabilities. It is used to gauge customer demand.

- Google Trends is a free trend analysis tool. It's reliable but not all that relevant at the level you will need to assess trend compliance.

- Pulsar Platform {https://packt-debp.link/KjSXa7} is an audience analytics tool that leverages social media data to track conversations in order to learn about community sentiment about competitors.

- CB Insights {https://packt-debp.link/lSxGyj} is used to provide competitive intelligence information based on company financial data. Technology competitive analysis is the focus. The CB Insights database provides information on trend analytics.

- Trends.Co {https://packt-debp.link/PgXjAQ} is a company that produces market reports on various trends by community. Start-up leaders use this to harvest new ideas from the signals available in the tool.

Other non-technology focused tools {https://packt-debp.link/3UbXmd} may be organized by domain, such as the following:

- AI trend analysis tools

- Big data analytics tools

- Competition analysis tools

- Customer journey mapping tools

- Expert advice

- Google Trends

- Industry reports

- Market research tools

- Sentiment analysis tools

- Social media monitoring tools

- Trade shows and conferences

- Your own research

Technology-focused tools are applied to get into the technical identification of data trends that are intrinsic to the data values being processed in your data factory. These are citizen data scientist tools, such as the following:

- Altair Rapidminr {https://packt-debp.link/YMZ0ZE}

- Alteryx

- Alteryx (Trifacta Wrangler) {https://packt-debp.link/kjaH7P}

- Microsoft Power BI {https://packt-debp.link/RdluoB}

- Qlik Sense Enterprise {https://packt-debp.link/mRPViY}

- Quest Toad Data Point {https://packt-debp.link/oukXsc}

- SAP Data Intelligence Cloud {https://packt-debp.link/haDtM0}

- Tableau {https://packt-debp.link/sp4re9}

- Talend Data Fabric {https://packt-debp.link/6hIVq0}

Gaps

When it comes to data profiling to detect gaps (missing expected values in a timeframe), you will want to look at *broken trend* anomalies in the dataset. This means that you have to have designed and built a data anomaly detection capability first. That requires special heuristic and deterministic logic to specify various data trends within a data class and for a given time period (the data's calendar contract). Then comes the task of justifying why you think a data gap has arisen when the data source is supposed to be trusted (expected to meet its contracted obligations).

In the past, the supposed gap could be tested by filling the gap with synthetic data and rerunning the anomaly detection logic to pass any credibility tests. Then you could, via numerical analysis methods, iteratively try to explore the edges of the data gap's boundaries until you can get a lower and upper bound on the values that would allow the synthetic data to be acceptable and still not break trend. Then, lastly, you need to select the mean values between those bounds as a best guess for the missing data values. After selecting this nominal set of gap filled data values, you need to create metadata to identify these values as synthetic data. Processing in the data factory can proceed normally until a data restatement arrives (from some source) that definitively replaces the values in the filled gap, or the synthetic data could remain in perpetuity. The synthetic data used to fill the gap will never be 100% correct data, but it will be good enough for downstream processing use. The gap-filled data values will make ML, AI, and analytics applications and insights correct. This type of gap/error handling will be music to the ears of the data scientist, and it will cause your auditors to have a fit. Be ready to render results with and without the synthetic data in order to satisfy them.

Overstatements

Just as gaps are problems in datasets, **overstatements** (data points received multiple times in a given timeframe) are also anomalies to be corrected. They also can be detected since they break trend, as does the presence of a gap. Just as a gap is a valley in the data, an overstatement is a mountain! Both have to be flattened for the dataset to be credible and processible downstream. How the flattening process takes place should be through any number of deterministic or heuristic algorithms that layer change after change over the raw data until it is smoothed to fit the expected trend curve.

The data adjustments should be stored as deltas to the raw data values so that when all corrective layers are munged together, the overstated data values form an effective value that is correct. Earlier, we alluded to the need for data lenses to be created and applied to data values so that adjustments could be made. These lenses also provide adjustments needed to produce the effective value that does not change the underlying raw value. These adjustments can be created by complex statistical processes. Some are as simple as a single instance run of overstatement correction code or having the data *fit a trendline in an expected swimlane.*

Adjustments to overstated raw data values should be factors to be applied over the data, such as adding a negative value, dividing by a number, or applying any other reducing factor/function that could be envisioned. An immediate goal we have for you is to visualize in your mind the need to create a stacked data point object. That object has the raw value and all the layers (deltas) leading to the interface's effective value aggregate return as a stacked data point. These layers are all retained in the metastore for lineage tracking purposes. The added benefit of this approach is that if one of the adjustment algorithms has to change or rerun to produce different adjustments in the future, it can be run without affecting other algorithms' prior adjustments. Being able to yank out and replace these adjustments without upsetting others is essential so that the effective value can always be trusted for use in downstream processes.

Summary statistical deviations

Just as gaps and overstatements can be detected by anomaly detection logic, so can statistical deviations arising from real time analytics applied over the dataset, as additive and non-additive measures, metrics, and statistical models are applied over any scope of data. Being able to create alerts when baseline anomalies are first detected is essential for data contracts to be maintained. Any derived data should also be entered as metadata and the original raw data identified as the derived data's context. This derived metadata is now subject to trend analytics quality checking, just as the primary source data was. Data is still data! It is not a mystery to anyone that statistical metadata could also have gaps, violate trends, or be overstated. Treat the metadata as data.

When statistical deviations arise, a programmer or statistician will have to get involved to create new data corrector lens(es) in order to account for *true positive* anomalies. Resolving the status of an anomaly can be labor-intensive, but data should *never* be manually tweaked. Always write new data corrector lens(es) after understanding the anomaly fully. Filters and factoring algorithms are run to produce adjusted stacked data points in order to resolve the anomaly as part of normal data governance practices. False-positive or false-negative anomalies are also easily handled in this manner by adjusting the anomaly detection logic itself, which means entering a judgment or *retaining the ML model.*

Calendar trend deviations

When processing data that has to be segmented by date, you first want to define the time buckets representing the equivalent dates over some longer timeframe. The granularity has to be set so that the number of day units represented in a bucket is comparable from year to year. As an example, buckets by month require special processing since each month has a different number of days, and in leap years, February will have an extra day added every four years. Running a query for sales volumes from month to month will have to be offset by the number of days in the month to make a real comparison. Note that the number of weekends also varies from month to month; this affects buying habits. This all means that the trend by month for a given calendar year compared to prior years' trend may not align. An analysis could lead to erroneous results and misleading insights if data calendars are not in alignment. You will learn of two core calendaring issues in the following section: **data calendars** and **event calendars**.

Types of data calendars

Obtaining retail insights is where data calendars have always played a key role in defining segments for apples-to-apples comparison. Oracle® Retail Insights {https://packt-debp.link/DIJwXL} has support for five dimensions of calendaring, as follows:

- Business calendar
- 4-5-4 calendar
- 13-period calendar
- Gregorian calendar
- Time of day

Business calendar: This is also called a time calendar. It is a fiscal calendar and set according to the financial calendar and not the Gregorian calendar. A business calendar can be based on some variation of the 4-5-4 calendar (see next) or the 13-period calendar (also defined shortly). The allocation of exactly seven days to every week is required. This is different from the Gregorian calendar.

4-5-4 calendar: The 4-5-4 calendar can be implemented as *4-5-4*, *4-4-5*, or *5-4-4*, depending upon your needs. You can determine the day of the week on which each week begins and ends. The quarters contain thirteen full 7-day weeks, two 4-week months, and one 5-week month. This 4-5-4 calendar is subject to the 53-week-year problem because each business calendar year is shorter than a Gregorian calendar year. The 53-week period of such a year should be associated with a 4-week or 5-week period depending on the business requirement, resulting in a 5-week or 6-week period, respectively. What's interesting is that when performing an analysis on data segmented by a 4-5-4 calendar, alignment between the business calendar and the 4-5-4 calendar should be automatic. Fiscal periods have the same number of days/weeks from year to year and the calendar mapping from the current year to the prior year has to be shifted. In an unshifted calendar, each year's first 52 weeks are aligned with each other, and the final 53rd week is omitted from year-over-year value comparisons. In a shifted calendar, the 53-week year is restated for the previous year's comparisons, such that weeks 1 through 52 of the following year align with weeks 2 to 53 of the 53-week year. Week 1 of the 53-week year can be used as the last week of the prior year for that year's week 53, or it can be omitted.

13-period calendar: A 13-period calendar year is divided into 13 periods of four weeks (28 days). Every fifth or sixth year, there are 53 weeks. The calendar has a 28-year cycle of six years, five years, six years, six years, and five years.

Gregorian calendar: The Gregorian calendar is our normal solar calendar that, as you know, is based on the length of the Earth's revolution around the sun. It is divided into 11 months of 30 or 31 days, plus February, which has 28 or 29 days, depending on whether the year is a leap year (every four years). A Gregorian year is either 365 or 366 days.

Time of day: Our standard day of 24 hours is broken down into periods of 60 minutes and 60 seconds. Standards have been created to normalize this time called **Coordinated Universal Time** (**UTC**) {https://packt-debp.link/JTUbkg}.

What you need to know is that if your data can only be analyzed correctly in the context of a data calendar, support for *calendaring segment analysis* will have to be built into your solution. Building this capability from scratch without any third party tool support can be exhausting, but it is a necessary part of making the data usable and not misused in downstream use cases. Our suggestion to you is that you don't leave it to the data scientist to figure it out!

Types of event calendars

Standard calendar buckets are great, but how does one create an event driven segment definition that works like a bracket across a segment of data in a dataset that defines some business event that took place during that timeframe? Examples of event calendars are as follows:

- Marketing campaigns
- Geopolitical events
- Corporate events/actions
- Other notable events

Marketing campaigns: Marketing campaigns run from one moment to another and can be used to drive an entire analytics scenario to gauge the efficacy of a price promotion.

Geopolitical events: Geopolitical events may occur that define a period of abnormal sales volume or financial sales activity that, when known, defines how this anomaly data is to be factored out or segmented out of the dataset or else it skews results.

Corporate events/actions: Corporate events/actions have to be marked in the company's general ledger for audit purposes. This notation helps an auditor explain abnormalities in the financial record. The notation also has to be available for segment analysis because those actions set the context for interpreting any data collected during the corporate action's timeframe.

Other notable events: Noting any external event that could result in an analysis segment has to take place when the data's context is set. These ad-hoc events may be free form text retained as metadata to be used later to create formal event types if and when new segments are proven to be creating anomalies in the data trend.

You can see that processing data by calendar buckets or for ad-hoc events requires some thought. Our suggestion is that you do *not* leave this capability out of the solution design. The data scientist will not appreciate having to figure out later how to segment your dataset by calendar, especially if the segment information was not gathered when the data was first ingested. Just because a software engineer supplied a very granular *jiffy* - level {https://packt-debp.link/BAricy} timestamp on the ingested data row, that does not mean the data is easily useable. Calendar and event segment enrichment of the data with decorative metadata enables the consumer to use the solution's data as intended.

ML-driven anomaly detection

Being able to apply machine learning to the task of detecting anomalies and other statistical outliers has been shown to be of great cost saving value. The cost of manual data anomaly detection or of creating deterministic, rule-based program anomaly detection can get very high. Manual anomaly resolution workflows do not scale as the volume grows and even with rule-based detection, well, you will see anomalies you did not expect (that's why they are anomalies); so, there are never enough rules to comprehensively cover your detection needs.

Training a machine learning model to detect and correct an anomaly pattern is a semi-supervised ML problem for the **support vector machine** (**SVM**) {https://packt-debp.link/ArgUFg} class of ML algorithms. Unsupervised ML model varieties such as **Latent Dirichlet Allocation** (**LDA**) {https://packt-debp.link/aD5xk2} could be used to detect patterns of features after they are found to cluster together. Once the pattern is known and then named, it could be used in the development of a training set of an upgraded SVM model, which can then detect the recurrence of the anomaly pattern in future arrivals. The pattern could also be used to drive data correction processes, so anomalies are automatically handled and corrected in the future. Doing this over and over enables the ensemble of ML models to cover all expected data behavior over the course of time. Anything not matching known patterns is, therefore, an anomaly.

It sounds simple, but I assure you it is not! Creating these models can be difficult, and they are subject to *drift* {https://packt-debp.link/DAdKBy} and need constant maintenance. Our suggestion is to segment incoming datasets and apply the previous ML approach to those segments. Gauge quality ruthlessly and in real-time. It is a good idea to study what others have done in this area and to look at third party toolsets. You can see what Dataiku {https://packt-debp.link/MqRT1r} has built as a data anomaly detection solution for banking (insider trading and so on), retail (QA minimizing faulty items), and healthcare (for earlier detection or treatment of disease). Do not try to boil the ocean! By this we mean, do not think that a 100% ML driven anomaly detection capability is possible. Go for the easy and valuable high-worth wins first!

Gathering and applying judgments

It may be a non sequitur from our elaboration on machine learning based anomaly detection, but we want to stress that no matter how good the stochastic approach is, it will always need to reflect subjective human judgments to be correct. These judgments have to be applied if and when a prediction made by the anomaly detection models is just *wrong*! 95% correct is great, and 100% is best, but in reality, you will implement ML based anomaly detection and measure its quality... and then you will see it degrade/drift over time, no matter what you do! Unless human judgments are entered and factored into the solution in such a way as to boost the effective quality to at least 96%-98%, the solution will lose senior management support because data ingestion TCO costs will rise. A 98% efficacy score is a stretch for many ML ensembles to attain. It is possible with some thought and careful measurement. After measuring quality, you will want to enter and maintain the quality metrics in a repository, as well as the human judgments needed to get to that level of quality in order to do the following:

- Direct the training of the models

- Reject *false-positive* outputs that are wrong

- Promote *false-negative* outputs to be true anomalies if they are determined to be real data errors

Erroneous data anomalies must not stop the data factory's pipeline flow, but be set aside to be included as part of the data correction process for which your data engineered solution is designed to handle. Once you have cleansed the datasets in your factory by eliminating anomalies and annotating acceptable values, you will want to turn your attention to the recording of the dataset's metadata in order to make the data catalog useful to consumers.

In the next few sections, we will elaborate on the options and recommendations for creating a metadata strategy. We will highlight some of the divergent thinking in this area as well so you can decide what is best for your organization. We will try to hold back on our prior positive biases toward the creation of a true semantic model so that you can objectively gauge the state of the art available and dialogue with various vendor and cloud service provider **subject matter experts** (**SMEs**) in a meaningful way.

Metadata strategy

What's metadata? The setup for this answer was provided in *Chapter 3*. Much emphasis was placed on metadata's three types: *descriptive metadata* (to catalog dataset contents), *administrative* or *lineage metadata* (the manner in which the dataset was curated as well as cloud topology and configurations), and *structural* or *semantic metadata* (formal information as semantic knowledge). Also, in *Chapter 4*, a number of principles were laid out to bring the issue into focus. The need for metadata in your solution is aligned with these principles:

- **Principle #10**: Metadata is associated with datasets and is relevant to the business

- **Principle #11**: Dataset lineage and at-rest metadata is subject to life cycle governance

- **Principle #12**: Datasets and metadata require cataloging and discovery services

- **Principle #13**: Semantic metadata guarantees correct business understanding at all stages in the data journey

- **Principle #15**: Implement foundational capabilities in the architecture framework first: zero-trust data security, test first design, data profiling, metadata design, auditing, and machine learning anomaly detection

The metadata topic also appears in *Chapter 5*, where your solution's *Conceptual Architectures Glossary* should reference various metadata classes across the architecture.

Our elaboration in this section will enable you to lean in on the issues that will be pushing you toward a **metadata design** solution needed to implement an end state strategy. Providing coverage for these metadata areas will consume a lot of your time. The goal is to preserve semantic integrity and make datasets discoverable and useful for the data scientist, analyst, and business user. We'll begin with a clarification of the divergent thought processes so you can converse with vendors and cloud providers without whipping up techno-religious fervor, and if you think this topic does not bring out the passion in a technologist, you will quickly find that is not the case.

The analyst, and particularly the data scientist, wants to see clean data. Once they have access to that cleansed data, they will want to know how good it is. That is going to be a subjective answer to the quality question with data transparency brought out through the data engineer's three types of metadata. Ask a cloud provider about metadata and you will get a lot of answers related to *administrative* or *lineage metadata*. This will involve an audit trace of the PaaS service calls made as data is transformed through various orchestrated flows. You may even get tools such as Microsoft Azure Purview to track service lineage calls made over time and be able to reflect on the metadata repository for discovery purposes. We'll leave it to you to determine whether the product today or in the future will address the data engineer's administrative or lineage metadata need, but you need to know that the cloud provider has implemented complex distributed services that obfuscate data lineage. Without a tool such as Microsoft Azure Purview in Azure, you must build lineage as a greenfield implementation. You do get coverage for most (not all) PaaS and then some SaaS, as well as third party integrations through OpenLineage that can be integrated with Purview. If you want access to the underlying metadata data store, you will be out of luck. If you also want multiple parallel versions of your metadata store mapped to your *blue/green* deployed processes, you are also out of luck. This last point is a serious consideration, since your solution is going to be released as sets of PaaS integrations and you want the metadata partitioned the way you want it partitioned and not as some tagged hairball of a metastore entry. Then, you will not be able to reflect/query those entries since the metadata was not treated as data. Similarly, security metadata is kept in Microsoft Graph and not integrated with Purview (at least, it was not before we asked Microsoft that it be integrated in 2021 when it was being developed). Purview is a great cloud product, and it takes Microsoft light years ahead of others. I'm sure it will evolve as time goes on because the community of users will weigh in on the value needed and future enhancements will be driven by customer need.

Now, if you ask teams outside the Microsoft Azure cloud team to define metadata support, they will respond from the data analytics and data warehouse perspective. In the next section, you will learn a lot about the cloud software developer's perspective.

Azure cloud provider perspective on metadata

Many cloud software product engineers see that metadata is data about data; however, it requires a systems integrator (programmer) to get that metadata to be useful. Microsoft software and data engineers think about metadata in terms of the table and object relational models they need to expose to the consumer. They use terms such as these:

- **Tabular models** in **Azure Analysis Services** (**AAS**)

- **Power BI Desktop**

- **Tabular Model Scripting Language** (**TMSL**)

- **Analysis Management Objects** (**AMO**)

- **Data Analysis Expressions** (**DAX**)

Tabular models in AAS run an in-memory database or in Microsoft DirectQuery mode AAS connects to backend relational data sources. The AAS **VertiPaq** analytics engine delivers access to tabular object model data via business intelligence applications (such as Microsoft Power BI Premium).

Power BI Desktop can create reports and when it does, it creates semantic model metadata in accessible **PBIX** and **PBIT** files (note that legacy PowerBI versions retained metadata in a closed proprietary format). Automatically, PBIX files are updated with new metadata when opened by the analyst. Power BI Desktop (and its APIs) are aligned with the table models in AAS, which conforms with the **Tabular Object Model** (**TOM**). Microsoft engineering considers this the *enhanced semantic model metadata* feature and views it as strategic and foundational. It's just not as semantic as you will want it to be! It does contain information about the data, but it does not contain any real structural or semantic metadata as we have defined the need. It is more of a tightly coupled data catalog. We are not trying to take away from its core value at all. We are merely trying to point out that there is too much technobabble in the interfaces and structure of these Microsoft developed semantics in order to make it at all useful to the data scientist. In fact, it frustrates them even more!

TMSL is the command and object model definition syntax for Microsoft's tabular data models. TMSL communicates with AAS through the XMLA protocol, where the `XMLA.Execute` method accepts JSON-based scripts in TMSL- and XML-based scripts in **Analysis Services Scripting Language** (**ASSL**) for XMLA. Microsoft points out that all future Power BI functionality will be built upon its metadata.

AMO is a libraryof objects that enables applications to manage AAS instances. AMO supports AAS, **SQL Server Analysis Services** (**SSAS**), and Power BI Premium tabular models. TOM is used for later versions, and you can see the Microsoft product teams dropping proprietary metadata approaches for standard ones. TOM is the extension of the AMO client library needed for this unification.

DAX is a formula language used to create custom calculations in AAS, Power BI, and Power Pivot (within Excel). DAX formulas will contain functions, operators, and values to perform advanced calculations on table columns. DAX is core to AAS tabular models created via Microsoft Visual Studio.

Microsoft's view of metadata leaves room for third party metadata tools such as AtScale {https://packt-debp.link/9teCPO} to arise (among many others).

A metadata tool provider's perspective on metadata

With AtScale and other such companies, metadata encompasses semantic truth about the data with an emphasis on administrative or lineage metadata and the data catalog, similar to the **Databricks Unity Catalogue** {https://packt-debp.link/Rft3K5}. The vision is that you need to design and build a semantic layer. See *Figure 13.2*:

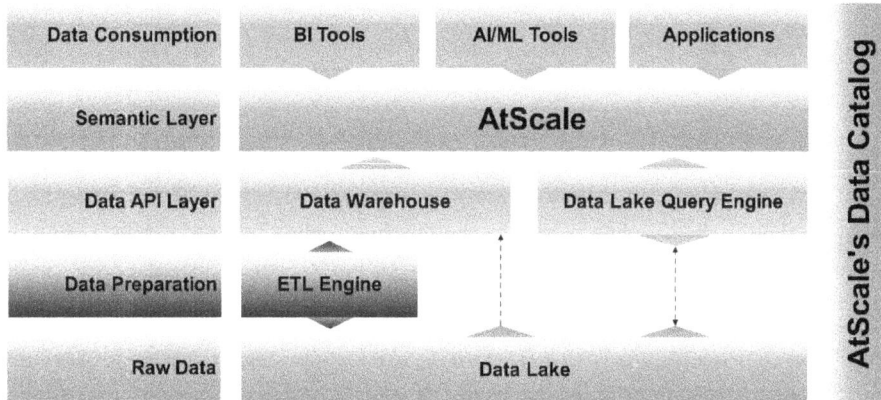

Figure 13.2 – 2023 AtScale semantic layer in the analytics stack

With AtScale, the semantic layer provides the intelligence needed to get to the data required.

> *"Experian reports that six in 10 companies believe that high-quality data increases business efficiency, 44% believe it raises consumer trust, 43% conclude it enhances customer satisfaction, 42% believe it drives more informed decision-making, and 41% report that good data cuts costs"*
> *(Experian via AtScale, {https://packt-debp.link/DFoVOe})*

The capabilities are shown in *Figure 13.3*, where integration, governance, and performance are not omitted from the list of essential needs!

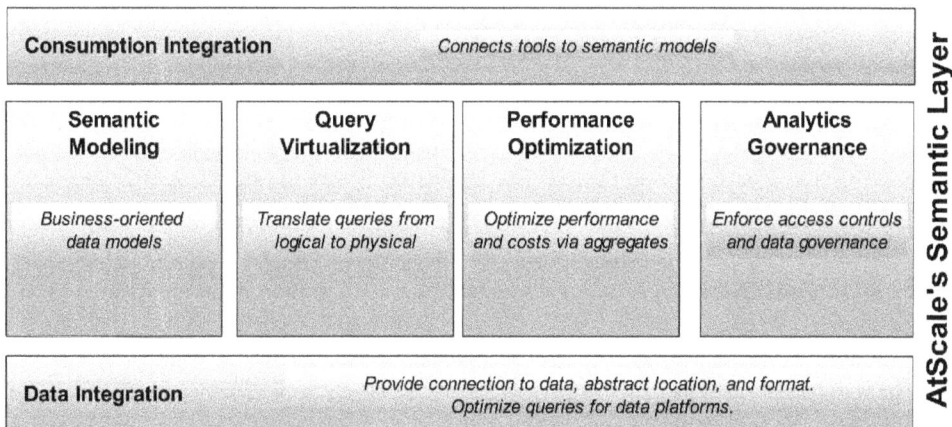

Figure 13.3 – 2023 AtScale semantic layer components

In *"AtScale's technical overview"* {https://packt-debp.link/Gec8uH}, you can find a brief definition of AtScale's Query Engine, which is an abstraction layer or query interface for business intelligence, AI/ML tools, and custom applications. Integrations connect through one of the following protocols:

- ODBC/JDBC (SQL) where AtScale looks like a Hive SQL warehouse

- XMLA (MDX or DAX) where AtScale looks like an SSAS cube

- Python where AtScale looks like a web service

- REST where AtScale looks like a web service

What we find most interesting is the semantic metadata layer's integration coverage that, in turn, makes the analyst and data scientist's jobs easier. AtScale is generally available today and appears to play well in the cloud provider service environments; however, if you want to extend it via Open Lineage and integrate it with some not yet supported products, you will have little recourse but to petition the vendor for product enhancement.

Now to address the final perspective, which is the need for formal metadata modeling when the dataset is complex with lots of semantic nuances. A few vendors now support this approach *"at scale"* (pardon the pun). For that, you want to explore the knowledge engineering approaches in the next section.

A knowledge engineering provider's perspective on metadata

For many, the mere mention of semantic modeling (the formal definition of at-rest metadata) and the implicit complexities of a formal OWL-RDF model for a business domain bring chills up and down the back of the data engineer. It's a niche technology and few are trained in the lingo, the techniques, and the best practices to make it work as expected. That is often left to a knowledge engineer {https://packt-debp.link/oJRDVR} to handle. This is why they are often called knowledge engineers.

Since it's not for the faint of heart, as a data engineer, move forward with caution. However, know that you will be rewarded with the clearest and most accurate definition of your solution's data semantics possible if you adopt this approach. Others, including cloud providers, punt on the topic and depend on custom coded **labeled property graphs** (**LPGs**), which are topic models that enforce semantic relationships. Examples are Microsoft Bing and Microsoft Graph Security. However, we'll continue with the semantic knowledge graph approach in this section in order to define a semantic model.

With a formal semantic model, you will get forward and backward inferencing capabilities without additional coding (this is called up via the SPARQL syntax query itself). You can get formal semantic data validation at ingest time for any instance of data associated with the **OWL** model you created and control, via leveraging SHACL or SHEX. These standards define the shape of data in such a way as to enable it to fit into your knowledge base (instance data and semantic ontology graph combined) without additional software coding. This is because validation comes from the model's enforced constraints. Note that vendor support is now available for semantic modeling, and the tools have matured (beyond just the first tool, Stanford University's Protégé) to support huge datasets at scale. We can't get into all the details of semantic modeling and how to work the technology into your designs, so we'll point you to a few vendors that have made great products in this field: Stardog, Ontotext, Cambridge Semantics, and PoolParty {https://packt-debp.link/nclKxX}.

These four knowledge graph vendors, among others, are worth studying. We've worked with many of these and other vendors in past lives and noted that Stardog is very impressive based on its cost, broad functionality, and willingness to provide professional service consulting to help get you started (as does Ontotext). Cambridge Semantics stands out based on its unique dual graph technology (OWL-RDF and LGP) as well as its professional industry knowledge. At this point in the chapter, you will want to understand more, not through a lecture but by example. Afterward, you can take a deeper dive into the technology.

The following are some examples of use cases and technology options that have been developed for various industries:

- Corporate use cases

- Technology use cases

- Business use cases

Corporate use cases include the following:

- The **British Broadcasting Corporation** (**BBC**) {https://packt-debp.link/7QTtB2} uses OWL-RDF as part of a linked data initiative that enriches content across its data systems.

- **Google** {https://packt-debp.link/qVhsHe} uses OWL-RDF in its search knowledge graph to organize information about entities and relationships.

- **International Business Machines** (**IBM**) {https://packt-debp.link/2Ud8O7} utilizes ontology standards in its information management and data governance products.

- **Oracle** {https://packt-debp.link/jGGFrx} uses OWL-RDF in its data management products, enabling semantic data integration and reasoning capabilities.

- **Siemens** {https://packt-debp.link/xr84Ct} leverages ontology standards in industrial automation and digitalization efforts for data interoperability and knowledge representation.

Technology use cases include the following:

- **Cambridge Semantics' Anzo platform** {https://packt-debp.link/SOGmGK} is a knowledge platform for many industries such as life sciences, financial services, manufacturing, government, and others.

- **EA Essentials Project** {https://packt-debp.link/o5cAnA} is a system with an extensible ontology for any technology enterprise.

Business use cases include the following:

- **Dbpedia** {https://packt-debp.link/DkijVm} is an ontology mapping to Wikipedia data in RDF format. It is a knowledge base of entities, concepts, relationships, and so on.

- **Dublin Core** {https://packt-debp.link/aFa7TS} is an ontology used for describing metadata of digital resources on the web. It is a vocabulary for information about resources, titles, authors, dates, subjects, and more.

- The **Financial Industry Business Ontology** (**FIBO**) {https://packt-debp.link/t8BSf3} is an ontology that defines financial business application interests and relationships. It is hosted and sponsored by the **Enterprise Data Management Council** (**EDMC**) {https://packt-debp.link/XjJKxz}.

- **Friend of a Friend** (**FOAF**) {https://packt-debp.link/XvBucF} is an ontology that is used in a social network context for people and relationships. It is a vocabulary for personal info such as names, email addresses, interests, and relationships (such as friendships, affiliations, and so on).

- **Gene Ontology** (**GO**) {https://packt-debp.link/IZbU9n} is an ontology used in molecular biology and genetics. It is a vocabulary for describing genes (and functions), cellular components, biological processes, and more.

- **PROV-O Ontology** {https://packt-debp.link/9LOiCu} is an ontology used for representing provenance data representing the origins, processes, and transformations of data and entities. It is a vocabulary to describe the data lineage of digital resources.

- **Retail Ontology** from the University of Toronto {https://packt-debp.link/IwW2go} is an example of a retail ontology.

- **Schema.org** {https://packt-debp.link/Hga7zE} is a search engine vendor that collaboratively developed a shared vocabulary for structured web markup data. Classes and properties describe various entities, organizations, events, products, reviews, and more.

- **Semantically Interlinked Online Communities** (**SIOC**) {https://packt-debp.link/K3ac1Q} is an ontology representing online community information such as discussion forums, blogs, mailing lists, and more. SIOC services content and metadata aggregation between online community use cases.

If you have the time, you may want to go through the endnotes for each of the vendors, companies, and tools identified in the previous section. Your research could get you into a rabbit hole of ontologies and use cases, so it's going to be an interesting journey. Then, you'll formulate an opinion about semantic modeling and how it could be used in your vision to implement a future-proof system. A key takeaway from this elaboration is that data has to be clean and smart and preserve its context for it to be repurposed in the future. This is part of the capability needed to enforce data contracts. The ontology approach to semantics is not a simple approach. It involves a lot of predicate logic and OWL-RDF modeling that some will find too difficult to master or get right for production within time and budget. Others have been successful, so it's doable and effective, but it will take very clear, data engineered design and buy-ins from senior leadership.

Metadata perspectives summary

After many years of working with data and semantic technologies in big data systems, we recommend that you put more effort into the development of semantic metadata than most data engineering efforts have required in the past. If you have lots of data and a relatively defined business domain with clear schema, then you may not want to embrace the complexities of formal OWL-RDF ontology modeling or build out your metadata as Microsoft has, using LPGs for their Bing, Security Graph, and Cosmos DB 2 products. However, if you have lots of intricate knowledge to manage, such as with finance, genetic cell, retail, or diagnostic health data, you will want to explore and then use semantic ontologies such as FIBO for your data and metadata storage.

After you choose your metadata-at-rest direction, you will want to handle lineage metadata using a data catalog that is search enabled. That enables entity discovery and drilldown to determine how values were curated into their current state. You'll also want to make sure the catalog covers cloud PaaS that will interact with your data along the journey. Additionally, you'll want to make sure that your SDLC and release plans can put logical brackets around your metadata groups so you can obtain multi version support across your catalog. This capability is often missing in third party and cloud provider metadata toolsets.

You will also want to make sure that metadata can be actionable in production. You may want to use the metastore and catalog to form a semantic layer (like AtScale does) for the automatic creation of skeleton **data access object (DAO)** interface code to acquire data facts and entities that are not directly exposed through a custom API. These are buried in many RDBMS tables and may even be accessible only through stored procedures. This facilitates the marshaling of data into analyst data spaces:

- MOLAP {https://packt-debp.link/mn0jUI} (cubes)

- ROLAP {https://packt-debp.link/PGDfAt}

- HOLAP {https://packt-debp.link/eazkXZ}

This easy way to populate data is invaluable to the data scientist and analyst. The transparency of data lineage and semantics would truly be a benefit to the organization and make your solution indispensable.

Summary

In this chapter, you were exposed to a number of areas for consideration in your designs. For data profiling, you want to really know your data, find data areas that are anomalies, and be ready to smooth the data, fill gaps, and if necessary, create temporary adjustments with synthetic data until real data can be supplied. When implementing a data factory, you need to know that the data is of the highest quality and that includes the core data, its metadata, and its trends. You should also include its data profile in that list.

You also learned that raw data is messy and has legitimate gaps and illegitimate (erroneous) gaps that need correction. It is important that you also know the shape of the data as defined by its profile so that it can be semantically maintained in downstream processing and not cause data to be misused.

You also learned about data calendars and that data has to be interpreted in context, one facet of which is the data calendar. This is essential to process data for analytics, and it can't be overlooked when processing data, so you want to systemically set this up in the design stage before jumping into a design that later has to be redone.

As a takeaway from this chapter, you will want to keep a sharp eye out for technology disrupters to your designs. These will appear as niche technologies with questions that will require answers. You will need to address these proactively and well in advance of them becoming destabilizing. Even the most thorough architecture can lose its future-proof status if it were not built to adapt and change.

In the next chapter, we will elaborate on data delivery considerations for your engineering designs. We will focus on data streaming, pub/sub, data flows, handling huge volumes, data organization, and confidential compute feature best practices.

14

Key Considerations for
Data Delivery Best Practices

This chapter will address important data delivery areas for consideration in your solution, such as the following:

- Data streaming

- Publish/subscribe

- Data flows

- Huge volume implications

- Efficient data organization

- Confidential compute

In this chapter, you will be made aware of the data delivery options that need to be considered when you design your solution. We begin with an overview of the topics that will be discussed. The goal of the overview is to level-set you on some effective approaches to deliver curated data as a product to consumers. Later, we will dive into the impacts of the selected delivery method since some consumers require different capabilities than others. Although one size does *not* fit all, consistent data is required for all data processing use cases. You will want to know that the data the use cases are consuming is complete and semantically correct. In this chapter, we will dive into some recommendations for data flows and organization, leading to effective delivery.

Data delivery best practices overview

Each cloud platform provider will provide a unique perspective on the data delivery problem with mechanisms best suited for their cloud service offerings. They will offer a promise that with their best practices, a best-in-class and least costly solution is possible for your business needs. With this promise comes the expected cloud training and resulting cloud vendor lock-in. The cost to switch after implementing data flows will be high unless you implement IaaS services (Infrastructure as a service) and avoid most if not all PaaS (Platform as a service) or SaaS (Software as a service) services that result in specific cloud lock-in. You can and should architect your data flows to enable your organization's procurement group some leverage to negotiate with cloud providers; but that requires you to focus on the capabilities, features, and interfaces of the data flows you intend to build into your data solution. These should be modularized and itemized in an exit plan with cost estimates as part of your business continuity efforts. You never know when such a plan will be needed. Often, plans are needed to recover not from natural disasters, but business disasters! You will never know when a cloud provider, key service provider, or platform vendor is acquired by a competitor and subsequently shut down. In a hyper-competitive technical field, the data engineer is foolish not to design what-if scenarios. Over the years, we can't begin to tell you how many times these contingency plans (with their impact-assessed architectural designs) had to be pulled off the shelf, dusted off, and made effected as an emergency response to critical business events. Only once did a large storm (a derecho) hitting **Amazon Web Services** (**AWS**)'s Northern Virginia data center {https://packt-debp.link/7Tj4TJ} cause us to implement the business continuity plan as a result of a natural disaster. All the rest were caused by business events, vendor bankruptcies, platform competitive acquisitions, or zero-day security-related events.

Delivering data into a dynamic environment is like throwing a ball at a receiver who is always moving. You have to predict how it will be received and, as a result, launch the data on a vector that can be received effectively. The mechanisms chosen must address the business need. At the core of that need is being able to keep your released datasets – from what you curate to what they (the receiver) can process/consume – logically in sync. There must also be mechanisms to initially provision and occasionally reprovision the entire data corpus, not just the sequence of released data changes that affect the state of the receiver's logical image of the data corpus. This way, consuming systems can be brought up from a green field (with no data) to a fully populated data state. What affects this goal's achievement is data volume and its change rate.

Technology has not yet reached quantum processing speeds. A change takes time to be made effective and propagates to a population of receivers. Too much change, and a backlog of change queues up and then overflows, forcing a full reprovisioning request to be issued from a receiver. Likewise, too many receivers and too much change within the curated consumer dataset results in a demand that just can't be cost-effectively satisfied. A balance has to be achieved by defining a queuing model so that your delivery mechanism scales under various real-world circumstances and remains capable of sustaining service level agreement (SLAs) when the ecosystem undergoes *stress*.

So, what are the key recommendations you will want to consider adopting as part of your data delivery mechanisms? Firstly, you will want to address the data service platform needs discussed in *Chapter 12*. Then, you will want your pub/sub capabilities to be performant, scalable, and secure. The pub/sub mechanism selected should account for small data transmissions and be tunable up to supporting full bulk transmissions. All too often, the granularity of the data in a pub/sub system is too fine for the queue to be processed efficiently. It will be necessary to have the queue truncated or put into digest mode where the units of data transfers are grouped together to provide processing efficiency at the receiver end. The queue is turned into a series of micro-batches, or smaller bulk units of transfer. This enables the receiver to process a **change data capture** (**CDC**) organized pub/sub stream most efficiently as either a bulk or streaming change mechanism at the discretion of the receiver. This adaptive approach will save you untold time when implementing auto-scalable features supported by a cloud provider's auto-scale service.

For on-premises data processing where both sending and receiving processes are going to be running with more constrained resources than in the cloud, building an adaptable pub/sub mechanism also works well. This approach also allows you to throttle data rates and catch up on queued updates to work more effectively. Just make sure that you set up monitors for the observed data rates and system processing response times. You will see how often and how well the system has to adapt to real-world changes and where resources have to be brought online to handle data volume, scale, and performance.

When publishing data in bulk, as stream or hybrid bulk-stream updates, you will want to leverage some legacy processing lessons. One, in particular, stands out: the ability to publish once and read many times from a subscribed queue via the topic the message is sent on. You will also want to be able to control the queue by dequeuing and re-queuing messages if a dequeued message was *not* transactionally processed completely. Some queueing systems even provide a peek at what's next in the queue, which can provide some insight into how to set up for processing what's about to arrive. When a message was dequeued and not processed completely, then this is akin to the catcher in a baseball game having the ball pop out of their mitt! You will have to make sure the message is re-queued or saved in a pre-queue area and applied *before* the next dequeue from that same topic. Each cloud provider PaaS service, such as AWS Kinesis or Amazon Simple Queue Service (AWS SQS)/Amazon Simple Notification Service (AWS SNS), works differently, and all have quirky behavior when queues overflow or reach their configured cloud resource limits. Be forewarned! Know what you want to implement. Be ready to test your knowledge of the cloud provider system before testing those services in the context of the system you have designed.

Data volume, data change rates, and method of delivery (queuing characteristics) vary greatly from vendor to vendor. You want to disallow software or data engineers from developing divergent flavors of queueing. This only builds technical debt and makes the system fragile. Worse yet, the system will become queue-bound at some point in time or at some place in the operational design. Just build what is needed and make this queuing capability a bullet-proof critical part of your design so it remains future-proof.

In our early years, financial services network data technology mimicked the old ticker tape. Data was provided as a feed. These feeds were streams of data that a receiver could subscribe to and process without support from line protocol-level acknowledgment. No retransmissions were possible, only reading from the stream. If you lost a chunk of data because your system could not keep up, that was too bad – but on the next trade, that same data (or, even better, more current data) was going to be sent. If you tried to act on any bad or missing data, the resulting trade was going to be caught and corrected as part of a more complex financial transaction, where checks and balances were applied to correct the dumb data errors the technologists made. Only after these dumb feeds were replaced with verifiable, trusted communications protocols and modern data services (with provider-supported APIs), has the business learned to trust IT data engineers with their precious assets. Maintaining that trust is going to be essential. Selecting the correct design for your streaming, bulk, CDC, and data replication capability is going to be a challenge but a reality until the industry can embrace the concepts elaborated on later in this chapter.

A real issue is that there is just *too much data to keep moving it around*. You will be faced with this at some point in your career. That need is fast approaching because the volumes of raw data being collected today and then tossed over the wall for your processing will only increase. The greatest drivers of this growth are **artificial intelligence** (**AI**) use cases and new real-time analytics development.

Imagine it taking 30 or more days to resync a dataset if a consumer had to have a full data refresh. We have faced that reality many times in prior years! Recently, a cloud consultant discussed that a client could put all the enterprise's gold zone (over 300 PB) on a series of drives and drop ship the drive array to the cloud provider's location for loading. It sounded good even though it also sounded a bit odd given today's technology. It would be a one-off solution, we were told. When asked how long that data would take to load, the answer was a week … at least! Before the week was over, a new set of curated data would need to be applied against the original 300 PB, equal in size to the original. Catching that data change freight train was not going to happen unless a bulk load *and* streaming pub/sub mechanism were built, and it was built eventually. In this case, it was built with Azure cloud provider services. There are going to be several options to select from when synchronizing huge volumes of data.

When high-impact data is complex, big, and fast-changing, there can be a perceived drift in correctness if any transmission time delays are evidenced. When this happens, it is time to build a modern embodiment of the older datawarehouse-to-datamart synchronization method with your data services protecting the data sync processes. The consumer's correct and complete view of your data remains consistent even during periods of rapid change. This way, a consumer's query to the consumption zone's data is kept in sync with your released datasets that were built for deployment into your analytics data sandbox. The consumer should always have access to valid data that does not mix old and new within an analytical result set when processing OLAP (online analytical processing) queries, or transactional OLTP (online transaction processing) queries. The data services you build must be able to set up, break down, populate, and keep in sync these consumer zone's data sandboxes. After provisioning, these sandboxes are walled-off secure divisions of an overall data mesh that become your customer's trusted data source.

Various industries form consortiums using common platform technologies. SaaS vendors have established trust and privacy as part of a confidential computing initiative. Many realize that moving too much data is costly and inefficient. How does one enterprise trust another with its gold value data? Only recently have tools and processes begun to support the goal of creating a shared industry data mesh. In sections of this chapter, you will be exposed to the technologies and SaaS vendors that are leading these efforts to build confidential compute solutions. If the cloud providers embrace the effort and unify their internal architectures to support data mesh/data fabric architecture with confidential compute, then you will benefit. You will have much less to design, and you will minimize data movement. Until then, you will be faced with real choices and cloud service integrations that will have to be implemented as this cutting-edge area matures.

Let's begin with a deeper dive into streaming data delivery in the next section.

Streaming data delivery

When you talk to the average person, streaming data is what you get from your internet provider when you watch a TV show. Well, as a data engineer, you know that this is true in a pure definition sense. Data is streaming and the sender need not know whether you have received all the data or part except that when you don't, then you request the whole again. What was not received is served up correctly again via your streaming application, which knows how to put pieces together and render them for your viewing pleasure ... or *not*, as the case may be. Various streaming protocols and compression techniques have been created to facilitate the sending, and processing of these massively parallel streams of data that are processed into your smart devices. Likewise, for data streaming, there are different tools and techniques used to stream enterprise data. Data may be produced as a continuous high-volume flow of updates. Changes are incremental and it is expected that a receiver will process data as quickly as possible with as low a latency as possible so that processing can keep up with the stream. Enterprise data sources produce change data capture (CDC) streams of various sizes, complexity, and frequency. The types of data streams available are as varied as the data sources themselves. Examples include sensor, IoT, event, system of record, transactional, analytic, customer, marketing, sentiment, and other data for offline as well as real-time analytics.

Delivering this data at scale can present some issues. AWS has produced some interesting articles on streaming since they have provided so much support for it in the past. See *What is Streaming Data?* {https://packt-debp.link/RCytTJ}. In Amazon Kinesis, you can find the kind of data streaming services that scale. The service makes it relatively easy to load and analyze streaming data. In Amazon Redshift's streaming ingestion, Amazon allows you to ingest data streams directly into the RedShift data warehouse from several Amazon Kinesis Data Streams. The advantages of being able to access the stream in a relationally accessible data warehouse are clear. This takes us a step backward in the evolution of the data architectures; since the future is in having the data in a data lake, a data mesh, or a data fabric. Having the data in a monolithic data warehouse defeats that purpose. We mention the Kinesis/RedShift design pattern because of the impressive performance that AWS has made possible with its introduction.

Microsoft Azure has its unique spin on stream service offerings as well. You can read about these here: *Azure Stream Analytics* {https://packt-debp.link/sZbTfE}. Likewise, Oracle Cloud has a data stream processing capability as defined here: *"Oracle Cloud Infrastructure Documentation - Overview of Streaming"* {https://packt-debp.link/3RASUm}, and for Google, you have Google Datastream.

Data delivery with publishing/subscribing (pub/sub) methods

A pub/sub stream is established by a subscriber requesting a subscription to data that a publisher will provide. Often, the SLA and data contract are implicit in the service definition when it was created, and it is often fixed. However, a modern pub/sub data service that implements the best practices from *Chapter 12* will enable some of the data contract options to be specified when a connection is made. In its base form, the contract is an at-most-once or fire-and-forget instantiated connection from the subscriber to the data provider. In products such as Redis Streams, the contract is more than just an at-most-once because it also supports at-least-once or an acknowledgment that must be sent by the publisher. With Redis Pub/Sub, you only get at-most-once support. The Redis eBook *"Understanding Streams in Redis and Kafka - A Visual Guide"* {https://packt-debp.link/fOARDh}, provides an excellent overview of the subject matter in a downloadable format. We don't give Redis's offering much attention in this chapter; but these high-performing, cutting-edge services deserve your attention when designing ultra-fast pub/sub processing requirements such as in high-frequency trading analytics.

A few key points regarding pub/sub streams follow in the next sections.

Volume considerations for data streams

What happens when a consumer can't keep up with the rate of processing necessary to handle the volume or change rate of the data chunks being published? The subscriber resets the stream and requests a bulk resync.

What happens when a consumer suddenly stops taking updates and you as a publisher need to make sure that the consumer updates are persistently queued until the consumer starts dequeuing updates again? You have to retain the backlogs (up to some limit) for failing consumers until their update queues are drained, after which a bulk resync is required.

When should you spew non-dequeued updates into a secondary queue to be drained first; *after* the consumer reconnects and *before* any fresh updates are sent to the primary queue? This is to account for slow readers and queuing the spew. At this point, you are questioning the design. You need to create a bulk refresh process to resync a publisher with a consumer after a queue overflow or any long-term disconnect. After all, you need this on day one for provisioning purposes and then again if any disaster occurs; so, why not design it from the start?

Speed considerations for data streams

When the processing rate of one or more consumers slows, the resources of the pub/sub system become stressed. The queues grow and they can't be unbounded; so, at some point, the subscribing consumer's slowness will degrade the publishing system and the consumer will be marked as failed, and have its subscription revoked. Monitoring and alerting will help with data contract maintenance as well as SLA compliance; however, your design will have to be adaptable within constraints and not be so fragile as to toss consumers out when they dequeue updates slowly at times.

Flow considerations for data streams

When does a publisher indicate the data is fully published and 100% consumed by 100% of the subscribers? This logically makes it possible to move on to the next published dataset. It sounds simple but what if your system rarely attains that state? You want to *look at your requirements* and *find the right balance between effective publishing and effective subscription* for your classes of data being made subject to pub/sub synchronization.

Be very careful about pushing data back to the publisher that you have derived from any subscribed data. This causal loop could become a feedback loop and, like all feedback loops, will cause more than just auditory pain! Likewise, building chains of microservice dependencies via queueing can lead to an equally bad feedback loop. We have seen both done with a mix of Kubernetes microservices and Kafka queues in the past. The unraveling of the resulting Gordian Knot was a six-month journey of outages, queue overflows, and sleepless nights. All could have been avoided with a good data flow engineering design without the loops.

So, in the following sections, we will provide a lot of good information to be used when you design your streaming data flows.

Data delivery streaming with examples

Delivering data via streams is a theoretical topic unless you can get your hands on the technologies and try a few things out first. We applaud that desire, but we want to caution you that you must first be level-set on what others have done. Note that this mouse trap {https://packt-debp.link/XWe32o} has already been invented and you will unnecessarily reinvent another unless you do some background research. Let's start with what AWS has done, then move on to Azure, then some third-party and platform developments, and round out the section with a write-up regarding Google Dataflow offering.

AWS Kinesis, Apache Streams, and Kafka Stream (confluent) examples

The capability to build your custom streaming application with AWS cloud service offerings is well documented; however, you have to know that there are many peculiarities when working with the AWS Kinesis service, especially at its configured limits or when it scales up. The three services available to you are Amazon Kinesis Data Firehose, Amazon Kinesis Data Streams, and Amazon Managed Streaming for Apache Kafka (Amazon MSK).

- With **Amazon Data Firehose**, the AWS Firehose is built to gather and store streaming data into Amazon Simple Storage Service (Amazon S3) or Amazon Redshift. By loading directly to AWS S3, many of Amazon's analytical services are enabled for real-time use cases using a business intelligence tool of your analyst's choice.

- **Amazon Kinesis Data Streams** {https://packt-debp.link/oKIs5v} provides a capability to process vast amounts of data. Each configured shard can support up to 1 MB/sec or 1,000 records/sec write throughput or up to 2 MB/sec or 2,000 records/sec read throughput from various sources. Kinesis supports several stream processing frameworks, such as Amazon **Kinesis Client Library** (**KCL**), Apache Storm, and Apache Spark Streaming. Some examples can be found here: *"AWS Kinesis Streams Example for Finance"* {https://packt-debp.link/lKK2nF}.

- In the **Amazon Managed Streaming for Apache Kafka** (Amazon MSK) service, AWS provides a managed service that makes it possible for you to create Apache Kafka applications supporting streaming data use cases.

- There are also other streaming solutions possible in AWS, but they require **Amazon Elastic Compute Cloud** (**Amazon EC2**) and **Amazon EMR** (previously called **Amazon Elastic MapReduce**) to create a custom stream storage and processing capability. This IaaS approach is flexible in that it enables various data storage layer options as well as streaming designs using Amazon MSK, Apache Flume, Apache Spark Streaming, Apache Storm, and others such as *Apache Streams* {https://packt-debp.link/W3awdD} and *"Activity Streams Example"* {https://packt-debp.link/TbgJ0D}.

Integration of third-party tools such as *Kafka Stream (Confluent)* {https://packt-debp.link/z8fAj1} with some examples can be found here: *"Confluent Examples"* {https://packt-debp.link/w22Kg8}. These could also be considered for your architecture since the CDC capabilities of this integrated approach enable you to run it on AWS, Azure, and GCP.

The core capability for streaming data processing is extremely important to your design. You will want to make it the centerpiece of your architecture. Give it much thought and build it in such a way as to enable its robustness, ease of integration, and trustworthiness. Now, we will take a look at a couple of other streaming solutions.

Azure Stream, Apache Flink, and TIBCO Spotfire data stream examples

Microsoft Azure has its unique spin on stream service offerings as mentioned earlier. Details are located here: *Azure Stream Analytics* {https://packt-debp.link/yeBRES} with examples here: *"Azure Stream Analytics Examples"* {https://packt-debp.link/lCkZEm}. It's worth taking a look at Azure's capabilities in depth since, like AWS, its cloud offers great scalability and Azure has a lot of integrated analytics service offerings once the data is curated into an analysis area or made ready for purpose in an accessible Azure Data Lake.

Using **Apache Flink** {https://packt-debp.link/jHreOT} can lead to some great solutions if your solution requires high throughput stateful processing. Refer to the documentation and examples here: *"Apache Flink Example"* {https://packt-debp.link/PeeHz6}.

If you want to leverage a leader who has built highly performant stream processing systems, you will want to leverage *TIBCO Spotfire Data Streams* {https://packt-debp.link/zl4obf} with their examples here: *"TIBCO Spotfire Data Streams Examples"* {https://packt-debp.link/sBxd2v}. After you dive into the detailed capabilities of these third-party offerings, you will find that although cloud provider offerings are broad and integrated, they are not as rich in features as some third-party offerings. In fact, they offer a lot more than what the cloud providers have glossed over. We'll take a deeper dive into the operations of StreamSets and Tealium in the next section.

StreamSets and Tealium examples

The **StreamSets** {https://packt-debp.link/YJI6Lu} product can offer you some great features in a complex hybrid and multi-cloud environment. Some examples are here: *"StreamSets Examples"* {https://packt-debp.link/w7sXkt}. Some integrations include GCP, AWS, Azure, Cloudera, Snowflake, and Databricks. The number of awards it won in 2023 is outstanding. Some of these are Big Innovation, Top 50 Infrastructure Products, and Impact 50, to name a few. The inbound and outbound prebuilt connectors will also be attractive to those wishing to integrate many data sources and destinations quickly.

The **Tealium Customer Data Hub** {https://packt-debp.link/d5wJzR} also offers many integrations (>1300) to connect your systems with data. Providing real-time customer insights is a primary goal. Some examples are here: *"Tealium Customer Data Hub Examples"* {https://packt-debp.link/sJ78f7}. The building of a **customer data platform** (**CDP**) is the primary focus of the product. Tealium had the following to say about it:

> *"Our research finds that the most experienced CDP users show greater confidence in their decisions, better ability to manage pending, healthier budgets, and most importantly, a bigger return on their marketing dollars than organizations with little or no CDP experience. Having a single, consistent view of customers enables the entire organization to make more focused investments, attract better talent, and plan with superior accuracy."*

> *"State of the CDP: 4th Edition"* by Tealium; 2023 (https://tealium.com/download/state-of-the-cdp-2023/)

With combined technology and business products such as Tealium, the vendor has created a compelling set of capabilities that should be studied. Even if you do not use the product, you will learn what goes into building a CDP business solution. Our advice is that you leverage that learning for your design efforts.

Google Pub/Sub Dataflow examples

With Google, we let their offerings stand apart from others. Like AWS, this cloud provider really groks streaming services and has produced a great **Google Pub/Sub Dataflow** {https://packt-debp.link/KlzfXd} capability. Some examples exist here: *"Google Pub/Sub Dataflow Examples"* {https://packt-debp.link/IZeFry}. You may find the Googlish techno-babble too much to handle; but if you want to geek out, then take a dive into the literature and find out how much control you can have over your data flow with Apache Beam. What is great about the open source foundation stack is that Google has stuck to the core. If you wish to run an Apache Beam pipeline using Google Dataflow, you need to perform these steps:

- Via the Apache Beam SDK, first define and then configure/build a pipeline. Google notes that you can also select a prebuilt pipeline from a published set of Dataflow templates. Apache Beam programs consist of `pipeline` objects, which form the basis for the pipeline's datasets. Google points out that the `pipeline` object represents a single, repeatable job. The pipeline data consists of `PCollection`, which is a distributed, multi-element dataset. Pipeline steps are transformations that use the `PCollection` objects as inputs and outputs. `PCollection` can hold a fixed-size dataset or an unbounded-size dataset. If the dataset size is unbounded, it comes from a streaming updating source.

- Call Google Dataflow to run the configured pipeline. Google Dataflow will set up the required pool of virtual machines needed to run your pipeline jobs, handle any code deployments, and then orchestrate the running jobs. An Apache Beam `transform` represents a processing operation on data in the pipeline. `transform` operates one or more `Pcollection` inputs and produces one or more `Pcollection` outputs. `Sink` and `TextIO transform` write to an external data storage system and individual files respectively.

- Google Dataflow optimizes the Google service backend to allow your pipeline to run as efficiently as possible. This will cause pipeline steps to run in parallel. The Apache Beam SDK `ParDo` operation is invoked as `PCollection`. `ParDo` performs parallel processing operations to collect zero or more output elements into a target `PCollection` output. Apache Beam I/O connectors provide the mechanism to bring data into and out of your `pipeline` object. The `Aggregation` processes enable you to compute a value from any number of input elements. An `Aggregation` example is grouping elements with a common key and time window together and outputs a summary value for a defined calculation or measure. You also can configure your custom user-defined function(s) as `transform`, in the `ParDo` or `Combine` operations.

- Google Dataflow management capabilities exist in order for you to keep tabs on your pipeline and its job execution progress. `Runners` constitute the software that accepts and executes a `pipeline` object. `Runners` will function as adapters to big-data parallel execution systems forming the configured Google Dataflow ecosystem.

Along with the Google Dataflow and Apache Beam advanced concepts, *Event time*, *Windowing*, *Watermarks*, and *Triggers*, the technical stack creates a robust data stream processing capability. Refer to the Apache Beam Capability Matrix {https://packt-debp.link/oUNCKm} for the stack's coverage with Google Dataflow.

The outcome of your stream data pipeline processing is the production of a consumable data at rest repository to hold any curated data release sets. The next section of this chapter will dive into that topic further.

Consumable data delivery as a repository

With many legacy data engineering approaches, the output was just a report, then over the years the report needed to be used to drive decisions, and as a result, **decision support systems** (**DSS**) were developed with MOLAP (multidimensional online analytical processing), OLAP, and ROLAP (relational online analytical processing) data mart technologies in the backend. The days of PowerBuilder and Crystal Reports became the days of Cognos, Hyperion, Micro Strategies, SAS, Power BI, and many others. Even these Data Mart-centered architectures are considered legacy approaches today.

In the future, you will want to focus on the creation of a secure data mesh or data fabric as the delivery target for curated data. With these patterns, we overcome the data volume issues present in legacy DSS systems without sacrificing observable analytic query performance since exabytes of data just do not load into tools such as Power BI. This is an evident truth because cloud providers such as Microsoft have implemented their Azure DirectQuery {https://packt-debp.link/osyfzd} interfaces, which are required to access datasets from a data warehouse such as Azure Synapse by Power BI. This query mode is more time-consuming than the indirect mode first requires copying the source data into the Power BI or **Azure Analysis Services** (**AAS**) data space. The SLA impact on the analysis scenario could be severe when performing non-additive measures, which require hundreds of thousands of direct query calls for fast changing source data. The data design has to address the data volume, scale, and analytics performance issues in light of cloud provider/tool provider capabilities, offerings, and limitations. We recommend that you consider the need to publish, stream, and maintain in real time an analytics space supporting your customers' analytics use cases. The approach needs to scale, retain data confidentiality, be secure, and be trusted/reliable with verifiability.

In this next section, we will assess recommendations provided by cloud providers, platform vendors, and third-party tool providers to try and illuminate the playing field, which is covered with the blood of those who insisted on riding on the cutting edge of technology without the armor of success we wish to provide.

Custom-built analytics sandbox with Confidential Compute

Delivering data with data services for an analytics solution while preserving security and privacy is needed in any future-proof data-engineered design. If data is to remain private (in other words, not leave your organization), it may best be protected with a castle defense. For additonal information refer to the following:

- *"Layered Security: Think Castle Walls and a Moat Full of Alligators (Castle Defense)"* {https://packt-debp.link/xVpQR5}

- *"Juniper's IDS/IPS"* {https://packt-debp.link/40QFM7}

- *"Zero-day vulnerability"* {https://packt-debp.link/iV4vPS}

Within the high walls of a cloud provider's access security model, you will be able to preserve data security and privacy with a defensive approach that leverages **role based access controls** (**RBAC**) and the native databases' group ID and row-level security, all within the cloud provider's **virtual private cloud** (**VPC**) with access further limited to those with a **virtual private network** (**VPN**) that you have provisioned. The enterprise analytics spaces that you publish are varied in size and capabilities, but they remain in *your* control. They are *yours* to provision. They are made accessible to requesting data consumers by *your* configurations. They can be implemented as single-tenant or multi-tenant data, in warehouses, marts, lakes, fabric, or mesh designs. They are standalone and carry a significant cost to operate. But provisioned enterprise analytic spaces are unable to support **confidential computing** {https://packt-debp.link/eg2y7f}. They are internal enterprise assets unless provisioned as an external target deployment of releasable datasets into a customer's published area; where legal contracts preserve: fair use, monetization, redistribution rights, and data privacy.

As an alternative to publishing into separate analytics spaces the concept of confidential computing, has given rise to a line of thinking that involves not having to bring data to consumers but rather having the consumers' algorithms meet the data in a protected **Data Clean Room** (**DCR**) {https://packt-debp. link/G5ASrC}. This way, each consumer, and even the host cloud provider, can't access your provisioned DCR-hosted data since it is kept separate by security controls and encryption. Yet, all data remains usable by accredited algorithms that are allowed to run against the DCR data. This way, insights can be gathered from data across the data's domain without having to know the data steward's private details or violating the legally protected data owner's rights. It is this concept that has given rise to an organization that promotes confidential computing: the **Confidential Computing Consortium** (**CCC**) {https://packt-debp.link/xOJr2R}. Taking a deeper dive into the related discussions that the (CCC) consortium publishes enables you to become well rounded on the topic and see the value the approaches discussed could have for your solution.

An example of a third-party product that has embraced CCC concepts is **Habu** {https://packt-debp.link/XgCAwr}, which, in 2018, was quick to provide support for DCRs and, as a result, produced technology to host DCRs in the cloud or on-premises. There was and still is a good focus on marketing and brand analytics, which is attractive to solution providers who combine technology with domain data and resell that data with a technical solution as a service to consumers. Habu has since been acquired by LiveRamp, a SaaS provider that allows for sharing data across several industries, but they are not purely a technology platform like Habu. Google initially developed DCRs called Walled Gardens. In 2017, Google created the **Ads Data Hub** (**ADH**), where YouTube advertisers were able to gain various measurements and insights from Google. ADH has replaced DoubleClick Data Transfer for brands to access data. Facebook Analytics, Amazon Marketing Cloud, and others have developed their own DCR environments.

You can develop DCRs for your own solution if required. You can leverage Habu software or implement similar designs from their architecture patterns. However, note that this is a cutting-edge area for cloud providers and the offerings are hyper-competitive. To future-proof your designs, be aware that third-party vendor offerings are also going to be in flux. Some vendors will be gobbled up like Habu was by LiveRamp. Our advice to you is that you build capabilities that are needed, and not just depend on a third-party vendor that may not even be shipping a **general availability** (**GA**) product by the time you go to production with your DCR-based analytics solution. Focus on the capabilities and features of a foundational set of building blocks required and be ready to shift gears as changes are needed. Be ready to replace components of your design when vendors fail, or cloud providers shift focus (and they will). Let's look at what a few big players have done in this area in the next section.

On Snowflake with Decentriq

Although Snowflake {https://packt-debp.link/WqG4J4} is a leading provider of data warehouse capabilities, it natively lacks (but stay tuned for changes) the deep confidential compute capabilities for true DCR operations necessary to support first-party data. The vendor sales folks may disagree, and so we apologize in advance, yet truth wins out. This gap can be remedied with the integration of products such as **Decentriq** {https://packt-debp.link/j81XVm}, whose sales team may applaud being sighted in this best practice book.

> *"First of all, a Data Clean Room without encryption in use is not a Data Clean Room, it's just another trusted party."*

> *Decentriq; 2023 (*https://www.decentriq.com/article/from-aws-to-snowflake-data-clean-rooms-are-trapped-between-privacy-and-progress*)*

None should get too excited about sales prospects yet! You are the one to decide whether or not the offering fits *your* needs. Knowing that some of the integrated capabilities are available today should be of great value to your design efforts. Decentriq with Snowflake implements a spot-on definition of the confidential computing concept:

> *"Confidential computing describes the entire act of computation using the Trusted Execution Environments (TEEs) the process of verifying the attestations reproducibly."*

> *Decentriq; 2023; in "What is Confidential Computing"* {https://packt-debp.link/eg2y7f}

There are many business drivers for confidential computing. The industry has not universally adopted the position that data could even be shared without moving it first to reside within *their* castles after being curated for consumption. How to make the best use of sensitive data while protecting it has been a *build it … and they will [MUST] come [to ME]!* philosophy, and with that, various technology tool providers have also built tools to support what the industries have asked for in the past. You are uniquely positioned to share data with proven confidentiality and verifiability going forward with confidential computing. That's what is needed for the future and if you build solutions that conform to that vision then your solution will remain relevant long after you make it generally available (GA). Here are two examples for you to consider:

- The messenger app called Signal {https://packt-debp.link/TVaQ91} uses confidential computing to find which of your contacts use the application. This happens without having to give Signal access to your contacts – you may want to ask how! See the *"Signal Apps Research Paper"* {https://packt-debp.link/tXI29z} for more details on how they accomplished what they did and why it should be a model for others to study.

- In another example, Edgeless Systems {https://packt-debp.link/HbAcT7}, built a confidential computing solution for training AI models that keeps sensitive data encrypted during the entire model training pipeline. This broke down the walls that financial service business competition and corporate governance privacy police erected over the years.

At this point, you will want to step back from your reading and explore some of the references and case studies related to building confidential computing solutions. Afterward, you will be level-set on this topic and appreciate what is needed going forward. Today, data is still too siloed; you recognize it needs to be protected from competitive and privacy perspectives. Various business data processing efforts have not been served well in the past, which you cannot fault the data engineer for since they had to work with what was available. In the future, you should envision a path forward where you can stream your data into an analytics solution supporting confidential compute. While you're at it, bring some friends and their data along. It just may change the world as momentum builds toward an implementation of the vision where *you bring algorithms to data and not data to algorithms*.

However, centralized confidential compute may not always sell in some industries, or your organization in particular. For that, you will want to study alternatives. In the next section, we'll take a deeper look at Habu.

On Habu with federated data stores

Habu is a technology provider that can be leveraged when building a custom DCR. Habu offers several *"Habu Playbooks"* {https://packt-debp.link/daPPDB} that fast-track the implementation of DCRs for specific industries. Since Habu is a technology platform, you can build your own DCR with its capabilities, but note it does not move data into a central place; it connects to decentralized data sources. It is quite different than what Decentriq with Snowflake on GCP built. When trying to proverbially: *"herd the cats"* and when you just can't get your data providers to trust a centralized SaaS architected confidential compute design, you may want to consider this federated Habu approach.

What may be raised as a red flag in your design is the niche vendor lock-in that will come when working with bleeding-edge technology. If that risk is too big to surmount in your organization, you will want to look at some of the biggest players, such as Databricks and Azure AAS/Power BI Premium as alternatives to confidential compute, as presented in the prior sections. We will dive deeper into this in the next few sections.

On Databricks with Azure Confidential Computing (ACC)

Databricks announced support for Azure Confidential Computing {https://packt-debp.link/E22FJU} on November 16, 2023. This is a great addition to Databricks and Azure's capabilities. Read the following statement:

> *"With support for Azure confidential computing, customers can build an end-to-end data platform with increased confidentiality and privacy on Databricks by protecting data in use, or in memory, with AMD-based Azure confidential virtual machines (VMs). This type of data protection complements the protection of your confidential data using existing Azure Databricks controls such as customer-managed keys for data at rest and private link with TLS encryption for data in transit."*
>
> *Kelly Albano (Databricks)* {https://packt-debp.link/bhnM7z}

You can leverage three technologies together for your confidential computing design as part of the recommended Databricks/Azure/AMD transformation {https://packt-debp.link/UWvBpd}.

> *"Azure Databricks on confidential computing VMs is our first choice for the robust protection of confidential customer data across multiple industries. Our successful collaboration with Microsoft and Databricks enables our customers not only to unlock significant value from their data, but it also emphasizes data privacy and ownership throughout the large-scale data analysis of sensitive information."*
>
> *Lasse Jenzen, Senior Consultant, ORAYLIS GmbH* {https://packt-debp.link/H8LJcp}

The private preview of the integrated Azure Databricks confidential computing offering stated that with Azure Confidential Compute (ACC), Azure Databricks customers can build an end-to-end data platform on the Databricks' Lakehouse architecture that has increased confidentiality/privacy through data encryption. Before this integrated offering, **customer managed keys** (**CMK**) were only available for encrypting data at rest. Databricks identified a few common use cases that require confidential computing.

These use cases include the following:

- **Anti-money laundering (AML)**: This provides for enhanced data protection measures that are required to process sensitive banking transactions while combating money laundering.

- **Adverse drug event detection**: This results in a fast response to adverse drug detection events driven by data to save lives. Secure processing and analysis of patient information while preserving privacy is required when detecting adverse drug reactions. Real-time insights are required as outputs for this type of healthcare application so that responses to these adverse drug effects are immediate.

- **Fraud prevention**: This requires the safeguarding of sensitive data during fraud detection processing. Fostering trust between data providers and consumers while preventing data tampering is part of that processing. The creation of Azure Databricks Confidential Computing based machine learning fraud detection data pipelines is facilitated with this Azure Databricks and **Azure Confidential Computing** (**ACC**) integration.

With all the new capabilities coming from the integration of two juggernaut vendors: Microsoft and Databricks, you can expect more to come in the future. However, beware of the marketware each produces since these vendors exist to sell, and they do that very well!

Our advice to you is that you map the capabilities identified by the **Confidential Computing Consortium** (**CCC**), to any multi-vendor integration to see if all the boxes are checked on your list of needs. Refer to *"A Technical Analysis of Confidential Computing"* {https://packt-debp.link/qyPzIf}, which is a publication of the Confidential Computing Consortium for an overview of confidential computing that stresses the need for protecting *data-in-use* from other owners and even the hosting cloud provider.

On Azure AAS/Power BI Premium

Microsoft has tried to leverage Power BI Premium Platform in the Azure cloud (a semi-IaaS level offering of the legacy product). It would be a natural fit for a full-blown cloud service if 100% redesigned for Azure but unfortunately, it has not been. AAS is the cloud replacement, but it has key capabilities missing that make it less attractive than a redesign that would preserve all of the important Power BI Premium Platform features. As a legacy platform, Power BI Premium Platform has always been the subject of a lot of administration. In the past, it has been the subject of many operational heartaches regarding scale, especially scale down operations. **Azure Analysis Services** (**AAS**) shares a lot of basic functionality with Power BI Premium Platform and it does not have scale or administration issues. You'll want to assess both and POC any of your advanced analytics needs before deciding. Expect more changes in this area over time. Look specifically at **calculation groups** and **direct query** call capabilities, as well as the size of the analysis RAM footprint, which is limited to around 400 GB or less. Today's analysis scenarios vastly exceed that RAM footprint, and you don't want to learn that you are at a dead end in regard to what can be put in memory at a time when you want to roll out your solution to many thousands of analysts.

Microsoft has not stated much about ACC and its plans for integrating/connecting the offering with AAS. However, I'm sure with the rate of innovation taking place, it's only a matter of time. It appears that they are on a good track. You will need to know that ACC's confidential computing **trusted execution environment** (**TEE**) {https://packt-debp.link/mOJVtl} design is accessible and usable by analysis services.

In the next section, you will learn about what industry data aggregators have been doing to service their data customers and encourage them to provide their trusted data into a processing environment for enrichment.

Using a third-party aggregator for analytics

Because trust is a hard commodity to come by in hyper-competitive industries such as retail, finance, or manufacturing, sharing data and being able to leverage value from aggregated statistics remains elusive. Trusted confidential compute architectures may enable an industry to trust some solutions over time once proven effective and trustworthy. Even then, trust may never be possible without a legal entity enforcing that trust, absorbing indemnification risk, and supplying audit services to ensure that the trust is maintained. The LiveRamp SaaS vendor provides this assurance.

Via LiveRamp and the CDP/DCR competition

When your whole business depends on your trusted data remaining verifiably trusted and confidential due to extreme competitive risk, you will want to look at third-party aggregators such as LiveRamp {https://packt-debp.link/7NEbaO}.

> *"One of the biggest reasons we felt like LiveRamp was the best solution for us is the scale. LiveRamp is one of the most robust identity solutions in the market and they go above and beyond with consent, which really aligns with our views on customer privacy. LiveRamp has hundreds of relationships with data providers, so that has really unlocked some partnerships for us."*
>
> *Katie Bell US Marketing, Customer Data Science, McDonalds (https://liveramp.com/our-platform/)*

LiveRamp is a platform provider for aggregating data in a Data Clean Room (DCR). Refer to *Figure 14.1*:

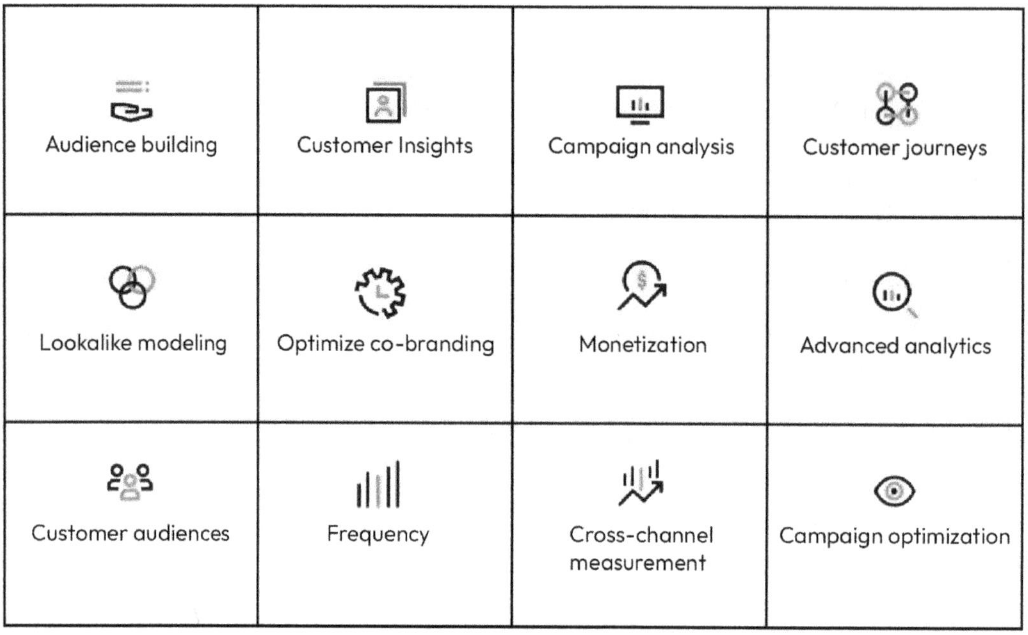

Audience building	Customer Insights	Campaign analysis	Customer journeys
Lookalike modeling	Optimize co-branding	Monetization	Advanced analytics
Customer audiences	Frequency	Cross-channel measurement	Campaign optimization

Figure 14.1 – 2024 LiveRamp features

There are four product portfolios in the LiveRamp Platform offering:

- **Live/Identity**: This is the first of four LiveRamp products that enable you to build identity infrastructure as part of an identity framework with rules that protect privacy. Advanced identity resolution together with a knowledge base enables you to create an accurate connected customer view.

- **Live/Access**: This is a product that enables the deep customer understanding needed to reach valuable new audiences and custom segments. It provides you with access to a marketplace to buy and sell third-party data. The capability to add value with new second-party collaborations is identified as a key advantage.

- **Live/Connectivity**: This enables an identity-matching capability to reach authenticated audiences at scale and deliver personalized experiences. Mixed mode channels are available, such as browser, mobile, and **Connected TV** (**CTV**) that all share a common `RampID` identifier for connecting LiveRamp data.

- **Live/Insights**: This enables you to discover insights. Measurement and collaborative analytics functions are provided. LiveRamp's DCR capabilities drive this key advantage.

LiveRamp implements a cloud-first strategy and as a result, works well across many cloud provider offerings without moving your data around (similar to Habu). Being able to resolve entities with a unique `RampID` in the cloud platform of choice (AWS, Google, Snowflake, Azure, and Databricks) is an important platform advantage.

Looking at this platform and its competitor's offerings could get you on the right path if you wish to add value to others' proprietary data or leverage their data to add value to yours in a DCR. Direct competitors are identified in the following as leaders, major players, and niche players:

- Leaders with those also identified by IDC {https://packt-debp.link/7NEbaO}:

  - **Epsilon** {https://packt-debp.link/tfUgjy} ships a PeopleCloud product to discover more in-market customers across the web, and to drive purchases and value.

  - **Habu,** with its playbooks, was defined in the earlier part of this chapter and the Habu section: *On Habu with federated data stores.*

  - **LiveRamp** was also defined in the earlier section of this chapter, *'via LiveRamp …'*

  - **AppsFlyer** {https://packt-debp.link/ZrXsKi} is a leader in marketing measurement, analytics, and engagement.

- Major Players as identified by IDC {https://packt-debp.link/7NEbaO}:

  - **Optable** {https://packt-debp.link/V7mE2l} is a provider of a DCR solution that focuses on the secure transmission of audience data. It offers secure and decentralized data collaboration.

  - **InfoSum** {https://packt-debp.link/75KfaD} (formerly Cognitive Logic) is a decentralized data collaboration platform provider.

  - **Amazon Web Services** {https://packt-debp.link/oKIs5v} see the section earlier in this chapter, *"AWS Kinesis, Apache Streams, and Kafka Stream (Confluent) Examples".*

  - **Snowflake** {https://packt-debp.link/WqG4J4}: See the section earlier in this chapter, *On Snowflake with Decentriq.*

  - **Transunion** {https://packt-debp.link/7WIcZ3} supports data collaboration in its TruAudience Data Collaboration Solutions, which allows you to resolve all consumer data and connect consumer data across devices, households, and channels using identity resolution processes. You can collaborate with data partners, brands, publishers, and walled gardens without sharing personally identifiable information or requiring data movement.

  - **Decentriq** {https://packt-debp.link/j81XVm} is a SaaS DCR provider. We get into more depth in *On Snowflake with Decentriq.*

  - **Acxiom** {https://packt-debp.link/wZ0eXW} IS a value-added SCR consulting firm. They built a *"Customer Intelligence Cloud"* product. This offering is a suite of services and an integrated marketing cloud platform with data, identity, and analytics capabilities for customer onboarding, retention, and penetration.

- Niche Players with a CDP or DCR focus:

  - **Claravine** {https://packt-debp.link/XC9eGw} (formerly Tracking First) is a provider of data integrity software. Their Data Standards Cloud connects standards across the company's ecosystems.

  - **EDO** {https://packt-debp.link/4Ul4xb} is a provider of business services (rather than data) for measuring the impact of television advertising in the advertising and media industry.

- **Karlsgate** {https://packt-debp.link/zWkpZ0} is a data collaboration solution that enables data owners, brands, publishers, agencies, and technology companies to share consumer insights while obfuscating consumer IDs.

- **Lotame Solutions** {https://packt-debp.link/YUfZri} is a global customer data management/identity solution provider with a data platform called Spherical for marketers to access first-party customer data.

- **MediaWallah** {https://packt-debp.link/AD9qA8} has a data connectivity and identity resolution offering.

- **mParticle** {https://packt-debp.link/0RuDb9} is a customer data and AI platform company that combines real-time data quality and governance protections.

- **Pyte** {https://packt-debp.link/PXxcGY} is a data-sharing platform to access data from different organizations.

- **Reltio** {https://packt-debp.link/XIp0U9} is an AI-powered data management cloud-hosted SaaS platform to unify, clean, and deliver data from multiple sources in real time. Industries serviced are life sciences, healthcare, financial services, insurance, retail, high tech, consumer packaged goods, and travel and hospitality.

- **Throtle** {https://packt-debp.link/zBLJlq} focuses on deterministic matching and identity resolution of entities to provide for targeted individual marketing. Matching is known to be of very high accuracy.

- **Zeotap** {https://packt-debp.link/aRyVHd} helps brands monetize customer data in their customer data platform. A universal marketing ID integrates data for targeted marketing purposes. The primary customer types are marketing and program advertising. Note that Zeotap's (CDP) partnership with Decentriq (as DCR SaaS provider) is worth evaluating as a viable way of handling the absence of third-party cookie data for analytics.

Data delivery into cloud service provided areas

At this point, you, as a reader, may be experiencing some reading fatigue. There are so many design options, vendors, and potential requirements that you can easily imagine bailing on the research effort and just starting coding. Please avoid that reaction!

Study what the cloud providers have started to build, which is intended to take away complexity and simplify choices, but also remove many needed capabilities. It may be that raw cloud service integration is good enough for you to get started, but you will have to handle security, privacy data collaboration, and data clean room designs. The business environment will demand it in the future. Think about the drivers pushing the CDP and DCR architecture designs for analytics and data collaboration. You will see that data analytics drives business decisions. If that data was web-, mobile-, or API-acquired and that data is no longer possible/available now due to GDPR and other regulations, then you are forced by circumstance to rethink the backend data-matching needs. Collaboratively processed data is now necessary to provide these replacement analytics services to your business. Does the cloud provider have these advanced services? Not yet! You get to build those capabilities at great expense or buy them with all the data that could be shared if collaborative data platforms are built.

That being said, the cloud providers are catching up. Note that their product focus is still on cloud services, not the data serving as a product in your solution. You can play with the cloud services provided today to get started and trained and to prove what can be built is minimally viable. Later, plan to get re-aligned to your vision and mission. So, let's go that route in the next few sections and show you what's available regarding a few of the cloud analytics offerings.

Using cloud provider's offerings

Microsoft Azure has produced a fine offering for analytics, and to use them you will be drinking the Kool-Aid so get used to it. Terms such as these are core to the cloud platform:

- **Microsoft Data Analytics**: Azure Analysis Services (AAS), Power BI, DAX, VeritPaq, Calculation Groups, direct query calls, and a host of other cloud services

- **Microsoft Databases**: SQL Server (Azure), CosmosDB (1 and 2), SQL Server 2022, data warehouse, developer tools, and so on

- **Microsoft Azure Services**: Synapse Link for Azure Cosmos DB, Synapse Link for SQL, Arc-enabled data services, Synapse Analytics, Databricks, AI, MLFlow, and machine learning

- **Microsoft Data Services**: Fabric, Microsoft Purview, and Security Graph

All of the preceding will have to be in your learned quiver of expertise to some degree. Even if you do not leverage any one particular service, you will be challenged to answer the question: why not? The answer: because! Or, just a *look* on your face which indicates your confounded state of mind. This does not instill confidence when challenged, and you will be for sure.

Amazon Web Services (AWS) Analytics {https://packt-debp.link/ccLWO9} has its own analytics stack with corresponding steep learning curve of its own in the following:

- **Amazon Web Services**: Athena, CloudWatch, DataZone, EMR, FinSpace, Glue DataBrew, Glue, Kinesis Data Firehose, Kinesis, Managed Service for Apache Flink, Managed Streaming for Apache Kafka, OpenSearch Service, QuickSight, Redshift, S3, and SNS/SQS

- **Amazon Blueprints**: Clean Rooms, Data Exchange, Identity Resolution, Glue, and Lake Formation

- **Amazon Integrations**: With Databricks, Redis, Stardog, and many more

Expect the learning curve to be a year or more and, of course, involve cloud certifications to make sure your staff minimizes the risks inherent in building complex cloud platform designs. All this is needed to get good at all the technologies and to learn all the various peculiarities and how to avoid them. Then you will find it does not all work as expected, and you will need tier 1 consulting assistance. Unless you first start with consultants, you must then pick others who know more than the first group. Expect all of it, and plan for it. You will always get the bug confused with the feature; Microsoft or AWS will remind you of this at every turn of your head. We're not down on the vendors, just realistic. We all know by now the 80/20 rule's meaning, but we are faced with vendors that implement 20% and try to make that look like 80% … and then call it all done! It does get this bad at times! At least agile processes, code refactoring, heavy doses of caffeine, and sleepless nights are still in vogue. But now, we have to take it all in as it is.

What we can do is put together some guidance on how to build the analytics space with Microsoft Azure and Amazon AWS.

On Microsoft

You want to first assess your requirements (for example, query latency and data sizing requirements) and then assess the rate of change for that data. There is no single reference architecture available other than the one that you can envision as a future-proof solution where data is always present and capable of supporting your needs. But first, you need to collect your needs and then propose a strawman reference architecture. Then, you need to be brutally honest. You need to assess yourself and your team's skills. Lastly, please get the skills required or hire a team of subject matter experts who can adapt a cloud provider's well-architected framework to your needs. If you are constrained to only an on-premises hosting environment, you will be limited in your ability to future-proof your solution. Note that in ten years, you can well imagine that all clouds will extend their capabilities (or reach) into your own data centers and be cooperative appendages to what today you may call your on-premises hosted environment.

You will want to look at some important requirements:

- Enormous exabyte scale for data, indices, and semantic graphs.

- In-memory performant analytics that requires data and indices to be optimized.

- Compute capabilities that are expandable as cloud resources.

- On-demand relational data, as well as a data mesh organized structure to currently support generally available (GA) tool integration (even though only a data fabric is available within some cloud provider offerings).

- The capability of supporting all data access methods in a confidential compute mode.

- The ability to implement nonfunctional requirements with SLAs that require 99.999% availability with less than 15 minutes **mean time to recover** (**MTTR**) and 1 year of **mean time between failures** (**MTBF**) when most cloud services are only at a 99.99% availability and leave the MTTR/MTBF SLA to your design goals.

- Data contracts that require all data to be semantically defined, metadata cataloged and dictionary-enabled across 100% of the data to define its 100% correctness.

- Data is to be released simultaneously across a release set (less than two seconds) that could be exabytes in size; so that a customer is not provided with time to act advantage over others.

- Machine learning algorithm enabled data that is fit for purpose when used for iterative model training and then model execution.

- An ability to know when digital twins (or parallel clones) of the same logical datasets are in sync across your environment and therefore ready for use by cooperative algorithms needed for AI, data mining, and insight generating applications.

- Know when to leverage semantic twin versions of your data mesh as graph data to account for knowledge engineering use cases requiring inferencing. Being able to reflect on data relationships enables data facets to be exposed by your data services.

- Build data quality, data lineage, and data at rest metadata into your system so that when data appears to be incorrect, the cause can be assessed and fixed.

- Provide the capability to collect and distribute all costs to consumers at a granular level, which includes a profit margin calculation for that data since data is a product, and knowing costs (TCO) and profit (ROI) is essential.

The preceding list looks difficult to achieve and there are some items you will wish to prioritize over others. Over time, cloud providers will continue to make these items easier to implement and more cost-effective for you, the engineer, as part of their vision to capture their customers' attention and gain an audience. You will want to be heard. We advise you to get a seat at their table and learn what they are doing and influence it if at all possible. Leverage the might of a tier-1 partner's existing seat at that same table. Look at this article to find the right partner: *"Microsoft: Find the right partner for you"* {https://packt-debp.link/FBBljm}.

On AWS

As with Microsoft, you can expect to be challenged. Rarely does the data engineer have the option to select an enterprise's cloud provider. That is often a C-level or executive committee-level decision based on due diligence and other factors such as risk, gross high-level cost, savings, business relationships, or security/trust. You will just have to go with the flow if directed to use a sole-source cloud provider. With AWS, you are cautioned to beware of security designs and the need to carefully manage costs:

- Regarding *security*, look at *"Zero Trust"* on AWS {https://packt-debp.link/3brhlf}. We will not get into exactly how cloud security should be set up; that is an entire book all by itself. Building on a zero-trust design makes sense for AWS solutions.

- Regarding *managing costs*, look to external cost reporting and analysis tools that know how to slice and dice the billing reports from the AWS cloud accounts. These tools should also be able to provide for alertable guidance when cost optimizations are required. AWS is not going to provide this guidance, but AWS Budgets {https://packt-debp.link/LfhE8E} can help. The article from CIO has a great assessment: *"Top 17 cloud cost management tools — and how to choose"* {https://packt-debp.link/wPx177}. We have worked with CloudCheckr and VMware Aria CloudHealth on past projects, as well as witnessed a FinOps manager's failure to correctly process raw AWS accounting cost logs before one of these tools was purchased.

With the AWS service stack, you will be faced with more open source technologies than with Microsoft's stack, and the number and type of resources are a far deeper and broader pool to drain. Look here to find Amazon partners: *"Engage with AWS Partners"* {https://packt-debp.link/7GEnkh}. As with any cloud service provider's offerings, you will have to become knowledgeable and avoid known pitfalls and weaknesses. Rest assured that you will discover anything that you failed to weed out before you engage in design implementation.

Regarding building out a high performing analytics space on the AWS cloud, you will want to think about high-memory instances {https://packt-debp.link/a35DPE} and big in-memory instances {https://packt-debp.link/nxeKGs}. This is an area where you will see a lot of AWS cloud service improvement in the coming years. There is also a lot of third-party competition for the prize of being the fastest, most scalable, and cheapest in-memory database for analytics use cases. The choices and costs vary but there is a lot of RAM in the cloud ready for you to put into your budget if the value statement you formulate for your solution dictates that it is needed.

In the next section, we'll walk through some of the trade-offs to consider when delivering release sets of data in bulk.

Bulk data delivery

Delivering data in bulk used to be the task of an File Transfer Protocol (FTP)/ Secure File Transfer Protocol (SFTP)/ File Transfer Protocol Secure (FTPS) outbound service that you crafted. For some older systems, this is still the case. Legacy data consumers can't deal with cloud technology and the variations possible when reading from Azure Blob storage, or AWS S3 buckets. For these consumers, an FTP facility is needed to push datasets out by FTP and/or post files for pickup. For both flavors (push or pull files), you will need to design a notifications capability so that the consumer is made aware of delivery or told that one is ready for pick up. You will also need to account for communication errors, and retransmissions with the potential that an SLA is exceeded in the worst case. The FTP facility works like a workflow manager, which requires constant monitoring and issue remediation tasks to get spun up automatically. You'll also need to create a packing list for what is in a release set and compress many files into a single unit such as `.zip`, `.gzip`, or other compressed binary file formats, or split a release set into many files. You can build or buy this capability and integrate it. A maximum file size of 2 GB used to be the limiting factor for file splitting, but that is artificial now given modern file system maximums and larger communications bandwidths possible that support more than 10 G ethernet connections. Also, there were FTP service limits on the receiving or sending side which limited file size to more than 2 GB or fractional line usage throttling. Most important is security for the communications traffic that would be in the clear if FTP were used. So, SFTP or FTPS is minimally required. An outbound and inbound FTP facility can be built, and, in the past, we have done just that: we built a service wrapper. However, one can be purchased as well.

Using various cloud provider offerings

AWS has an AWS Transfer Family {https://packt-debp.link/FjBl3R} of services and connectors for this purpose. Similarly, Azure has an Azure FTP Connector {https://packt-debp.link/xszFhd} for your use but it requires you to still build that FTP facility. Third-party solutions vary in capability, such as IBM Sterling Secure File Transfer {https://packt-debp.link/45VkQG}. Alternatives exist, such as the following, which stand out as viable options from Sterling:

- GoAnywhere MFT
- Progress MOVEit
- Globalscape EFT
- Oracle MFT
- Citrix ShareFile
- SolarWinds Serv-U

A set of less expensive FTP primitives can be found in the next section.

Via FTP (other)

You can cobble together your own FTP service with these products:

- Cerberus
- Complete FTP
- Core FTP
- FileZilla
- Globalscape
- GoAnywhere
- IIS FTPS Server
- JSCAPE (recommended)
- OpenSSH
- ProFTPD
- Progress MOVEit
- Rebex Tiny SFTP Server
- Sysax Multi Server
- Titan Server
- Vsftpd
- Xlight FTP Server

There are many choices because FTP has been around since the dawn of the internet. How you put together your FTP facility to manage the flock of files that will be produced is going to be a challenge. The customer will always want that one file that you produced a year ago, which you have already purged. You are going to have to convince that customer that they really don't want that old data anyway. However, there are modern ways to send files (such as AWS S3 files in buckets or Azure blobs) that are secure and support better scaling. Acceptance of these cloud-based file delivery mechanisms is not guaranteed by your data consumers, so be prepared to make accommodations for their needs. AWS has! In the next section, we will get into that more.

Via AWS S3 (a best practice) rather than FTP!

When transferring cloud storage (aka blobs or files within a bucket on AWS S3), different approaches can be implemented. You will want to consider first how to transfer from one S3 bucket to another since many consumers will have AWS accounts or be able to procure one to sync files from your transmission bucket. You must create a legally binding data contract with SLAs for data deliveries and build monitors for compliance so that there is a clear line of separation between what is produced and sent, as well as a receipt provided for what is received. You must turn on file versioning and storage classes' life cycle management to archive infrequently used files to reduce costs. Look at this article for some AWS approaches depending on your use case: *"What's the best way to transfer large amounts of data from one Amazon S3 bucket to another"* {https://packt-debp.link/lAtmxI}:

- Execute uploads in parallel using the AWS CLI

- Use the AWS SDK

- Use replication (cross-region or same-region flavor)

- Use batch for Amazon S3

- Use AWS S3DistCp {https://packt-debp.link/F4IRuD} with Amazon EMR

- Use AWS DataSync {https://packt-debp.link/a8hh0E}

Given the number of file sync options possible on the AWS cloud service platform, and the diversity of customer systems you wish to sync to, you will find AWS DataSync a great option. In all of the approaches, you need to create a command and control system similar to the FTP facility discussed in the prior section. This command and control capability drives the transmission of release sets that your factory is producing as they become available to a consumer. An example of a complete file transfer and command and control capable solution for media artists is Amazon Nimble Studio {https://packt-debp.link/FmV4GO}, where **digital imaging technicians** (**DIT**s) and content creators can transfer files without having technical AWS expertise.

Summary

This chapter addressed important data delivery practices to be considered as part of your data-engineered solution. When starting out, we stated that the data consumers' data processing use cases always require data to be consistent, complete, and semantically correct. This chapter elaborated on that goal with best practices and examples of choices that you have to make as you implement your data solutions.

You were exposed to data streaming considerations and how bulk operations can be treated as streaming operations of smaller bulk (or micro-batch) sizes. This enables you to tune the entire system for the best performance given the technologies being used. Best practices for publishing and subscribing to data were outlined and then elaborated upon. Data flow was discussed with a study of Google's implementation in detail with Apache Beam.

We also explored how best to organize huge volumes of data while making the output of the data factory available for real-time analytics with confidential computing capability in DCRs.

Lastly, we explored the legacy, and yet still current, file sync/transfer mechanisms possible with today's cloud service providers. The goal you will need to achieve is to create a command and control facility necessary to maintain data contacts and SLAs for released datasets delivered or received.

In the next chapter, we will elaborate on some important data considerations, such as measures, calculations, data restatements (of derived data), and data engineering designs for data science (in notebooks).

Other Considerations – Measures, Calculations, Restatements, and Data Science Best Practices

In this chapter, we will elaborate on important data considerations, such as measures, calculations, data restatements (of derived data), and various data engineering designs for data science, as well as the notebook tooling needed for these designs.

We'll begin with a metaphor. Think about making a loaf of bread. When a baker prepares, they begin by assembling all the materials needed so that everything that will be required is present, in equal if not greater amounts than required for the finished product. This careful planning makes it possible to not have to scurry around looking for an item (tool or ingredient) when it is needed in the recipe. All these items are placed on the table or at the edges of the work area in a place that can be easily accessed. This way, the item may be added to the mixture in just the right amount and at just the right time. The French culinary term for this is *"mise-en-place"* (Refer to *Cooking for Geeks, 2nd Edition* by Jeff Potter (https://www.oreilly.com/library/view/cooking-for-geeks/9781491928110/) since it presents the flip side of our metaphor by suggesting that cooking is like programming).

Miss an item or a processing step and you'll have a mess in the end! The process of baking is part art and part science. It is that chemical process that has been practiced by mankind since the first kernels of wheat were ground into powder and man figured out that it could be transformed by adding a few ingredients, such as salt, yeast, and honey, followed by kneading, rising, kneading again, and finally shaping the semi-finished product before subjecting it to heat. That baking process results in a finished product that is far greater in value than the original ground kernels.

What we can learn from the baking process is that processed data is far richer than its original raw form. Along the baking process, chemical transformations take place. Starches and sugars in the wheat are transformed into carbon dioxide and alcohol, giving the lump of wet dough the unbaked consistency of bread that has risen to a much larger size. Likewise, with data, you overlay metadata, create derived data, and retain it all in some semantic preserving structure, so that when the data is put under its pressure tests and the effects measured, new information is evidenced. Only the data scientist (akin to the baker) who knows what might happen under the right process and with just the right data mix can direct the creation of these data recipes.

Lots of trial and error goes into baking, and if you think it is easy, just ask the finest bakery chefs how they become as successful as they are. Genius, luck, training, perseverance, and perspiration are the types of answers you'll get. First, a process is defined, and then it is perfected. Then, a recipe is produced in enough clear detail to capture all the environmental conditions: ambient temperature of the room, the baker's table temperature, the altitude of the bakery, the alkalinity (pH) of the water used, the type/amount/temperature of yeast used, and the humidity of the rising and then baking ovens. All these factor into the success of a superior baked product. Furthermore, bakery chefs must innovate in response to changing market demands – vegan, gluten free, and low carb have all become significant market segments that require new processes.

Once a data scientist has an epiphany, where data has been magically transformed into information and a new insight created, a data engineer will be brought in to observe the process. Only then can the engineer elaborate on how to design a solution from the science. Maybe the engineer is even allowed into the scientific process early to take notes describing the elegant outcomes that were just demonstrated by the scientist. The data engineer has to be prepared at this point in time to ask good questions. The creation of an effective data engineer engagement model enables the engineer to gather the list of ingredients used, the process for success, the environmental conditions, and finally, the outcomes. These are all placed into a data notebook, which works like a cook's recipe. The only difference is that the data engineer often does not have the skills needed to pull the recipes out of the mind of an excited data scientist who is probably jumping around the room as if they have just discovered the winner of the next three-star Michelin award {https://packt-debp.link/QpU2Kk}, or a way to turn iron into gold!

So, the data engineer calmly writes down the recipe as dictated by the data scientist with all the proper context added by the data engineer that makes the experiment reproducible. Then, the data engineer reproduces the data recipe to prove that the math works. The language of the data scientist is math, and the data engineer has to be well versed in that language, especially when replicating stochastic algorithms, probability/statistical methods, and linear algebraic formulas that are all outputs of the data scientist's work. Imagine a sous-chef not being able to understand the command of the master chef in a high-end restaurant's kitchen! Hide all the sharp objects that can be thrown if that happens! Likewise, the data engineer cannot be ill equipped and ill trained to work with a data scientist.

Regarding our bread metaphor, croissants are notoriously difficult to make. They require a lot of time, patience, and skill. The dough has to be laminated with butter, which requires multiple rounds of rolling and folding, and then proofed and baked in order to achieve what the patron experiences as a flaky, buttery, layered masterpiece of modern baking. *Laminate, rolling, folding,* and then *proofing* are key operative words for the chef, and anybody trained in the art will know what they mean. For the data engineer or their team, the brilliance of the data scientist must be made mundane and consumable by mere mortals. To say the data engineer has to remain humble is an understatement. But just like the sous-chef, nothing happens without the data engineer leading the notation and then execution of the chef's recipes, therefore leading to success.

For the data scientist, the data science workbench is the table on which all the data magic takes place. This is like the baker's bread making table. The data engineer has to build that workbench so that all the tests, hypotheses, and observations the scientist conducts are defined, measurable, and repeatable. Too often, a data scientist is trained as an engineer first and then forgets the root of their success. The truth is that obtaining a great result requires a team effort, so communicating the way the success was obtained is just as important as the success itself. This is essential because a solution must be repeatable and proven to always be repeatable in the future under the same conditions. We're reminded of a great chemical research scientist we worked with in the past who told us of a story where a difficult chemical process resulted in a reaction that was unimaginably fantastic. What was created was a one-pot process that saved costs by eliminating a vast amount of environmentally unacceptable intervening chemical by-products of drug manufacturing. But the problem was that the scientist's process definition was missing something because *nobody* could reproduce the results! In scientific publication, if the results cannot be independently validated, then peer reviews would result in negative findings and the results labeled as being fabricated. The chemical research scientist was desperate. Observers were brought in and even under this observation, the process inventor could not produce the same result; that is until the evening when he flipped on his classical music. The harmonic caused by the music set up a crystal resonance and that allowed a critical crystal lattice to form. The experiment was reproduced, the process validated, and the awards granted *only* after this critical context: *sound* was identified as the missing ingredient. In this case, the *sound* was an environmental catalyst for the entire reaction.

If the data engineer has the right human interaction, language, and observational skills, that engineer is the data scientist's symbiotic partner for success. It is this relationship that you as the data engineer must foster with the data scientist. Do not think you can step into those shoes without the training and rigorous discipline science demands. The result of your engineering efforts will be the data scientist's **Minimum Viable Product** (**MVP**) entering a **Generally Available** (**GA**) product form. The scientist has to be reminded of the value you (the engineer) bring to the table. When you do have this conversation, just knock on the table to which you have been granted a seat, smile, and say, *"It's a solid workbench, no?"* It will be solid only after you are able to create that platform for the scientist to use and afterward very much appreciate.

Overview of other data engineering problems for consideration

In this chapter, we will dive into some key data analysis needs. First, we'll look into the data engineer's statement of value in order to remind the data scientist of this key role. Then, we'll address the capabilities needed in a modern data science/analysis workbench with features that include support for additive measures and non-additive measures at a petabyte in-memory scale. With this capability comes the need to support calculations within on-the-fly defined groups (aka **calculation groups**) and understand how this feature adds value beyond the well understood relational GROUP BY operations. Then, we'll get into notebooks and open source as well as commercial tools that compete to be the best pencil sharpener (tool) for the citizen data scientist or analyst workbench. We'll also define the process of change management for data as a product. This is the ability to manage **data restatements** and assess forward impacts as well as backward looking reasons for an observed change. This aids in carrying out an impact assessment for a data change that has its effect on your analysis' outcomes, official reports, and previously validated insights.

Data engineer statement of value

What should a data engineer bring to the table? The answer to this question is going to be based on who is asking the question and their preconceived expectations. A data scientist will expect the engineer to build tools and facilities that can be utilized in the process of creating models and analyzing data. The data processes need to be productized after a series of scientific experiments are run. One or more of those experiments will be proven to produce truthful results. That experimental process will need to be operationalized. This step requires the skills of the engineer who will do the heavy lifting required to create a business solution around the insights the science uncovered. Sometimes this is easy, but often it is not.

The scientist is expected to develop hypotheses, carry out experiments that prove or disprove the hypothesis, and record all processes and observations along the way in order to create data and analytical models that can be proven when independently reproduced. The engineer has to know the language (math) of the science being used, understand what's happening in the experiments (especially the few that produce positive results), and be ready to transform successful processes into a scalable production-ready product. The data engineer will also tease out of the solution any assumptions and environmental contexts that could have been overlooked by the data scientist. Often, problems arise when the required assumptions drift from numerical truth, economic feasibility, or business reality. This occurs too often with the following:

- The estimated cost of creating supervised training data for machine learning models

- The volume of manually curated judgments needed to preserve a model's correctness

- The inability to constrain an input dataset to fit a given model

- When constraining the dataset to run against a subset of the domain data for which the model is most effective

This last problem happens when the engineer is often not given a proper or complete understanding of the model's tuning dials by the scientist. As a result, the engineer can't create a map of the data corresponding to the ensemble of machine learning algorithms that have to be applied across the input data in order to cover 100% of the data domain and not just a niche segment. As a data engineer, you must be clear on the expectations and objectively show where the data scientist has not given you the handle on their work, which is to say – provide a way to make it possible to transform the science through the engineering process into a solution. This makes the science workable in a business production setting. The engineer must be practical and, in a real world setting, be able to discover unrealizable assumptions or unclear environmental contexts that the scientist leveraged in the development of data science model outcomes. The approach the engineer should take when accepting new data science innovations is a framework approach.

The data engineer framework is grounded in an understanding of technologies, processes, and people for which this book was written. This was explained in the first five chapters. The architecture will constrain the choices possible for your solution's design, as well as enable many more that the data scientist will want to see in the production business solution. There will be some costs and technology choices that will just not be possible or easy to implement. The data engineer could become a bearer of bad news to the data scientist when it is not feasible to operationalize the data science methods due to cost and effort. If the engineer was not trained in soft people management skills, then communications and the effective working process can break down. Fitting the tools and technologies into the data solution will require training and communication that flows both ways between the engineer and the scientist. The cost and effort to operationalize any data science method has to be transparent. Together and collaboratively the two roles have to refine a solution until science can be made fit for its intended purpose within the engineered solution. This ebb and flow of collaboration is essential. It has to be started on the first day of the data science effort. The data engineer builds tools, and those tools are placed on the analyst's workbench. The value statement of the data engineer is to be clarified for the data scientist and acknowledged by the data scientist.

Modern data science/analysis workbench

Building an analyst workbench is the best way to bring structure and utility to the data science effort. This is especially true today with the rise of new classes of machine learning algorithms that implement stochastic data processing beyond legacy supervised (and unsupervised) machine learning models or the second generation of deep learning approaches of the past. Generative AI algorithms are being integrated into production daily, and their data engineering necessities are not fully being taken into account. Missing these necessities results in poor outcomes such as proprietary sourced training data being output directly by the model (see the *New York Times lawsuit against OpenAI* {https://packt-debp.link/sFVXOJ}) rather than the targeted semantically summarized output that the model intended. How knowledge graphs can be used to produce embeddings that drive semantic correctness in generative AI models is a technique the knowledge engineer (an advanced data engineer) can provide to the data scientist. The creation of automated judgments against a generative AI models output model can also lead to the prevention of hallucinations. The data engineer wants to inject a healthy dose of practical thinking into the hype that generative AI raises. Any ML/AI solution that does not adapt to a domain's knowledge is going to cost too much to maintain. The cost of creating manually created judgments (we can also call them exceptions or tweaks) will drive a model to become specialized or broken down into submodels that are scoped to fit into an ensemble of models. The dials (or metrics) that need to be watched are the model quality management methods that the data engineer creates. The data model's quality dials will be the success indicators for the business leader. When asked: *Is it working ?* The unacceptable answer is *"It looks fine,"* whereas it should be *"See for yourself!"* The quality measures were built in, in advance. to monitor efficacy.

In the next section we'll explore the capabilities of the **analyst's workbench**, which has all the needed capabilities required for the analyst and data scientist to perform their tasks. This workbench enables the engineer to have an organized and constrained set of technologies, and processes to leverage when operationalizing a future-proof solution. The analyst's workbench provides the framework needed to govern what can and should be constrained early in the development of the solution when the scientific methods are first created. It is then used later as these methods are scaled out so that they perform in a real-world setting.

What are the capabilities of a modern data science/analyst's workbench?

The answer to the question, *"What are the capabilities of the analyst's workbench?"* will be based on your budget and tolerance for pain in *not* having these capabilities. There will be many more one-off solutions getting created by the data scientist's team without building a workbench. The capabilities you want should be divided into two basic types:

- **Tool-based capabilities** include the development and integration of tools that render data into a neutral analytic form. Microsoft has made this easy in the Power BI platform; however, alternatives exist that can rival this commercial cloud offering, but only if you have software development staff able to enhance and maintain home grown tools.

- **Quality based capabilities** maintain the highest quality when the analyst curates data as effortlessly as possible. This way, **model creation** and then **model back testing** are made equally easy when rigorously applying the standard quality measurements discussed in *Chapter 4, Architecture Principles* {https://packt-debp.link/5iMKHG}.

The tools and quality capabilities will need to be unified; today, this is often done in notebooks. A question will arise, and that will be whether you should operationalize notebooks (by putting them into production as is) or refactor the notebook logic into a more maintainable and efficient form.

Why not operationalize notebooks?

So, why not operationalize notebooks? Notebooks are sometimes operationalized because they can be, but whether you do this or not will be the chief architect's call, and often, that is just *no*! Successful vendors such as Databricks advocate for notebook operationalization for many use cases. The decision not to operationalize notebooks will have an impact on the data engineer who must recode the analysis workbench process flows. In the notebooks, these are knitted together with various scripted code snippets that must be unwound. A process flow engine capable of scaling at a reasonable cost will be required to knit together the data cell flows that are inherent to notebooks. **Amazon Web Services** (**AWS**) has gone so far as to make it easy to operationalize their AWS SageMaker notebooks; you can read more about it in their machine learning blog post, *Operationalize your Amazon SageMaker Studio notebooks as scheduled notebook jobs* {https://packt-debp.link/gWBDo2}.

We'll explain some of the arguments that one could raise to operationalize notebooks versus why the best practice is *not* to put them directly into production in order to implement the most cost efficient and high performing solution.

Using notebooks in production

There is a great degree of expedience that could be obtained by operationalizing notebooks. This is a clear benefit when taking a technically complex model as is and not having to refactor it into other technologies. As ML processing becomes mainstream and ensemble machine learning use cases arise, these complex processing steps will become very difficult to maintain if the model execution environment exists only in the notebook. When making the decision to put a model's training tasks into a notebook, there may be some opportunity to use notebooks for this type of task; but even then, there will be lots of wrangling, categorization, and data segmentation tasks that need to take place within the coded steps. Only if cloud and tool providers support an efficient operational environment can your crafted notebook steps be made operationally efficient and metadata supportable. Without this, the software and data engineer's maintenance efforts will dissolve into chaos as the number and versions of notebooks grow over time. You may want to review *Should You Use Jupyter Notebooks in Production?* {https://packt-debp.link/dQrR7o} to assess what others say about this topic. Contrast the engineering versus the science positions and appreciate why this is a hotly contested topic.

What are some of the notebook choices available?

We have had many heated discussions on notebook selection in the past. The features of these notebooks cut across many categories; some overlap, though many do not. This is to be expected as each tool is intentionally positioned as being different from its competitors. In the past, in order to reduce or eliminate this type of differentiation, we have always limited running notebooks in production. A few notebook choices are as follows:

- Anaconda {https://packt-debp.link/3aQYtN}
- Amazon Web Services (AWS) SageMaker Notebooks
- Apache Zeppelin {https://packt-debp.link/OyX98F}
- Microsoft Azure Notebooks {https://packt-debp.link/YTLsc4}
- Binder {https://packt-debp.link/dWgeo6}
- Databricks notebooks {https://packt-debp.link/CnTDmF}
- DataRobot Notebooks {https://packt-debp.link/Lhoxjo}
- Google Colab (Vertex AI Workbench) {https://packt-debp.link/iBcDJo}
- CoCalc {https://packt-debp.link/Yh87ur}
- Count.co {https://packt-debp.link/t9OXg2}
- Deepnote {https://packt-debp.link/OQUDII}
- Fanchise {https://packt-debp.link/yPzYh5}
- Hex {https://packt-debp.link/0eaiTM}
- IPython {https://packt-debp.link/bf1qzi}
- JetBrains Datalore {https://packt-debp.link/qQMBH9}
- Jupyter {https://packt-debp.link/y6J0ii}
- Kaggle (kernel) notebooks {https://packt-debp.link/XUXJjj}
- marimo {https://packt-debp.link/N11Wyk}
- mljar Mercury {https://packt-debp.link/DWIxjv}
- Noteable {https://packt-debp.link/3HnrMz}
- Paperspace by DigitalOcean {https://packt-debp.link/oYVAKp}
- Paperspace Gradient by NVIDIA {https://packt-debp.link/4gMaLE}
- Polynote {https://packt-debp.link/n935eJ}
- R notebooks {https://packt-debp.link/JjPcLk}

- SageMath {https://packt-debp.link/SE4v84}

- SAS notebooks {https://packt-debp.link/EcAqKt}

- Spyder {https://packt-debp.link/BVTZLc}

- Starboard {https://packt-debp.link/4smTjM}

It is the best practice to pick a robust way of quickly turning a notebook into production. The data quality processes in your analytics workbench have to be good enough to provide enough reason to use a refactored approach to putting notebooks in production. See the source article of the following quote for a walk-through of the arguments:

> *"The goal should be to empower data scientists and their entire delivery teams to come together and build software that delivers the required business functionality while still retaining the ability to experiment and improve."*

Don't put data science notebooks into production {https://packt-debp.link/IYf4Mu} and {https://packt-debp.link/hAkrDx}

Some of the challenges of using notebooks are that there is a lot of effort that goes into setting up data source definitions and reliably accessing that data over time, since the data access formats and methods drift. Then, after getting access to data via the notebook, the data loading itself becomes tedious and time consuming. Cloud or on-premises resource limits do not always scale as needed within the operational cost constraints available when running the notebook. You can also lose track of all data lineage via metadata traceability that the data engineer requires to access any data issues (without the extensive use of markdown tags). The nature of the data scientist or analyst's work is to iterate quickly, and that often leads to sloppy code remaining after the work is marked as done! Additionally, the analyst does not always adhere to data or software engineering rigor and will copy data values where expedient rather than creating normalized, efficient, and maintainable data flows with abstracted data access objects. A serious issue with the use of notebooks in production is that code is not modularized and can result in data operations that are out of order, operationally inefficient, or sloppily coded. This creates potential errors in system state processing that software engineers are trained to avoid. Equally important are data security issues that arise with data being accessed with the opposite of the zero-trust principle, with the hope being that trust will be layered on after the analysis is complete. Often, this reapplication of correct security does not take place and the data becomes insecure/non-private after processing through notebooks. Notebook access management often lacks the granularity required to preserve enterprise security standards, making it difficult to reapply security once it is stripped off during notebook processing.

So, there is a need to recode, refactor, and re-envision the output of a data science notebook effort. And that will be a key capability of the analyst workbench's tooling. The data engineer has to be thankful structured notebooks exist to provide a good definition of the science that needs to be refactored for production.

How does the engineer operationalize the value contained in notebooks using Integrated Development Environments (IDEs)?

There are a number of IDEs available to software engineers. These are also available to the data engineer. They can even be used by a trained data scientist (if willing). Tools such as these exist:

- Android Studio {https://packt-debp.link/keeT9s}
- Apple Xcode {https://packt-debp.link/dum0f6}
- AWS Cloud9 {https://packt-debp.link/sryM3h}
- Azure DevOps {https://packt-debp.link/pkL7nK}
- Eclipse {https://packt-debp.link/MC0SSC}
- JetBrains:
    - IntelliJ IDEA {https://packt-debp.link/996rX8}
    - CLion {https://packt-debp.link/mQMtxW}
    - PyCharm {https://packt-debp.link/k42jPS}
    - Rider {https://packt-debp.link/YWjYr6}
- OutSystems {https://packt-debp.link/6thKWT}
- NetBeans {https://packt-debp.link/VcINxN}
- Microsoft Visual Studio {https://packt-debp.link/d3caLY}
- Qt {https://packt-debp.link/0cb2QN}
- R Studio {https://packt-debp.link/zjbfp7}
- Spyder {https://packt-debp.link/eQ92RD}
- Xamarian {https://packt-debp.link/leswcj}

And there are many others to consider.

These IDE tools have important capabilities, such as the following:

- Autocompletion
- Ability to find methods and their parameters
- Integrated help (with coding assistance to jump to class definitions)
- Code highlighting help
- Code refactoring help
- Metadata and annotation features
- Version control system integration

At this point, you as a data engineer will think we are preaching to the choir! We should be telling this to the data scientists! Well, you are right! But it's going to be your job to tell them that the nice extensions that can be added to notebooks just don't work the same as a properly engineered solution. It will also be your job to refactor the code that these extensions have made possible for the data scientist. Their life may have been made easier, but your work is made harder. Refactoring will not be a trivial task. The notebook enthusiast will tell you that these extensions exist to facilitate development and that leads to the justification for operationalizing notebooks:

- Bookstore {https://packt-debp.link/TKZAVf} is used to enable versioning and storage.

- Code Prettifier {https://packt-debp.link/nat2DR} is used to beautify code according to PEP 8 style standards.

- Commuter {https://packt-debp.link/hTbOVb} is used for storage and access control.

- Git {https://packt-debp.link/Q6VcY5} is used to add a check-in capability and review existing notebooks in GitHub.

- Hinterland {https://packt-debp.link/Y4kf2J} is used for code completion.

- nbdime {https://packt-debp.link/eq7LBd} helps a **diff operation** on notebooks (aka line-by-line level file comparison).

- nbdev {https://packt-debp.link/pQMqO3} is a framework for exporting identified notebook cells to a Python library, and it automatically generates documentation using function signatures and notebook cells. Lastly, it is used to run notebook cells as unit tests.

- nbgather {https://packt-debp.link/5iTP6e} is used to log cell executions.

- nbTranslate {https://packt-debp.link/VNJC7l} is used to translate notebooks.

- Papermill {https://packt-debp.link/j4Es0k} adds parameters to notebooks (mostly for parallel notebook version support).

- RISE {https://packt-debp.link/AYXayU} makes notebooks presentable as slides and PDF reports.

- scrapbook {https://packt-debp.link/7XfX7v} to save drafts from notebooks.

- The **Snippets** menu {https://packt-debp.link/jV7sV7} exposes various Python libraries, such as the following, as ready-to-use code inserts:

  - Matplotlib {https://packt-debp.link/MzeWJY}

  - NumPy {https://packt-debp.link/zBvZ11}

  - SciPy {https://packt-debp.link/5rwwkS}

  - And so on, as ready-to-use code inserts

- Table of Contents {https://packt-debp.link/fiMpxv} is used to help navigate in the notebook.

- Variable Inspector {https://packt-debp.link/TZbOVC} is used to collect defined variables.

There is a lot that can be done with notebooks and the notebooks' ecosystems (desktop or cloud-based versions). If you do decide to operationalize notebooks, you will have to mitigate the risks that violating best practices entails. You also may choose to operationalize some and not all of them. That being said, we await a maverick tool vendor or cloud provider that will provide cost effective ways to accomplish this feat. The task is akin to writing an AI driven compiler for a given notebook that emits efficient code that runs in a cloud provider's engine. Is it possible? Yes. But has anybody done it yet? No! It could be a great start-up idea for the next generation of entrepreneurial data engineers.

In the next section, we'll focus on the capabilities the tooling needs to deliver for the analysts, particularly the data rollup mechanisms and data access methods needed when performing selective on-the-fly iterative operations across vast datasets.

Calculations and measures

Let's look at some of the calculations and measures you will have to do as part of this process.

What features should be included to support additive and non-additive measures at a petabyte in-memory scale?

Let's begin the answer to the question posed by defining what a **non-additive measure** is. They are measures that do not lend themselves to be easily aggregated across any of the dimensions used as input. Additive measures are derived from direct fact or dimension calculations, such as COUNT, MAX, MIN, or SUM. The fact that they cannot be logically aggregated between fact rows means that an algorithm decides how aggregations are to be made during the calculation of the resulting measure. A non-additive measure is characterized as requiring it to be formed of a ratio, percentage, or rate that is derived from other facts. Some examples are as follows:

- Average

- Count distinct

- Conversion rate

- Customer satisfaction score

- Median

- Profit margin

To throw a twist into the discussion, there is also a **semi-additive measure** that is a hybrid of an additive measure and a non-additive measure. The semi-additive measure uses calculations such as SUM to aggregate over a set of dimensions and another aggregation method over other dimensions. The choice is driven by some segmentation rule, such as time or a business operating location. SQLBI {https://packt-debp.link/TJXxfU} (a leading consulting firm for Microsoft's Power BI and DAX technology) produced an article titled *Semi-Additive Measures in DAX* {https://packt-debp.link/S7OU3z}, which is a good source for understanding how Microsoft handles semi-additive measures.

The reason why the topic of measures appears in this chapter is that when building out big data processing facilities, you will be faced with building measure and calculation capabilities into the analyst workbench with or without Microsoft **Azure Analysis Services** (**AAS**) or Power BI Premium capabilities. There are memory limits to that tools/services memory size. This affects your solution's scalability, and when looking for alternatives that do scale, you will be faced with tool limitations that affect the performance of your implemented measurement feature. Nevertheless, DAX supports a large set of analysis patterns that can be studied and leveraged. Many examples appear in DAX Patterns {https://packt-debp.link/4zzPa7}. Rather than calling the patterns out explicitly, it is worth studying the features most commonly available in Microsoft's analysis solutions and providing equivalent functionality in your solution if your scale, performance, and data functionality exceed Microsoft AAS's capacity.

To handle non-additive measures, you will have to include additional values in the normal data model and leverage set analysis such as:

```
SUM(AGGR(MAX(Value),Month))
```

But you can also create separate tables in your data model that are associated with your primary model, in order to reflect a specific aggregation level. Another way to implement the measure is in the data access method (or the visualization GUI) where a ratio can leverage the numerator and denominator from the data model and the ratio retained in a cache. The cache must be big enough to handle huge data volumes, so a tool such as Databricks Delta can help in this regard. When working outside of the supported capabilities of a single analysis tool, you will have to craft specific solutions to accommodate your use cases. Expect that you will be asked to integrate one or more tools into the analyst workbench so that the features are easy to use by the analyst. This way the analyst's outcomes/goals can be implemented without overly complex technical workarounds.

An alternative way of handling non-additive measures is to have some injected code executed during the rollup processing or being able to have auto-generated SQL run on the fly during query execution. Holistics {https://packt-debp.link/aVH3NL} has an auto-generating SQL capability, which can be useful when having generated code execute for your non-additive measure. Using **dynamic SQL**, either auto-generated or pre-generated and called by rules, is an approach that we have leveraged in the past. This has resulted in great success as determined by the solution's scale, performance, and functionality criteria.

How should generic calculation support features be built?

There are choices you will have to make when building support for calculations in your analysis workbench. One of these will be how you support on-the-fly algorithm-defined groups versus alternatives that can scale. **Calculation groups** {https://packt-debp.link/jE74wO} in Power BI allow for a single calculation to be run across multiple measures without having to explicitly create additional calculated columns or measures. It is a functionality that is unique to Microsoft via Azure and Power BI. This saves development and analysis time and makes your data model more manageable. The concept of a calculation group is clear and necessary for an analyst's workbench. Microsoft built this capability based on user feedback and years of practice. Kudos to them! But if you can't use Power BI Premium or you have a data scale issue that blows through the Power BI Premium memory limits, you will want to know alternatives to implementing this capability.

The article *An Alternative to Calculation Groups in Power BI* {https://packt-debp.link/lCx5V1} contains an explanation regarding how you can replace the DAX calculation group functionality with the deployment of the following:

- A new version of the data model.

- A many-to-many approach that requires configuration table structure and referential data (to define period granularity if, for example, a time series were to be used in the calculation group). This approach is needed if the analysis engine does not provide support for calculation groups.

We mention these approaches because you will be faced with choices when weighing the trade-offs between the CPU-intensive calculation group implementation versus the data-intensive many-to-many alternative that can scale better for larger datasets. This is due to the massive parallelism available for data-intensive operations.

Difficult analytics features

In order to process data into an analytics workbench, as you will know from experience, data comes in various formats and has to be ingested in order to be made useful. It is then transformed by your data factory and made fit for purpose along the journey. All this processing leads up to the point where data can be explored via analytic capabilities. Data's purpose is to be transformed into information, knowledge, wisdom, insights, and, ultimately, value to the business. All data is *big data* in time since the volume of data being collected is always increasing. Even when limited referential data is joined with big data, the sum becomes huge data. There will be many dimensions across a multi-dimensional dataset that will exist. All this data is subject to operations of the analytics workbench.

Key capabilities in the analytics workbench

You will need a number of key capabilities in the analytics workbench to be effective in your mission. Some of these are as follows:

- Data explorer

- Data inspection and data mining for value with free-text search

- Semantic search

- Semantic model explorer

- Semantic facets

- Faceted semantic search

- Filters

- Data lineage exploration

- Handle time-series analysis

- Decision support

A *data explorer capability* enables you to discover datasets that could be analyzed. Data may be kept in various raw formats but normalizing them into the standard for your platform analytics workbench is a data factory best practice. Think of what would be needed to process data in the following raw formats if not already set into a normalized form:

- Facts and dimensions in **Data Warehouse (DW)/Data Mart (DM)** format.

- Relational structure with indices (for example, RDBMS).

- Tabular rendering (for example, DAX).

- Semantic graph model (for example, RDF/OWL2).

- **Labeled Property Graph (LPG)**.

- Various serialized data forms: JSON, JSON-LD, Turtle, N3, Parquet, Avro, Thrift, **Protocol Buffers (Protobufs)**, custom object binary formats, and so on. The data explorer feature enables datasets to be identified from raw ingested form to its fit-for-purpose analytics form.

Being able to *inspect data and mine it for value requires a free text search* capability. You can first drill down past all data structures and discover a key string that may be a clue as to what context was added to that raw data earlier. Learning what data exists and what was done to that data is essential. Iterative free-text search discovery is a key data mining capability.

A *semantic search capability* enables inferred semantic relationships to be associated with retained facts and analyst search terms. The **free text (keyword) search** feature gets a boost in efficacy, with an added semantic search capability. This way, data's free-form text will match the search terms concept. You can search for concept matches in this manner. Meaningful results are returned even when free-text matchings would not have found any results.

The *semantic model explorer capability* enables the analyst to explore the subject matter expert's semantic rendering of the knowledge domain. We've modeled activity streams (for web analytics), small molecule chemistry (for predictive chemistry), finance (for financial news search), and retail (for search and anomaly detection). This gives the analysts an understanding of an industry's structure and brings clarity to the analyst's effort. The populated domain's knowledge base will be assessed for data sparsity or richness in the context of the domain's modeled knowledge graph's correctness. As an example, if a brand's shoes come in 10 colors and your knowledge base has 5, the knowledge graph will show the full enumeration of colors and that 5 are missing from the dataset undergoing analytics. If the semantic model is flawed, incorrect, or eye openingly more expansive than the analyst anticipated, then the semantic model explorer has been of value.

Regarding *semantic facets*, in the semantic model, the graph can be viewed as an aggregation of directed graphs. Each directed graph can be walked to extract facets that can be used later in faceted search solutions. These two-dimensional views of the semantic graph have great value as stand-alone taxonomies. By using facet intersections to quickly reduce search result sets, this value becomes evident. An example is going to a shopping website and identifying an item to purchase across thousands of items by intersecting these facets to reduce the options to what you are looking for (each facet is a tuple):

- `type=shoe, gender=men`

- `gender=men, size=11`

- `size=11,color=black`
- `color=black,heel=low`

Building a generic data **facet extractor** feature and leveraging that feature in your **faceted search** capability is essential to your analyst workbench. The directed walk criteria over an **ontology** (or knowledge graph) produces a taxonomy that should be considered the generated facet to be extracted; hence, we use the term facet extractor.

A *faceted semantic search capability* leverages the power of faceted search and semantic search together to carry out insightful data mining operations. The plain text that represents the taxonomy's nodes will also define concepts that are defined as collections of matching objects (with textual descriptions) that have inferred relationships to other objects. These objects have a discoverable **semantic distance** from the concept identified in a search string. A resulting found item may be brought into a search's result set based on relevance (also called a **semantic match**) and then ranked.

Regarding *filters*, the capability to reduce a dataset through the iterative application of filter criteria is known as a **data reduction** capability. Being able to scope the analytics data in order to shape it for its intended purpose is important. Insights need to be found from this data and they should be of the highest quality. By applying filters, the data will be in the right format for numerical analysis. Filters can be regular expression based, logical, relationally validated (with alt-data lookups), or subject to statistical smoothing so data can be fit to reality (and not reflect dependent gaps that may be present in underlying data).

Regarding *data lineage exploration*, the analyst must have the ability to answer the question, *What is this data and how did this data obtain its values?* The answers come from being able to look at the data's at-rest metadata, and the data's evolution through the factory (**metadata lineage**). This capability is exposed when you build a metadata/data lineage explorer capability into your solution.

The ability to *handle time series analysis* is something you should consider supporting. This type of data requires a unique data model and often a database capable of supporting time-series operations. The problem being solved with predictive analytics is to predict future values based on previously observed values. **Time-series analysis** is used to predict the future values of a time series, based on its past values. In *Machine Learning for Time Series, Second Edition* by Ben Auffarth {https://packt-debp.link/q6B7uS}, the following time-series machine learning use cases are identified:

- Forecasting
- Decomposition (separation of data into parts to identify seasonality, repeating patterns, and so on)
- Classification/Regression
- Anomaly detection
- Clustering
- Drift detection
- Smoothing

It's worth reading Ben's book for yourself if your analytics use cases require time-series analysts.

Lastly, a *decision support capability* enables you to try out data and analyze algorithm changes in order to see the changes' effects. This capability also enables you to compare many change outcomes in order to select an optimum result driven by some leading cause (data or algorithm change). What is often not built into **decision support** systems are linear algebraic capabilities to select an optimum set of input values given a multitude of variables that can affect an outcome. This should be considered when you build out your decision support capability. You do not just want one single alternative when making a decision; you want to build a system that automatically optimizes a final result given a multitude of inputs.

The **citizen data scientist**, via the analyst's workbench, brings data, metadata, algorithms, and data into focus for the business. Building the best capabilities for that class of user is essential for success. If your time and budget allow, you will want to build the GUI, the APIs, and the supporting code for your analytics patterns into a set of tools. Then, you want to place these tools on the analyst's workbench for repeated use.

Next, we will look into the issue of handling historical data correction processing and then some analytics tool capabilities. Some are open source tools and others are commercially available. Vendors will compete to be the best tool on the citizen data scientist or analyst's list. Your job will be to select and integrate tools for your data customers and enterprise analysts.

Historical data correction processing

In this section, we will define a real data change management process for data as a product. The capability to manage data restatements and look forward for a change impact assessment or backward at data lineage for a change effect is required. The outcome or end state of a prior analytics scenario can change if the facts change. Sometimes that is exactly what happens! Special attention is to be given to the assessment of a change's impact. The analysis outcomes, official reports, and previously validated insights may or may not yet have been acted upon. Change impacts can be as severe as restating corporate earnings, quarterly forward looking estimates of business revenue, or a consequential business case's justification disappearing. All as a result of a piece of critical input data being corrected; therefore, causing the business to issue an impactful restatement.

Your ability to explain the impact of a raw data change or algorithm change begins with your metadata, particularly your data lineage capability. You will want to have a systems environment for forensic analysis of data problems and any forward prediction of data change impact. This way, the analytics workbench **Decision Support Services** (**DSS**) have a place to look for old versus new outcomes as a result of data restatements or underlying changes. This environment can be the same as the data quality management environment, so long as the analysts can be given sole access to the environment during the impact assessment phases of their forensic journey. This is needed before any change is allowed to propagate into the *production* environment and be accepted by the analysts. Once you assess the impact of change, you have to know who used the data in their analytics, reports, or other business communications so that they can be notified. It's not good enough to make a change and hope the subscribers will not notice. They will!

The assessment of impact can be driven by subjective or objective factors. As an example, IBM Databand {https://packt-debp.link/izbvjB} is an observability product for data pipelines and warehouses. It can automatically collect metadata, build historical baselines, detect anomalies, and coordinate alerts to remediate data quality issues. One product reviewer is quoted as follows:

> *"What do you like best about IBM Databand? I like its ability to capture metadata automatically, set historical baselines, and use the alerting system to warn them of irregularities and rule breaches.* **What problems is IBM Databand solving and how is that benefiting you?** *My Organization is using this tool for various purposes in data observability and transformation specifically from my perspective by using triggers and alerts the tool assists in observing data flow from source to destination and also offers insights for future decision-making, and warning users of anomalous dataflows."*

> *Vishal S. (Software Engineer at Citi US Tech) via G2.com* {https://packt-debp.link/KNUbef}

There are alternatives to IBM Databand, as follows:

- Bigeye {https://packt-debp.link/UuhnUf} (formerly Toro Data Labs {https://packt-debp.link/fm9GG4}) is a data observability tool vendor with capabilities that include data quality monitoring, ML-driven anomaly detection, and root cause analysis.

- data.world {https://packt-debp.link/VfbYgC} has a platform to find data from various sources, manage file formats, and understand data's meaning prior to enhancement. Data triggers promote collaboration.

- Datameer {https://packt-debp.link/msC4O6} is a data transformation platform for data preparation, cleaning, combining, and organizing. It is often used to maintain data quality.

- Explorium {https://packt-debp.link/jsWd3P} is a data cloud platform used to connect data sources. It provides solutions such as lead scoring/qualification, market analysis/segmentation, revenue solutions, and feature discovery/generation.

- Monte Carlo {https://packt-debp.link/2jIhmP} is a data reliability and observability platform that detects, resolves, and prevents data downtime, providing end-to-end visibility and interoperability between data tools.

- Soda (https://www.soda.io/platform) provides a data monitoring platform. Its collaborative real-time capabilities inform users of important data events. The Soda platform for data quality testing makes it easy to test data quality in your development and production pipelines. Soda makes it possible for you to catch data problems before they impact your business's downstream processing.

The goal of creating an actionable set of steps for any data analytics restatement that is driven from a data restatement can be difficult. You will want to select a platform tool partner from those identified previously, or others. This problem has become very visible in the data engineering field, so you have been informed – just be ready to handle the downstream impacts of data changes.

You will want to minimize false positives and focus on highly impactful changes. You'll also want to automate the detection and communication processes to avoid the system becoming too noisy / squawky. There is going to be a fan-out effect for any data restatement; so, expect the notification workflows to produce business communications that in turn will cause some consternation, because nobody likes to change. Change that literally can pull the rug out from under an analyst's key insight can be career impacting.

Data restatements versus release set changes

The process of change management for data as a product also involves the creation of justifications for a data restatement. Was the restatement due to a software bug, a storage failure, late data ingestion caused by transmission delays, a poorly crafted statistical fitting algorithm, a bad statistical factoring configuration, or just a human entry mistake? The causes of a data restatement are endless, as well as the potential justifications for allowing it into your data factory. It's best to be transparent!

You want to let everyone know that there is a time window during which data restatements can take place. This needs to be accounted for in the data scientist's (or analyst's) work schedule. Once this time window closes, restatements are no longer possible; after that, a data change process becomes fully effective. The technical implementation for both is the same; however, data changes (not restatements) have to be handled on the data's next release cycle, when a new release set of data (not a restatement) has to be prepared. This change is natural and should account for some acceptable degree of change in the data of a prior cycle. We've used +/- 4% deviation as an acceptable level of change between release cycles, but that can be adjusted based on the type of data being released and the frequency of your data factory's release cycle.

Impact on downstream analytics and insights generated

Being able to look backward in order to assess why a data change had an impact using a data lineage tracking-capable tool is essential when assessing why and how a data change impacted a previously generated insight. This inward look into the workings of a data factory works like an X-ray feature that is essential when resolving data issues. In *Chapter 5, Architecture Framework – Conceptual Architecture Best Practices*, in the capabilities table section, the need for a metadata repository was identified. In that table, we did not go into much detail on each tool vendor in the metadata space. But for data lineage (aka metadata management for data in transit), we do need to call out a few key players. The Alation, Octopai, and Collibra products have extensive data lineage support, as do these additional products:

- Atlan {https://packt-debp.link/NcdYqI}
- Dremio {https://packt-debp.link/qWFOS8}
- Keboola {https://packt-debp.link/zPulRu}
- Kylo {https://packt-debp.link/Dxi0FW}
- OpenLineage {https://packt-debp.link/vD5IQY}

What you do *not* want to do is reinvent a better mousetrap for data lineage! It is a tough problem to address and using a tool vendor is advised. Kylo is worth evaluating, as is OpenLineage.

> **Note**
>
> You will want to be able to gather metadata fast, and when working with specific cloud services integrations, you can keep an eye on Microsoft's Purview (formerly known as Azure Purview) as it matures to support a more open metadata store with comprehensive versioning. The ability to easily knit together PaaS, SaaS, and IaaS services used in your data pipelines is going to drive you to use an extensible tool that you can integrate. Alternatively, you can use an already integrated cloud tool such as Microsoft Purview for your data lineage X-ray use cases.

When building out your solution's data lineage capabilities, you also want to give some thought to how pipeline data with metadata lineage is going to be released as release sets. A capability to support this need is absent in most tools today; you will want to access the various tools' metadata stores or at least get support from a vendor for versioned lineage trails aligned to your release sets. We advise you to not let the tail wag the dog! Downstream impacts for processed data changes need to be assessed and impacts fully unearthed before allowing data restatements or release changes to propagate through your data factory. Leverage the analyst's DSS capabilities to make use of data lineage metadata (since it is just data after all). This way, you can see a change's effect before that change is allowed to adversely affect data in production and the published analytics insights that depended on that data.

Next, we'll dive into notebooks and some key skills, processes, and requirements that the data engineer needs in their working toolkit.

Notebooks

Where the data engineer meets the data scientist will be at the negotiating table. This is where the science and the art get transformed into a solution that can operate and be maintained for a high-quality delivery. And on that table, now manifested as the analyst's workbench, the tools and processes to make the data science or analyst magic take place. This way the art and science come together to be implemented in the engineer's solution. In the center is a notebook that allows the insights to form. Modern data analytics approaches make the manipulation of data in the notebook easy for the data scientist. Data is brought to life with step-by-step scripted recipe transformations. Data manipulations across small and big datasets via Python, Java, Scala, R, C#, and so on make it possible to code the core of an MVP's features in order to prove the viability of the art that will have to be made production ready. This key point is often a bone of contention in the field. Using notebooks to their greatest effect in order to operationalize models and effect changes to operational thinking is what the data engineer is challenged to perform.

Notebooks can grow over time. We have seen flocks of notebooks become full-blown data processing system pipelines. This is not good! The data engineer could become disenfranchised and have to rip the notebook apart to make it ready for production. A smooth notebook refactoring process is desired.

We've worked with many types of notebooks over the years. You will see that the industry has not settled on a consistent approach to markdown and metadata annotations within notebooks (refer to the *Babelmark project's FAQ* {https://packt-debp.link/Qqx1us}), which leads to notebook lock-in once a selection is made by the architects and engineers. However, the data scientists will exert a lot of pressure to use features provided by alternative notebooks. This will confound your data strategy. Just try to pry a favorite tool from the clutches of a data scientist! The only way to avoid tool changes it to not have to make those changes later. Choosing the right tools from the beginning is essential. Your due diligence is very important from the start of the project. Make sure a clear direction is available regarding the use of enterprise standard notebook options and then communicate this along with the selection's justification. You just can't have all these different notebooks and their various versions in your solution, but you can select one or two.

In the next section, we'll go over some of the various notebooks' strengths and weaknesses.

Notebook technologies

Before we begin to recommend a notebook for you, you will have to itemize what you view as valuable capabilities and then decide which notebook is best for you. This technology area is very competitive and the choices will be in flux for a long time. The vendors will be pushing their supported integrations for their tools and/or their cloud platforms.

Zeppelin versus Jupyter – a comparison from a different perspective

Many analysts, scientists, and data wranglers prefer to use Jupyter notebooks, and when collaborating, they use JupyterHub {https://packt-debp.link/O25cQ3}, which allows you to log in to a hub server and write Python code with the web browser; there is no need to install software locally. JupyterHub is not the same as the cloud-based services for running Jupyter notebooks, such as Google Colab, Microsoft Azure Notebooks, and Binder. Note that Zeppelin comes with multi-user support without the need for a separate hub or other hosted integration. Zeppelin is not as popular as Jupyter, but they are competing in the same space and some see them as trying to solve the same problem.

There are many similarities in capabilities between the Zeppelin and Jupyter notebooks. What stands out with Zeppelin are the many integrations that are possible and the security model. Access restrictions for viewing versus using a notebook should exist for both notebook types. With Zeppelin, those configurations are available so that you can create flexible security configurations. Regarding first-class integrations, Zeppelin comes with Spark, Angular, Python, Pig, SAP, Groovy, Elasticsearch, and many others. Many have identified Zeppelin as being a gateway tool for those wishing to work on big data directly from a notebook. Zeppelin clearly has an advantage when it comes to working with technologies supported by one of its interpreters. Additionally, Zeppelin has a built-in scheduler, so for data scientists, this capability does not require additional extensions like Jupyter; however, some enterprises mandate the use of Apache Airflow or other schedulers that obviate this advantage. Note that Databricks notebooks take this capability to a whole new level, and they are compatible with Jupyter.

The degree of open source support, and the number of **kernel engines** {https://packt-debp.link/5xllcO}, for Jupyter far exceeds Zeppelin's. As of 2024, more than 100 kernels exist for Jupyter! These kernels define what languages can be used within a Jupyter notebook. A great place to look for Jupyter projects and add-ins is here: *Best-of Jupyter – A ranked list of awesome Jupyter projects.* {https://packt-debp.link/Sh1pPH}.

> **Note**
>
> The Jupyter kernel engines are available to Zeppelin through the Jupyter interpreter for Apache Zeppelin {https://packt-debp.link/trGZMK}. Knowing that you can get the two notebook engines to work together provides a great complementary set of capabilities for each. One important reason that we have liked working in Zeppelin is that it is easier to share visualizations. In Jupyter, you have the Plotly {https://packt-debp.link/qhb2FT} library, which can output a chart into the notebook, but Zeppelin has native Matplotlib {https://packt-debp.link/3GUQn3} support for two-dimensional plotting with outputs that can be rendered as HTML.

Zeppelin had issues in the past when you tried to run it on Microsoft's Windows 10/11, but with Microsoft's added support for the **Windows Subsystem for Linux** (**WSL**), you can run it just fine now. For instructions on how to accomplish this, you can check out *Install Zeppelin 0.7.3 on Windows 10 using Windows Subsystem for Linux (WSL)* {https://packt-debp.link/akzbb0}.

For additional comparative thinking regarding Zeppelin versus Jupyter, you can check out *Zeppelin v.s. Jupyter — a Comparison from a Different Perspective* {https://packt-debp.link/DOSCYj}.

Jupyter Notebook best practices

Jupyter Notebook is an open source web application. As an open source tool, you have a wealth of extensions, plugins, and facilities for **Exploratory Data Analysis** (**EDA**). With the tool's power also comes the ability to really cause your engineering staff operational and maintenance headaches without taking into account a few pointers. Some best practices are needed such as tagging standards and templates to enforce compliance. When working with these notebooks, you will want to know whether they can be ported to popular vendors such as Databricks. This vendor tightly controls its notebook ecosystem. Databricks supports the import and export of `.ipynb` files. This facility allows for your work to begin with Jupyter notebooks, move to Databricks notebooks, and then move back to Jupyter notebooks {https://packt-debp.link/pQsFoo} later. Databricks notebooks are similar to but not the same as Jupyter notebooks! Databricks also supports a number of core open source Jupyter libraries within the Databricks Machine Learning Runtime {https://packt-debp.link/xyjwx4}. What is really nice with an ecosystem such as Databricks is that you can interface Ray {https://packt-debp.link/tYzHLd} directly from the Databricks notebook, in order to build Databricks jobs that scale and perform well when building models.

Regarding best practices for Jupyter Notebook, take a look at the following recommendations:

- The notebook should not become a notebook of notebooks
- Organize your notebooks
- Use Markdown text

- Notebooks should be presentable at all times

- Execution flow in the notebook should be logical and sequential

- You do not have an infinite amount of memory or storage

- Modularize your notebooks

- Test your notebooks

- Capture data lineage where data in the notebook persists in the data lake, data mesh, or data fabric

The notebook should not become a notebook of notebooks or a monolithic gargantuan notebook. Each notebook should solve a particular problem and be able to stand alone. One way to determine how to split up analytics objectives is to focus on the target of the result, and when that shifts, carve off a new notebook. This must be balanced because sub-goals in analytics can cause tight integration of one step with another. When this happens, the notebook should remain combined. Just avoid a chain of these dependencies; otherwise, the notebook grows to be the monolith you want to avoid.

Organize your notebooks by first identifying the hypothesis/goal you are going to bring out when analyzing any ingested data. Make sure at the end of the notebook the **quod erat demonstrandum** (**QED**), or rather that which was to be demonstrated is clear and supports or refutes the hypothesis/goal defined at the start. You also want to make use of the notebook's **Table of Contents** (**TOC**) capability (with or without the inclusion of a notebook extension).

Use Markdown text {https://packt-debp.link/l5ectK} when building up a notebook. Use it because formatted text adds clarity. The notebook will try to render data tables and all text as code. Comments with interspaced code need to be rendered as markdown text to add clarity. A secondary goal of the data scientist has to be to leave a breadcrumb trail for the engineer. It will be frustrating to the data engineer if the algorithm or data recipe is not clear. In the rush to generate an insight, the data scientist often omits this step. The story contained in the notebook must be clear to the engineer. Examples of text that should be rendered as markdown text are headers, anchor links, math equations, and inline code. Note the following:

- The R language has its own form of markdown that it supports called R Markdown {https://packt-debp.link/6GMpgC}.

- Jupyter notebooks support CommonMark markdown and MyST Markdown {https://packt-debp.link/FRwaRn}.

Notebooks should be presentable at all times. They are always subject to review. A picture is worth a thousand words, and charts, tables, and graphs help illustrate the message that the notebook is trying to convey.

Execution flow in the notebook should be logical and sequential. Too often we've seen notebooks drop data all over the cells and reuse that data much later without so much as a comment as to why. Execution of cells out of order takes place but can cause errors when the cell data required is not available when needed. Deleting cell data can cause issues especially when cells are executed out of order. Once gone, data is not available for repeated runs or rollup calculations. Our recommendation is that you build up the output from one cell to another sequentially. When running cells out of sequence, beware of running cells that update content that is needed later. Duplicate data in a cache to avoid corrupting downstream analytics. Think transactionally; when cells depend on other cells, they need to execute as a group. It's best to just restart and run all – but this can be very time inefficient.

You do not have infinite memory or storage and have to think about this when defining your notebook. Execute your notebook with sample data (or the first number of rows if sampling is not easy), then scale it back up to real data later. You will also want to convert data types from strings to their numerical integer or float downcast sizes. Using enumerations (aka category data type) also reduces memory. What you will also want to do is think of ways to aggregate data at higher levels so that when lower granularity is not needed, the summary aggregate data becomes good enough (think of organizing the datasets at the month and quarter levels versus the raw data's daily level). Lastly, delete data that is no longer needed, and do this as often as possible to free up memory.

Modularize your notebooks. This means creating functional modules. It makes the notebook readable and promotes the reuse of those functional modules. This involves the creation of custom Python modules. It is not a notebook feature, but a Python feature.

Test your notebooks. This means creating a data science life cycle and then aligning that with your **software development lifecycle** (**SDLC**) and then adding test support in your notebook. The details are complex, and you will appreciate the challenges of trying to get a data scientist to think like a software and data engineer (aka developers) where unit, system, and regression testing is required to maintain a solution. The following are some pointers regarding notebook testing:

- You can read a good article on how to implement testing in your notebooks here: *How to Test Jupyter Notebooks with Pytest and Nbmake* {https://packt-debp.link/snExCm}.

- You'll also want some notebook code review support to keep the work evergreen over time. Refer to **ReviewNB** {https://packt-debp.link/zEMDu7} for this kind of side-by-side notebook comparison support.

- Lastly, you will want to really focus on unit testing, especially if notebooks are to be run in a production setting. It's not that we are suggesting that you do this *but* you may be forced to do what you *know* will not be good in the long term in order to just be able to move on and focus on winning other battles. The main issue to be worked around is that you can't directly import functions defined in a notebook into a testing module. A solution is that you can use **TestBook** {https://packt-debp.link/6JAi4g}, which is a unit testing framework to test code inside of the notebook. Also, you can use the nbval {https://packt-debp.link/KoDe2p} (an ipytest plugin {https://packt-debp.link/cHLGOy}) to compare a notebook's stored outputs to its current outputs.

Capture data lineage where data in the notebook persists in the data lake, data mesh, or data fabric. As intermediate data is created in the notebook, make sure it is defined by metadata and the transformations are made clear in a data lineage tracking tool. This capability will have to first be built by the data engineer and then used by the notebook's analyst.

> **Note:**
> You will want to define and then use a structured set of tags in your notebook, which contains the metadata for the entire notebook, the cell, or the notebook's output.

These three levels of metadata are set as follows:

- You can set the *notebook-level* metadata by going to the Jupyter notebook's **Toolbar | Edit | Edit Notebook Metadata** setting and structuring your tags to align with your data lineage design.

- You can also set metadata at the *cell level* by going to the Jupyter notebook's **Toolbar | View | Cell Toolbar | Edit Metadata** setting and each cell will become annotatable once the button appears on each cell.

- You can, lastly, set metadata at the output level by leveraging Python as follows: `IPython.display.display(obj,metadata={"tags": [])`. This way, metadata for your notebook's outputs can be set, which is going to be critical when meeting your data lineage engineering goals.

As identified previously, you will want to follow these pointers when working with Jupyter notebooks.

You will also want to remember that structuring your notebook and adding a TOC (take a look at the TOC2 extension {https://packt-debp.link/TP75cg}) will provide a lot of value, especially for you, the data engineer, as a third-party reviewer. Using the Collapsible Headings {https://packt-debp.link/Qte1SQ} extension allows you to hide sections of the notebook with cells that are less important. This allows the reviewer to focus on the notebook's important flows. When you have all your best practices, tagging conventions, metadata strategy, markdown style, and other conventions clearly documented, then make that enforceable by creating a template in order to make future compliant notebooks easy to create. Read the article *Set Your Jupyter Notebook up Right with this Extension* {https://packt-debp.link/pqD6dY} so that you can learn how others have addressed frustrations with inconsistent notebook structure.

When running notebooks, you will want to know whether the notebook or some key processes are running long or are never going to complete (infinitely looping). The tqdm library {https://packt-debp.link/l3vrbG} provides a progress bar as a pacifier to those anxious to see results or that progress toward a result is taking place within the notebook.

Use Git! Your notebooks need to be reviewed and then checked in, but not with bloated data, so you need to have that data stripped out by using a Git pre-commit hook. Use **nbstripout** {https://packt-debp.link/r2Mqs2} to remove notebook output. Commits, reviews, and notebook diffs will be much more readable. Notebooks should be run in a dedicated environment. You will want to store the notebook's `requirements.txt` file in the Git repository with your notebooks and any modules created. This is how you can reproduce all workflow needed to facilitate making the logic production ready.

Summary

In this chapter, we elaborated on some important data considerations, such as measures, calculations, data restatements (of derived data), and data engineering designs for data science (including notebooks). We explored in depth some of the various notebooks available in detail. You can find that information in the chapter's linked references for your research.

In the various tools, extensions, and approaches presented, you can see a short list of what is possible in your solution. Many will be used in your development and production environments. The criteria you use when selecting one tool rather than another will be based on the capabilities your solution needs, which in turn drive the required features. Those features then drive the selection criteria for tooling. We can't explicitly tell you what the best tool is for your solution, but we can guide you in making the best choices possible given your development constraints. With that guidance, you are equipped in the best possible way to create something future-proof. The best practices that were identified in this chapter focus on one key decision: *Will you operationalize your notebooks directly into production?* We provided pointers on how to code the notebook logic into efficient software while emphasizing the role of the data engineer as that individual takes an insight/algorithm from a Data Scientist's Minimal Viable Product (MVP to its Generally Available (GA) product form.

In the next chapter, we will dive into the topic of machine learning (ML) pipeline best practices and processes.

Machine Learning Pipeline Best Practices and Processes

At this point in the book, you are equipped with a fine understanding of how to produce a data factory pipeline and render data into releasable sets in a consumable area. After making them clearly available as a knowledge base with transparent metadata and lineage services, you can provide analysis capabilities in your analytics workbench. What we have not yet covered is the iterative cycles needed to tease out insights from your quality information via machine learning algorithms. This involves minimizing the technical effort, implementing a high degree of objective quality, organizing flows and optimized models, while integrating cutting-edge technologies. All this must take place to bring a production-ready solution to the market. Not all of your machine learning solutions are going to implement artificial intelligence! Be satisfied that you are adding value to the businesses and/or customers that you serve by reducing the manual time it takes to operate data-intensive tasks, as well as increasing the overall quality of the results.

Machine learning (ML)/artificial intelligence (AI) overview

In this chapter, we will address the best practices for classic **machine learning** (**ML**) operationalization and more advanced **deep learning** (**DL**) and **generative AI** (**GenAI**) requirements. This technology field is evolving rapidly! It is an uncomfortable proposition to try to future-proof your engineering designs on the shifting sands of developing reliable, high-performing solutions based on probabilities and fuzzy logic. You will always feel uncomfortable as a result. Get used to it. But what you *can* feel comfortable with is your ability to know the constants and build accordingly. Since change is going to be your constant, accept that as it is and build accordingly. What does not change is that the truth within a subject matter area that begins as a hypothesis over gathered facts is proven over time, and elevated as a truth within a domain of knowledge. This progression is what your data factory has created if you have taken the preceding chapters to heart. Now, it is time to apply the intelligence to new raw data and use the current state of machine-represented knowledge to create a new hypothesis, and then feed that back into the knowledge base so that a cycle is created. It is an iterative process, and it leads to high-quality machine intelligence. *Figure 16.1* should give you some idea of all of the different concepts that vie for your attention at the moment. It is a technology word cloud for what tools, technologies, and vendors you will be exposed to as you proceed with your ML solution:

Figure 16.1 – A 2024 ML technology word cloud

The current and future state of AI

In early 2024, Katja Grace released her study entitled *"Thousands of AI Authors on the future of AI"* {https://packt-debp.link/2DnAIw}. The question posed was: *How soon will* **High-Level Machine Intelligence (HLMI)** *be feasible?* This resulted in a response showing that the predicted advent of HLMI had fallen back by 13 years between 2022 and 2023. You may ask *What is slowing progress?* The study provides some interesting insights.

Participants estimated that *halving the drop in costs of computing* would have had the greatest effect on AI progress over the last decade, while halving *researcher effort* and *progress in AI algorithms* would have had the least effect. So, progress is being slowed by IT-engineered operating costs. Additionally, AI systems suffer from uninterpretable reasoning, which is considered a significant AI risk factor, leading to outcomes ranging from unjust biases in the treatment of people to the active pursuit of harm by capable human agents. Participants predicted that AI system risks in 2043 will carry specific traits that we see today, such as the following:

- Alignment {https://packt-debp.link/iGQrkQ} (for details, see *"Pearson Artificial Intelligence: A Modern Approach, 4th Edition"* {https://packt-debp.link/zAbCJy})

- Trustworthiness {https://packt-debp.link/mzgUsT}

- Predictability {https://packt-debp.link/hEKTdD}

- Self-directedness {https://packt-debp.link/cSRXvc}

- Capabilities {https://packt-debp.link/vphXp7}

- Jailbreakability {https://packt-debp.link/uziO7g}

Respondents were also asked, *"How likely it was that at least some state-of-the-art AI systems in 2043 would have each of eleven traits [in the following list]?"* The responses are ranked from top to bottom, according to their potential for occurrence:

- AI makes it easy to spread false information (for example, with **deepfakes**).

- AI systems manipulate large-scale public opinion trends.

- Authoritarian rulers use AI to control their population.

- AI systems worsen economic inequality by disproportionately benefiting certain individuals.

- AI lets dangerous groups make powerful tools (for example, engineered viruses).

- Bias in AI systems makes unjust situations worse (for example, AI systems learn to discriminate by gender or race in hiring processes).

- A powerful AI system may have its goals set incorrectly, causing catastrophe (for example, it develops and uses powerful weapons).

- AI systems with wrong goals become very powerful and reduce the role of humans in making decisions.

- People interact with other humans less because they spend more time interacting with AI systems.

- Near-full automation of labor makes people struggle to find meaning in their lives.

All eleven traits were considered to have a high chance of existing in AI systems from now until 2043. The first one, that AI makes it easy to spread false information, can be countered with AI solutions that can detect deepfakes, and it is essential when you engineer your solution. You can't assume all information is truth unless it is fact-checked, and you can only do that when you have the truth codified into a knowledge base. You also need to be able to know how your AI system produces its output. In this regard, many of today's AI researchers have little to offer because many respondents considered it unlikely that users of AI systems in 2028 will be able to know the true reason for AI system choices, with only 20% giving it better than even odds.

As a data engineer, you have to be able to be true to your goal, which is aligned with your vision, mission, and architecture, yet be able to adapt to practical patterns aligned with technology offerings and trends that are beneficial. These directions cannot be based on hyped-up research ideas with assumptions that will never be allowed to make it to production. Specifically, you need to know why an AI system produces any given output. Human intelligence can apply backward reasoning, justifications, and explanations as to why you make a particular statement (for example, utter a human conclusion to a query). With that explanation comes the listener's realization that either a fundamental truth was considered (or not) when the reasoning is dissected.

An observer can state that a fact is true based on experience, knowledge, and perceptions of reality. These will differ between subjective and even objective observers. A fact's truthfulness should be taken into consideration when processing data, but it may be discovered later that the assigned truth metric is, in reality, a mistake. This affects downstream processing and requires correction processing. Human judgments could be retained in the truth metric for any given fact, causing the basis to be rejected. The assumed truth, that the false fact needs to be discarded, reformulated, and re-asserted by your solution's data factory system. In legacy data processing systems, this means that data is adjusted and the software has to be rerun. It is important to know that some facts are absolutely true because they have been proven mathematically. Most facts are formulated from a human or machine hypothesis that will change over time and cause the status of that fact to change as well. This causes prior facts to be designated as false in the future. Remember that before Nicolaus Copernicus, it was thought that the sun revolved around the earth (geocentrism), and only later was it proven to be false with heliocentrism.

Your future-proof data engineering solutions have to account for a sliding scale from a true fact to a false fact, and vice versa, when storing information. Holding only true facts in storage is the implied goal of past legacy data systems. Future-proof systems have to be able to account for a variation in the truthfulness of those facts. An explanation is required to separate a false fact from a true fact. How that is retained, and then made use of by the information consumer, will be an important discerning task. This ability to explain the reasoning behind an outcome can't be omitted from your engineered solution, even if it is a hard problem to address. And if it is omitted, then you will *not* have input to produce objective judgments later, necessary to identify or correct any AI errors that will appear. This fact-checking is an essential part of human learning, and not having it in your AI system will reduce its efficacy. It is easy to confuse AI correctness with problems caused by the underlying truth used to train and later tune those models. Eliminating information errors by leveraging the retained truth status of that information is essential to maintain quality. Rarely is this highlighted in current data engineering designs. It will be your task to model your metadata to account for this need.

Systems that can detect deepfakes in images, text, and videos are arising, but a more general problem has to be addressed. That problem is to detect non-truths by any ML output. This requires advanced ML models trained to detect anomalies in truth. It's good to anticipate the need for a representation of a domain's truth, semantically modeled in your knowledge base. This came from the best practices previously discussed in this book. Engineers will need to address the AI cost issues to train a model and execute it at runtime.

Lastly, architecture teams need to address the selection and placement of new approaches and algorithms as technology choices arise, enabling them to be leveraged in enterprises' data science endeavors. This chapter will focus on the best practices to reduce costs and address the reasoning issues. First, we will begin with **machine learning operations** (**MLOPS**) {https://packt-debp.link/yOElF0}.

Machine learning pipeline

MLOps is a core process needed to operationalize ML engineering efforts. It is a cyclic process to assess a model's quality, tuning parameters, scoping domain applicability, and assembling model ensembles. The goal is to produce a final solution for a business domain that is scoped by assumptions that are preserved and validated by the domain's subject matter expert. All this results in a production-ready and maintainable, operable, and monitored solution. MLOps is a team effort, and it involves data scientists, data engineers, DevOps engineers (system engineers), and software engineers (IT).

This chapter will focus on the best practices needed to operationalize various types of machine learning models to build out the final piece of the puzzle where insights are vetted to produce true wisdom. Through the data engineer's effort, data is progressed from its ingested raw form to generate insights.

But even this insight has to undergo a gauntlet of challenges before being acted upon. The ability to train a model, weigh insights, and create targeted actions, responses, and outputs that are correct for a given domain is part of the vision for a successful AI implementation. Doing this repeatedly and under various contexts is a sign of intelligence, which becomes wisdom if the truth is delivered at the right time. This chapter will provide best practices to make this vision a reality and assist you on the journey.

The *MLOps Manifesto* {https://packt-debp.link/hp1D1c} states that the machine learning model has to be *reproducible*. It must also be *accountable*, which means the model has to be tracked back through provenance checking to the model's data sources, with its assumptions, configuration settings, and so on. The model also has to be collaboratively developed asynchronously with other developers, and software development processes should be applied, such as the use of git processes with pull requests. Also, the model has to be put under **continuous integration and continuous deployment (CI/CD)** so that it is rebuilt, tested, and deployed in an orchestrated manner. Lastly, developed models need to be cataloged and then registered like metadata for your data release sets.

We'll begin with an explanation of the ML pipeline and how it fits in as the last pipe segment in your data factory. With machine learning model development comes an iterative cycle, often shown as a figure eight diagram, as depicted in *Figure 16.2*. This cycle is not the framework for running models but the development of the models themselves.

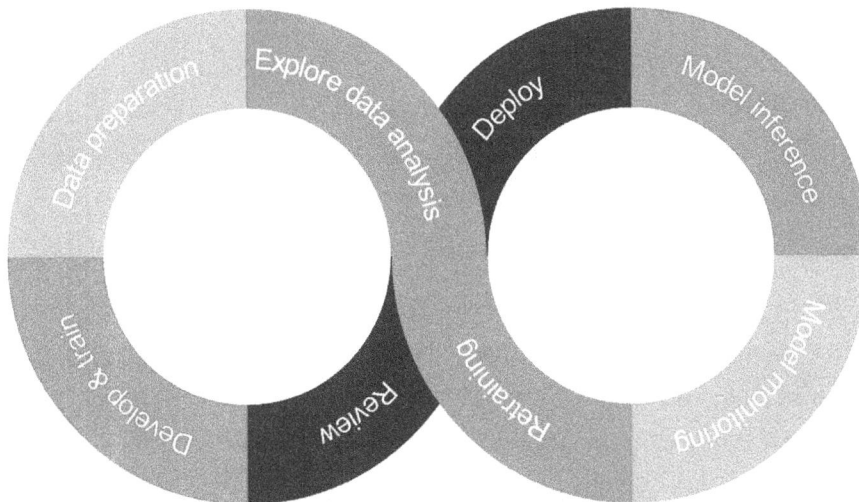

Figure 16.2 – Machine Learning Operations (MLOps) flow

We agree with the *Databricks definition of MLOps* {https://packt-debp.link/E2EGmf} as having the following life cycle stages:

- Data preparation
- Development and training of the model
- Review and governance

- Deployment

- Model inference

- Model monitoring

- Retraining

- **Exploratory data analysis** (**EDS**)

Data preparation involves the selection of features for model training by evaluating the profile of the dataset for which the model (or models) was created.

The development and training of the model can take place using many **hyperparameter optimization tools**. **Hyperparameters** are parameters that determine how an algorithm behaves when it creates the executable model. Hyperparameter optimization is the process of selecting the best-performing combination for your model. Manual optimization is still possible, but it is a very old-school approach. Using an optimizing tool cuts time and effort and increases model quality. The following are tool choices from *Top 10 Tools For Hyperparameter Optimization In Python* {https://packt-debp.link/3a4Bbq} and others:

- AWS Sagemaker {https://packt-debp.link/HOnGSL}

- Azure Machine Learning {https://packt-debp.link/xKPNvd}

- Ax {https://packt-debp.link/OZOJGq}

- **Bayesian Optimization** {https://packt-debp.link/rtUoZQ}

- **Bayesian Optimization HyperBand** (**BOHB**) {https://packt-debp.link/uQTwgf}

- Google Vizier {https://packt-debp.link/w3El4V}

- GpyOpt {https://packt-debp.link/RGtw1x}

- Hyperopt {https://packt-debp.link/HjEe26}

- Microsoft's **Neural Network Intelligence** (**NNI**) {https://packt-debp.link/zdOy6F}

- MLMachine {https://packt-debp.link/5sza3K}

- SHERPA {https://packt-debp.link/RShppM}

- scikit-learn {https://packt-debp.link/O4egrX}

- scikit-optimize {https://packt-debp.link/SRGiet}

- Optuna {https://packt-debp.link/eZWPsF}

- Polyaxon {https://polyaxon.com}

- Ray-Tune {https://packt-debp.link/dVVOKE}

- Talos {https://packt-debp.link/6oaDEx}

Configurations for your model can make your algorithm perform well or result in rubbish outcomes. Capture these essential settings as if they were code.

Review and governance enable model lineage tracking, model versioning, and model artifact life cycle transitions (model lineage, similar to data lineage).

Deployment can be performed simplistically using the same CI/CD processes used in software solution deployments via DevOps processes, with the model steps codified in scripting. However, this can also be orchestrated in an MLOps framework, which will be discussed later in this chapter. Deployment is now part of the MLOps process.

Model inference makes it possible to manage the thresholds and timing of model regeneration. Many factors go into deciding when a model needs to be regenerated, and inference logic is needed to make those decisions.

Model monitoring shows when a model drifts in quality and requires retuning and then regeneration.

Retraining happens when a model has drifted from its quality baseline.

EDS is the same work that took place in the Analyst Workbench of the prior chapter. You can prepare, explore, and share analytic solutions. You can also assess all metadata associated with a model or its source data to implement the MLOps Manifesto.

Next, we will examine the necessary features of the MLOps process that require an engineering build and support effort as part of your solution.

Model governance and compliance

The capabilities you will need for your ML pipeline to have some component MLOps life cycle processes:

- Version control
- Continuous integration/continuous deployment (CI/CD)
- Model management
- Scalable infrastructure orchestration
- Monitoring with alerting

These life cycle processes require some elaboration.

Version control is necessary to guarantee that a model can be reproduced exactly as it was intended. Often, notebooks are checked into git with data preparation steps, as well as an important set of recipes required to maintain the model over time. The model should always remain auditable. It also needs to maintain a change history and a record of all ML experiments run in its development. Docker images preserve much of the infrastructure needed to run the model. So, git and Docker are essential to track changes, collaborate effectively with peers, and reproduce all experiments.

Continuous Integration/Continuous Deployment (CI/CD) is needed to automate all building, testing, quality, monitoring, packaging, and deployment steps necessary to create a model in its ensembles, as well as a manager that will run in production. Catching errors via testing and regression testing builds a quality solution and increases development velocity.

Model management will use tracking and management subsystems to keep track of versions, quality metrics, and a model's other descriptive metadata. The types of frameworks used in model management include MLflow {https://packt-debp.link/89seVH}, Kubeflow {https://packt-debp.link/wGyCZV}, or Neptune.ai {https://packt-debp.link/MTzyfc}. Model management involves model versioning, experiment results tracking, and a registry of models you will need to create a discoverable repository for reuse. Later in this chapter, there is a section on the need for an ML model catalog, which has distinct capabilities over ML model registry tools.

Scalable infrastructure orchestration is needed so that a model can be created/trained in a cost-efficient and reasonable period of time. This DevOps capability will use cloud providers as well as other container managers, such as Kubernetes, to automate cloud provisioning needed when regenerating the model (such as retraining, testing, and packaging). Scalability will also be needed when executing the model against a real input stream.

Monitoring with alerting is needed to guarantee that a model's recipe works as expected during model regeneration (tuning and training). Deviations from quality metrics aligned to the model's assumptions (for example, data profile tests), will pick up concept (or data) drift and any degradation in efficacy from one regeneration to another. Many monitoring and alerting tools were explored earlier in this book, but Prometheus, Grafana, and DataDog stand out as being full of features well suited for model development and alerting. Other MLOps monitoring alternatives that can be integrated were mentioned in *Chapter 11* {https://packt-debp.link/e6T4Xr}.

Why are models rarely deployed?

The 2022 KDNuggets article by Eric Siegel entitled: *"Models Are Rarely Deployed: An Industry-wide Failure in Machine Learning Leadership"* {https://packt-debp.link/JPYPIU} states the following for the 114 respondents to the question *How many models get to production?*

> *"The majority of data scientists say that only 0 to 20% of models generated to be deployed have gotten there. In other words, most say that 80% or more of their projects stall before deploying an ML model."*
>
> Source: Eric Siegel (KDNuggets), January 17, 2022.

These facts have not changed much since 2022. Operationalizing your ML models and refreshing those same models will cause you to face and overcome many difficulties. Maybe 0-20% of the models you develop will make it to production. Often, these failures are caused by quality issues in the model or the dependence on model assumptions, which are unrealistic in the real world. Also, a prior released model's quality will drift over time as a business changes or the input data quality matures. Models (aka specific algorithms) will have to be abandoned if they do not scale to support the volume of data that the business needs. Many have faced technical challenges getting sparse matrices implemented on any of the large third-party big processing engines. As an example, Apache Spark does not support **sparse matrices** {https://packt-debp.link/eYPejX} for many of its algorithms. Using a sparse matrix will slow the processing algorithm at the expense of memory consumption, which then limits the size of the matrix to something manageable for the Spark engine to handle. We have been faced with these underlying problems. They become severe when you do not have access to modify the source, as in Databricks, or are too difficult to modify, as in Apache Spark. This can halt progress. The mandatory re-coding of an algorithm's implementation, changing a pre-budgeted runtime estimate, or requiring an unacceptably large amount of memory can happen with **Latent Dirichlet Allocation** (**LDA**) and **Latent Semantic Analysis** (**LSA**) ML models when the matrices become *big*. The model has to fit your MLOps life cycle and not the life cycle of a particular model. Some would have you think the reverse is true.

What is the MLOps model life cycle?

It is essential that you look realistically at what machine learning can provide for the business. Ensure that the entire team drives toward getting a solution into production, rather than some niche algorithm working for a segment of the data domain served by a specific model. Your effort often requires many models, orchestration, and continuous change and model retraining. This process resembles the flow in *Figure 16.3*:

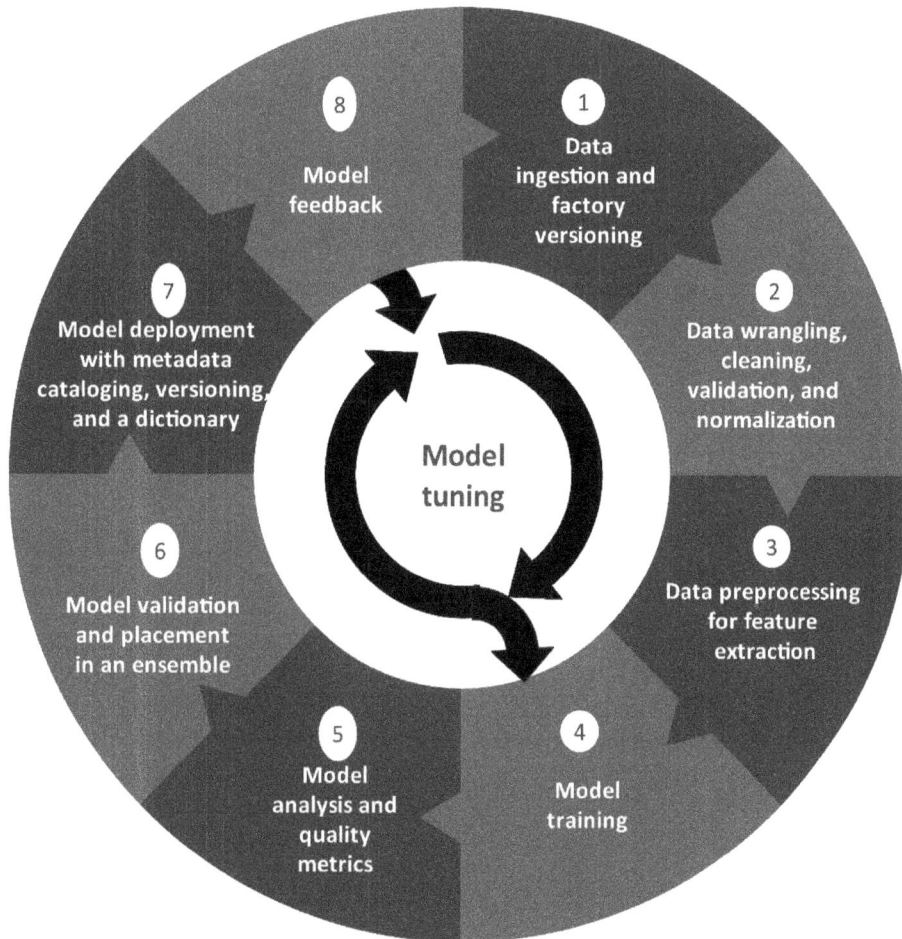

Figure 16.3 – A MLOps model life cycle

In any ML model life cycle iteration, you will have to envision a number of steps that drive a model's maturation to its final readiness state. Then, even if it is ready on day one, it will require retuning in the future. Automation is key to preserving the model's efficacy. Even if your data factory creates the most normalized, cleansed, and semantically aligned data, it still needs the following four steps before it can be used in model training:

- Ingestion
- Validation
- Feature extraction
- Training

After the model is created from the training set (if semi -or fully supervised learning models are being created), then the tuning process begins with these four additional steps:

- Analysis, leading to a model's quality assurance
- Validation
- Deployment
- Feedback

These four steps provide objective as well as subject assessment of a model, which leads to the creation of **judgments**. These have to be factored into the centrally positioned model tuning cycle. That cycle's output can affect ANY stage of the life cycle and cause each step to be rerun. This is why automation of the MLOps model life cycle is required.

There are just way too many configurations, feature selection criteria, and parameter combinations that affect the final quality of your solution, which may include the creation of **ensembles** (if that approach is being leveraged). A great source for building this ML pipeline (flow) is *"Building Machine Learning Pipelines"* by Hannes Hapke and Catherine Nelson {https://packt-debp.link/HFiNPl}. A lot of effort will be put into building your MLOps model life cycle. That effort will go a long way toward getting a larger percentage of your models into production with the least effort.

Note that building data flows for decision support pipelines, as covered in *Chapter 10*, is similar to that for machine learning pipelines discussed here in *Chapter 16*, since there are iterations required for both. It's just that the volume and degree of change that affects the MLOps model life cycle flows require special attention to detail. What is often forgotten in both life cycles is the intermediate and final representation of a model's metadata. Databricks and others have created a great metadata catalog tool for big data processing in the Unity Catalog. But if you want to register your model and capture your model's metadata, you will want to extend those tools, but you can't. Only with OpenTelemetry support can this be built by you relatively easily (but still not without some effort). Neptune.ai has a complete platform that can address many of these issues if you have the budget and desire to start with a third-party tool. However, note that Neptune.ai does NOT have OpenTelemetry integration, even though it was built to reduce the cost of the W&B Platform {https://packt-debp.link/UsWaNu}. Both are worth a look if you wish to get started quickly and have the budget.

Technology necessities

Technology makes machine learning algorithm implementation a possibility. There are different technologies and subject matter lingua franca {https://packt-debp.link/o28ZEb} for each of the following ML/AI categories:

- Classic predictive ML algorithm development, including supervised, semi-supervised, and unsupervised (clustering) algorithms.

- DL algorithms, which form a subset of ML approaches to conduct data analysis via artificial neural networks. They mimic the human brain's structure and function.

- GenAI algorithms, which implement **large language models** (**LLMs**) with the ability to generate human-like output. With semantic embeddings, they can be tuned, or they can be guided by prompts. With **Retrieval Augmented Generation** (**RAG**) techniques, a generic GenAI LLM model can be tuned to handle your domain's concepts, vocabularies, and data.

Each of the machine intelligence approaches requires an understanding of configuration options, parameters, and model training techniques before you, a data engineer, can engage in a meaningful solutions dialogue. In the next few sections, we will establish some of the basic knowledge needed to effectively collaborate with your data scientists.

What is MLflow?

Mflow enables you to build models and generative AI (**GenAI**) apps on a unified, end-to-end, open source MLOps framework. However, it is also integrated by many cloud provider offerings as their MLOps tooling, so you will encounter it in many places. There are four functions in Mlflow, as follows:

- **MLflow Tracking** is used to store a model's code, data, configurations, and output.

- **MLflow Projects** is used to package model sources.

- **MLflow Models** are used to deploy and manage ML models in a model's execution environment.

- **MLflow Model Registry** is a model store with versioning, stage transitions, and annotations, used to manage ML models (note that an *ML model registry* is not the same as an *ML model catalog*).

There are specific capabilities in MLflow provided to track LLMs. With MLflow's LLM Tracking {https://packt-debp.link/AqhRei}, an enhancement to MLFlows's ML and DL support, you get artifacts, parameters, tags, metrics, lineage, and quality tracking, which provide predictions for generative models.

Before we leave the MLflow topic, let's touch on the AutoML tool as well. AutoML's {https://packt-debp.link/4Nfxzv} tools are used to automate ML processes. These tools help a non-expert develop ML models for real-world problems. Integrating MLflow with the **AutoML framework** {https://packt-debp.link/h7tnEs} enhances the MLOps life cycle by providing ways to track, manage, and deploy models, whether they are created by a novice citizen data scientist or a fully accredited data scientist.

What is Retrieval Augmented Generation (RAG)?

Retrieval augmented generation (RAG) {https://packt-debp.link/903e8Z} is a process for optimizing output from an LLM, aligned to your data sources. RAG makes a generic LLM work with your data. Your data factory best practices advocate for the creation of a knowledge base, which contains information that's well suited to provide data context for use in the RAG process. The LLM costs ~$1M to train, and you can't expect to fully retrain it completely with your data. So, you have to tweak the LLMs because they are trained for general-purpose use across a very large corpus of information, with many billions of parameters. LLMs are well suited for multi-language natural language processing. With RAG capabilities, the LLM may be augmented with your domain-specific knowledge, rather than having to expend funding to regenerate the entire model.

Amazon Web Service's GenAI RAG approach

With **Amazon Web Services** (**AWS**), you can access a managed service called Amazon Bedrock {https://packt-debp.link/LtY7RP} that provides access to a number of high-performing **foundation models** {https://packt-debp.link/Jql7l9}, which are API-enabled LLMs. They do not have any special domain knowledge but can produce output that appears coherent, giving them AI capabilities. Your vector knowledge bases, hooked through Amazon Bedrock, are stored in a way that abstracts your data from the LLMs' internal mechanisms and data formats. The conversions, retrievals, and resulting improved output from the foundation model are handled automatically in the Amazon Bedrock service. In AWS, you also have an alternative approach that can be leveraged for a RAG implementation – that is, if you wish to manage your own RAG. There is Amazon Kendra {https://packt-debp.link/Q2SAfq}, which is, in essence, an enterprise search service that is powered by machine learning. This can reduce your time to develop any enterprise search effort considerably. There is also an optimized Kendra Retrieve API that gives you access to the semantic ranker {https://packt-debp.link/0x3ImL}, which provides 1 to 100 relevantly ranked semantically aligned enterprise search result passages, with up to 200 words each. Amazon Kendra also integrates security by filtering result set access to only those results that a user is entitled to.

The Microsoft Cloud's GenAI RAG approach

With Microsoft Azure, you also have RAG support via the cloud service's Azure AI Search {https://packt-debp.link/7bl2a4} capabilities. As you would expect, Microsoft has several implementations to use Azure AI Search with your RAG solution:

- Azure AI Studio can leverage your vector index for retrieval augmentation when calling various foundational models, such as the following from Azure AI Studio's model catalog {https://packt-debp.link/GysuZ6}:

  - **Facebook AI Similarity Search** (**FAISS**) {https://packt-debp.link/K1BghP}

  - **ChatGPT 3.5** {https://packt-debp.link/tStk9Y},

  - **ChatGPT 4** {https://packt-debp.link/paeADA}

  - **Llama-2** {https://packt-debp.link/8gEMeJ}

- In Azure AI Studio (for OpenAI) {https://packt-debp.link/50rMO6}, you can use a search index {https://packt-debp.link/6tNT1x} with or without vectors.

- In Azure Machine Learning {https://packt-debp.link/GNkq96}, you can use a search index as a vector store {https://packt-debp.link/DDUhym} in a prompt flow {https://packt-debp.link/DpL6A5}.

Oracle Cloud's GenAI RAG approach

You can read more about the Oracle Cloud's offerings in *"Oracle Artificial Intelligence - Generative AI - What is Retrieval Augmented Generation (RAG)"* {https://packt-debp.link/R6Asku}. Oracle is a relative latecomer to the field, but late does not mean that its offerings are not worthy of consideration.

Google Cloud's GenAI RAG approach

You can read about **Google Cloud Platform** (**GCP**)'s support here:

- *"Build enterprise gen AI apps with Google Cloud databases"* {https://packt-debp.link/nKMNVv}

- *"Building AI-powered apps on Google Cloud databases using pgvector, LLMs and LangChain."* {https://packt-debp.link/QZM7Pi}

With Google, you want to focus on the enterprise integration aspect of their capabilities.

GenAI RAG conclusions

In order to grok what you have just read, you will really need to dive into the referenced materials and gain a good understanding of the options and approaches available from the cloud provider's offerings.

We suggest that you focus on how a RAG solution is tested. We recommend that you use tools such as RAGAS {https://packt-debp.link/rAcHFv} to generate synthetic test data to evaluate an RAG solution's performance. The imperative here is to *evaluate* your solution. Manually creating hundreds of question-context-answer samples from documents is a very time-consuming effort. The complexity of manually curated imports will affect the quality of the assessment itself. Using synthetic data generation can reduce developer time by 90%. Once you can test effectively, then gather all your test metrics and make informed decisions while iteratively improving. For a great how-to article, read *"Evaluating Naive RAG and Advanced RAG pipeline using langchain v.0.1.0 and RAGAS"* {https://packt-debp.link/zSwsg6} by Plaban Nayak {https://packt-debp.link/HugcBf}.

This field is rapidly evolving. The novel GenAI technology is sucking all the air (and funding) out of the room from traditional search and information retrieval projects, so, as a data engineer, you have to be prepared to support and contribute to these efforts as they arise. Expect a great degree of improvement in the RAG, embedding, and prompt engineering areas as this field moves from its hype stage into mainstream acceptance. We'll begin with an explanation of prompt engineering.

What is prompt engineering?

Prompt engineering {https://packt-debp.link/culiPd} involves the iterative adjustment of text prompts for an LLM so that the most accurate and relevant responses are generated. This is an engineering task that requires the application of a few best practices. In this chapter, we will discuss these and provide endnotes that will be useful. The use of these prompts when leveraging the general capabilities of an LLM will allow you to effectively tune a solution's overall output.

First, you will want to assess your current LLM model's state by gathering the quality assessment from MLflow and any other metrics you have available to capture. Particular attention needs to be given to metrics driven by your domain's business data semantics. Then, you want to segment and categorize the domain areas that the LLM struggles with. Be ready to create fixes for these subject areas through the use of prompts. Then, you want to create an iterative quality step in your MLOps model's life cycle so that you can focus on an acceptable and maintainable level of quality for your model. This way, the areas that the prompts have been created for will have been tuned. This means you want to work with the most defined domain segment first. Then, categorize your prompts into simple versus complex buckets, and refine each bucket for the targeted domain segment from the previous step. Newer LLM tools provide templates and lessons learned from others, such as the following:

- **LangChain** {https://packt-debp.link/dDCgNp} with its LangChain Templates provides a collection of deployable reference architectures.

- **LlamaIndex** {https://packt-debp.link/w54l2s} with LlamaIndex.TS {https://packt-debp.link/v8Xbns} is used to help build an LLM-powered application with your enterprise's data factory's consumable data, in order to produce a RAG GenAI solution.

You will have to implement RAG to get a GenAI solution into production with your factory's knowledge base. Earlier attempts to operationalize LLMs without RAG caused models to produce all kinds of errors, hallucinations, and off-topic output. As with supervised ML models, tuning and training can become onerous tasks, so this effort has to be automated. The collections of prompts (with LangChain) have to be applied and aligned with your knowledge base's extracted embeddings (via LlamaIndex.TS) and then refined over time. Databricks recommends the use of the DSPy {https://packt-debp.link/onk7ex} framework for this purpose because it algorithmically optimizes LM prompts and weights. The DSPy framework makes the normal set of **language model** (**LM**) tuning steps automatic, since every time a data pipeline changes or your LM is adjusted, all your prompts will also need to be re-optimized. DSPy runs optimizers that are LM algorithms themselves.

Microsoft's take on prompt engineering

Microsoft has its own take on prompt engineering in its training material (refer to *"Getting started with LLM prompt engineering."* {https://packt-debp.link/okoH7C}). Microsoft does not embrace the architectural principles driving the need for formal semantics of a knowledge base (dependent on a formal semantic knowledge graph). Leveraging a knowledge base is what we propose that you adopt. The inability to align prompts (fine-tuning) with embeddings (implementing transfer knowledge within or between knowledge domains) is a very intensive effort. Think of all the one-off software developed LPG or vector database schemas that need to be created as an alternative to a unified knowledge base, driving a comprehensive tuning capability for your LLM.

What are embeddings?

So, *What are embeddings?* {https://packt-debp.link/Korrxj}? (Please read the reference links provided in this chapter.) Embeddings are values, objects, text, images, or audio that are designed to be consumed by machine learning models. They are translations of information into a numerical form as per some defined factor, trait, or attribute that each piece of information embodies. The information's context within a given subject domain is a very important feature of an embedding. A knowledge base is a great source for embeddings, since the information is already codified as an object, and its relationship to other information is formalized by a semantic model, with object instances aligned to the model. Semantic context is provided. Embeddings enable the ML model to discover similar objects. For example, by ingesting a photo or a sample document, as input to an LLM, you could find similar documents.

Note that an embedding is a representation of a piece of information, such as an example, a piece of text, a document, or rich media (images, video, audio, and so on). The goal of an embedding is to store semantic meaning for the embedded concept so that it can be used later by the data consumer. Most embeddings are objects or numeric representations of the concepts in a document; therefore, these can be used by ML algorithms when turned into a vector space. As a vector space, the distance between embeddings can be used to determine how the meaning of two sentences, paragraphs, and documents match. Embeddings can be created for an image and used in mixed-modal applications to compare an image to the text describing the image. On a more general note, embeddings can be generated using a class of open source libraries called **Sentence Transformers** (**SBERT**) (Refer to https://www.sbert.net; **BERT** stands for **Bidirectional Encoder Representations from Transformers**). It is a bit frustrating to learn that most open source embedding tools are not driven from a knowledge base with its underlying knowledge graph. As such, many data and ML engineers are led awry when implementing semantic embeddings without a clear vision of information semantics. It is important to drive engineering from a clear vision and not the expedience of free open source tool offerings. Feel free to innovate with those tools and share your results with the world!

Embeddings cause recurring relationships between words and other objects to be highlighted with meaning. In LLMs, embeddings includes the addition of context to every word, as well as the word itself, during model creation. Sentences, paragraphs, sections, chapters, and book meanings can be analyzed with high quality. Even the context of queries can be retained as embeddings, in order to speed up performance or reduce resource consumption for subsequent queries.

Data annotation

So, you want to know how you, as a data engineer, can help a data scientist? You can begin by curating your data into indices that are semantically aligned with the object in your knowledge base. These, in turn, are correctly modeled for your business domains with the current known truths that are relevant for your enterprise. We bet you thought we would say, just build a vector store and make it available to your LLM, using the cloud provider of choice's tool. However, you will fail to have that model propagate to production, since it would suffer from many of the failings of untuned GenAI models (such as hallucinations, quality errors, and the inappropriate exposure of training source text in model output). If you did as some of the cloud providers propose and built your RAG with an embedding, without semantic structure, you would be creating several iterations in your factory. Your vision should be to implement a knowledge-aware semantic form for embeddings. But let's get real! Not all enterprises have a knowledge base, and it would take a lot of knowledge engineering expertise and effort to create a semantic model as a knowledge graph, and then have it populated from your corpus. To do that in an automated manner requires the kind of GenAI model we are trying to enable with the embeddings we want to create. It's a chicken and egg {https://packt-debp.link/DpYkz1} scenario of which one comes first, and we need one before the other. Previously, we have first looked at using the techniques mentioned in the *"Embeddings Learned By Matrix Factorization"* presentation {https://packt-debp.link/M1WIpm}, where various approaches are evaluated for use when you create your corpus embeddings. The categorization of any embeddings, context setting, and semantic expansion from tokens to objects can be added, as your knowledge base is populated from semantically aligned embeddings. The semantic knowledge graph also undergoes evolution as the recurrence of concepts (or topic) occurs with greater frequency, thus driving your enhancement efforts to become focused on areas of greatest recurrence. This does not mean the user is not going to search for, or ask a question about, a concept that has the least occurrence! In fact, that will be like searching for the proverbial needle in the haystack, which is where there is the greatest value. As an example, what diagnostician does not want to find the cause of a rare disease, with symptoms that are in the knowledge graph and knowledge base but infrequently written about in an enterprise's corpus? Embedding that knowledge into a model will provide semantic value.

When you build embeddings aligned to a knowledge base and you do the same when you develop prompts, they will be aligned with each other! The same vocabulary for the semantic objects, the objects' semantic context, and the dictionary form the basis for effective retrieval.

Sometimes, you will just have too much document information and it all just needs to be annotated manually, or semi-manually with processes that call niche tools and algorithms (some of which are based in machine learning). This is possible, and to do that you want to leverage third party data-annotating software or build this capability yourself, but only if there is development bandwidth. The human costs also have to be kept manageable.

In the article *"21 best annotation tools for documents, video, and more"* {https://packt-debp.link/2CBzep}, Filestage references 21 annotation tools for various information types. You may be interested in this list, as there are many tools that can help you with the initial mix-modal document annotation requirements for your solution. Initially, building your solution with one or more of these tool solutions into your MLOps life cycle will enable you to bootstrap development efforts to improve your GenAI embeddings. This happens after the knowledge base is built up from the corpus annotations. It takes effort to build up a knowledge base. That effort involves processing free text, performing document analysis to harvest patterns for concepts aligned to the knowledge base, or detecting new word patterns (clusters) that will be considered new concepts, which need to be added to the knowledge base. Various low-level GenAI and non-GenAI ML approaches will be applied in the creation of your knowledge base. The efficacy of your GenAI embeddings, extracted from the knowledge base, will improve as the knowledge base improves over time.

Note that success is not measured in the MLOps quality stats for a single model but, instead, over the entire data-engineered process that gets refined through life cycle iterations. The pipeline leverages many tools, models, and execution steps so that the quality of the GenAI RAG solution eventually improves. It can now be said that your GenAI solution is given a real learning capability.

The ability to extract features from your curated released data is the next capability we willexpand upon.

Sampling the data lake

There are numerous choices you can make when selecting data from your data mart, data warehouse, data lake, data mesh, or data fabric, with the best choice being the one that your cloud provider advocates. Not all gravitate to the same architecture today. Unifying your model's feature extraction methods is a best practice, implemented hopefully in your analyst workbench. This involves accessing release sets (curated and consumable datasets) through an interface that is aware of that data source's metadata. You can even go so far as to produce access methods that are auto-generated from the metadata defining the factory's released data. This makes it possible to extract different data from various sources in a feature-neutral form and have it annotated for use in training and testing ML models.

In addition to extracting features in bulk, you should build extractors that sample the core data in smart ways. As an example, you may want a standard distribution of data based on algorithmic criteria, a Gaussian or flat random distribution, or a flat random distribution over some distinct selection criteria. As the data engineer, you need to create these feature extractors and catalog them to make them available to the data scientist. The library should be based on cloud technologies, scalable, produce derived data in cloud hosted data areas that are secure and cost attributable to a data scientist's line of business. Note that there are real costs to running feature extractors.

After feature extraction is complete, the features should be cataloged and stored. This data needs to be identified as data at rest in your factory's metadata dictionary and source data lineage metadata in the metadata repository, with any decorations and annotations required. This makes the model regeneratable and testable, since it will undergo the MLOps life cycle. You may be counting the repositories needed and asking why the cloud providers have not made this process easy. Some have efforts underway, while others have services that are still in development.

Managing data annotation

In the article *"21 best annotation tools for documents, video, and more"* {https://packt-debp.link/2CBzep}, cited previously, the Amazon SageMaker Ground Truth service was identified as an excellent choice for data annotation tasks. The service offers a set of human-in-the-loop capabilities to create human judgments (or feedback) in your ML life cycle. This improves the quality and accuracy of the models being developed. Data generation and data annotation features provide the capabilities you will need to review, customize, evaluate, and QA your models. **Reinforcement learning from human feedback** (**RLHF**) {https://packt-debp.link/vl2Idd} is a machine learning technique that uses human feedback (judgments) to optimize a model, so it self-learns. Generic **reinforcement learning** (**RL**) approaches provide training inputs to create model outcomes that perform better. Aligning model outcomes with the necessary human feedback helps bring overall performance into line with human outcome expectations. RLHF is a needed capability to tune an LLM's system from good performance to excellent performance, thereby bringing business value. Using Amazon SageMaker Ground Truth, you can implement RLHF processes via a service without the heavy development normally required.

Even if you decide to not use **Amazon SageMaker Ground Truth**, you need to study its capabilities, since it provides a solid foundation to operationalize automated data annotation. The concept of a ground truth (a term used in statistics and machine learning) refers to the correct or true answer to a specific problem. It constitutes the gold standard when comparing and evaluating various model results. Creating high quality training datasets will improve your model's accuracy. This is done by applying subject matter expertise at the right place at the right time, as orchestrated by an on-demand workflow. AWS can manage a workforce for you or leverage your internal workforce, and Amazon SageMaker Ground Truth offers options and flexibility. Amazon **SageMaker Ground Truth Plus** also lets you do the following:

- *Hire and manage a domain-expert workforce* directly through Amazon SageMaker Ground Truth to perform tasks such as labeling audio files or translating documents into a specific language.

- *Manage your existing data operations team's in-house workflow* to correct and/or augment annotated data. Your team is going to have the best subject matter expertise regarding your enterprise's information.

- *Coordinate a preferred vendor's annotation work* by leveraging AWS Marketplace to complete assigned tasks.

- *Execute crowdsourcing* through Amazon Mechanical Turk, which can be cost-effective and scalable. Note that quality and delivery have to be assessed for your domain because not all mechanical Turk participants will have the skill levels you want them to have.

Whatever path you take, be aware of the operational human costs of producing annotations and judgments for your ML solutions.

Real data versus created (synthetic) data

So far, we have focused on the creation of models from the training data that you have at hand. But what if the data can be created synthetically to drive the training of your models? This is possible! It can be formed from a domain's knowledge base. You can leverage a class of tools that, until recently, did not exist. Gretel {https://packt-debp.link/ViHTTb} can generate these artificial datasets with characteristics that are similar to real data. This way, you can develop and test several AI models without having to worry about data privacy issues. The top alternatives to Gretel are as follows:

- K2view {https://packt-debp.link/b794Y1}

- Mostly AI {https://packt-debp.link/cG2tdR}

- Tonic.ai {https://packt-debp.link/77FkPa}

- Syntho {https://packt-debp.link/01tS4i}

- Test Data Manager {https://packt-debp.link/3GyBQo}

There are also many other Gretel competitors, and they meet many mixed-modal data generator niche needs:

- AI.Reverie {https://packt-debp.link/0x0JCb}

- Aindo {https://packt-debp.link/3VIw3y}

- Anyverse {https://packt-debp.link/IauQ4J}

- Benerator {https://packt-debp.link/liinTq}

- Betterdata {https://packt-debp.link/v7PbjS}

- Cvedia {https://packt-debp.link/2SwmYG}

- DataGen {https://packt-debp.link/4xN3Pn}

- Datomize {https://packt-debp.link/5AwrUe}

- Datprof Privacy {https://packt-debp.link/bQZMmg}

- Dedomena {https://packt-debp.link/PYLG4B}

- Delphix {https://packt-debp.link/V5XhqY}

- GenRocket {https://packt-debp.link/0MZUHY}

- Hazy {https://packt-debp.link/xrtTDH}

- KopiKat {https://packt-debp.link/O2CwAr}

- LexSet {https://packt-debp.link/NUacQh}

- MDClone {https://packt-debp.link/VtWucQ}

- OneView {https://packt-debp.link/uM7xbn}

- Protopia AI {https://packt-debp.link/B1Ca4L}

- Replica Analytics {https://packt-debp.link/DDTBQE}

- Sarus {https://packt-debp.link/2053xP}

- Statice {https://packt-debp.link/J6HS60}

- Synthesized {https://packt-debp.link/xe6GVz}

- Synthesis AI {https://packt-debp.link/JfsARD}

- Synthetaic {https://packt-debp.link/mgh9TL}

- Trūata Calibrate {https://packt-debp.link/GmvbGL}

- Tumult Analytics {https://packt-debp.link/0O92UA}

- Ydata {https://packt-debp.link/EkiIRI}

Sometimes, you just do not have enough real data to drive a model's training effort, but synthetic data makes that possible. Plan to augment your data factory's curated data with synthetic data to preserve privacy, or for subject areas that require a dataset to exist for complete model training. When you know that some data area should have representative data and you discover that it is absent, then it needs to be generated so that the model is trained with complete scope. You can tell what you don't have regarding data by running queries against your knowledge base, allowing you to see the subject areas where object instance coverage is sparse. The knowledge graph will give you insight into what should be present in the knowledge base when the count of object instances for a class is low, or when instance data is completely missing but the model indicates that data should exist.

The effort needed to create synthetic data that is best suited for your solution can be daunting. You can the problems Tesla faces (see the later example in the *ML asset deployment* section) when trying to account for real-life driving experiences, with incomplete multi-modal data being unavailable for all situations. You have a fleet of integrated tools, from the preceding list, to help with the synthetic data curation necessary to train your solution. Many tools will be applied to the creation of systematic data and the replacement of it with actual collected data, as a system runs in production. The processes required to create, train, and then replace synthesized data over time will be part of the glue that binds your MLOps processes together.

Model training

When training models in the MLOps life cycle, it can become operationally expensive in terms of IT! You want to make use of as many cost optimizations as possible in your cloud infrastructure configurations. As an example, you want to use **spot instances** in order to manage cloud costs. You also want to scale down immediately after model training. If you use tools such as Databricks, you want to especially tune scale-down, since third-party tools want you to stay subscribed even when you aren't using them, so that they can incur costs for doing nothing.

We have learned many lessons over the years regarding classic ML model training. Some of these best practices are the following:

- Your ML models need to be clearly defined

- Avoid confusing ML learning and skill-building efforts with real model experimentation

- Understand your data and its metadata

- Build your data quality metrics

- Beware of SaaS-hosted ML solutions/tools

- Define your model to be robust

Each of these requires some more elaboration.

Your ML models need to be clearly defined by goals, parameters, and metrics. On average, today, only about 13% of all models produced make it into operation. This problem can be reduced in severity with proper planning, as well as by working to make sure the assumptions and scope of your training data are profiled and then aligned to a model's goals.

Avoid confusing ML learning and skill-building efforts with real model experimentation. We might all have great ideas and want time to play with the data and the technology, but at the end of the day, it's just that – play! If you want to learn using your employer's environment, then get permission and set limits on a task. Then, get back to real work! Follow the accepted computer science and data scientific processes. Create the hypothesis, experiment, measure, conclude, iterate, and accept (or refute) the hypothesis, and finally, complete the science effort before engaging in any production rollout.

Understand your data and its metadata, and make sure it is complete for the subject matter domain. If not complete, get the necessary data or be ready to create accurate synthetic data.

Build your data quality metrics as well as your model's quality metrics in advance, and use these to measure model performance through the MLOps life cycle. Collect experiment outcomes, recall partial experimental successes, and reset the MLOps life cycle to some last know good state if, or when, the MLOps outcomes diverges from your expectations.

Beware of SaaS-hosted ML solutions/tools that cause request traffic or data to leave or enter your cloud service provider's environment. This incurs iterative exit or egress data charges from your cloud provider. The complete **total cost of ownership (TCO)** of your ML solution can rise precipitously if these costs are not considered in your planning.

Define your model to be robust, which has to be one of the vaguest terms used in this book! But you get the general idea. The model has to be valid for a scope of data, it has to pass objective and subject output quality checks, it has to be measured by a set of ML metrics from iteration to iteration, it has to be assessed for drift, and so on. It also has to have many tests applied and evaluated in an automated manner, results tracked, trended, and alerts raised when deviations are detected.

Since this section ended with a testing topic, let's dive into that a bit more in the next section.

Model testing and evaluation

ML model evaluation does not simply mean reviewing training accuracy metrics. Real models need multiple test metrics to be certified, in order to ensure that they are safe to deploy. ML testing processes will evaluate and then validate many acceptance measures to make sure that a model subjectively meets its goals. Some important categories are correctness, accuracy, performance, reliability, ethical, and robust. In addition, there are objective model measurements that need to be created:

- A measure of accuracy and sensitivity

- A measure of resilience to change

- A measure of reliability

- A measure of bias

Let's look at these points in more detail.

A measure of accuracy and sensitivity is a model's *accuracy* (or *precision* {https://packt-debp.link/630pDP}) versus its *sensitivity* (or *recall* {https://packt-debp.link/Nh1BjW}), which provides the model's historical trend to adhere to the data used in training and the outputs aligned to that training across the full domain of data. Predicted outcomes, from running a model against its intended dataset, can be measured to produce an **F1-score** {https://packt-debp.link/br5uPj}, reflecting an even optimization of both precision and recall measures. This tends to produce precision and recall results that are contradictory to each other (in other words, one goes up, while the other tends to go down). Your engineering goal is to get both precision and recall metrics up at the same time, and that happens in classical ML models where semantic features are leveraged in the model's training. The F1-score is often criticized, and if you wish, you can leverage the **P4-Metric** {https://packt-debp.link/hoemJH} instead.

A measure of resilience to change. Since data is always changing, the concepts that the data represents will *drift* over time as it evolves. ML testing will continuously ensure that ML models are as effective in the future as they were in the past, even when new data is introduced that is aligned with drifting concepts. Being able to sustain a model's correct predictive capability is essential when putting a model into production, ensuring it remains future-proof. Seldon {https://packt-debp.link/BYSuFF} presented an interesting approach to handling drift in this presentation: *Seldon Presents on Context-Aware Drift Detection at ICML 2022* {https://packt-debp.link/44zifN}. It would be a good idea for you to dive into the way drift was handled by Seldon, using context/categorical variables {https://packt-debp.link/yF8JZI}.

A measure of bias. An ML model can tend to predict outcomes that favor one class of outcome over another. This leads to unfair outcomes or **bias** {https://packt-debp.link/ImTHlm}. Testing can reveal and measure the degree of bias. The root cause of bias is often found in the training data used or the ML algorithm itself. The solution requires the model to be retrained, with different features selected from the source dataset or the section of a new ML model better aligned to the class of data being processed. You can build your own tools to check for bias/fairness or leverage third-party tools to inspect models. Here are six of the many tools for detecting bias/fairness:

* The Aequitas Tool {https://packt-debp.link/rJESPK}
* Google's What-If tool {https://packt-debp.link/nlzlFK}
* IBM's AI Fairness 360 {https://packt-debp.link/InraXA}
* scikit-fairness/scikit-lego {https://packt-debp.link/Pex0cE}
* Fairlearn {https://packt-debp.link/pLtXmY}
* FairLens {https://packt-debp.link/AgGMNS}
* FairML {https://packt-debp.link/wkz5d3}

These tools are rapidly evolving, and the list grows each year because this issue has a severe impact on your ML solution's performance. These tools are part of the **Trust, Risk, and Security Management (TRiSM)** {https://packt-debp.link/00xrhG} AI governance framework, which should form a core part of your solution. TRiSM products include AI auditing, AI monitoring tools, and AI governance frameworks.

A measure of reliability. Comprehensive and well-architected testing procedures strengthen ML system reliability. Keeping a model performant and at a production-ready level of quality will be a data engineering task. The task of knowing your data will be tested when building reliability into your ML models. When implementing cross-validation testing approaches, you will determine how well your ML models handle new, previously unknown data and still produce correct predictive outputs. Cross-validation takes a dataset, creates several subdivisions, and then trains the model on them. Model testing takes place by processing the rest of the dataset through the ML model. Cross-validation will give you a solid level of confidence that the model *may* perform well for future datasets of similar content. That will be based on how new datasets are aligned conceptually with the original dataset used to train and cross-validate the model. Another aspect of an ML model's reliability will be in testing its degree of robustness. When the ML model is presented with carefully synthesized, curated, or discovered data that is contrary to the original datasets used to train the model, the model may become confused. This **adversarial testing** {https://packt-debp.link/dXDtbH} approach reveals a lot of good metrics regarding the model's robustness.

Your testing approach for ML models requires you to define objectives, iteratively take measurements, collect metrics, and report results to a test manager. This way, postmortem meetings (meaning after test failure is reported) can be conducted and result in collaboratively developed meaningful ML model changes. Any subjective quality assessments can be used to change the objective test suite or process flows. We have seen many ML solutions produced by consultants that looked good when developed but failed when put under the stress of a production data workload. You have the best chance to get your ML models to production with a solid ML model test process.

If you base your ML model testing processes on one of these well known ML test frameworks, you will most likely be satisfied with the outcomes:

- **TensorFlow**: **TensorFlow Extended (TFX)** {https://packt-debp.link/9k8CIo} is an open source framework developed by Google. It offers a solid set of capabilities for ML pipeline testing. Additionally, there is TensorFlow Data Validation {https://packt-debp.link/mQBOTR}, which is used to test data in ML, and TensorFlow Model Analysis {https://packt-debp.link/TMFFVo} for in-depth model evaluation.

- **PyTorch**: PyTorch is also an easy to use open source ML framework that produces a computation graph {https://packt-debp.link/Lqwjbv}. PyTorch comes with several tools for model evaluation, debugging, and visualization. Read this article for a tutorial on how to apply it to DL: *How to Evaluate the Performance of PyTorch Models* {https://packt-debp.link/mIJGxj}.

Other ML test frameworks exist, such as scikit-learn and Fairlearn, which we have already covered.

Now, let's address the testing needed after you think you have finished testing. The test cycles never end when developing ML models and getting them ready for production. We need to create some smoke tests.

Smoke tests

What is a smoke test in the context of ML model testing? The answer is, a simple type of testing that is implemented when you wish to make sure that integrated code and a ML model run successfully in a production environment.

There are many libraries, packages, and cloud configurations needed to run, scale, and then degrade resources during ML model execution. Updates have to be backward compatible, and have access to data and results, updated metadata, and effectively collected measures and metrics. You need to catch any adverse effects within your system before deploying changes. Tests are designed to make sure the boundaries of acceptable functionality and performance are sustained from release to release. Even data changes, due to processing new data factory release sets, can kick off ML test detected drift anomaly alerts, which are best handled before outputs are visible to end users.

Small changes to dependent packages may *not* be backward compatible. Sure, even ML model code should undergo the same engineering discipline requiring unit, integration, stress, regression, performance, stress, acceptance, and finally, ML model testing. These tests take place well before the smoke test is conducted; however, the successful execution of a suite of smoke tests is the last hurdle before going live with a solution after deployment, but before it is activated. It is important to note that business predictions should be included in any smoke tests you produce to provide a good assessment of a model's performance, even when an F1-score improves. This applies to other model metrics' scores as well. A proper metrics result can still cause a model to erroneously predict an outcome for a critical case.

Dependency management (which includes the validation of software contracts and data contracts) has to be implemented as configurations and not be hardcoded in scripts. Just as with other types of released builds, ML pipelines should include smoke tests to quickly catch problems. In the article *Smoke testing for machine learning: simple tests to discover severe bugs* {https://packt-debp.link/576Cjt}, the authors created definitions for 22 smoke tests related to classification algorithms and 15 smoke tests for clustering algorithms, as well as a naive approach with linear complexity for combinatorial smoke testing that considers different values for all hyperparameters. These tests included a case study for Weka (Frank et al. 2016), scikit-learn (Pedregosa et al. 2011), and Spark MLlib in order to test 53 classification and 19 clustering algorithms. In that testing, 11 bugs were found and reported to the developers.

Additionally, a proof-of-concept proved that the smoke test approach works with deep learning and the TensorFlow framework. Finally, two best practice recommendations were suggested that can improve the quality assurance of ML libraries:

- *Define smoke tests*, with input data and with various combinations of hyperparameter combinations, should be used to test algorithms in three ways:

  - With input data coverage of extremely *large values*, extremely *small values*, and finally where all features are set to *exactly zero*.

  - With *empty classes*, or near empty classes, when implementing classification smoke tests.

  - With *empty category inputs*, for categorical data.

- *Use combinatorial testing* for basic model functionality that is to be applied with a wide set of hyperparameters. You want to make sure you have equivalence class {https://packt-debp.link/u9jO7u} coverage. This can be achieved with a grid search {https://packt-debp.link/vdd5LM} over hyperparameters. Testing each of the equivalence classes once using default values for all other hyperparameters can also give you good test coverage. This latter approach scales better and produces fewer invalid tests.

In summary, with proper smoke testing, you can catch errors that were missed because you never thought to test assumptions that you never anticipated being changed. Over time, anything can change, and it's the unexpected change that will degrade your system. We have always applied the first principle of paranoid programming in such matters – it's better to test than be sorry later:

"What can go wrong will go wrong, and it's going to happen to ME!"

Murphy's Law (sort of!)

This is an application of Murphy's Law {https://packt-debp.link/RufZpq}, made more personal for a data engineer or data scientist. Take it to heart. Smoke testing will help catch many errors.

ML asset deployment

As a data engineer, you will be tasked with building out an ML model deployment process. This way, science can be made part of a solution and then operationalized. Even if that deployment is to a **user acceptance testing (UAT)** environment, from the perspective of the data scientist, it is operationalized. So, the first configuration option you have to support is the deployment to a target environment. You must be able to un-deploy (rollback) any ML model as a package with equal ease. You need to then follow a zero-footprint rule – that is, leave no trace after removal.

There will be many scripted and parameterized configurations (or steps) to be taken when deploying ML models. You want to build a deployment framework or use a proven third-party tool. Being able to call the deployment process from a notebook is also really useful. Some tools and MLOps frameworks used for model deployment are as follows:

- Amazon SageMaker {https://packt-debp.link/Ke4CyN}
- Azure ML {https://packt-debp.link/AV2RjX}
- Banana {https://packt-debp.link/4bfUQT}
- CognitiveScale {https://packt-debp.link/skCvgL}
- Einblick {https://packt-debp.link/fp5GgB}
- FlyteInteractive (Flyte) {https://packt-debp.link/z290Hv}
- Google Vertex AI {https://packt-debp.link/bvzOiT}
- Inferless {https://packt-debp.link/8calAi}
- Modelbit {https://packt-debp.link/OiBAfv}
- ModelOp {https://packt-debp.link/ccoEvO}
- Modzy {https://packt-debp.link/tMUF0u}
- Nod {https://packt-debp.link/mFudq9}
- OctoML {https://packt-debp.link/zCLoSH}
- Oracle ML deployment {https://packt-debp.link/PLkZZ8}
- Wallaroo.AI {https://packt-debp.link/ja2tH0}

We'll let you look into the capabilities in more detail; those of you who wish to rapidly develop and operationalize ML models will want to examine the third-party frameworks closely to get a head start.

A/B testing

A/B testing is a great class of experiments that you can perform when testing your model's performance and quality. The concept of A/B testing was created by those wishing to compare website versions and be able to toggle customer traffic between versions. Web analytics collects results, and Optimizely and Adobe Analytics lead the field in the coordination of such efforts.

With ML models, you can confirm that your model works better with real data than another model, or compare performance with that of a human. Different models can be compared or tuned variations contrasted. All recommendation ML systems or predictive ML models can undergo A/B testing. This type of testing follows a predefined series of steps:

- Deploy the ML model.

- Split the model request traffic into A/B streams.

- Set assumptions that will be common to both A/B runtime model deployments (such as time, location, and other text contexts).

- Measure, gather metrics, and baseline any data into a test catalog.

- Analyze metrics and prediction outputs to create conclusions regarding the performance of the A and B model outcomes.

A/B testing can produce inconclusive results if you do not fix parameters and variants when testing. So, when you design A/B tests, you want to categorize the input streams for each and compare apples-to-apples metrics for the outcomes associated with those categories. Test scenario design and management are critical. The design will always be hypothesis-driven, since this is still a test practice and you have to test to prove or disprove your test hypothesis.

First, you want to create your test groups. You want to profile the universal test group and select participants aligned to test invariants, such as age. Then, you have to design the test with this select group profile for both A/B test streams. Other invariants affect the model's scope. As an example, you want to fix the context of the dataset, such as a medical diagnostic category such as colon cancer. This way, the subjects and subject matter are aligned between A/ B test invariants. Also, make sure A/B subjects are really separate and do not talk, mix insights, or share observations. Even knowing about the other group will skew results. This A/B testing has to be conducted as a 100% blind test effort.

Lastly, measure any random fluctuations in the A and B models for the scope you have defined. This results in an error factor being calculated and later used to calibrate metrics and results, so as not to assume that small differences between A and B are statistically significant. Some truth will be lost in the noise, but you need to know that the noise exists and handle results outside that inconsequential band. The noise error can be measured by having both A and B tests run by the same group. The results should be the same, and if that is not the case, then the test will be inconclusive. If so, redesign the test (see *"A/B Testing in Machine Learning"* {https://packt-debp.link/7xRoNK}).

Handling regressions

In ML development and MLOps, regression testing will provide a safety net that can prevent bugs from reoccurring. The period after retraining a model is when your regression testing will catch issues that you thought were already fixed. Your datasets will drift over time, and they will suffer from success as a business grows. A prior dataset will be augmented, with new twists on those edge cases that undo the patches placed in your ML model. Regression testing is a best practice way of maintaining model performance over a long period of time and a number of life cycle iterations as your models improve.

It is advisable that as problems arise with your model, you categorize the issue that the regression test addresses as low, medium, or high complexity (or any other way that makes sense for your future reuse). The issue priority is going to drive the regression tests that you need to produce. The more complex the issue in your model, the more regression testing will be needed. Those tests will require data, expected outcomes, and measurements to be retained. This way, a baseline and trend can be gathered from one iteration to the next cycle.

An example of the type of issue that regression testing can detect is seen in the following example, where the goal was to create a computer vision ML model to detect when a runner stays in their lane during a track competition. On cloudy days, the system performance metrics indicate great performance, but when the sun is out and at certain times of the day, the image shows a runner's shadow. That shadow appears outside the runner's lane. This caused the model to predict that a runner had left the prescribed lane and was disqualified. This bug has to be fixed, and you can do so by collecting similar shadow-containing images, generating a regression dataset, and then reevaluating your model. So, you fixed this poor model's outcome by collecting additional training data! With regression testing, you will continuously be on the lookout for bugs by reevaluating your model on normal test data but, now, with the new regression test data as well. This way, when shadows appear, they are handled, and you can be sure that this type of bug doesn't reoccur, unless an outsider jumps on the track and crosses the lane boundary, but that is a case for another regression test, so only registered runners are considered important.

In another example of regression ML models, in 2020, Tesla's director of AI, Andrej Karpathy, explained how Tesla employs large-scale regression testing to verify the performance of its autopilot system. Tesla has created extensive custom MLOps and test infrastructure, allowing them to automatically create test sets for some uncommon scenarios by mining data that was collected from their fleet. Tesla implements regression testing retroactively and proactively to probe a system for correct behavior. Andrej's video, quoted here, expounds on the challenges faced at Tesla:

> *". . . we [at Tesla] curate not just the training sets. We curate the test sets and spend just as much time on those, if not more, because you want to make sure that your evaluation is really good. You can do arbitrary things on the training set. But you must have a really good evaluation because that gates your release into the world. So, basically what I'm trying to get across is that it's a very difficult sort of domain because of the complexity of it [self-driving cars]. So [below] are slightly outdated numbers now:"*

Tesla's **HydraNets** are ML models that handle objects, traffic lights, and markings for frontal analysis, as well as a shared **Convolutional Neural Network (CNN)** for backend processing. There are 48 networks (for 8 cameras), 1,000 distinct predictions (output tensors), and 70,000 GPU hours required to train the models:

> *"… We maintain roughly 48 networks in production (but)it's actually more now. They make roughly a thousand distant predictions. None of these predictions can ever digress. All of them must improve over time. It takes a long time to train these networks if you were to train the models from scratch. Of course, we can get away with a lot of fine-tuning and things like that, and we do … But if you were to train this stack from scratch, then it would train for a long time. So, on one node this would require an entire year …"*

> *Andrej Karpathy in "Tesla's Andrej Karpathy in CVPR 2020: "Scalability in Autonomous Driving Workshop"* {https://packt-debp.link/3zg6cH}

Andrej also mentioned the problem of finding what he called *needles* (as in the adage, a task is like finding a needle in a haystack). The needles are images of rare but key moments that needed to be considered by the Tesla HydraNets. All image information was collected from across the world, and that makes it like a haystack. These interesting scenarios captured events such as a chair falling out of a pickup truck on a highway going at speed. This is a rare occurrence, but it is important that it is handled correctly in the Tesla models. Likewise, any potential accident-causing scenario is a needle, since autonomous vehicles have to be smart enough to handle these edge cases. He stated that finding enough of these needle images is very hard because they are associated with rare events. A training data set can't be built on these edge cases today from real data images.

We hypothesize that the creation of synthetic training data can be effective if driven from your knowledge base, which can be used to create synthetic but credible training images to augment any real training data collected. With this approach, you will have to pay special attention to automated regression testing associated with issues posed by the needles. This way, you can monitor an ML model's performance as more and more real data is collected to supplement or replace the synthetic data.

Since the MLOps life cycle is just that, a cycle, there will be times when you want to undo what was done and get back to a previous good state, relating to an ML model's configuration, trained state, and execution environment. We'll discuss that next.

Rollback

In order to roll back, you need to create a snapshot of your analyst workbench. Some physical science researchers actually record videos of their lab workbench areas, allowing them to get back to a last known good state or recreate a failure. Often, the hard sciences' physical notebooks do not contain enough detail to recreate an experiment. Data science is really no different. The number of variables can be enormous. You want to build the equivalent of a flight recorder for your analyst workbench, gathering timestamps, configurations, steps, metrics, probes, training data, input data sets, and outcomes, as well as any other environmental contexts. Then, you want to create checkpoints and resets for them. This way, rollback efforts are made easier. However, sometimes, you will need to just partially roll back, and not to a previous checkpoint. That will need to be reflected in your flight recorder.

There is no magic bullet to slay this monster. You need to be able to roll back work that you hoped was going to perform as expected. The reality is, you are going to be doing this many times. As an engineer, you want to look for common ways to do the uncommon.

Reviewing training data

Models and the training datasets used to create or tune them have to be cataloged in an ML model catalog, with descriptive metadata. The catalog needs to reflect model versions and be discoverable. A catalog has many capabilities beyond those of a model registry, since it needs to be able to do the following:

- *Share model assets*, making them reusable by exposing the model's metadata – who, what, where, how, with what quality, training inputs, feature extraction logic, sample expected outcomes, calibration methods, outcome usage, and so on.

- *Support model governance* rules, tests, and auditable outcomes that are necessary to comply with ethical, legal, and societal mandates. Enterprise compliance teams can define rules for ML models and their output and approve guardrails for PII, security, and publishing rights management (when training on third-party data).

- *Sustain model performance* with usage metadata that should be cataloged with the model, in order to supply performance prediction trends and end user usage. Models can exist for a short time and no longer afterward. You'll want to purge or archive them to reduce your technical debt load. Objective and subjective assessment information should also be retained in the model catalog.

- *Support streamlined deployment* workflows, since they are not fixed between models. They can be as flexible as custom scripted steps, or as structured as configuration in the selected MLOps framework configurations. CI/CD tools can also be used, but they need to be integrated to report progress steps in a catalog and the final entry in an ML model registry. Git check-in is a best practice for any notebook (without data, but with the dataset location linked to the ML model catalog).

Verta {https://packt-debp.link/ZtL7qM} has an ML model catalog {https://packt-debp.link/dOTzWz} that can provide advanced capabilities. There are a couple of alternatives to Verta:

- Metaphor {https://packt-debp.link/oRSDIM}

- DataGalaxy {https://packt-debp.link/Hgg6mM}

The cost of maintaining many ML models in production will produce a lot of drag on your development velocity. That will be felt by the data engineer, since the data scientist will be off to the next new challenge as soon as a solution goes to production. The maintenance task for sustaining a suite of high performing ML models is high. The cost and effort can be lowered with the best practices you have read about in this chapter.

MLOps frameworks

There are going to be a lot of evaluations and a final choice to make regarding the selection of an MLOps framework. You could develop one yourself, but you need to weigh the costs/benefits before trying to build these capabilities yourself. A summary of existing MLOps frameworks follows (with reference links for any additional information not previously cited):

- Amazon SageMaker

- Azure ML

- Banana

- CognitiveScale

- DataGalaxy

- Einblick

- FlyteInteractive (Flyte)

- Google Vertex.AI

- Inferless

- Kubeflow

- Metaphor

- MLflow

- Modelbit

- ModelOp

- Modzy

- Neuromation {https://packt-debp.link/oXRPJy}

- Nod

- OctoML

- Oracle ML Deployment

- Verta

- Ydata

- Wallaroo.AI {https://packt-debp.link/ntsNJV}

Note that many in the preceding list were mentioned earlier in the chapter when the purpose they serve was elaborated upon. Also, note that cloud providers have produced credible MLOps offerings, but these are not always as capable as some of these specific third-party tools and platform offerings in their respective niches. You will want to ask: *Which tool do I choose?* The answer will be driven by the due diligence criteria: the weights that you put on one capability and its derived features versus another when you formed your architecture. If we were to say that one tool is best for you, then you could find reasons to refute that position, as could your enterprise peers. If we were to objectively give you the mechanism to structure your criteria and then select and communicate the tool selection to your peers, *then* you would have a solution that could remain in production over time.

Summary

Information Technology has evolved a lot since machine learning and artificial intelligence have replaced natural language processing as effective approaches to handling unstructured human communications. The real world is unstructured, and the task of IT is to bring order to confusion. With cloud computing comes the commodification of resources that were unheard of 25 years ago. With the future advent of quantum computing, the previous pattern of software, processes, and algorithms preceding the advent of hardware is again playing out. When that key processing capability comes online, you can expect the best practices and vision that are aligned with that future to remain effective for a long time.

Since many MLOps life cycle best practices apply lessons learned from DevOps, you can expect these processes to become more unified in time. As with many cutting-edge technologies, you will see a bloom of third-party vendor offerings, open source solutions, and cloud provider unification of these approaches into cost-effective bundled service offerings. This chapter has identified many of the options. You can research them, understand their capabilities, and see how they can help you in your effort. Just don't be oversold. You can expect these trends to continue, so do not get too married to a niche tool. Be ready to shift gears when the cost/benefit equitation warrants a change.

We began this chapter with a caution about change. Just remember:

> *"Change is going to be your constant, accept that as a given and build accordingly."*

In the next chapter, we will summarize the book and leave you with a few parting thoughts.

17

Takeaway Summary – Putting It All Together

In this chapter, we will summarize some best practices. We will try not to rehash what has extensively been presented but rather to leave you with a few key takeaways that we think you will want to use on your journey.

In this summary, we will recap some of the practices from other chapters and we'll begin with how important *people, technology, and processes* are to your engineering effort. All too often an engineer will become focused on some interesting feature of a capability within the architecture that is exciting and challenging. Losing sight of the big picture is a constant threat to your success. The data engineer has to remain mindful of the big-picture view of the entire effort, even while tackling detailed challenges.

This chapter will recap the challenges that cut across all capabilities of your solution designs: business, security, privacy, knowledge engineering, and artificial intelligence. The goal of this summary is to underline the need for you, the data engineer, to conceptualize your work as a big requirements matrix where each cell in this multidimensional matrix has a design requirement that you have thought out completely. When challenged, you will be sure you have not missed anything essential and provided yourself enough slack to accommodate future changes. If Agile purists point out that this is not aligned with the current Agile processes in vogue, you can point out the fact that, without a common vision, the solution will perish! It does not mean we are suggesting your solution be built outside of an Agile framework; rather, you want to rationalize the business processes needed with the people and technologies on hand while optimizing the cost of developing a future-proof solution.

People, technology, and processes

When reflecting on the goal of the book, which is to provide you with data engineering best practices, we focused on people, technology, and processes first, understanding the answers to the question *why?* For example, *Why* did you design it like that? *Why* did you not consider these options? *Why* did you go in that direction, rather than that direction? All these choices are driven by your enterprise and IT mission and vision. Otherwise, you will have no ground to stand on when challenged and, as an engineer, expect to be challenged frequently. Together, the mission and vision enable you to craft an architecture that organizes your thinking and provides guardrails for your design choices.

Working with other people

Transparency is required to meet the challenges of the skeptics you will encounter. *"Build it and they will come"* is just not going to happen without a healthy dose of social engineering before you attempt to engineer a solution. When we were younger, earlier in our careers, we heard this phrase often quoted and abused:

> *"… it's not what you know but who you know"* that will bring you success.

> G. P. Bush, L. H. Hattery, *"Federal Recruitment of Junior Engineers"*, in *Science*, volume
> *114, number 2966, page 456 (1951)*

It was made in reference to the summary sentiment of 84 students who referred to political influence as a major disadvantage of federal employment. It meant that:

> *"There are too many political connections necessary – in spite of apparent merit systems."*

> *"Federal Recruitment of Junior Engineers"* (see previous)

The fact is, it's not going to be your superb engineering that brings success. The people governing funding, your project's staffing positions, and the choices made while you form your solution will also be critical to success. Consider all those you will encounter while traversing the mountain of success. We've seen rock star engineers climb to untold heights only to crash and burn without the support structure needed to maintain those lofty heights of success. We've also seen companies with cultures that departed from successful engineering ways in order to just ship a product. So many crushed eggs (people and careers) were destroyed in the process just to get a product out the door. These cultures promoted aggressive internal competition. They even created internal process dead ends in order to challenge future leaders to think out of the box and then rewarded many who were non-compliant with enterprise policy since, after all, they achieved a successful **generally available** (**GA**) product release. We'll leave those company names out to protect the innocent, but you may also have had some of these sour encounters in the course of your career. Those leaders had to depart the company even after the successful release of their product! There was just no going forward and no way they could go back into the development herd. How you survived those types of situations had a lot to do with the soft skills you developed. The psychological and social engineering skills that you used in your interactions were and still are vital for your data engineering success.

You're never going to be working in a vacuum, and as such, many others will have to contribute constructively. Anarchy will reign if there is no structured framework for your designs to be mounted on. These designs need to be augmented with input from others, and the structure makes that possible. Then comes the need to make engineering choices, with justifications, in order to create your physical architecture. This is based on the high-level concepts you defined in the conceptual architecture and the logical components broken down in the solution's logical architecture. These artifacts provide the framework for business and technical communication as well as governance. There is a tapestry of logic that needs to be communicated to others, which is visible only from a higher perspective of your architecture. That picture has to be a simple, crystal-clear communication vehicle for you to use as a flag to marshal troops – you will need to marshal all the key players toward the goal. In your mind, that will be the vision of the end-state implementation, and the way you get there will be your mission. You will also encounter many types of would-be adversarial contributors, detractors, intransigents, dinosaurs, naysayers, opponents, and possibly competitive enemies in the course of your development. Deal with them as you would a sick puppy. Understand them and why they are acting as they are, isolate them and their support structure, and then leave them in capable hands. Reward and protect your helpers and get them to *"help you help them."* Avoid firefights that will zap you of your strength and your established support at critical times. Fight battles that you can win and avoid those that you will lose! *The Art of War* has shaped competitive business landscapes for generations. You can learn a lot from reading this ancient treatise, but I want to posit Abraham Lincoln's (the sixteenth president of the United States) thoughts when confronted by an old woman who rebuked him for his conciliatory attitude toward the South, which she felt should be destroyed after the American Civil War. Abraham Lincoln replied:

"Madam, do I not destroy my enemies when I make them my friends?"

Abraham Lincoln, 1864

Your attitude will flavor all your dealings with others. If you are smart, you will overcome technical challenges, and if you are wise, you will win political ones, and the hearts and minds of those supporting you on your journey. If you implement best practices, you will be both good and wise.

Know yourself and your limitations

Know yourself and your limitations! Often, technical folks, especially engineers, are not great communicators. We think we are. But we're not. There are a lot of excellent, expressive marketing types that will run circles around an engineer and leave others bewildered with options, marketing speech, and general confusion. Yet, the target will remain optimistic that what they were sold on will really work out well in the end. It often does not, and the marketeer will not be around when that kind of vaporware solution fails to deliver. The marketeer will be off selling the next widget to the unknowing and unaware. The engineer will be left picking up the pieces of those projects, and even if there is some degree of success, credit will be taken by the same leadership that caused the catastrophe that you fixed. If this starts to happen, withdraw or be determined to fight, but let your actions be driven by the principles you adopt. Not all projects are successful; ideas come to market, nor do all well thought out solutions remain in production. Your goal is to create something that outlives even your career, as a future-proof solution should. This goal could be obtained if you can simplify the complex and effectively communicate a clear message.

A human being can only hold so much in their mind at once when considering actions. We know some geniuses have the capability to balance seven concepts at once and come to optimized conclusions intuitively that are stellar. Others are limited to being able to balance three concept variants. It's not that some are smarter; it's just that some people have thought limitations and can't balance the fourth to seventh variables.

Every individual is different

Everyone can handle three contrasting ideas with ease – beyond that it gets harder. You will encounter this in your future project dealings. Don't judge someone when they say:

"I don't follow …"

Every individual is different. Key stakeholders need to buy into your solution, and you need their agreement. Craft versions of your message to your target audience and tell them a story that makes sense to everyone. The message must fit each listener's perspective and capability while maintaining a clearly articulated set of well-organized facts that do not change. These artifacts will back up your claims. Use an architecture tool rather than building a house of cards for solution documentation and keep it evergreen. Special attention is needed to clarify your solution's capabilities, and the features that support the architecture and will appear in your later designs. Tell everyone the data story where the hero wins, all are happy, and there is no end to the business's success. Who does not like a true and enduring journey where the outcome is success and where you can see yourself sharing in that success?

When going for those ambitious, lofty heights, avoid climbing greased flagpoles! You'll come down fast when headwinds pick up. The solid footing that your best practice engineering process makes possible will also make success possible. That success will be repeatable over time, without any back-sliding. You will be able to point back to a long trail of breadcrumbs so that others can follow behind. As an engineer with a vision, on a mission, you will often outpace others, and others need to follow your path of thinking. Being an engineering leader means you want to foster a learning culture where tacit, self-actualized, self-guided learning takes place. In this manner, you will build a mountain of support. *"Aha!"* moments will be encountered by others (your managers and teammates) as a result of your kind, servant leadership style. Who does not want lessons from a non-credit-seeking leader and teacher?

Develop your solutions' technology with clear processes

You will want to develop your solutions' technology with clear and consistent processes. Technology should be developed in a way that others who follow after can see what you have designed, why it was designed the way it was, and how it was delivered so that they can add value when maintaining your solution. Keep your technical artifacts organized and maintained, even when it can cause you hours of work without knowing if or when that documentation will be read. Do not think that the project's stream-of-consciousness documentation anchored in Agile tickets will be sufficient! The project's twists and turns will be at best confusing for someone trying to follow along based on poor documentation methods. Build and maintain the current state (or what-is) document base from the perspective of the architecture framework. Being able to drill in and out of a component located at any level is very important. This is the best practice for everyone involved in your engineering solution. Let the JIRA documentation be used to color how the architecture came to be as it is, not to define the what-is, because there will be lots of misleading information that frustrates rather than communicating the current state.

Since we are summarizing, on the topic of Agile development, you will recall the *Prevent Agile from being fragile* section in *Chapter 9*, where we elaborated on the need to create an enterprise version of the Agile Manifesto. There are just too many individuals who will want to twist Agile concepts and best practices to their will. Often, they will not share the intent behind those modifications. In many companies, the introduction of the Agile methodology hits road bumps where yearly budgets, funding, and inflexible product expectations hit the reality of cost and time to deliver, with the data and software engineers in the middle. The product development process must not be omitted from the enterprise's published Agile manifesto. This process cannot be picked off the shelf and used by your enterprise. Generic Agile processes have to be formalized, published, and governed for compliance. If this is omitted, then the service-level agreements, Agile delivery timeframes, costs, and the product's data contract will all be in jeopardy. When building processes, measure and report on key progress so all can see from month to month how much progress is being made. Be very careful about reporting a red status since the shoot-the-messenger reaction to the one who brings bad news is still a practice that is in vogue. This means that when you find a process violation or a failure to deliver, the culture of the company should be to accept those failures as areas of improvement and not ammunition to fire the blamed offender. You best know your enterprise's culture. Beware of this aspect of Agile development, or else it will naively become fragile.

Develop technology that operates well

As you develop, and the process is running smoothly, you will want to see iterative progress regarding the creation of features associated with the capabilities your solution's project team has committed to deliver. You want to build a great quality process that has **service level agreements** (**SLAs**) (software contracts) and data contracts at its core. These are to be monitored, trend analyzed, audited, and reported against when violations take place. They need to be refreshed at least each month for critical review. Building a quality dashboard will go a long way toward establishing trust that the solution you develop will not melt down under adverse conditions. The dashboards will help you assure leadership that the technology product will be of the highest quality. You will be in charge of your success, and you need to be open and honest about how that is measured. Then, you need to clearly report on your success. One class of project SLA constraints is going to be your IT operating costs! Measure cloud costs particularly carefully. Set your various teams' usage boundaries and alert on team and individual cloud usage deviations on a daily basis. You will get badly burned if you do not take this recommendation to heart! All these cost details have to be in a clear summary report with a drilldown into the detailed IT operations report. This transparent high-quality reporting goes a long way toward establishing trust with senior management for your delivery. Any **objectives and key results** (**OKRs**) obtained will be your highlighted successes.

Build a technical solution with a focus on quality

Your test processes and technology choices used in software and data product development will tell everyone how well you want to retain a high level of quality over time. Regression testing will be your safety net when you accelerate development after a minumum viable product (MVP) is created and you build toward GA rollout. You do not want to have any previously developed features degrade as you add new ones. Automated, efficient, effective, and parallelized testing within your CI/CD pipeline is critically important. This quality approach needs to be clear when your project is funded. Inefficient testing can cost 40% of your budget if not optimized. You also want to remember the key best practices for testing and implement **shift left testing** (**SLT**) in order to have testing as close to the developer at the time of development as possible. Some will advocate for a separate testing team, but if that team was a center of excellence with developers rotated in and out, your testing framework would be recharged with fresh talent. The value of testing would then get pushed into your core developers' psyche before the temptation to over-produce and under-test takes hold. Remember the quote from earlier chapters:

"In all things, consider the end!"

The value in your data engineering will be the new insights the processed and then analyzed data paints for the business.

Focus on the process of information value creation

Focus on the process of information value creation, with the goal of teasing out new insights. As you plan your data product solution, you want to take a walk around the data, information, and knowledge to assess possible insights in order to ensure that they are actionable insights. IT has always been data processing. It's time to stop moving bits and bytes around and focus on knowledge and how a domain of knowledge can be represented as part of the universe of understanding. This means software systems should be built to operate around a central core of data and improve that data over time. This makes that data ever more expansive and correct, while refining the derived information, knowledge, and insights that can be added to the source raw data. The traditional flow of processing input, processing it to form output, is to be replaced. Even the more current request/response or persistent pub/sub model is altered so new processes can exist. This is where your data services define a user's goal, represent the user's context, and then use both to set up a multi-faceted view of the knowledge domain. This is the new paradigm. Reducing data movement is the new data service model's goal. Then, the logic backing up those new data services will become analytic processing services. This is required to execute complex analytics scenarios to accomplish various business-defined steps leading to the goal. Today, analysts have to structure those steps in notebooks. In the future, notebooks will be supported by data service backend systems that run in the data source's consumption zone. IT will not require data to be pulled from a data warehouse, data lake, data mesh, or data fabric – it can and should be operated on where it resides.

Ensure that the technology of your data solution scales

This will make big data analytics tasks scale and reduce data movement considerably. It will enable citizen data scientists to work more effectively since statistical analytics tasks will be commoditized in data services called from enhanced notebooks focused on analytics steps rather than data wrangling. Rethinking data services as an orthoscopic manipulation of data to expose the domain's knowledge as a diamond is essential to this vision. The various facets are cut from the user's definition that was set up via the data service's analytics context. This is a good picture to keep in mind as you implement a key goal, which is to reduce data movement to improve your analytics' time to insight metrics. With this new approach, confidential computing is also made possible, with much less IP leakage than traditional analytics approaches. Data, information, and knowledge would be available for insight and ML/AI use cases where, today, only the raw form is available.

All too often, we bemoan the fact that we are still moving bits and bytes around after years of technological advancement. You have the opportunity to build out a centralized knowledge base and build the system services to revolve around your data, transforming it and adding value to it in order that the next system enhancement does the same. AI systems then have a mechanism to store the knowledge and use that evolving knowledge base for true learning. The AI system can become an expert system with this approach. Just as Wikipedia and DBpedia grew from crowdsourced contributions, it will take a visionary (maybe you!) to create a GenAI-contributed version of DBpedia to show the value of linked data in a universal data mesh.

Data in a mesh – or fabric if you implement that pattern – has to be wrangled in order to be made ready for use in analytics. A data factory approach makes that possible, and this has book elaborated a lot on how to go about accomplishing that goal. This is a pragmatic approach since you are going to be treating data as a product. This can be implemented today with various cloud services offerings, to make data fit and ready to meet analyst expectations. This book has given a lot of attention to overcoming niche analytics problems such as time series data, gap filling, data smoothing, data profiling, metrics, additive and non-additive measures, calculations, data calendars, trend analysis, data anomaly detection, data cleansing, and a host of other issues related to data quality. There is no need to go into detail again about these in this summary. The details are where your effort needs to be applied. A key takeaway is that a data engineer must not trivialize the effort needed to create information from raw data that can be made fit for purpose.

Build your technology to the data factory pattern

In your efforts to make data fit for its use, you will want to make the transitions between your data factory processing zones as easy and painless as possible by implementing standard **Change Data Capture (CDC)** design patterns. This formalizes what would have been developed as a number of one-off pub/sub mechanisms the software developers will want to build. Cloud providers make this anti-pattern easy, so you will have to explain that just because you can build something you may not want to, given the mission you are on together as a team. Using standard patterns for tough tasks also enables standard metrics and reporting to be created across the data factory. Overall backlog tracking and queueing metrics are needed to keep the factory system humming smoothly.

If it exists, it should be measured – this includes your solution

On the topic of metrics, we advised you to let your technology choices and solution designs always be validated by metrics. Let due diligence be guided by objective assessment criteria that are weighted by importance for your solution. Then, the decision to buy a tool versus building a solution for a given capability can be summarized for all to review. Otherwise, decisions will be driven toward subjective criteria, and third-party tool marketers will have a field day with your budget. Also, let your IT operational metrics be reflected in dashboards. Always be able to tie those metrics to historical trends and use them with your analytic reporting prowess to determine whether the operational aspect of your solution is working as expected. To create a future-proof implementation, non-functional requirements must be treated as equally important as the delivery of your OKR functional requirements when driving toward implementations that are future-proof. Knowing how well you are doing should not be based on a subjective assessment but based on hard numbers that you have designed to be objectively collected. You will be accountable for maintaining service-level managed agreeable trends and data contracts now and into the future. Proving that you are in compliance will help you build a mountain of success without backsliding. Building a trustworthy system begins with the definition of trust, followed by the ways to verify that trust over time.

Build a trustworthy solution

Trust is also supported by your **data quality management (DQM)** processes. Knowing your dataset's status in the data factory pipeline will go a long way toward building trust in the final data product's delivery as a release set. If trust is required – and it is – then you need to prove it with the collection of metadata for the data lineage transitions that take place between stages where data at rest stabilizes. Some of your most important data is your data's metadata and its transparent lineage from raw source to curated final version. Cataloging that data and important attributes, including its quality assessment, version transparency, and semantics, implements a clearly verifiable data quality transparency capability for the analyst. You can't let the PaaS cloud service's complexity interfere with this goal of implementing curated data as a product, even if the cloud provider makes anti-patterns easier to deliver for your developers. The data engineer will be the one to guide these development prerogatives. Misuse of data and a lack of clear metadata would make it possible for data consumers (analysts or big data algorithms) to reach incorrect conclusions. You will want that kind of error to be designed out of your solution by making it impossible for that to occur. It will be your data engineering design for a great business solution that channels correct processing within the guardrails established.

Other important challenges

There are a number of important areas of your engineering effort that cut orthogonally across the capabilities required by your architecture, as mentioned in *Chapter 5* when the conceptual architecture was elaborated upon:

- Business realities
- Security and privacy
- Knowledge engineering
- Artificial intelligence

Business competition, intellectual property rights, and fair use of data are very real drags on data engineering efforts. It would be great if all data were free and open source, but that is not the case. The monetization of low-value data into high-value insights will require lots of your own data as well as other data. Some of that data has to be bought, integrated, and metered for fair use. That data, and its derivative data, needs to be secured and remain subject to privacy constraints that are auditable and maintained in perpetuity.

Knowledge is the outcome of information processing of raw data, and artificial intelligence is a technical way to keep knowledge true and evergreen over time. The approaches of ML and AI will evolve and continue to be as disruptive as GenAI was to traditional ML and DL approaches. A machine becomes intelligent when it learns and can retain what it learns in a knowledge base that improves over time. This ability to retain AI outcomes in a semantic way will allow your solution to refine truth from mountains of raw facts. Keeping this cyclic process in mind gives you a clear perspective on what you will need to do as you make your data engineering solution future-proof.

Business realities

There are business realities you need to consider. This means data will need to be accessed, used, and reported on in a confidential computing manner. In the future, you will not be able to move titanic blocks of released data around the internet and between cloud providers! However, businesses are not fully ready to bring their valuable algorithms to the data until trust, intellectual property protection, rights management, source attribution, and fair use can be linked to the data as a product being shared alongside a competitor's dataset. This challenge is being addressed in various industries (for example, retail analytics). If successful, this will break your data free so it may be enriched far beyond what is possible today. This paradigm shift is something you want to be ready to adopt as it becomes a reality in your engineering designs. With confidential computing comes the requirement to establish trust that the data contract is being enforced.

Security and privacy

Some of the highest barriers on your guardrails will be the security, privacy, governance, compliance, and data protection constraints you put on developers. Implementing zero-trust is a goal and you will be challenged to implement a cost-effective and fully capable solution given the state of current technology offerings. Not all third-party tools play in the sandbox well with each other and you are going to have to make compromises until they do. Expect lots of changes as nefarious actors are always going to try to worm their way into your castle and take, hold hostage, and exploit your curated data for their benefit. Assume the following:

"What can go wrong will go wrong and it's going to happen to me!"

So, plan accordingly and do not compromise on your architecture's security capabilities. If forced to integrate a tool with an anti-pattern (and you will be forced), then wall off access and protect the design's exception with secure data services. Build a security command and control console capability, with an alerting/metrics dashboard, and be ready to shut down internal access to datasets if nefarious penetration is detected. This way, the system's trust will be assured and remain auditable. Even with all this, have an insurance policy ready, and an external security group contracted to take charge if a major incident arises. We have seen this work well in multiple companies when catastrophic attack profiles were detected and only with the help of external experts was the attack thwarted. Expect anti-virus, anti-hacker, and anti-nefarious-AI solutions to arise with the capability to know your system and automate defensive actions on your behalf. It is with knowledge of your data and the solution architecture's metamodel that this type of automated response capability will be possible.

Knowledge engineering

Adopt the principle that the data engineering field is morphing into knowledge engineering. Data is getting smarter and massively growing in volume and it needs to be moved less and remain where it is curated. A knowledge base tunes a GenAI model and outcomes as insights should be collected as hypotheses and stored in the same knowledge base. That knowledge base should be linked to other knowledge bases. This way, an AI system really learns. The AI system is then able to retain truth, infer truth, and discover new patterns from existing ones. The capability to retain and evolve the correct understanding of this truth via reasoning from many GenAI algorithms is where artificial intelligence is headed. As a data engineer, you need to embrace a higher level of strategic thinking in order to build toward that goal. Note that the advent of quantum computing will accelerate GenAI but there is no common data engineering solution yet to make use of that revolutionary capability, without considering the strategy outlined in this book.

Artificial intelligence

Machine learning (ML), **deep learning (DL)**, and **generative AI (GenAI)** have taken all the air (and funding) out of the room in regard to the project topic of conversation. They're all the rage and, as you'll know, when the air leaves the room, the topic falls flat. That is not going to happen with this topic. It will evolve with some common-sense thinking and third-party tools that are arising. What we summarized before regarding the output of GenAI, keeping the outcomes, and how these outcomes are rationalized as truth is very pertinent for the future. If, let's say, 90% of a GenAI model's outcomes are objectively and subjectively determined to be correct based on a subject matter expert's review, who is going to fix the 10% that is incorrect, and how? Do you think 90% is good enough for your retirement portfolio's recommendations? Is 90% acceptable for a health diagnostic and treatment plan? Is the potential for your self-driving car to ram into an overpass support structure in a rainstorm, okay? You would say no, and heck no! Getting quality outcomes from AI is essential, and knowing when to override its output with a judgment that feeds back into the model is a necessity. Build designs to address this required capability. Do not settle on accepting the GenAI model's errors until others can fix the core model. Fix it yourself! You can and must make your solution work at a much higher level than the default out-of-the-box quality of the technology.

When it comes to implementing non-generative models, you want to also think about the niches that each model services and make sure that each is measured for quality and remains performant under the assumptions and constraints that you have stipulated. These were contracts that you took on when the model went live. All models will have to undergo iterative ML lifecycle re-tuning as the data drifts and evolves over time. MLOps processes have to be built onto the end of your data factory's pipeline with a lot of care and focus. Under the MLOps umbrella of capabilities, you are going to have to deal with the friction between data scientists and data engineers. To preserve détente, you want to work all the necessary tooling and processes into the analytics workbench. Don't be distracted by letting a group clandestinely adopt a cloud service provider's easy fix or all-in-one tool that breaks the peace! Some will try. Govern the notebooks to be used, any operationalization of those notebooks, and processes for engineering a scalable version of the science that needs to become part of your enterprise's analyst workbench. Establish MLOps process defaults that make it easy to get models into production without omitting the CI/CD-enabled quality checks for metrics that are critical. Analytics regression testing, A/B testing, measurement/metrics reporting, the ML model catalog, and the ML model registry are all pertinent, just as they were in the creation of a data release set from the data factory.

Summary

When we began this book, we defined the data engineering landscape as being an environment that is an ever-changing technological landscape. The following question is still going to be on the tip of your tongue and a constant companion throughout your journey:

"Are we there yet?"

The answer is soon! *Soon*, there will be even more powerful computing environments (there are always new hardware innovations). Quantum computing will supply the horsepower to drive developments in the data engineering field as its potential becomes reality. *Soon*, there will be evidence that your future-proof data engineering efforts will bear fruit, as software algorithms are run on the new cloud hardware technologies aligned to your chosen architecture. And, very *soon*, you will be done reading all about these best practices and begin applying them in your daily work.

We wish you much success with your engineering solution designs, and great fulfillment as you build what will withstand the test of time.

18

Appendix and Use Cases

This appendix will discuss our experiences with a few use cases that cover the assertions made in this book. Each use case will drive home the need for raw data to be transformed into information, and for that information to be curated into knowledge and retained in a highly dimensional storage engine, or a **knowledge base** (**KB**); so that insights and, ultimately, business wisdom are outcomes. Once, in conversation with a CEO, I asked how much an insight generating engine would be worth to a retail business. The answer was *immeasurable*!

Use cases overview

The use cases in this section are built on **knowledge graph** (**KG**) technologies for 20+ years in anticipation of **generative AI** (**GenAI**) **large language model** (**LLM**) **Retrieval Augmented Generation** (**RAG**) prompt engineering and embeddings. The root of today's current technology advances began with **natural language processing** (**NLP**) and the lack of algorithms and processing approaches required to make AI useful.

Technology innovation and processing power and costs have made it possible for the AI field to mature. We are ready for the *a-ha* moment where human productivity takes a great leap forward with AI assistance! The beginning of that process is happening now with LLMs and advanced approaches to tuning those LLMs with prompt engineering. You can muster data architectures and data engineering solutions to exploit the opportunity via DataOps and MLOps (or **AIOps** if you are working with GenAI models). Working with key partners that have knowledge graph capabilities to structure your data and especially your semantic metadata to drive RAG is going to be your focus. Vendors such as Stardog are truly grokking the semantic metadata needs, and they can be a key enabling partner as you build out your solutions capabilities.

All AI involves learning cyclically, always moving toward better performance. It's a refinement process that is meant to produce *true facts*. The processes' capabilities are iterative and often recursively called on to produce outcomes. AI outcomes are not yet truth until the hypotheses are vetted. What drives correct performance or alignment to truth is your LLMs tuning. This is driven by domain truths that are inferred, implied, vetted, and rationalized as well as retained as semantic metadata. All this appears in knowledge base (KB) that represents the knowledge needed to tune the LLM model. What you will need to do is to retain key outcomes of the LLM model and create processes for auto-tuning what today requires **human in the loop reinforcement learning** (**HITL RL** or **RLHF**) to auto-correct false facts that are generated with the highest quality. What we are saying is that GenAI is still immature and being oversold without putting an engineered ecosystem in place for the combination of the following:

```
KG ➜ KB + RAG + LLM + Prompt Engineering ➜ KB
```

In this chapter, we will begin by discussing the foundation for the high-level use cases to be elaborated upon. This way, you have the technical context needed as a foundation for the use cases. The foundation requires the data engineer to step out of a pure data role and into the knowledge engineering role. That is going to be your future-proofing key to success. As part of GenAI's AIOps processes, you will be using GenAI to create a KG from traditional metadata and then populate it with objects from your enterprise data and craft GenAI queries in technologies that mere mortals can easily use. You'll be storing outcomes in KB-accessible locations and vetting them for truth. This is the AIOps ecosystem that will replace many deterministic processing software efforts today. The focus on quality outcomes and adherence to the AI vision where performance needs to exceed human correctness capability needs to be clear. That means that all unstructured information in the future can be processed and assessed for its truth (today, much of what is published is nonsense) and then only the new truths are retained as hypotheses and over time vetted and iterated over till all error is eliminated. This is the human learning process and an AIOps ecosystem needs to mirror it. We just have to get ready for that future.

We'll dive into health sciences, life sciences, and social media analytics processing high level use cases, which all require KBs aligned to ontologically curated knowledge graphs. All three areas suffer from information complexity and volume issues. So, they are well suited and difficult enough to illustrate the points made in this book. Each use case is real and has produced working systems and shown enterprise business, and societal value. We hope that you also see that as we explore the details and see where the best practices presented in this have been applied or could be applied if they were around when the use cases became part of GA products.

We'll begin with some background details on currently available technologies and vendor products that will be used to set the stage on which our high-level use cases depend.

Technology background deep-dive

This book was written at the intermediate data engineering level. The technology choices and use cases may stress your comfort level and require some advanced knowledge and deep dives in the hyperlinked endnote materials of this chapter. References will be supplied for you to take deeper dives into the technical.

As we discuss the necessary technology background, you will note that even this elaboration will contain additional references that contain lots of supporting examples and drill down use cases. After all, this is the use case section of the book, and all the deep concepts will need to have solid examples to enable you to grok the materials. Feel free to skip over details if you already are trained in the art! When you take deeper dives into the engineering, you will appreciate the assertions that are going to be made in the high-level use cases to follow.

Some of you will already have experience with graph technologies, property graphs, and knowledge engineering technologies that were touched on earlier. This is not a traditional data engineering skillset. This will now be leveraged, so you may be at a disadvantage. If you do not have that experience, you will want to study a bit more. Take a look at *"A Data Engineer's Guide to Semantic Modelling"* {https://packt-debp.link/L4441W} to get started if you are not already informed. Dive deeper into the materials provided as referenced links will expand your basic understanding so you can approach the examples with just enough understanding to get the points now but afterward go back and gain that deeper appreciation needed when you bring all your data together for repurposed use in training and augmenting GenAI models. We'll try to guide you as we set you up for success.

The first technology that you will want to know more about is the technology needed to build a knowledge base with its ultra-high dimensionality and why you should select a graph structure for that purpose. Alternatives exist, such as a snowflake designed relational database (not *SnowFlake*, the vendor, but rather the design pattern known as the **Kimball snowflake** {https://packt-debp.link/CzhtLx}). We've been there and done that twice in past lives and regretted it sharply! For a data warehouse, the snowflake pattern works well. However, it has a serious query complexity problem that requires a lot of expert data engineers query tuning and optimization to get it right. It is often also the case that you don't dare change the RDBMS schema after this effort has completed or your tuning is trash. This is a fallacy since the schema often must change to address data drift, thus trashing your customer efforts over and over again. For ultra-high dimensional data, a graph schema works better. However, operationally, this data organization paradigm can run considerably slower than a relational organization as data volumes increase. If the wrong graph engine is selected or the wrong semantics model is created, your entire solution can fail to perform. One also can't store massive quantities of data in a graph without it all being in one massive memory instance. This avoids the cost of exercising queries across partitioned graphs (as a serious performance issue). Stardog and Ontotext's GraphDB have overcome many of these limitations, as have others such as TigerGraph (but for its **labeled property graph** (**LPG**) design). Using massively large RAM-configured cloud instances is one way to solve this problem. An alternative is to virtualize large volumes of data via links from the KB to a Spark-enabled data store such as Databricks. These vendors have solved the scale and performance issues that plagued semantic graphs in the past. They also have matured their tool offerings to support the modeling ecosystem beyond the original RDF-OWL/RDFS* tools such as Stanford's Protégé or TopQuadrant. This way, semantic modeling is made easier when supporting SPARQL queries or implementing semantic data validation via SHACL. Property graph access methods have also improved, but they are far from unified in the approaches taken by Tinkerpop Gremlin, Cypher, GSQL, and so on.

We have in this book and still do recommend that you develop formal data semantics in an ontology or set of ontologies that may be linked over time. This implements your metadata strategy perfectly. This metadata conforms to the knowledge graph which is analogous to the schema in a relational database that you may have modeled with erwin, Quest Toad, or other relational modeling tools. This legacy approach worked well and still does for many traditional IT data engineering needs, but if you want to structure semantics into your data to make it smarter, you need to add the rules also as metadata together with the data's structure so that metadata semantically fits into your ontology. Data instances (also known as objects) go into the knowledge base. The effort to do all this will pay off over time, but initially trying to get your teams to adopt the technology approach will be an uphill battle. Fight it anyway! Show the value over time! Use the graph database vendor's professional services consultants to assist you. Mentor the team along the way! Guide them to an effective solution that operates well. But also, avoid getting trapped into using immature technology provided by some graph engines. Note that proper object versioning, backup restore, checkpointing, or ACID transaction compliance will be difficult. Know what is needed and build around the weakness. You must not assume that any graph technology has all the capabilities of the RDBMS you require. Do not underestimate the power of the new capabilities; you will need to leverage these in the future.

You may still be wondering why you need to have a graph structure for RAG and prompt engineering. We're not forcing you to! We are just pointing out that you will want your data at rest's semantic metadata to be used as you gain success. This will be evident when you want more and more enterprise data to be fed into your GenAI solution. Without semantic metadata alignment, you will have to refactor your RAG solution at a very bad time. This is when you are trying to scale it up. You do not know what is going to be needed to properly set the context of all your enterprise facts until all your data is available to the GenAI LLM via RAG. You also don't know what enterprise data is needed for tuning, nor the effort required to get the necessary data available when it is determined that it is needed for later re-tuning. This will put IT (data engineering particularly) in the MLOps/AIOps loop at a critical time; that is when the data scientist is trying to improve AI/ML performance. Having your business domain data and metadata ready and accessible in a clear semantic model exposes what is today in production, and it also exposes what in your datasets is not populated fully. It also allows information context to be implied, inferred, and contextualized, with any degree of separation (semantic distance) from other facts, thus removing information ambiguity. Reducing this error goes a long way toward not propagating repairable error factors into an AI system that multiplies those error factors. All this benefit comes from the KB backed by the semantic metamodel (or the knowledge graph or ontology).

So, we want to elaborate further on prompt engineering frameworks and some prior vetted art required to select data to tune generic LLMs for your business needs. You will want to take away a solid understanding as to why a **Graph-of-Thoughts** (**GoT**) framework works best and how that pattern can be used with your knowledge base. This is a best practice approach as you craft quality prompts.

Prompt engineering frameworks with examples

In the article, *Prompt engineering frameworks designed to enhance LLM reasoning* {https://packt-debp.link/bnECmp}, you can see how Graph-of-Thoughts (GoT) is assessed as the best way to approach your engineered prompts. It is important to note that a GenAI LLM does not equate to *"human thinking,"* or *"reasoning."* Note that we have added a lot of context pertaining to the need for semantics in this book's prior chapters. This along with semantics there is this a lot of discussion regarding false fact/true fact maintenance requirements, which are hypothesis driven. All this is provided so that you can formulate your vision/strategy to create an ever improving repository of knowledge that in total simulates human thought, retention, and reasoning processes. The ML/AI algorithms enhance the reasoning and knowledge retained in the KB, and that in turn enhances the quality and performance of the ML/AI algorithm.

GoT versus Chain of Thought (CoT), Tree of Thought (ToT), and others

The Graph-of-Thoughts (GoT) Framework should be understood by you, the data engineer. You have other choices, and understanding each alternative is an academic exercise that we leave to you to explore. The benefits are driven home in the following paper: *"Graph of Thoughts: Solving Elaborate Problems with Large Language Models"* {https://packt-debp.link/lkWsAc} by Maciej Besta, who asserts that GoT advances prompting capabilities in large language models (LLMs) beyond those offered by paradigms such as **Chain-of-Thought (CoT)** or **Tree-of-Thought (ToT)** methods.

> *"When working on a novel idea, a human would not only follow a chain of thoughts (as in CoT) or try different separate ones (as in ToT) but would actually form a more complex network of thoughts. For example, one could explore a certain chain of reasoning, backtrack, and start a new one, then realize that a certain idea from the previous chain could be combined with the currently explored one, and merge them both into a new solution, taking advantage of their strengths and eliminating their weaknesses. Similarly, brains form complex networks, with graph-like patterns such as recurrence. Executing algorithms also expose networked patterns, often represented by Directed Acyclic Graphs. The corresponding graph-enabled transformations bring a promise of more powerful prompting when applied to LLM thoughts, but they are not naturally expressible with CoT or ToT" (Maciej Besta et al. (see previous link)).*

In *Figures 18.1-18.4*, you can see a comparison of Graph of Thoughts (GoT) to other prompting strategies. The comparison contrasts these alternative approaches:

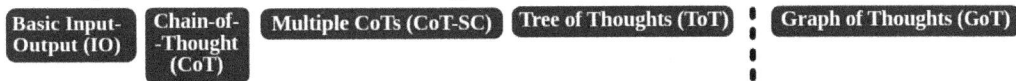

Figure 18.1 - A comparison of GoT with other prompting strategies (source: Besta et al. (2023))

In *Figure 18.2*, and *Figure 18.3*, you'll see a few comparative approaches. Then, in *Figure 18.4*, the GoT approach is shown, which we and Maciej Besta advocate.

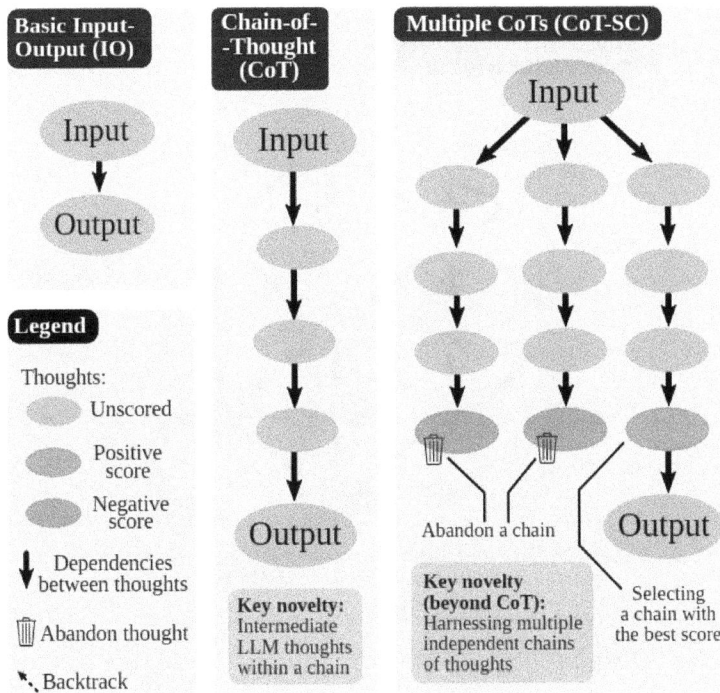

Figure 18.2 - A comparison of basic input-output (IO), Chain-of-Thought (CoT), and multiple CoTs

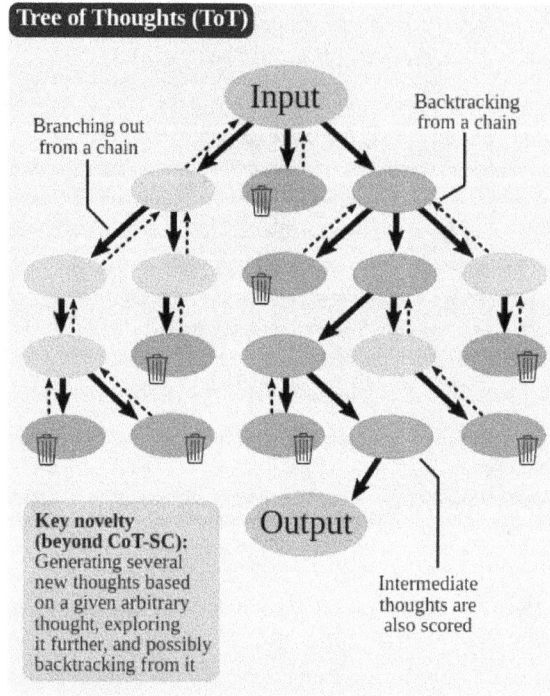

Figure 18.3 - Tree-of-Thought (ToT)

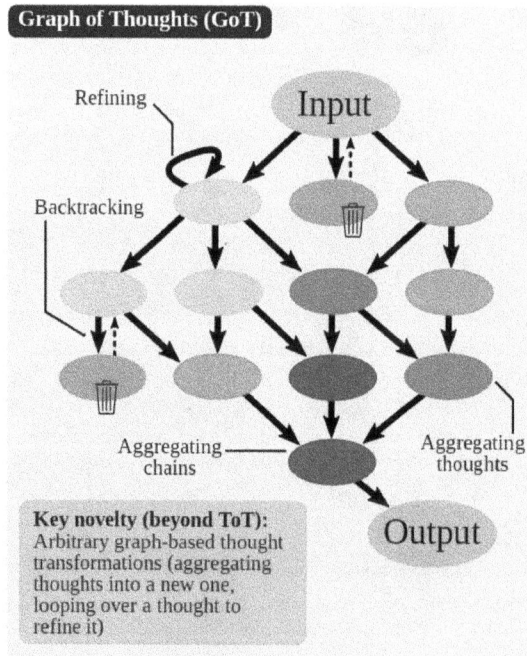

Figure 18.4 - Graph-of-Thoughts (GoT)

In the paper, a few use cases are implemented with the GoT framework.

GoT use cases

In the first use case, the LLM is shown to be able to implement `merge sort`. This is done by first decomposing the sorting task and the graph operations. This is central to implementing and executing any GoT workloads. The author considers sorting numbers 0–9 with duplicates. LLMs are unable to sort a sequence of such numbers correctly beyond a certain length consistently because duplicate counts do not match! The details of the approach are worth learning.

In the second use case, the *set intersection operation* is shown. Set intersection operations are recurring problems in genome, or document comparisons and general pattern matching. The LLM handles set intersection for two sets in a similar way to how it handles sorting. The second input set gets split into various subsets and the intersection of those subsets is assessed for intersection with the first. Then, all intersected sets are aggregated as the completed result. This is also worth studying.

Additional use cases for *keyword counting* and *document merging* are also examined in the paper and are likewise worth your study.

The paper concludes with a clear statement of value for the GoT framework:

> *'The overall goal when conducting graph decomposition is to break down a task to the point where the LLM can solve it correctly for the majority of time using a single prompt (or with a few additional improvement steps). This significantly lowers the number of improvement/refinement steps needed during the later stages of the graph exploration. Furthermore, as indicated by our results, combining or concatenating sub results is usually an easier task than solving large task instances from scratch. Hence, the LLM is often successful when aggregating the final (or end state) solution"* (Maciej Besta et al.)

But now, you will want to know how to learn about knowledge graph creation, since you have already learned that it is not that easy a technology to work with and it comes with a steep learning curve. This is part of the journey towards your effective prompt engineering designs.

Learn about new knowledge base generation tools with examples

Because building a KB begins with the formulation of an ontology that is operationally hosted in a graph database as a KG, you have a lot of technology to master. Even the precise terminology will be abused by vendors and cloud providers that often call the KB (which is the ontology model with object instances) a KG. Maybe this is because the KG must have several objects present in the base model to be able to enforce object constraints, as per modeling standards in OWL. This KG/KB confusion is understandable. However, when you scale out all objects aligned to the ontology, the technology hosting the KB must be able to scale, and those who have ever worked with Protégé know that the limit on your desktop will be exceeded quickly. So, vendors such as Stardog have realized that they are successful when you, the customer, can use the technology that they offer. The handcrafting of the ontology is a problem since RDF/OWL/RDFS/RDF* are acronyms that are not commonly known. Can a GenAI solution craft the model for you and create SPARQL queries without the steep learning curve? The vendor promise is yes, but once generated, you will still need to grok the underlying syntax and semantics created. But wait … Didn't we say we wanted to use the KG for prompt engineering and embedding and now we want to use GenAI technology to help generate the KB? The answer is yes, and there is a solution for that!

Stardog Voicebox

Refer to *"Stardog Voicebox Customer 360 Use Case Demo"* {https://packt-debp.link/39oRUe}, where the Stardog Voicebox product can talk to your data so you can deliver on the AI's promises and produce accurate explainable insights. In the demo, Sarah, the fictitious head of customer experience, has acquired a major competitor. Post acquisition customer loyalty, seamless customer experience, and growing the business are top priorities. Sarah's boss asks her to find the top ten customers across both organizations who signed up for their reward board accounts in December 2023. Sarah has multiple tools, from both companies, available for her use with lots of dashboards. But none provide the answer to this question. Worse yet, she gets conflicting information from the reporting systems about many of the customers. So, Sarah reaches out to her data analyst who states, *"They need to engage IT to curate new data to answer the question,"* but the IT representative explains that they don't have direct access to all this underlying data and would need to load a fresh set from the data lake. IT has a project backlog and can curate a model, ETL the data that Sarah needs, and have an answer for her in a *month*!

Stardog Voicebox can get timely accurate explainable answers using an enterprise KG and conversational AI. With its LLM powered user interface, Sarah can chat with the enterprise data to discover hidden connections that are accurate, timely, secure, and hallucination free. Sarah can ask her questions and get verifiably accurate responses. Voicebox gives her insight into *all* her enterprise data. The underlying data is easily traceable, and, as pointed out before, it is hallucination free. Sarah can see data and its lineage, so every answer is explainable. Without code or dashboards, answers and insights are clearly evidenced in Voicebox outputs.

It is clear that the Stardog vendor groks the use of a KG to understand enterprise data and then uses that understanding to get answers from a tuned LLM. The vendor understands this so well that they have come out with this presentation foretelling the future of data management: *"Voicebox FAQ: How LLM, Generative AI, and Knowledge Graphs (KG) are the Future of Data Management."* {https://packt-debp.link/Vu0HlF}. The product can best be understood in the context of this Stardog Voicebox presentation {https://packt-debp.link/n1hkkp}, which is used to craft your KGs or at least provide much of the heavy lifting work needed to create one. The tool does not directly provide the RAG LLM tuning or prompt engineering interfaces. Those still have to be built once the KG is created.

In *"Demo Day: LLM-Powered Stardog Voicebox and Knowledge Graphs"* {https://packt-debp.link/99rkJk}, Stardog provides you with a lot of insight into how a credible KG is created from any data source and populated with object data to constitute a KB. Refer to *Figure 18.5*:

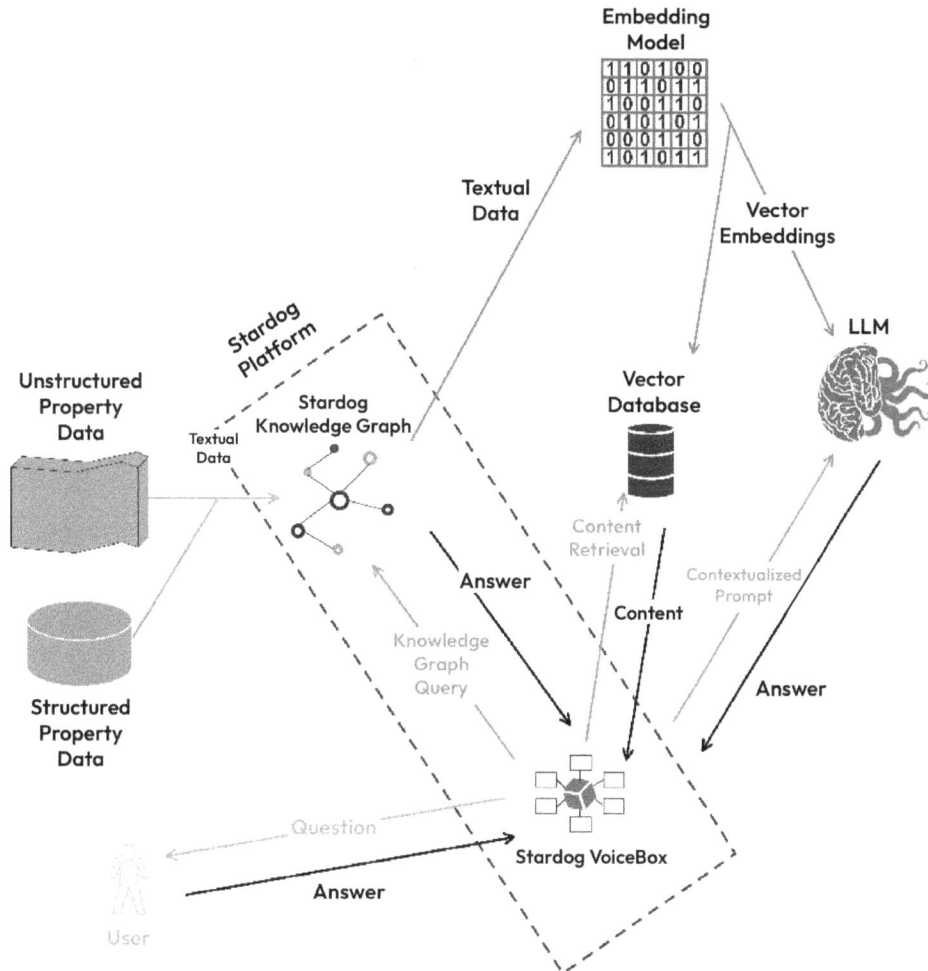

Figure 18.5 - Stardog Voicebox architecture

In *"Voicebox: End-to-End Example of Knowledge Graph Creation"* {https://packt-debp.link/WDSfyi}, you can see a brief but complete example of a GenAI created Knowledge Graph (KG).

Stardog is addressing core data engineering problems

Lastly, in *"Knowledge Graphs, Semantic Layers, and Data Democratization, Oh My! (from TDWI Executive Program)")* {https://packt-debp.link/etEVM7}, several core data engineering problems are addressed by Knowledge Graphs. The technology makes the marshalling of enterprise data for GenAI LLM tuning easier, so that the promise of AI may be realized in your enterprise. Problems addressed by the Stardog solution are as follows:

- **Current data management systems can't keep up** with growing data complexity and insight demands that are hidden from data consumers. This information is buried in siloed databases, catalogs, and analytic applications.

- **Conventional graph and relational data architectures lack the access**, context, and inferencing required to meet the demands of innovating and monetizing advanced analytical solutions.

- **Data catalogs provide an inventory of information assets**, however, if the catalog is disconnected from the rest of the enterprise data, you are left with a meta-data silo. This causes data consumers to be constrained by architectural limitations when performing scalable, discovery-based business analysis.

- **Data fabrics have emerged as a modern solution to address these needs and free data, but how?** Without a KG a key enabling ingredient to the data fabric, the capability to enable the data mesh self-service vision is impaired.

The knowledge graph is a semantic data layer that weaves together your core datasets. With a knowledge base you may connect to data where it's stored, populate, and use your data catalog, and leverage data and analytics teams with the data, information and knowledge data consumers need.

High level use cases

Now that the KG, KB, and GenAI background technology review has been set with many examples, let's get into the high level use cases, starting with the **Health Sciences Knowledge Graph** (KG).

Health Sciences Knowledge Graph (KG)

When working at Elsevier, the world's largest provider of medical information to scientists, clinicians, doctors, nurses, and researchers, the core problem was that this corpus was locked up in text form. It and any newly published research had to be freed from its text form and the knowledge distilled for reuse. Doing more with less was and still is imperative when the value of insight can save a life or lead to discoveries. It was an exciting problem to tackle! Having that much information available is a great opportunity, but the challenge is how to parse it out of the text since the knowledge to understand it is a niche specialty of select subject matter experts. ML/AI algorithms train on big data with lots of inputs to be able to harvest meaning. That is not present across the health science domain spectrum. The medical information problem that this high-level use case is trying to solve is crystallized in the work by Peter Densen in *"Challenges and Opportunities Facing Medical Education."* {https://packt-debp.link/yfy9fT}.

Elsevier's solution to the information problem is to craft a KG to organize the concepts and then use that to harvest new facts that have to be added to the semantic model so that the KG improves as well as the KB gaining depth of coverage over time. The evolving KB was called the Elsevier Healthcare Knowledge Graph and it was used to facilitate clinical search services as part of the Elsevier product **ClinicalKey**. Being able to capture search efficacy enabled the graph to mature and be more correct over time. The medical corpus indexing correctness also improved. The Elsevier Healthcare Knowledge Graph enabled better search results due to improved index quality, faceted search alignment (created from directed acyclic walks of the KB), and query term autocomplete, spell correction, and term expansion. The result was and still is a success. The clinician and diagnostician were able to see tangible benefits through their search experience. Later, the semantic modeling inherent to the graph organization made it useful as a standalone data product. The Elsevier Healthcare Knowledge Graph contains many millions of medical concepts, dictionary terms, synonyms, and concept expansions. All is correctly contextualized for medical, clinical, and diagnostic purposes. Today, it is useful in the GenAI processing of research papers, text, and diagnostic tooling that Elsevier is envisioning for the medical professional.

In *"Text Snippets to Corroborate Medical Relations: An Unsupervised Approach using a Knowledge Graph and Embeddings"* {https://packt-debp.link/5Xmwsr} knowledge graphs are shown to significantly improve search results. Subject matter experts have to normally keep up to date with medical literature for search indices to remain relevant. Dynamically identifying text snippets in the literature that confirm or deny knowledge graph triples is a differentiator between trusted and untrusted medical decision support systems. The article describes Elsevier Clinical Solution's approach to mapping triples to medical texts. A KG is the source of triples used to find matching sentences in reference text. The unsupervised ML approach uses phrase embeddings and cosine similarity measures and boosts candidate text snippets when key concepts are found to exist in the text. The approach accurately maps semantic relations within the KG to text snippets with a precision of 61.4% and recall of 86.3%.

Further elaboration on the KG technology appears to be in the works from Elsevier's lead development manager, Mev Samarasinghe, in *"Elsevier Knowledge Graph that Captures the Worlds Medical Concepts"* {https://packt-debp.link/Uoxqkr} and is further elaborated on by Maulik Kamdar {https://packt-debp.link/VIeTMr} in *"Elsevier's Healthcare Knowledge Graph: An Actionable Medical Knowledge Platform to Power Diverse Applications"* {https://packt-debp.link/ryWPeg}. Refer to *Figure 18.6* to gain an appreciation for the volume and scope of medical data covered:

Figure 18.6 – Elsevier's Healthcare Knowledge Graph, 2021

In *"The Knowledge Graph Conference - Web-Scale Data Integration in Life Sciences & Healthcare Through Knowledge Graphs - Optum Health"* {https://packt-debp.link/4aAw44}, Maulik states the following:

> *"So, there's definitely a lot of different challenges and opportunities … but this is a very popular diagram that you must have seen several times, but this diagram [referring to Figure 18.6] is so true in biomedicine in healthcare and Life Sciences because the biomedical knowledge is continuously evolving and expanding with new research and discoveries, so you* **need to think about automation associated with your enterprise** *knowledge graphs.* **Knowledge must be trustworthy and accurate for use in products,** *so you need to think about your visualization interfaces for subject matter experts. And knowledge can arise from several different heterogeneous sources … with machine learning we can see how it helps with knowledge extraction." (Maulik Kamdar, 2023).*

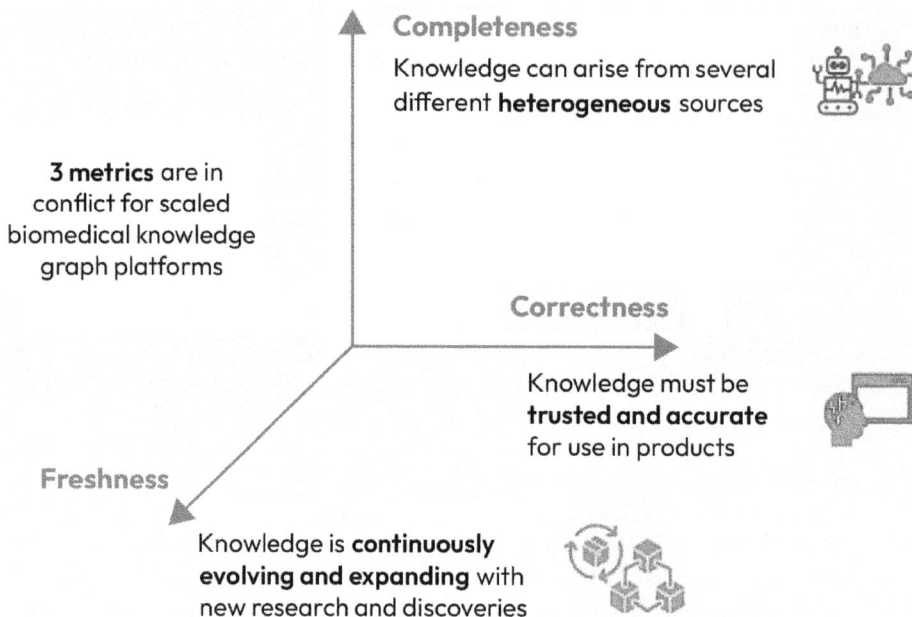

Figure 18.7 - Completeness, correctness, and freshness of biomedical knowledge

The success of Elsevier's *"Healthcare Knowledge Graph"* played out again with the more recent development of the *"Biology Knowledge Graph"*, in the **EmBiology** {https://packt-debp.link/pTyckb} product. You can read more about the product from *"EmBiology Fact Sheet"* {https://packt-debp.link/Erbdcu}, where millions of facts are stored to help visualize and discover biological relationships to find new opportunities that a researcher may have overlooked. The biology domain's information semantics in the knowledge graph leds to less errors in processing as stated in the following quote:

> *"Using EmBiology could result in a 20-25% global improvement on rate of success."*
> *VPGT (Vice President, Gene Therapy)* {https://packt-debp.link/Z29WwW}).

The power of the knowledge graph in healthcare clinical search and discovery as well biology research is very clear given these use cases. The organization of knowledge and metadata in a semantic manner enables data to be re-envisioned for future uses. Both graphs are used in Elsevier Labs AI research, and we can all expect some great new developments as GenAI methods replace current ML text mining methods that have already benefited from the KGs today.

Life Sciences Semantic Information Engine (ELSSIE)

Before Elsevier's Healthcare Knowledge Graph was conceived, architected, or designed, there was Elsevier's *Semantic Information Engine (ELSSIE)* {https://packt-debp.link/adMGXn} which proposed to create an engine to service various linked knowledge graphs across the research domain; therefore, moving forward with Tim Berner Lee's vision of the internet. At the time, Stardog and Ontotext graph databases were not available to the team; so, what was done was we built on Gridgain (or Apache Ignite) with the work that originated as project Quetzal {https://packt-debp.link/DEa9YP} from IBM Research. The SPARQL

ARQ parser's backend needed to be recoded (no easy task) to work with an in-memory database holding semantic graph data. The modifications were made with consulting vendor Knoldus' assistance. The life sciences modeling focused first on small molecule chemistry with the goal of being able to predict chemical outcomes via model reactions based on properties already collected in the Reaxys {https://packt-debp.link/B37IeE} system. The project produced what it set out to accomplish and became the Elsevier Entellect {https://packt-debp.link/5iBVYZ} project after we moved on to build with the team creating Elsevier's Healthcare Knowledge Graph and Elsevier's clinical systems search functionality.

What was envisioned originally is the seed for what Elsevier later built. ELSSIE is a prime candidate for revival, because of the enormous value in being able to reduce the time and cost of drug discovery. Because their enormous value in being able to reduce the cost and time for new drug discovery. If a unified life sciences KG engine such as ELSSIE could be operationalized and used to support GenAI discovery, the societal value would be enormous. This requires a very deep understanding (or a model) of small and long-chain molecular chemical processes and properties, and the expression of those molecules in biological pathways. Elsevier and others have the right data. It's just not put in a way that can be processed well for GenAI use cases. This use case is still open on one end, and we welcome others to take up the challenge and build on the current work.

Microblog message analysis (Dataminr)

In 2009, while working at Bloomberg after implementing the architecture and designs with proof of concept for the enhanced and faceted Bloomberg News Search and then later many features of the Bloomberg Law Search, I was approached by the co-founders of Dataminr. They explained what they wanted to achieve, and I immediately thought of a way to help crystallize the ideas into an architecture that would accomplish what was needed. This came out later as the patent *"METHODS, APPARATUS AND SOFTWARE FOR ANALYZING THE CONTENT OF MICRO-BLOG MESSAGES."* {https://packt-debp.link/sIEdlu}. The key problems being addressed in **Dataminr** were as follows:

- **How do you process over 1.5 billion tweets a day and create a signal alerting system while the core data set is growing ferociously?** A Twitter micro-blog message could be clustered with similar messages, and that cluster could be determined as viral. Even if the viral threshold was not reached in a small time frame, the superset time frame could amass enough messages to trigger *virality* for a concept the cluster represents. The number of sliding windows of time in the original Dataminr system ranged from 1 to 10 minutes in length. What would be an indicator of an alert could be worth noting as a standalone event until it gained viral status after clustering and categorization. Virality is a strong indication of truth and worthy of action. Note that AWS had two services at this early date: EC2 and S3. All the PaaS services that we see today came afterward, and Microsoft's Azure alternatives did not even exist yet.

- **How is k-means clustering implemented across this volume of data using twee-English n-grams (this was the shorthand notation Twitter users used to stay within the 140 character tweet limits at the time)?** At the time, **n-gram** research focused on English n-grams. Note that in 2009, we were just at the start of the ML revolution and the Princeton Corpus was just about all we had to work with beyond the raw definition of **k-means clustering** and cosine similarity.

- **How are concepts that are core to the system's effective alerts organized?** We could cluster on token recurrence using modified k-means clustering approaches, but afterward, you still needed to assign a most like classification of the cluster and then categorize the cluster into the domain the concept aligns with. This is all driven by an ontology, and we need to evolve it over time, with the taxonomy being a walk (a directed acyclic walk of the graph) if the graph were an ontology. Note that semantic technologies from Stanford University in Protégé, OWL/RDF, and so on were non-existent or just forming and therefore untrusted. Also, there was a pressing need to buy the financial hierarchy and a brand-to-company associated dataset. These were expensive and partnerships were formed: D&B, Yahoo Finance, S&P CapIQ, and Xignite to name a few sources for third party data.

- **Additionally, an ontology was needed to become a source of truth.** The Oracle RDBMS was selected (much to my eventual pain) to build a highly dimensional **Financial Industry Business Ontology** (**FIBO**) {https://packt-debp.link/xSDwma} equivalent ontology. Note that this was the second of two times in my career that I regretted being part of the decision to use an RDBMS when a graph DB was required (the first use was at Bloomberg for the News Search ontology, which was far less deep than what Dataminr required).

The challenge was set, and the system was developed and iteratively improved over the years. I (Richard Schiller) was Dataminr's chief architect and CIO during the early days and employee number 3 or 4 (but who's counting?). We drilled into the technology with all our vigor and came up with a great solution to each of the problems. When it came to handling volumes, we built a highly available and yet scalable number of instance tirades. Three instances in one availability zone, and three in another, and yet others when scale was needed could be launched even in other regions. Later, after my time, AWS PaaS services were used. Any self-hosted message queuing and other communication services move to efficient AWS services. It's a good thing we had this when storm El Derecho hit Virginia and we lost all instances except three that we used to rebuild our system. Without a high availability cloud infrastructure configuration, we would have been lost. In this test, we were successful!

We also squeezed all we could out of the AWS system! The team massively parallelized all Twitter message ingest and message cluster processing. Dataminr was 1 of 10 Twitter firehose recipients worldwide, and all unfiltered messaging was ours to choke – or not, as the case may be. If Dataminr's system choked, we lost our candidacy for the firehose. We were very successful so that was not a real risk for us after a while! Special handling of raw data at scale under huge volumes was a success story. Then, we did the same for the classification and categorization logic of the clustered tweets.

The data structures for the ingested data looked a lot like time series data since we had to optimize processing within the allocated time slot, which was a rolling window that overlapped the prior window evenly. For example, 5 1-minute slots fit into the 5-minute slot, and 2 5-minute slots fit into the 10-minute slot, and so on. These were tunable but rarely changed in the early days of Dataminr. Then, there were the data implications of the Oracle ontology and knowledge base. The ontology was fixed but built on very early versions of FIBO.

The knowledge base was carefully curated from many sources because, as a start-up, there were no operations folks on hand to manually curate this important contextual data. It was essential to add meaning to the data being aggregated in clustered tweets. As an example, who talks over Twitter (now X) about a company or company hierarchy owning a particular brand? Nobody! People talk about the common name of a product that is implicitly associated with a company's brand. Sometimes, the brand name is forgotten and often it changes. Yet analysts want to see the effect of social media sentiment changes on their company, which is an aggregation of brand sentiments sometimes under a corporate hierarchy. Analysis systems should hide the sausage factory processing of data and bring context to light without exposing the complexity. Making the complex simple is what Dataminr excels in. The knowledge base pattern provided value way back when the system was brought online; and then over the years its importance waned. Lately, it once again has great value for GenAI LLM RAG processing.

I will not get too much more into the details of Dataminr's processing capability. You may be able to glean a bit more understanding of the original system from the patent. The company has changed a lot in the past 15 years, and it has grown to over 500 full-time employees. You can appreciate the data engineering work that went into its first build. Today, Dataminr is a very successful unicorn start-up and I'm proud to have been a part of it in the early days.

I hope that this use case provides background to the best practices of this book and that it brings home some ideas for you as you complete your journey.

Summary

In this chapter, you were exposed to a technology deep dive with lots of references to research and come up to speed on the topics of any blind spots. Then, we dove into three high level use cases. Although they may have lacked serious details, they have been portrayed descriptively enough for you to get the general idea. The challenge has always been to leverage your data in new ways and to organize it in a way that makes that possible. Data needs to be smart, fungible, and linkable.

Over time, the future will show us how trust can be enforced among the roles both within and external to your IT organization. Data silos will come down when the truth of the underliyng data is transparently preserved with formal semantics. Data consumers will know that the information presented is correct, complete, governed, and able to be used as intended. This makes the promise of an AI ecosystem a reality, because the errors are reduced during RAG processing, prompt engineering/tuning, or core LLM model generation.

Index

O

‹packt›

packtpub.com

Subscribe to our online digital library for full access to over 7,000 books and videos, as well as industry leading tools to help you plan your personal development and advance your career. For more information, please visit our website.

Why subscribe?

- Spend less time learning and more time coding with practical eBooks and Videos from over 4,000 industry professionals

- Improve your learning with Skill Plans built especially for you

- Get a free eBook or video every month

- Fully searchable for easy access to vital information

- Copy and paste, print, and bookmark content

Did you know that Packt offers eBook versions of every book published, with PDF and ePub files available? You can upgrade to the eBook version at packtpub.com and as a print book customer, you are entitled to a discount on the eBook copy. Get in touch with us at customercare@packtpub.com for more details.

At www.packtpub.com, you can also read a collection of free technical articles, sign up for a range of free newsletters, and receive exclusive discounts and offers on Packt books and eBooks.

Other Books You May Enjoy

If you enjoyed this book, you may be interested in these other books by Packt:

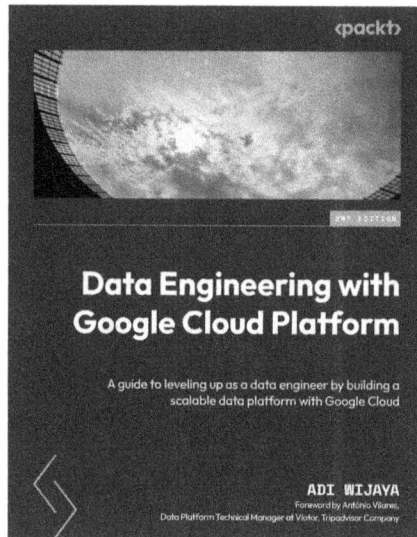

Data Engineering with Google Cloud Platform

Adi Wijaya

ISBN: 978-1-83508-011-5

- Load data into BigQuery and materialize its output

- Focus on data pipeline orchestration using Cloud Composer

- Formulate Airflow jobs to orchestrate and automate a data warehouse

- Establish a Hadoop data lake, generate ephemeral clusters, and execute jobs on the Dataproc cluster

- Harness Pub/Sub for messaging and ingestion for event-driven systems

- Apply Dataflow to conduct ETL on streaming data

- Implement data governance services on Google Cloud

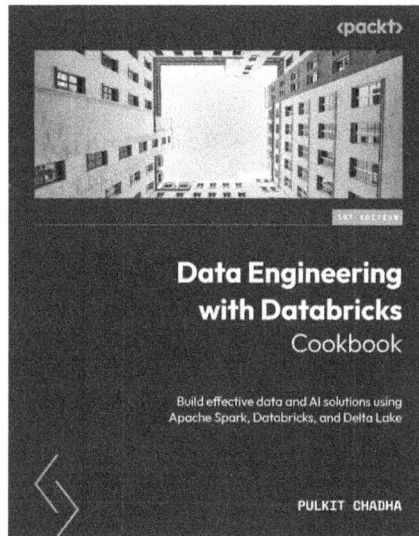

Data Engineering with Databricks Cookbook

Pulkit Chadha

ISBN: 978-1-83763-335-7

- Perform data loading, ingestion, and processing with Apache Spark

- Discover data transformation techniques and custom user-defined functions (UDFs) in Apache Spark

- Manage and optimize Delta tables with Apache Spark and Delta Lake APIs

- Use Spark Structured Streaming for real-time data processing

- Optimize Apache Spark application and Delta table query performance

- Implement DataOps and DevOps practices on Databricks

- Orchestrate data pipelines with Delta Live Tables and Databricks Workflows

- Implement data governance policies with Unity Catalog

Packt is searching for authors like you

If you're interested in becoming an author for Packt, please visit authors.packtpub.com and apply today. We have worked with thousands of developers and tech professionals, just like you, to help them share their insight with the global tech community. You can make a general application, apply for a specific hot topic that we are recruiting an author for, or submit your own idea.

Share Your Thoughts

Now you've finished *Data Engineering Best Practices*, we'd love to hear your thoughts! Scan the QR code below to go straight to the Amazon review page for this book and share your feedback or leave a review on the site that you purchased it from.

https://packt.link/r/1-803-24498-4

Your review is important to us and the tech community and will help us make sure we're delivering excellent quality content.

Download a free PDF copy of this book

Thanks for purchasing this book!

Do you like to read on the go but are unable to carry your print books everywhere?

Is your eBook purchase not compatible with the device of your choice?

Don't worry, now with every Packt book you get a DRM-free PDF version of that book at no cost.

Read anywhere, any place, on any device. Search, copy, and paste code from your favorite technical books directly into your application.

The perks don't stop there, you can get exclusive access to discounts, newsletters, and great free content in your inbox daily

Follow these simple steps to get the benefits:

1. Scan the QR code or visit the link below

https://packt.link/free-ebook/978-1-80324-498-3

2. Submit your proof of purchase
3. That's it! We'll send your free PDF and other benefits to your email directly

www.ingramcontent.com/pod-product-compliance
Lightning Source LLC
Chambersburg PA
CBHW081217220326
41598CB00037B/6803